Quantitative Geophysics and

Springer
London
Berlin
Heidelberg
New York
Barcelona
Hong Kong
Milan
Paris
Santa Clara
Singapore
Tokyo

Louis Lliboutry

Quantitative Geophysics and Geology

Published in association with
Praxis Publishing
Chichester, UK

Louis Lliboutry, Emeritus Professor, University Joseph Fourier, Grenoble, France

Original French edition, *Géophysique et géologie*
Published by Masson, Editeur, Paris, © 1999

This work has been published with the help of the French Ministère de la Culture – Centre national du livre

---

SPRINGER–PRAXIS BOOKS IN GEOPHYSICAL SCIENCES
SUBJECT *ADVISORY EDITOR* and *TRANSLATOR*: Philippe Blondel, M.Sc., Ph.D., FGS, Senior Scientist, Department of Physics, University of Bath, UK
SUBJECT *CONSULTANT ADVISORY EDITOR*: Kirill Ya. Kondratyev, Counsellor of Russian Academy of Sciences, Research Centre for Ecological Safety, St. Petersburg, Russia

---

ISBN 1-85233-115-1 Springer-Verlag Berlin Heidelberg New York

British Library Cataloguing in Publication Data
Lliboutry, Louis, 1922–
Quantitative geophysics and geology.
— (Springer-Praxis books in geophysical sciences)
1. Geophysics. 2. Geophysics—Mathematical models. 3. Geology.
I. Title. 550

Library of Congress Cataloging-in-Publication Data
Lliboutry, Louis, 1922–
Quantitative geophysics and geology / Louis Lliboutry.
p. cm. — (Springer-Praxis books in geophysical sciences)
"Published in association with Praxis Publishing, Chichester, UK."
Includes bibliographical references and index.
ISBN 1-85233-115-1 (alk. paper)
1. Geophysics. 2. Geology. I. Title. III. Series.
QE501.L578 1999
550—dc21 99-34589 CIP

Printed by MPG Books Ltd, Bodmin, Cornwall, UK

Cover design: Jim Wilkie
Typesetting: Originator, Gt. Yarmouth, Norfolk, UK

Printed on acid-free paper supplied by Precision Publishing Papers Ltd, UK

Contents

Preface

When I wrote *Tectonophysique et Géodynamique*, published by Masson (Paris) in 1982, I was hoping to contribute towards bridging the gap that exists between geologists and geophysicists. In France this gap is due entirely to different training: for geologists this is based on observations of the terrain that are literally far too naturalistic and for geophysicists the training is too abstract and removed from the complexity of the real world. Despite the entrenched attitude of many teaching departments, this gap is now partially bridged. More and more French universities are creating Earth-science departments, merging geology and geophysics, as is the rule in the United States.

However, at the same time some degree of ultra-specialization is inevitable to face the explosive development of research in these domains. The international book market now offers only very focused studies, even when written by several authors. Students starting Earth-science courses have already heard of plate theory, but in the original simplistic form in which plate tectonics first appeared. For them and for graduate students thinking about future specialization, a synthetic textbook like this one is more necessary than ever. For postgraduates there is the need for them to undergo a general update, because past teaching has mostly become obsolete.

It may well be that this is the last time that a textbook as ambitious as this – because of the multiple subjects treated – is the work of a single author. Had I not already written two books on these matters, and because my position as professor emeritus gave me some free time, this would have been impossible. It was indeed not only a matter of updating a 365-page textbook but of completing and enriching it so that it is still useful for the new millennium. The content and standard both aspire to the same aim: that the student, having assimilated this book, should understand related articles in the scientific literature.

This requires that scientists with a geological background make the effort to absorb the basics of mechanics, classical physics and mathematics in order to understand the modelling of geodynamic processes (which is the only possible way to really explain them). They must not take the results obtained through modelling in isolation, nor should they trust them blindly. Similarly, physicists interested in Earth sciences must make the effort to absorb the rudiments of petrology and geology. In this way they will not mistake the simplistic model of some geophysicists for the faithful and accurate representation of the real world. In both cases the reader should be advised to read traditional manuals of mechanics, physics or geology, which try to cover too much but sometimes miss essential points. These rudiments are introduced as needed in this book and require the reader to have studied science up to pre-university level. I decided not to relegate these rudiments to an appendix, which would have been contrary to the synthetic goal of the present book. Only those mathematical complements that were unavoidable and not clearly shown in current mathematics textbooks are grouped in an appendix at the end of the book.

Compared with the French edition of 1982, the order and contents of chapters have been changed to make them more homogeneous and gradual. The method adopted is to start from well-established facts and, as far as possible, end with a discussion of the most conjectural questions, such as continental drift in Palaeozoic and Precambrian times or mantle convection. Some of the discussions in the French edition are only of historical interest and have been suppressed, only the correct modern hypothesis is presented. Other discussions have started in the meantime. As I do not need to defend my own hypothesis and can stand back with the passage of time, I hope I have been reasonably unbiased in my choices and fair to my critics.

The following questions have been introduced or treated more substantially, because of the current interest shown by scientists:

- The constitution of the deep continental crust, which seems much less rigid than previously thought.
- Rock metamorphism, which provided geothermo-barometers.
- The different processes of rock creep, more numerous than thought. Structure and fabric formation and the anisotropy that may result from it have been introduced.
- The important role played by fluids in rupture processes occurring in earthquakes and large overthrusts.
- Decadal-scale measurements of plate movements obtained by different modern techniques and already presented in my recent book *Sciences géométriques et télédétection* ('Geometrical Science and Remote Sensing').
- The limitations of plate theory; for example, plate theory is no longer valid regarding the boundaries between plates – sometimes over large regions – and does not account for the sudden formation of basins off continents or for the occasional intense stretching of continental crust.
- The many tectonic mechanisms recently introduced, such as ablative subduction, magma underplating, post-orogenic subsidence of mountains, the exhumation of crustal rocks by retro-subduction, or the highly disputed delamination.
- The effect of the Moon, and its increasing distance from the Earth, on the rotation of the Earth – slowed down by tides – and the disagreement between computed and observed slowing rates.
- Palaeo-geographic reconstructions of post-Palaeozoic times, a domain to which

French geologists have made a substantial contribution, undoubtedly greater than that of the United States.

- The segregation of continental crust as distinguished from the formation of the first continents.
- Mantle heterogeneities and their mixing by convection currents. At this point, I introduce the modern notions of bifurcation, catastrophe, deterministic chaos and 'hard turbulence' in thermal convection.

Many of these problems have yet to find a solution that is unanimously approved. This led me to justify many affirmations by references to recent articles and, where possible, review articles. These references are only targeted at scientists willing to go further on a particular subject. They will then find the other necessary references (and sometimes hundreds of them!).

Louis Lliboutry

1

Rationale

1.1 GEOLOGICAL SCIENCES AND GEOPHYSICAL SCIENCES

The topics outlined in this book do not (yet) form a discipline *per se*, if this word is to be given its etymological, academic meaning. To each field of education one discipline would therefore be associated. This is because the development of science and technology has rendered completely obsolete and unsuitable our classification into separate disciplines, which is two centuries old. One could indeed separate from modern physics and chemistry a more concrete and naturalist portion, including physical mechanics, thermodynamics, and heat and state changes, and adding physical geography and geology (excluding palaeontology). This would make a very functional new curriculum, essential for many jobs, and which would include, among other disciplines, geophysics and geology. This idea is most likely utopian, and was only introduced to list the fields that will be needed throughout the present book.

Geodesy is the daughter of astronomy. Seismology, meteorology and physical oceanography are the daughters of rational mechanics, for elastic bodies or perfect fluids. Geomagnetism and aeronomy are the daughters of physics. It was therefore natural that the international associations of scientists from these different fields should merge in 1924 into the International Union of Geodesy and Geophysics (IUGG). It also includes the International Hydrology Association, which has the largest membership, and which is involved in the management of water resources and the related human implications.

The IUGG represents all *geophysical sciences* and is the counterpart of the International Union of Geology (IUG), which represents all *geological sciences* (mineralogy, petrography, sedimentology, palaeontology, stratigraphy, structural geology and geomorphology). The application of some geophysical sciences to the exploration of the sub-surface is called *geophysical exploration*. For many geologists, this is the whole of geophysics. This very restrictive meaning of the word is a source of great confusion in library catalogues and classified ads.

At first, geophysical sciences were completely different from geological sciences: in their objectives, in the techniques used (although more and more geologists used sophisticated physical instrumentation and computers), but mostly in their methods and philosophy. These differences have now blurred. New fields have emerged, with their own meetings and scientific journals, which cannot be easily classified as 'pure geology' or 'pure geophysics': radiochronology, magmatology–volcanology, palaeomagnetism, planetary sciences, etc. It is therefore better to group together everything in a single discipline: *Earth Sciences*. In France, only the prior education makes a clear difference between geologists and geophysicists.

The present book concerns mostly (but not exclusively) three domains:

- *Structural geology*, which describes the large structures of the Earth, such as mountain ranges, and aims at reconstituting their kinematics and their past movements.
- *Geodynamics* extends structural geology by looking for the mechanical or thermal causes, often far away or very deep inside the Earth. Geodynamics uses mechanical models, e.g. the model of lithospheric plates, but is not restricted to *plate tectonics*.
- *Tectonophysics* studies rock structures at a local scale, and aims at reconstituting the deformations undergone, and at inferring the stresses, pressures or temperatures to which the rocks were submitted.

1.2 GEOLOGICAL RULES REDUCED TO THE LAWS OF PHYSICS

The reason the traditional geological method is insufficient for a complete study of processes, despite all it brings to the knowledge of the Earth, will be shown for the specific example of structural geology.

Traditional structural geology faces three obstacles:

- It only looks at movements and strains, without estimating the forces needed for them (no mechanical models).
- It cannot explain the origin of these forces, hence the cause of the horizontal and vertical movements, whose first origin is at depths not accessible to the geologist.
- It cannot test contradictory hypotheses, which is often possible in geophysics through modelling and computations.

Geophysics enables the linking of the processes observed by the geologist to the fundamental laws of physics. It cannot exist on its own, but is a necessary addition.

Without the quantitative knowledge of causes, geologists needed to use the principle (or dogma) of *uniformitarian theory*. This principle was introduced by Charles Lyell in 1833. It states that the processes responsible in the past for the features currently observed by the geologist are always active at the surface of the Earth. This principle contradicted the theory of natural catastrophes, introduced by Cuvier a decade before. But many ancient facts contradict the strict uniformitarian theory (e.g. glaciers covering regions now temperate or tropical, the formation of craters by meteorite impacts). These processes have been established with certainty, although they have never been observed during historical times.

Some geologists have tried to establish *geological laws*, which would be peculiar to the discipline (e.g. for the evolution of geosynclines, see Chapter 14), similar to the laws in biology, economics, psychology, etc. These laws concern complex processes, whose understanding is far from complete and certain, and they always suffer exceptions, in varying numbers. They are in fact *rules*, masking immutable physical laws, of which they are only consequences if the process has been correctly and completely analysed. The Russian language makes the difference between the law (*zakon*), and the 'behaviour in accordance with the law' (*zakonomernost*), which can be translated as 'regularity'.

The uniformitarian theory must therefore be replaced, not by 'geological laws' *sui generis* which cannot be changed, but by the principle that *physical laws do not change with time*. In the few rare cases where this has been doubted (e.g. variation of the gravitational constant), the hypothesis has been proved false.

1.3 USEFULNESS OF PHYSICAL MODELS

In the Earth sciences, the construction of *physical models*, followed by their mathematical processing, replaces experimentation. A model is a *simplified* model of reality (which is always far richer and more complex). A model forms an *isolated system* (which can only be an approximation), such that the laws of mechanics, physics and chemistry may be applied to it. From an epistemological point of view, this is a *reduction* to basic 'hard' sciences, i.e. with firmly established laws.

Through this book, 'models' will always signify physical models. *Mathematical models* are talked about quite a lot, but cover completely different cases. They may concern a physical model, after solving the underlying thermo-mechanical problem (this solution generally consists of a computer programme). Hydrologists make use of mathematical models, which are only simple empirical correlations of regularities, and have not been inferred from any physical model. From the start, quantum physics is a set of mathematical objects and equations, but it is no mathematical model of a reality which can expressed otherwise. This reaches a domain which can only be described mathematically.

The common language uses substantives, verbs and other signifiers which are spontaneous models of the Real. Pure mathematics must be seen as an extension of the common language, clarifying the meaning of these spontaneous models, ensuring their logical consistency, and building other models from the first. It is an extremely evolved language. Applied mathematics consists only of techniques (without any derogatory connotation). Mathematics should therefore not be considered a proper science.

The simplest physical models are found in geophysical exploration, when solving what is called an *inverse problem*. Often, one knows the rocks that should be located at depth, their density, their electrical resistivity, or the velocity elastic waves should have. From surface measurements, one aims at detecting the presence, thickness and dip (deviation from the horizontal) of the different layers. The unknown thickness and dip variables are the *adjustable parameters* of the model. More often, the velocity of the elastic waves, and the density and the resistivity of each layer are also adjustable parameters of the model. Once the parameters have been adjusted, one should not claim to have 'solved the inverse problem', but merely to have found one possible solution. An inverse problem indeed has no unique solution, but an infinity; there are almost as many solutions as the number of physical models one can imagine.

However, the models are often very complex, because many processes intervene. This is, for example, the case in thermal convection in the Earth, in the triggering of an earthquake, or in a volcanic eruption. The design of a physical model to explain them is already in itself a work of synthesis worthy of being published, even if it has not been tested yet, i.e. confronted by new facts.

In any case, the division of the scientific thought process into two or three successive and independent stages ((1) design of a physical model, (2) mathematical processing of this model to adjust its parameters, and (3) tests of the final model), does not correspond to the actual scientific steps. There exists a constant dialectic between the modelling and the mathematical processing of the model. The model may be too complex for the mathematical solving to be achievable, even numerically. It must then be simplified, certain parts of the equations must be neglected. A researcher cannot be criticised for publishing a model without completing all the numerical computations; but to be taken seriously, he/she must have estimated orders of magnitude. The neglected terms must have been demonstrated to be indeed negligible, and the terms retained in the model must be shown to produce results of suitable orders of magnitude. In this way, modelling is an 'art', like medicine, economics or politics.

1.4 BACONIAN AND POPPERIAN APPROACHES TO REALITY

This book will not follow the method of the book *Plate Tectonics* written by Le Pichon, Francheteau and Bonin (Elsevier, 1973). The authors there first present the existence of plates as a fact, and thence deduce consequences.

First, lithospheric plates are not real objects in the way of an ocean or a calcareous massif. They form a model which unites several processes, which allows computations at the scale of a continent, but which is *per se* inappropriate to detailed studies. In particular, when two continents above different plates collide, the results may be very different from traditional subduction. Rigid slabs may pile up here, whereas beds may deform there. *The plate theory is but one part of the geodynamics of the Earth.*

Second, the above approach is a Baconian approach. It follows the method proposed by Francis Bacon (*Novum Organum*, 1620). According to Bacon, science is characterized by *inductive* reasoning. A few specific facts allow the formulation of a general law, completely *true*, from which consequences can easily be inferred. The scientist has, in fact, a much more pragmatic approach, as explained by Karl Popper

(*The Logic of Scientific Discovery*, 1959). The scientist uses laws and theories obtained through induction, as long as they produce exact results. But he or she is ready to modify them, or totally reject them, as soon as the resulting predictions prove to be wrong. Each use of the law or theory implicitly constitutes a test. A series of positive tests makes the theory more trustworthy, but in all logic, a new test may always question it again. Induction does not yield irrefutable truths. From this analysis, it appears that the attitude of some: 'Let us wait for the demonstration of the plate theory as true before using it' is not scientific. But the attitude of others, who believe as a dogma that the plate theory is immovable because it is the Truth, is no more scientific.

The analysis of the scientific process led Popper to establish a clear and objective criterion to distinguish which laws and theories are scientific, and which are closer to religion, ideology, or fake science. This criterion is that a theory is scientific if it can be submitted to tests which may eventually disprove it.

Popper applied his criterion to psychoanalysis and to the dogmatic aspects of Marxism, but it must also be kept in mind for Earth sciences. Thus *creationism*, which relies on the dogma of a revealed Truth, irrefutable in essence, cannot be presented as one scientific hypothesis among others. Only the relevance of its teaching may be discussed, but this is no longer a scientific question.

In a less passionate sphere, many beginning researchers have a Baconian approach to Science, when they adopt without discussion a physical model found in some previous article and consider it as true, although it has been very little tested, or not at all. They use it to investigate a similar case, or to make predictions. Very well, but they should be less affirmative. At best, their work makes it possible to study the *sensitivity* of the results to the values of the parameters entering the model, in order to orient future work towards a better determination of the important but poorly known parameters. At worst, this is a purely academic exercise, which enables them to perfect their numerical analysis skills Their only excuse is that research managers have demanded that they obtain results in the limited time allotted to our short Ph.D. theses.

2

Earthquakes and seismic exploration

2.1 TECTONIC EARTHQUAKES. FAULTS

Seismic events (also called earthquakes when they cause destruction) correspond to a shaking of the ground due to the arrival of elastic waves, coming from a place called the *focus* or *source*.

Most earthquakes, and all the catastrophic ones, are tectonic in origin. They come from the sudden release, by breaking of the rocks, of elastic energy which has accumulated in a place over decades, whereas the neighbouring areas and the underlying layers have been deforming continuously and plastically, without any increase in the stresses.

It will be explained later, but let us mention now, that a distinct discontinuity exists everywhere around the Earth, at depths varying between a few kilometres and 70 km, the *Mohorovicic discontinuity*, or *Moho*. It separates the surface *crust* from the *mantle* underneath. The layer between the Moho and another discontinuity at a depth of 670 km forms the *upper mantle*.

Seismic regions cover the Earth over more or less narrow belts, often in *subduction zones* where cold and brittle crust is engulfed at depth (see Chapter 7). Most seismic foci appear at shallow depths, either at mid-ocean ridges (see Chapter 5) or more often in the upper continental crust, between 2 and 20 km deep. In the large subduction zones, they may be found at all depths down to 670 km. Seismic sources can sometimes be found in the upper mantle far from actual subduction zones: down to 90 km below Tibet, between 50 and 150 km deep in the region of the Strait of Gibraltar.

Seismologists classify earthquakes according to the type of recorded seismic waves, which depends exclusively on the depth of the source. They call all those with sources less than 60 km deep *shallow earthquakes*, those originating between 60 and 300 km *intermediate earthquakes*, and those with sources deeper than 300 km *deep earthquakes*. This classification does not hold any interest for the geodynamicist.

The rock failure occurs without disrupting the surrounding terrain (because the

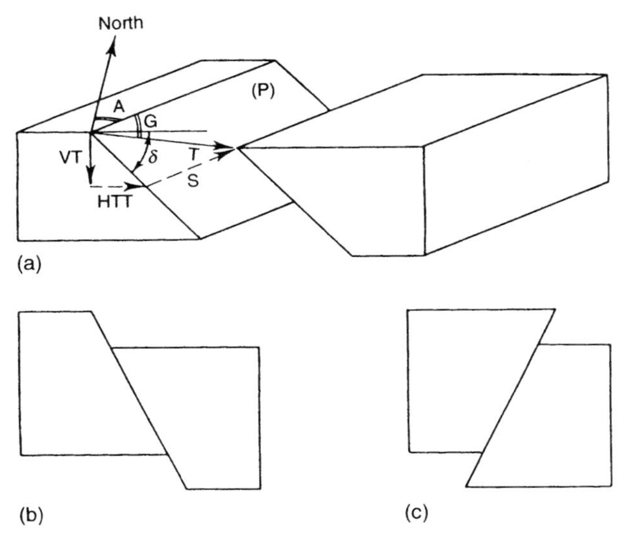

Figure 2.1. (a) Drawing of a normal fault; (b) normal fault; (c) reverse fault.

pressures exerted prevent it), and creates a *fault*, or more often reactivates an old one. Old faults may be hundreds of kilometres long, but only small portions play a role in the earthquakes. During a large earthquake, the fault may move at most by a few metres over a few tens of kilometres. This displacement decreases progressively until it becomes null at the end of the region affected by the seism, except in the case where the fault reaches the surface. The large faults which reach up to the surface cut the topography and the different geologic layers linearly.

In that case, the following values can be measured in the field (Figure 2.1):

- The vector, $\boldsymbol{T}$, corresponding to the *throw* (for geologists) or *displacement* (for seismologists). The angle with the edges of the fault is called the *slip angle* σ.
- Its horizontal component, $\boldsymbol{S}$, corresponding to the *strike–slip movement*. Its angle with the North is noted A and called *strike*, or *azimuth*. The movement is said to be *sinistral* if an observer placed on one side of the fault sees the other side moving to the left (Figure 2.1). The two forces which caused it were acting counterclockwise. Conversely, the movement is said to be *dextral* if the forces were acting clockwise.
- The perpendicular component of the movement is called *throw* by the seismologists. The angle of the vector with the horizontal plane is the *dip angle*, δ. It can be decomposed into the vertical throw (VT) and the horizontal transverse throw (HTT) (for geologists; seismologists call it the horizontal throw). According to the respective orientations of these two throws, the fault is said to be a *normal fault* or a *reverse fault* (Figure 2.1b and c). In a horizontal geological layer, a normal fault makes extension perpendicular to the fault possible, and a reverse fault makes compression possible.

2.2 INDUCED, VOLCANIC AND OTHER TYPES OF EARTHQUAKES

Induced earthquakes are triggered by the filling up of large dams (Rothé, 1968). A typical example is Boulder Dam, in Colorado, with a volume of 40 km^3. No earthquake had been recorded in the 15 years before its filling up. But in 1937, as water was reaching up to 130 m above the riverbed, a hundred seismic tremors were observed. Seismographs were then installed, which recorded thousands of tremors. The peak activity occurred some 9 to 10 months after the maximum water level (145 m) was reached, and seismic activity lasted another 10 years.

Kariba Dam, in Zambia, has long been the largest volume of retained water in the world (175 km^3). Its filling up caused a strong seismic activity, with a destructive earthquake of magnitude 6.2. At Koynia Dam (130 km south of Bombay, with a volume of 2.78 km^3), an earthquake of magnitude 6.4 occurred in December 1967. Even the relatively small dam of Monteynard (Vercors, France), with a volume of 0.275 km^3, caused in April 1963 an earthquake which moderately affected some neighbouring villages.

These earthquakes are not caused by the overall compression of the ground under the weight of water, but by the increase in pressure of the underground water in the cracks and joints of rocks. The friction between the sides of cracks is attenuated, and pre-existing elastic stresses become free to act.

Other seismic sources are underground explosions and implosions, whether natural or man-made.

Volcanic tremors are only felt locally, before or during volcanic eruptions. Before Montessus de Ballore, one century ago, all earthquakes were thought to be of volcanic origin. They are produced by the opening of fissures, or solidified magmatic plugs, which 'pop out' because of the high pressures created by the *magma* (molten rock). Decompression liberates the dissolved gases, which push on the 'plug'. Some small volcanic tremors may correspond to the collapse of the roof in magma chambers, or to echoes from the ramming of lavas propagating in the volcano's chimneys.

Neither volcanic tremors nor *phreatic eruptions*, resulting from the vaporization of groundwater entering into contact with magma, are indicative of an incoming catastrophic volcanic eruption. When half of Mount St. Helens (i.e. 4 km^3) blew up in the northwestern USA in 1980, the seismic activity of the preceding months had almost entirely subsided, and so had phreatic eruptions.

Underground nuclear tests have been important for the development of seismology. To detect them, the USA, the USSR and France developed extremely sensitive seismographs and worldwide networks of seismic stations. A long time, however, was needed before the records, and even the location of monitoring stations, became unclassified. These tests were important for seismic exploration because their timing could be known in advance.

Implosions are produced by the collapse of cavities: natural (in gypsum, for example), former mines, or underground quarries. In France, Yves Rocard had set up a monitoring network for the CEA (*Commissariat à l'Energie Atomique*, Atomic Energy Commission). When two submarines sank in the Mediterranean and imploded at a depth of 600 m, each time this network made it possible to know the exact location of the tragedies.

The detection of seismic signals is limited by a background noise, the *micro-seismicity*, of an amplitude that fluctuates widely with location. It comes from man-made sources (mainly lorries and trains), natural sources (wind, breaking waves, and mainly periodic pressure oscillations on the seafloor linked to strong swell).

2.3 MACROSEISMICS: SEISMIC INTENSITY IN A GIVEN LOCATION

In a given location, the seismic intensity measures the observed effects. For the same earthquake, the intensity will vary with the observing points, the distance to the source, and the geological structure. The site of higher intensity is the *epicentre* of the seism.

The intensities are denoted with Roman numerals so that they are not mistaken with the magnitudes, defined later. The intensity scales in use have successively been the Rossi–Forel scale (1873; 10 degrees, material destruction from degree VIII), the Mercalli scale (1888; 12 degrees, destruction from degree VII), the Mercalli scale modified by Cancani (1917; 12 degrees, destruction from degree VI). The MSK scale (1956), and the EMS 1992 scale which is given in Table 2.1, and specify the kind of destruction for different types of building, classified as:

- Type A: buildings made out of adobe (sun-dried clay) or unhewn rocks badly cemented.
- Type B: buildings made out of brick, timber and brick, hewn stones, or pre-designed concrete panels.
- Type C: buildings made out of reinforced concrete, well-built timber lodges.

Genuine aseismic buildings can resist earthquakes of intensity IX. An example is a modern hotel in Chimbote, which resisted an earthquake in 1970 whereas the surrounding town, mostly built from adobe, was destroyed.

Cost is a limiting factor; in the most seismically active region of Japan, regulations require that the buildings resist accelerations of $2.4\,\mathrm{m\,s^{-2}} = 0.24\,\mathrm{g}$ (i.e. intensity VIII only).

The Kobe earthquake killed 6,308 people on 17 January 1995, with a resulting destruction estimated at tens of billions of pounds sterling. The earthquake demonstrated that these regulations were not sufficient. Furthermore, in Kobe or Mexico City or other places, it was realized afterwards that contractors and architects had not followed the rules. Even in the absence of bribery, local governments are not stringent enough when faced with very fast urban growth.

The number of casualties depends not only on the intensity of the earthquake, but also on the local type of construction, on the density of population, and on the time at which the earthquake strikes. The most deadly earthquake recorded is the one occurred on 2 February 1556, which destroyed the capital city of China, Xian (830,000 victims). The next most deadly one also happened in China, on 27 July 1976, shortly before the death of Mao, at a time when China was closed to

Table 2.1. EMS-1992 scale for seismic intensities.

		Maximum ground acceleration (g)
I	Only recorded with seismometers	
II	Rarely felt (except by people at rest in upper storeys)	
III	Felt by few people (similar to a lorry passing nearby)	
IV	Felt by many people. Cracking, oscillation of chandeliers	
V	Many sleepers awake. Strong oscillation of chandeliers	0.012–0.025
VI	Small fissures in adobe and plaster walls	0.025–0.05
VII	Large fissures in type-A buildings, small fissures in type-B buildings; fall of chimneys	0.01–0.1
VIII	Partial collapse of type-A buildings; important damage in type-B buildings. Statues and tombstones are displaced. The water level in wells varies, lake waters get muddy; fissures appear in the ground	0.1–0.2
IX	Partial collapse of type-B buildings; important damage in buildings of type C. Landslides	0.2–0.4
X	Partial collapse of type-C buildings. Fissures in the ground can reach 1 m. Interruption of roads, railways, underground pipes and dams	0.4–0.8
XI	Similar damage, generalized	0.8–1.6
XII	Total destruction of the Earth's surface. All man-made constructions are destroyed. New lakes or waterfalls appear, riverbeds are displaced	>1.6

foreigners. This earthquake of Tangshan (province of Hopeh) officially killed 242,000 victims, but was rumoured to have, in fact, killed 655,000. In the course of the twentieth century, around 20 earthquakes caused more than 15,000 casualties. In Europe, the most deadly earthquake was the one in Messina (Italy), on December 28, 1908: 58,000 victims.

A shallow earthquake may cause a *subsidence* of the ground over a relatively large area. When this occurs below the ocean, there follows the propagation of a series of very large waves, much faster than the normal swell, which can cause great damage in coastal areas. This is a *tsunami*, improperly called a *tidal wave*.

Let us look at a recent example. On 12 July 1993, an earthquake of magnitude 7.5 occurred in the Sea of Japan, off Okushiri Island (southwest from Hokkaido). Linked to the subduction of the Japanese seafloor below the island of Hokkaido, this earthquake created a subsidence of 0.2 to 0.9 m. The resulting tsunami was characterized by waves of 15–30 m in the southern part of the west coast, 5 m elsewhere, although with heights of 10–20 m at the southern tip of the island, and 10 m in the north. The source of the earthquake was too close for the population to be warned in time, and large coastal villages were destroyed. There were only 114 casualties, the houses being traditional wooden buildings on one floor. The tsunami reached the Russian coast and Korea, travelling through the Sea of Japan at a speed of 670 km/h.

The countries bordering the Pacific Ocean have set up a tsunami warning system, but the predictions of amplitudes still need to be improved.

2.4 SEISMIC WAVES: GENERALITIES

Seismic tremors consist in elastic wave trains, infrasounds, but with speeds largely greater than in fluids (2 to 8 km/s depending on the environment). *Body waves* propagate along in all spatial directions, and may cross the Earth as *seismic rays* (the analogue of light rays), whereas *surface waves* propagate parallel to the surface of the Earth, and only affect it down to a certain depth.

The displacement at the source is not a periodic function of time. It can be considered as the sum of an infinite number of infinitesimal vibrations of any period, whose amplitudes depend on the period (Fourier integral). In the case of a sinusoidal, permanent vibration, the maxima and minima would travel at a speed called the *phase velocity* (or *celerity*), c. If T is the period, the wavelength is, by definition, $\lambda = cT$. As the predominant period is approximately 1 second for body waves and larger than 20 seconds for surface waves, they are also called short waves and long waves respectively.

The elastic wave train, of a finite total duration, propagates with the *group velocity* u. For any type of waves, elastic or not, material or not, the following relation can be demonstrated ($v = 1/T$ is the frequency):

$$u = c\left[1 - \frac{\lambda}{c}\frac{dc}{d\lambda}\right] = \frac{c}{1 + \frac{T}{c}\frac{dc}{dT}} = \frac{c}{1 - \frac{v}{c}\frac{dc}{dv}} \tag{2.1}$$

For example, the lows and highs of the swell propagate in deep water with the celerity $c = (g\lambda/2\pi)^{1/2}$. Therefore the wave train resulting from a passing ship propagates at the group velocity $u = c/2$.

The phase velocity of body waves does not depend on frequency, and there is no need in their case to make the distinction between phase velocity and group velocity. This is different for surface waves, for which the two velocities may differ by 10 to 30%. The sinusoidal vibrations making up a surface wave propagate at different velocities, and dispersion occurs. In a given location distant from the source, surface waves are very slow oscillations of large amplitude, which may cause panic but do not produce damage, unlike short waves.

The motion with time of a point submitted to seismic activity can be decomposed along three vectors perpendicular to each other. The same is possible for forces and accelerations. The fundamental equations of mechanics and elasticity can be written

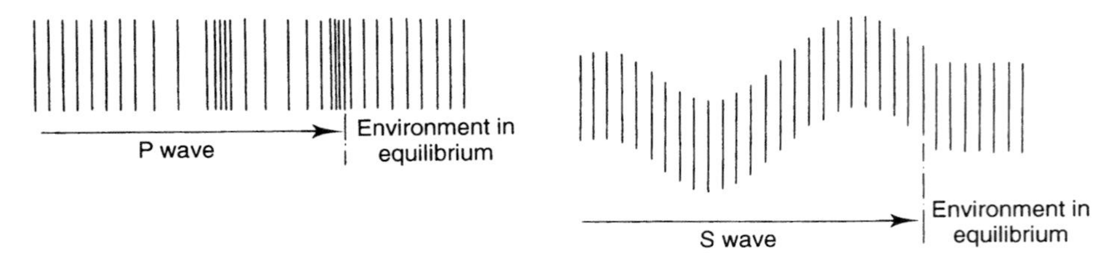

Figure 2.2. Displacements in an elastic medium of P and S waves propagating in the direction of the arrow.

along only one direction. In the absence of a free surface, one finds that three body waves, independent of each other, can propagate in a continuous elastic medium (see Figure 2.2):

- *longitudinal waves*, or *P* waves, correspond to movements along the direction of propagation. They create, along this direction, successive compressions and extensions. Their velocity v_p depends on the adiabatic elastic parameters k_s (bulk modulus) and μ_s (shear modulus), which will be defined more exactly in Section 10.6, and on the density ρ:

$$v_p = \sqrt{\frac{k_s + 4\mu_s/3}{p}} \tag{2.2}$$

- *transverse waves polarized vertically*, called SV waves.
- *transverse waves polarized horizontally*, called SH waves.

Transverse waves propagate with the velocity:

$$v_s = \sqrt{\mu_s/\rho} \tag{2.3}$$

Whatever the medium, $3k_s > 2\mu_s$, and therefore: $v_p > \sqrt{2}v_s$. During an earthquake, the P waves always arrive first (the letters P and S come from the Latin words *primae* and *secundae*). The S waves are generally the strongest and most destructive. Many people in countries seismically very active owe their lives to the fact that they rushed out of houses when feeling the vibration of P waves. People living on higher floors have to rush to the safest corners of their apartments, identified in advance.

During seismic activity, a given small volume of ground has a kinetic energy and an elastic potential energy, which continuously pass from the one to the other. For a sinusoidal oscillation of period T and amplitude a, the total density of seismic energy is proportional to a^2/T^2. The flow of seismic energy across a unit section, perpendicular to the propagation direction, is equal to this density multiplied by the group velocity u.

This energy has been radiated by the source, along all directions in space in the case of body waves, along all directions in a plane in the case of surface waves. The flow of seismic energy at a distance r from the source is therefore varying proportionally to r^{-3} for the latter, and r^{-2} for the former. This is true as far as the sphericity of the Earth is not taken into account. This is the *geometrical attenuation* with distance.

There is also a *physical attenuation* with distance, whose measure is a dimensionless parameter Q, called *quality factor*. By definition, a sinusoidal vibration loses a fraction $(2\pi Q^{-1})$ of its energy during one period (i.e. while the phase increases by 2π radians). It follows that the flow of seismic energy propagating through a ray tube decreases as $\exp(2\pi Q^{-1}r/\lambda)$.

There are two reasons for physical attenuation:

- When the medium is heterogeneous (e.g. in continental crust), seismic energy is diffusing all along its travel.
- Even if the medium is homogeneous, the deformations are not exactly elastic, and a small portion of energy dissipates as heat (this is the reason why Q^{-1} is often called *internal friction*). This process is called *anelasticity*, and does not depend very much on frequency. It affects S waves more than P waves. It should come from frictions on the grain boundaries, and differ from transient reversible creep (see Section 11.8). For a given pressure (i.e. a given depth inside the Earth), anelasticity increases with temperature. The determination of Q^{-1} therefore enables the detection of abnormal temperatures in the Earth's interior.

2.5 SEISMOMETERS

The ground movements are measured with horizontal and vertical pendulums (Figure 2.3). Let us note x the abscissa of its centre of gravity relative to the frame, fixed on the ground, ξ the abscissa of the ground in an absolute reference frame, and M the pendulum mass. The inertia force is $-M(d^2x/dt^2 + d^2\xi/dt^2)$. The return force, proportional to the elongation, is $-Cx$. The dampening force, resulting from electromagnetic induction currents, is proportional to the relative displacement

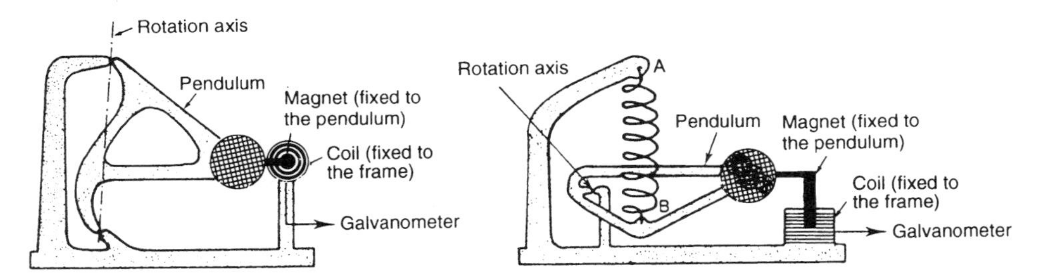

Figure 2.3. (a) Horizontal seismometer. The rotation axis is nearly vertical to increase the period. (b) Vertical seismometer. The indifferent (or astatic) balance is approached as much as possible to increase the period. Perfect equilibrium is achieved when the angle AOB is right and the spring has a 'zero length'. Such a spring, with joined spirals and a fixed length for loads less than F_0, has a length proportional to the load when it becomes larger than F_0.

speed dx/dt. The balance of forces reads:

$$M\left(\frac{d^2x}{dt^2}+\frac{d^2\xi}{dt^2}\right)+f\frac{dx}{dt}+Cx=0 \tag{2.4}$$

The intrinsic period of the pendulum in the absence of ground vibrations is approximately:

$$T_0=2\pi\sqrt{M/C} \tag{2.5}$$

A sinusoidal ground vibration with a period T causes a forced oscillation of the pendulum with an identical period T, and then:

$$\frac{d^2x}{dt^2}=-\left(\frac{2\pi}{T}\right)^2 x \tag{2.6}$$

The members $M\,d^2x/dt^2$ and Cx are therefore linked through the equation:

$$M\frac{d^2x}{dt^2}=-\left(\frac{T_0}{T}\right)^2 Cx \tag{2.7}$$

The intrinsic period of accelerometers is much smaller than the period of seismic waves. $M\,d^2x/dt^2$ is therefore negligible compared to Cx. After amplification, the value recorded is $q=Vx$. The quantities $M\,d^2\xi/dt^2$ and Cx are predominant:

$$q=-\frac{VM}{C}\frac{d^2\xi}{dt^2} \tag{2.8}$$

Accelerometers measure the accelerations of the ground, which cause damage.

In the design of seismometers (see Figure 2.3), some tricks yield a period T_0 greater than the period of seismic waves despite reduced weights and sizes. Cx can then be neglected when compared to $M\,d^2x/dt^2$, so that $d^2x/dt^2 \approx -d^2\xi/dt^2$, and therefore $x \approx -\xi$. The transducer also measures the electromagnetic induction force appearing between the extremities of a coil fixed to the frame, when a magnet fastened to the pendulum is moving. This force is proportional to dx/dt. After amplification, one records the quantity $q \approx -W\,d\xi/dt$, corresponding to the ground velocities.

The exact theory is more complicated and depends on the type of seismometer considered. In particular, amplifiers only amplify correctly a specific frequency band.

Over the years, recording was successively performed with a stylus on blackened paper, a light spot on photographic paper, a hot filament on heat-sensitive paper, and finally magnetic recording. Nowadays, digital recording and further processing has replaced filtering and analogue recording, so that no information is lost. A seismologist now needs solid background knowledge in numerical signal processing.

Three seismometers are needed to observe all three components of the ground movement. A complete seismic station is made of three seismometers with short periods ($T_0 \approx 1$ second), three seismometers with long periods (30 to 40 seconds), and an extremely precise time-base (quartz clock and reception of hourly time signals with automatic synchronization).

Seismology greatly benefited from the US implementation in the 1960s of the WWNSS (World-Wide Network of Standard Seismographs). In the last years, other regional or national networks have been installed, using more modern instruments with digital recording capabilities. Among them are the networks from Germany (GDSN, regional), the United States (GSN, worldwide), China (CDSN, regional), France (*Géoscope*, worldwide, and *Terrascope*, regional), Italy and Japan (regional).

Increasingly, seismographs are used as *arrays*, networks of around 20 seismographs over areas of a few square kilometres or a few tens of square kilometres, with remote transmission of signals to a single location. By appropriately mixing the signals, the background noise can be greatly reduced.

Seismic exploration uses mainly *geophones*, small and robust seismographs whose amplification and physical characteristics are poorly defined, and designed to exclusively record the vertical movements of the ground.

2.6 DISSIPATED ENERGY, SEISMIC EFFICIENCY, SEISMIC MOMENT

Chapter 12 will examine in detail the processes leading to rock rupture, to small earthquakes which always follow a major earthquake (*aftershocks*), and to the ones which often precede them (*foreshocks*). This will be accompanied by the inevitable question: can earthquakes be forecasted? Here, we shall concern ourselves with only one single earthquake. The energies at play and the concepts of moment and magnitude can be simply introduced. We need however to anticipate Chapter 9 to define what is called shear and shear stress in mechanics of continuous media.

Simple shear corresponds to a deformation such that all the points move parallel to an axis Ox of a plane xOz, with a displacement U proportional to the third coordinate, $y : U = \gamma y$. The value of shear is γ, dimensionless. This is a diffuse sliding, one layer above the other, without volume changes. In mechanics, cutting with shears, or the movement of a fault, are ‘shear fractures’, not shears.

A layer parallel to xOz undergoes the shear γ when its two faces are submitted to tangential forces parallel to Ox and acting in opposite directions, with values per unit area τ and $-\tau$. This value τ has the dimension of a pressure and is called *shear stress*. For an elastic medium, shear and shear stress are proportional:

$$\tau = \mu\gamma \tag{2.9}$$

The elastic parameter μ is the *Coulomb modulus*, or *shear modulus*. For rocks at the depths at which shallow earthquakes occur, its value ranges between 30 and 60 GPa. It increases with ambient pressure, i.e. with depth. ($1\,\text{GPa} = 10^9\,\text{Pa} = 10\,\text{kbar}$; and $1\,\text{Pa} = 1\,\text{N}\,\text{m}^{-2} = 1\,\text{J}\,\text{m}^{-3}$).

The elastic energy stored per unit volume is computed by supposing one side of the layer is fixed, and the shear stress increasing progressively and reversibly from 0

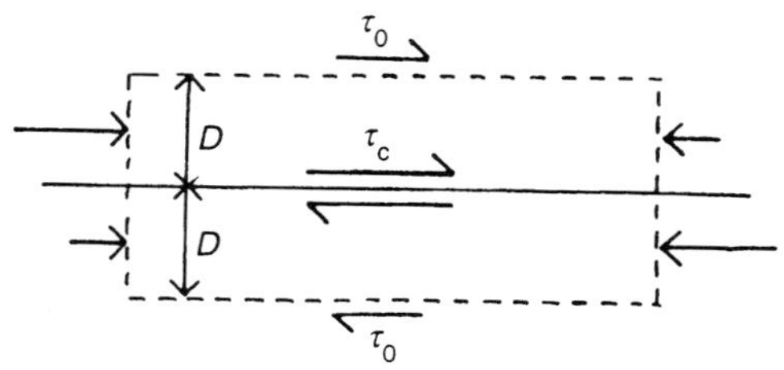

Figure 2.4. Shear stress (half-arrows) and normal stress (full arrows) around a fault.

to τ:

$$e_{elast} = \int_0^\tau \gamma \, dt = \frac{\tau^2}{2\mu} \tag{2.10}$$

Before an earthquake and close to its source, shear and shear stress are continuous functions of the coordinates. The fault's throw becomes zero at its extremities. But good global estimates can be obtained with the model of Figure 2.4. Just before the earthquake, the shear stress on planes parallel to the fault is uniform and equals τ_c over a distance D on each side of the fault, and over an area A equal to the area of the fault. The stress equals τ_0 elsewhere (in the real world, the transition from one value to the other is progressive). As soon as the fault becomes active, the shear stress is τ_0 everywhere. In particular, the friction between the two sides of the fault is τ_0. The throw, henceforth denoted R (from the French word: '*rejet*'), is uniform over the area A.

The initial model of *elastic rebound* postulated wrongly that $\tau_0 = 0$. The realization that the *stress drop* $(\tau_c - \tau_0)$ was in reality only a small fraction of the permanent stress in the area was a major discovery. This fact explains why the date of an earthquake is nearly impossible to forecast.

The stress drop during an earthquake is:

$$\Delta E_{elast} = 2DA \frac{\tau_c^2 - \tau_0^2}{2\mu} \tag{2.11}$$

A portion E of this energy has been radiated as seismic waves: this is the earthquake's energy. The remainder was dissipated as heat through friction between the edges of the fault, and equals the work of the friction force $A\tau_0 R$. Furthermore:

$$\frac{R}{2D} = \gamma_c - \gamma_0 = \frac{\tau_c - \tau_0}{\mu} \tag{2.12}$$

Therefore:

$$\Delta E_{elast} = 2DA\left(\frac{\tau_c - \tau_0}{\mu}\right)\left(\frac{\tau_c + \tau_0}{2}\right) = AR\left(\frac{\tau_c + \tau_0}{2}\right) \tag{2.13}$$

and

$$E = \Delta E_{elast} - A\tau_0 R = AR\left(\frac{\tau_c - \tau_0}{2}\right) \quad (2.14)$$

The *seismic efficiency* is the ratio:

$$\frac{E}{\Delta E_{elast}} = \left(\frac{\tau_c - \tau_0}{\tau_c + \tau_0}\right) \quad (2.15)$$

It is now known that the shear stress drop during an earthquake is of the order of a few MPa, whereas in seismic areas, the shear stress is permanently around a few tens of MPa. Seismic efficiencies are therefore less than 10%.

The drop in the tangential force between the two sides of the faults, multiplied by the average distance between these opposing forces, is termed the *seismic moment*:

$$M_0 = A(\tau_c - \tau_0)D = \mu AR/2 \quad (2.16)$$

Comparing with the expression of E in equation (2.14):

$$E = M_0(\tau_c - \tau_0)/\mu \quad (2.17)$$

Today, E and M_0 can be estimated at better than a factor 2, from the numeric recording of an earthquake on seismic networks. Seismic moments have been found to range between 10^{15} and 10^{23} N.m (i.e. 1 to 10^7 MPa.km^3). Energies can range between 10^{10} and 10^{17} J. As a comparison, the chemical energy produced by 1,000 tons of TNT (the kiloton, used as energy unit for nuclear weapons) is 4.2×10^{12} J. These values may vary by 7 orders of magnitude depending on the earthquake. This is not related to the stress drop, which varies at most between 0.1 and 10 MPa, but to the volume $2DA$, which can vary from less than 10 to more than 10^6 km^3.

2.7 MAGNITUDE OF AN EARTHQUAKE

In order to compare earthquakes, Richter introduced in 1935 a logarithmic scale of *magnitudes* (similar to the one used for stars). The magnitude of an earthquake should not be confused with the intensity felt in a given location, as is often done by journalists obstinately calling the magnitude 'intensity according to the Richter scale'.

The initial definition of magnitude was grounded on the response of a particular horizontal seismograph, the Wood–Anderson seismograph, 100 km away from the source, a situation which never happens. The reduction of observations to this standard case is doubtful, and yields diverse magnitudes for the same earthquake. Several operational definitions of magnitude have been suggested (see Madariaga and Perrier, 1991). Among them, the two following ones are of use for the earthquakes recorded at a distance Δ from the source between 25° and 90°. The amplitude a of the vertical displacement of the ground is in microns, the period T in seconds, and Δ in degrees.

For surface waves only, a magnitude is defined as:

$$M_S = \log\left(\frac{a}{T}\right) + 1.66 \log \Delta + 3.3 \quad (2.18)$$

For any source depth h, a magnitude m_B is drawn from the body waves by:

$$m_B = \log\left(\frac{a}{T}\right) + \text{function}(\Delta, h) \quad (2.19)$$

Gutenberg and Richter estimated roughly the seismic energy E (in Joules), and found that it was correlated to the magnitude M_S:

$$\log E = 1.5M_S + 4.8 \quad (2.20)$$

In 1977, Kanamori found that the computed seismic moments were correlated with the corresponding magnitudes M_S:

$$\log M_0 = 1.5M_S + 9.0 \quad (2.21)$$

Comparing equations (2.12), (2.14) and (2.9):

$$\frac{E}{M_0} = 6.3 \times 10^{-5} = \frac{\tau_c - \tau_0}{\mu} \quad (2.22)$$

The stress drop associated with a shallow earthquake should then be of 1.9 to 3.8 MPa. The estimations of seismic moments computed by the University of Harvard for all important earthquakes showed that the correlation (2.21) is not valid any more when $M_S > 7.5$. The following magnitude has been defined using the seismic moment:

$$M_W = \tfrac{2}{3} \log M_0 - 6.0 \quad (2.23)$$

It can be very different from M_S when the earthquakes are very strong.

The use of M_W prompts a new assessment of the strongest earthquakes in the twentieth century. The Assam earthquake (on 15 August 1950) keeps its magnitude of 8.6 but is no longer the strongest. It is surpassed by six other earthquakes:

21–22 May 1960	Lebu-Valdivia (Chile)	$M_W = 9.5$ (5,000 dead)
28 March 1964	Anchorage (Alaska)	$M_W = 9..2$ (130 dead)
9 March 1954	Aleutian Islands	$M_W = 9.1$
4 November 1952	Kamchatka	$M_W = 9.0$
31 January 1906	Ecuador	$M_W = 8.8$ (2,000 dead)
4 February 1965	Aleutian Islands	$M_W = 8.7$

Fortunately, four of them occurred in nearly desert regions.

The estimates of E could be greatly improved with the new seismic networks. Indeed, they provide data already digitized, and are more sensitive to frequencies between 0.1 and 1 Hz, where most seismic energy concentrates. These frequencies were partially filtered in the WWNSS, to better distinguish signals from the background noise due to micro-seismic activity.

Recently, Choy and Boatwright (1995) studied the 397 earthquakes of magnitude

$M_S \geq 5.8$, recorded worldwide between 1986 and 1991 (including 10 earthquakes with magnitudes larger than 7.5). They found that the correlation expressed in equation (2.20) uses a constant of 4.4 and not 4.8. This implies an average stress drop $(\tau_c - \tau_0)$ of 0.47 MPa. The correlations improve much when computed on seismic regions with one single type of focal mechanism each. The stress drop is less than 1.5 MPa in the subduction zones with reverse faults, and greater than 3.0 MPa in transform faults affected by strike-slip movements. The values obtained in the other cases are between these two end-members.

Further in this book, we shall only use the magnitude M_E proposed by Choy and Boatwright (1995):

$$M_E = \tfrac{2}{3} \log E - 3.2 \tag{2.24}$$

Let us remark that, if the equation (2.20) had been adjusted by Richter, considering M_S as a function of log E, and not the opposite, then the coefficient of M_S would have been larger than 1.5.

2.8 RAY REFLECTION AND REFRACTION; HEAD AND SURFACE WAVES

The points where the seismic vibrations are in phase form a *wave surface*. The energy trajectories are *seismic rays*. In an isotropic medium, the seismic ray is always perpendicular to the wavefront.

When interacting with another medium, the seismic wave produces reflected and refracted waves, whose respective directions are given by the Snell–Descartes law of optics. Assuming the surface between the two is horizontal,

- a P wave produces reflected P and SV waves. When the reflection is not total, the P wave also produces refracted P and SV waves (see Figure 2.5). The same is true for incident SV waves.
- an SH wave produces a reflected SH wave. When the reflection is not total, it also produces a refracted SH wave.

Giving the two media the indices 1 and 2 respectively, the Snell–Descartes law reads:

$$\frac{\sin i_1}{v_{P1}} = \frac{\sin i_2}{v_{P2}} = \frac{\sin j_1}{v_{S1}} = \frac{\sin j_2}{v_{S2}} \tag{2.25}$$

When one of the sine values is greater than 1, the corresponding ray does not exist. There are thus many possibilities of total reflection.

Head waves

As in optics, the Huyghens principle is valid. In order to know which waves propagate outside a wavefront (Σ), the source O can be replaced with an infinity of sources $F_1, F_2, F_3, \ldots$ in phase, located on (Σ) and transmitting waves of the same

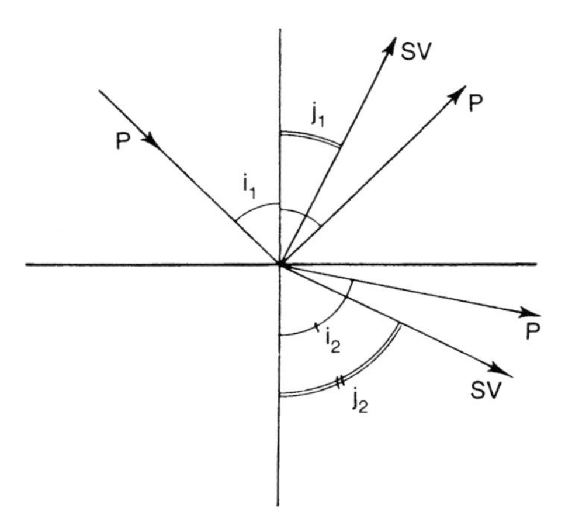

Figure 2.5. Reflected and refracted seismic rays at the interface between two media, when the incident wave is a P wave.

nature (P, SV or SH, accordingly), their amplitudes decreasing away from the propagation direction of the ray (Figure 2.6).

Let us apply the Huyghens principle to a spherical wave transmitted from O with a celerity v_1 in the medium (1), and arriving in medium (2) where the celerity is $v_2 > v_1$ (Figure 2.7a).

For incidence angles greater than i_0 (defined by $\sin i_0 = v_1/v_2$), there is no refracted ray. Further than the wavefront defined by the horizontal circle going through I, the wavefront in medium (2) reaches the interface between (1) and (2) perpendicularly. In this plane, it forms a horizontal circle whose radius increases at the speed v_2. According to the Huyghens principle, this circle behaves like an infinity of sources in phase, as an annular source growing with the speed v_2.

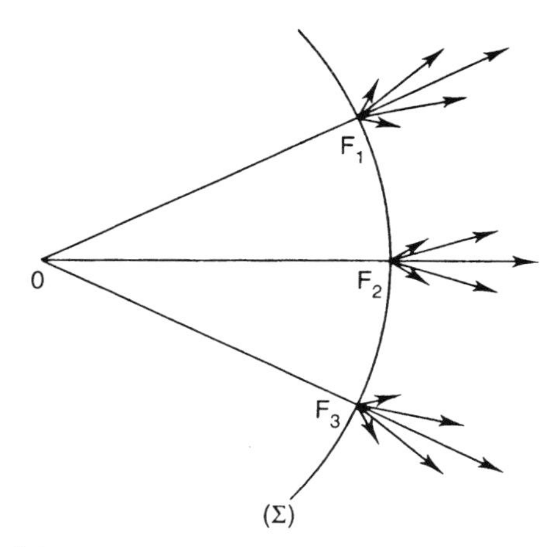

Figure 2.6. Huyghens principle.

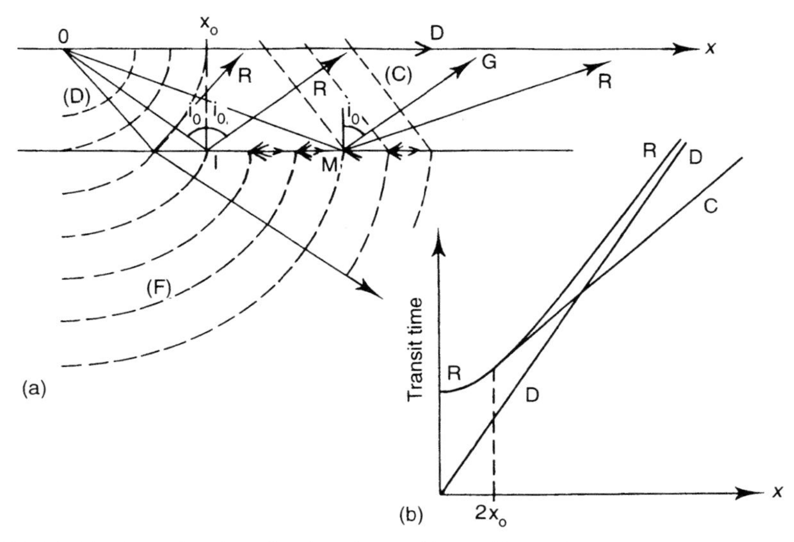

Figure 2.7. (a) Seismic exploration of a plane layer with a uniform thickness (*D*), *OI*: wavefront and ray for the direct wave; (*R*): reflected ray; (*F*), *IM*: wavefront and ray for the refracted wave; (*C*), *MG*: wavefront and ray for the head wave; (b) Corresponding time–distance graph.

In the first medium, this speed is supersonic. Consequently, the annular source produces a shock wave, called the *head wave*. Its transit time corresponds to *OI* at speed v_1, *IM* at speed v_2 along the interface, and *MG* at speed v_1. The trajectory *MG* is along the critical incidence angle, according to a *reversibility principle* analogue to the one used in optics.

Exploration geophysicists use the time–distance curve displaying the travel time as a function of the distance between the shooting point and the geophone: $OG = x$. Seismologists call the time interval 'propagation time', and its graph as a function of the angular distance between epicentre and receiving station is the 'time–distance curve' (see Figure 2.7).

In a layered medium with celerities increasing downwards, a head wave originates at each interface. As shown in the time–distance graph of Figure 2.8, at increasing distances from the shooting point, the first waves detected correspond to the head waves generated at deeper and deeper interfaces. The energy associated with a head wave is, however, very small, and it is often not recorded any more at long distances (or what is recorded is, in fact, a refracted wave, which penetrated a lower layer on its course).

Head waves of interest to us are the P or S waves propagating along the Mohorovicic discontinuity (Moho), which separates the crust from the mantle (P_n or S_n waves). In the terrestrial crust, v_p increases *grosso modo* with depth, up to values of 6.7–7.4 km/s at the bottom. But P_n waves show that it ranges between 7.8 and 8.2 km/s in the upper mantle close to the Moho.

When the focus of an earthquake is not too deep, some waves are propagated along the surface, at shallow depths compared to the distances travelled. They are

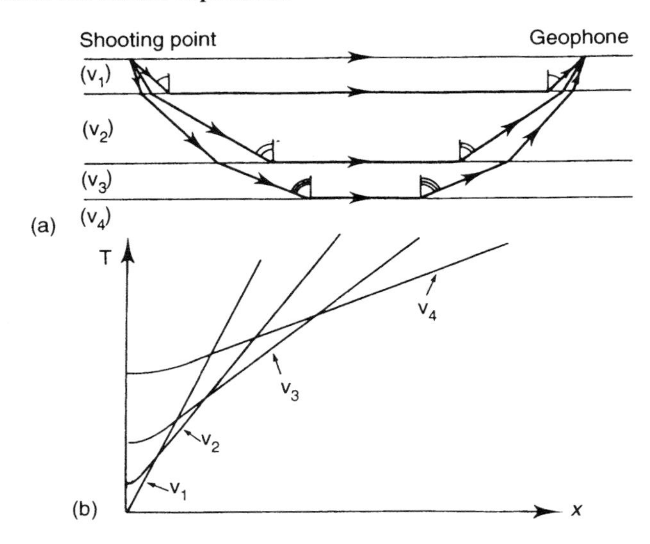

Figure 2.8. Seismic exploration of a layered medium: (a) energy trajectories; (b) timedistance graph.

called *surface waves*. We saw that in the presence of a discontinuity, P and SV waves on one side, and SH waves on the other side, behaved independently and were decoupled. The same happens here for the vibrations in a vertical plane, the *Rayleigh waves*, and the vibrations in a horizontal plane, the *Love waves*.

In a homogeneous medium, the Rayleigh waves are similar to the swell, the return force being elastic and no longer due to gravity. The layering of the Earth imposes that all surface waves, horizontally progressive, behave like stationary waves vertically. As any stationary wave, the surface waves have several *modes*. (Surface waves are a particular case of *guided waves*, like the centimetric electromagnetic waves propagating in a metallic 'waveguide'.)

Figure 2.9 shows the phase and group velocities as a function of period. As the period increases, surface waves penetrate deeper and deeper into the mantle. For periods larger than 200 seconds, the whole mantle is involved. Such waves are perceptible only when completely stationary waves are established on the Earth's scale, which requires very definite periods. These are the *eigenfrequencies* of the Earth, which are observed only during the largest earthquakes.

2.9 INNER CONSTITUTION OF THE EARTH AND SEISMIC PHASES

Below the terrestrial crust, the celerities of P and S waves depend more on depth than on the geographical location. They increase nearly continuously between the Moho and a major discontinuity located 2,886 km deep. The only decrease, very moderate, occurs around 100-km depths (see Section 9.15). This is the *mantle*, formed of silicates. The variation of celerities with depth increases at depths of 410 km, 520 km,

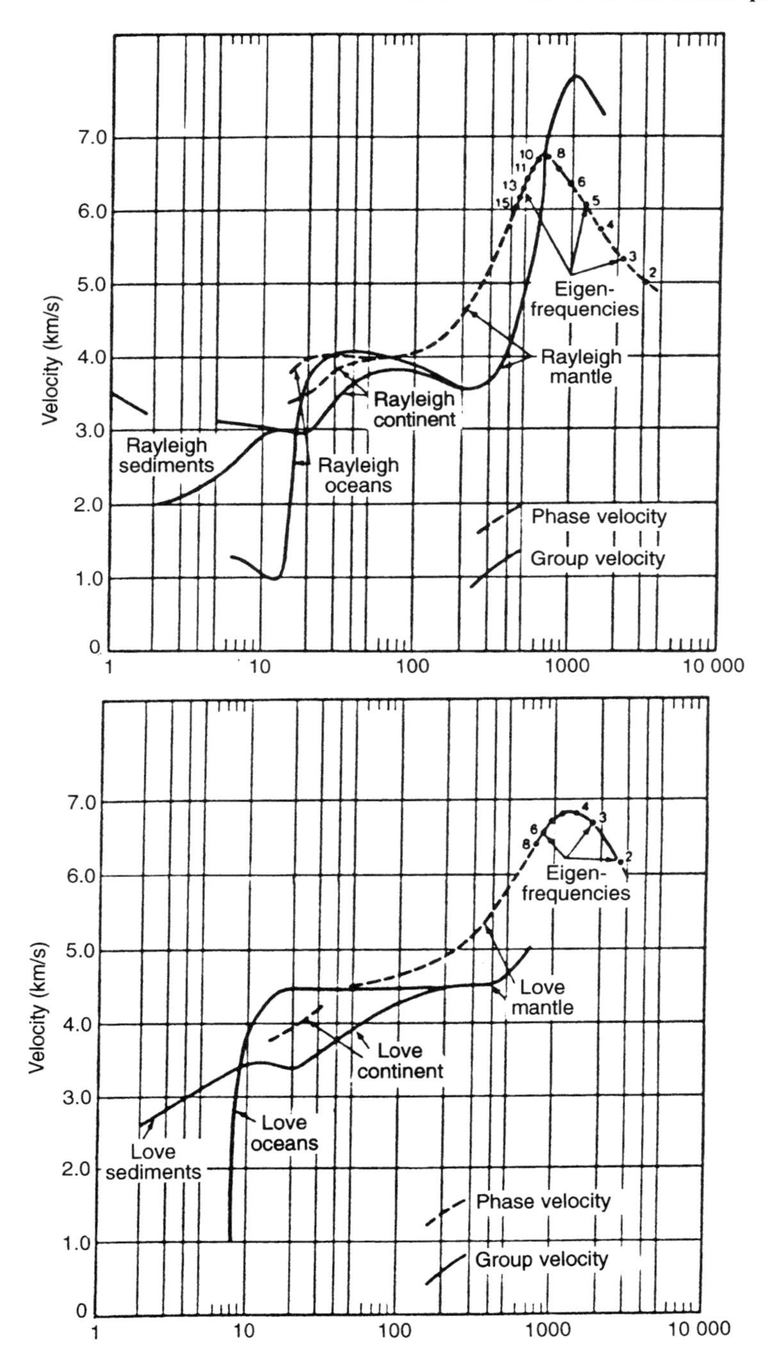

Figure 2.9. Dispersion of surface waves as a function of their period: (a) Rayleigh waves; (b) Love waves. From Bureau des Longitudes, 1977.

and particularly 660 km. This can be explained by the modifications of minerals, without the need to call for a modification of the global chemical composition (see Section 17.2). Only the last discontinuity is strong enough to reflect some seismic waves.

At a depth of 2,886 km (i.e. 3,485 km away from the centre of the Earth, give or take a few kilometres), the celerity of P waves drops abruptly from 13.7 to 8.2 km/s, and S waves disappear. One enters the *outer core*, which is liquid. In its centre, one finds a solid *inner core*, 1,220 km in radius. Anticipating Chapter 15, we can say that the core is made of ferronickel with 5% nickel, in addition to 5 to 10% of light elements (probably sulphur and oxygen) in the liquid part. In passing, let us remark that at the considerable pressures in the core, S and O have acquired the properties of a metal.

Dziewonski and Anderson (1981) proposed a model of seismic velocities depending only on the radius (PREM model; see Section 17.4). It can be roughly resumed as:

Mantle:	$v_P = 8.05\text{–}13.69$ km/s;	$v_S = 4.70\text{–}7.27$ km/s
Outer (liquid) core:	$v_P = 8.24\text{–}10.16$ km/s	
Inner core:	$v_P = 11.1$ km/s;	$v_S = 3.6$ km/s

The small velocity of S waves in the inner core should not come as a surprise. Because of the immense pressure, the density is very high, as well as the incompressibility measured with $1/k$. However, the parameter μ, which measures the possibility of shear without volume reduction, is not influenced by pressure.

The velocity in the mantle increases with depth, and therefore the seismic rays are curving upward. A PP wave is a P wave that has been reflected once on the surface of the Earth and remained a P wave. A PcP wave is a wave reflected on the core–mantle boundary, which remained a P wave (Figure 2.10). There are also waves denoted PS, SP, PcS, ScP, SS, ScS, etc.

A P wave crossing the outer core and reaching the surface of the Earth will be a PKP wave (or a PKS wave). If it crossed the inner core and re-crossed the mantle as a P wave, it could become a PKIKP wave (the inner core being crossed as a P wave) or a PKJKP wave (the inner core being crossed as an S wave).

The arrivals of large numbers of wave fronts or *phases* at a particular location after large earthquakes can thus be explained. Other phases occur when the earthquake is deep (e.g. pP waves). The different time–distance graphs are shown in Figure 2.11. In fact, seismologists use tables of propagation times: the tables of Jeffreys and Bullen (1940), corrected by Herrin *et al.* (1968) for the P waves, and by Randall (1971) for the S waves.

One of the current areas of study in seismology is the search for the small deviations of the propagation times observed from the ones given by the tables (*residuals*). They are used to deduce the *lateral* variations of velocity in the mantle, i.e. as a function of geographical coordinates, for a given radius. This method is called *seismic tomography*, as it studies the lateral variations on centred spheres, layer by layer (see Section 18.4).

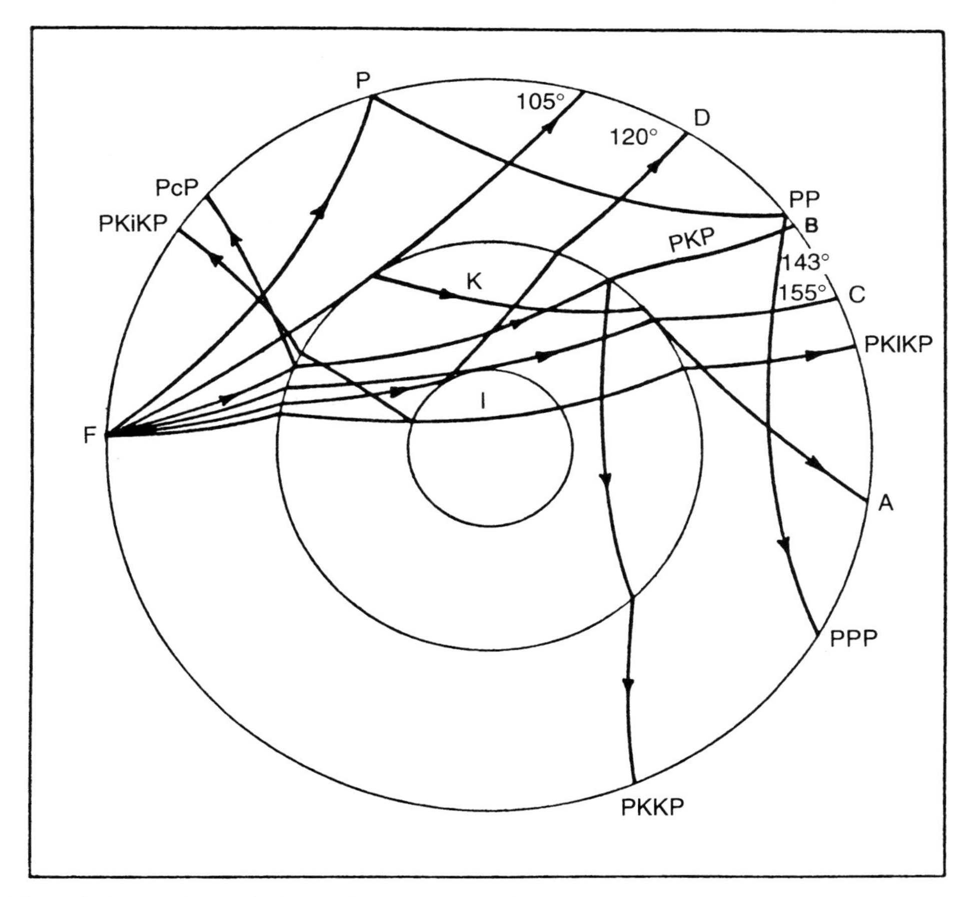

Figure 2.10. Seismic rays in the Earth. From Bureau des Longitudes, 1977.

2.10 TECHNIQUES OF SEISMIC EXPLORATION

Seismic exploration makes use of artificial earthquakes. The dropping a weight is enough to explore the subsoil over ten metres for civil engineering applications. At sea, one uses airguns, which expel compressed air, electric sparkers, or explosions in a pierced chamber (*flexotir* system). On shore, one uses breaking explosives, with no-delay detonator, which are emplaced in a drill hole which is later filled (*bulling*). To study structures as deep as 50 km ('experimental seismology'), several neighbouring holes are drilled, down to approximately 60 m, in a basement outcrop. The holes are filled to one-third with explosives, and two-thirds with sand for bulling. The explosions are triggered simultaneously at millisecond precision.

In order to avoid the formalities and the precautions necessary to use explosives, and to be able to work close to buildings which would be shaken by the explosions,

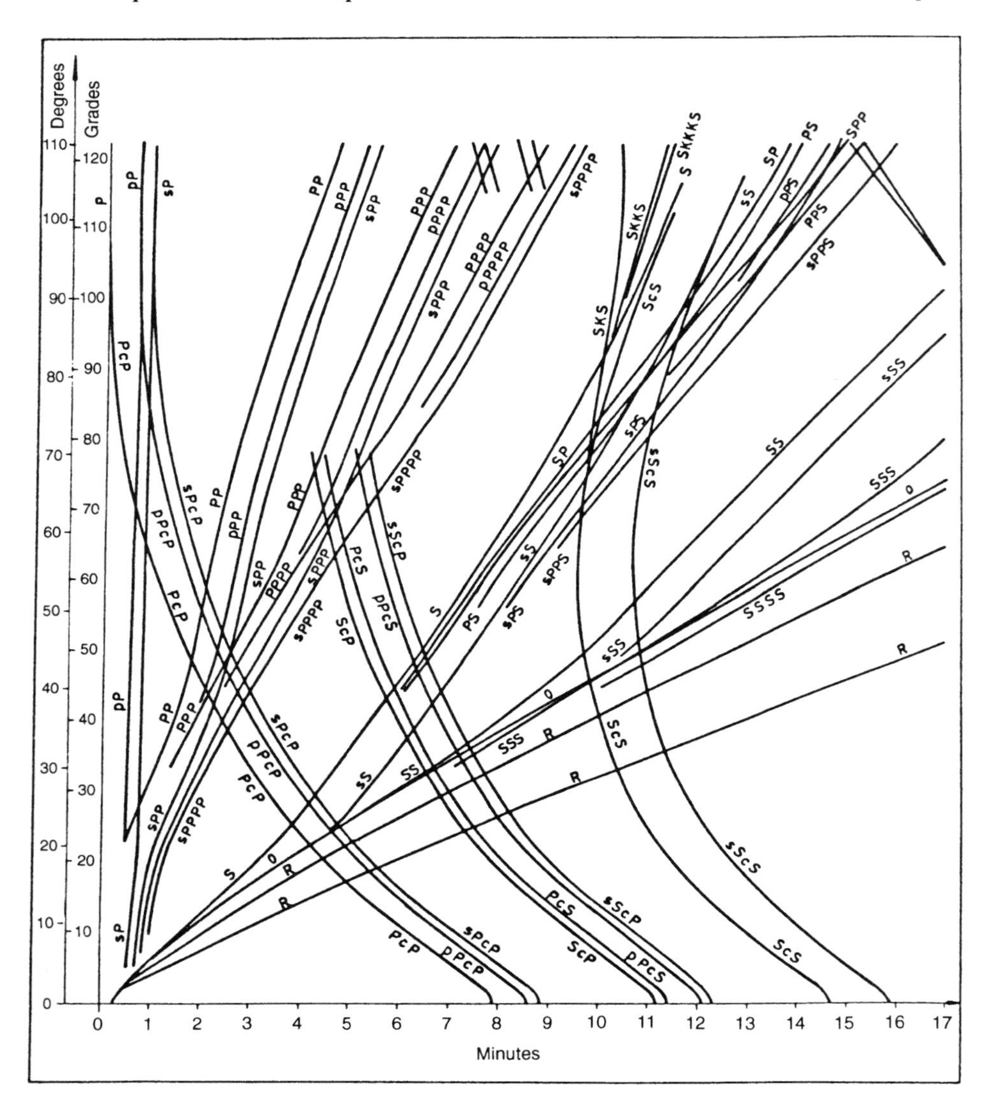

Figure 2.11. Time–distance graphs for a focal depth of 150 km. The X-axis shows the reduced time $t - t_P$. From Bureau des Longitudes, 1977.

the *Vibroséis* system is more and more used. This system uses maintained waves, and records the phase shifts.

The receivers used, simple geophones or portable seismographs in experimental seismology, are designed to record the short periods (P and S waves). The end of the recording is blurred by the surface waves (*ground roll*). The waves recorded are reflections from the underground discontinuities and, if the distance between the shooting points and the geophone is large enough, conic waves can be observed (they are also improperly called 'refracted waves').

In *reflection seismics*, the shooting points are close to the network of geophones, and the reflected waves (i.e. the echoes) are recorded. This method is the only one used at sea: one ship tows the geophones and the *streamer* (seismic cable linking them to the recorder) on one side, and the seismic source on the other. The resulting profile is continuous, where the different sediment layers and the rocky basement are visible.

In *refraction seismics*, the shooting point and the geophones are aligned on a long distance, and the conic waves are used. This allows the determination of velocities in the successive layers, as long as they are increasing. And these layers should not deviate too much from the horizontal either. By inverting the direction between the shooting point and the geophones, another dataset is obtained, which allows the determination of both the dip of the layers and their seismic velocities.

2.11 FOCAL MECHANISMS. SYNTHETIC SEISMOGRAMS

For a strong earthquake with focus *F*, several stations record the first arrival of *P* waves, also called *iP* (*impetus of P waves*). The path of these waves in the mantle being known, it is possible to determine the points where the corresponding seismic rays cut a small sphere centred in *F* (this is always in the lower hemisphere). These points are divided into two classes, depending on whether the *iP* is a movement toward the focus (meaning that some extension in this direction has disappeared during the earthquake) or away from the focus (meaning that some compression along this direction has disappeared). From the general stress theory (presented in Chapter 12), it can be predicted that the 2 classes can be separated by 2 planes passing through *F* and perpendicular to each other (*nodal planes*). One class occupies 2 opposite quadrants (Figure 2.12).

This study is now routinely performed by seismological data centres. It is called the determination of the *focal mechanism*, as early on it was thought this would yield the orientation of the fault plane and the direction of the throw. In fact, this determines *the stress drop in the focal region allowed by the earthquake*. It is only a few per cent of the preexisting deviatoric stress (the difference between the stress and a simple isotropic hydrostatic pressure). However, it may be assumed that both are proportional, and have the same planes of symmetry, so that the knowledge of focal mechanisms is of the utter interest in geodynamics. It indicates the orientation and the sign of the tectonic forces.

The computation of the ground movement in a given geographic location as a function of the underground geology, is a most important problem in seismic engineering. To give the location of the focus and the focal mechanism is not sufficient. The computation of such a *synthetic seismogram* is made by assuming a particular *model of seismic source*, i.e. either a particular throw, or a particular variation of the stresses at the source, both as a function of time. In theory, the model of the source could be computed from a fine study of the source processes and

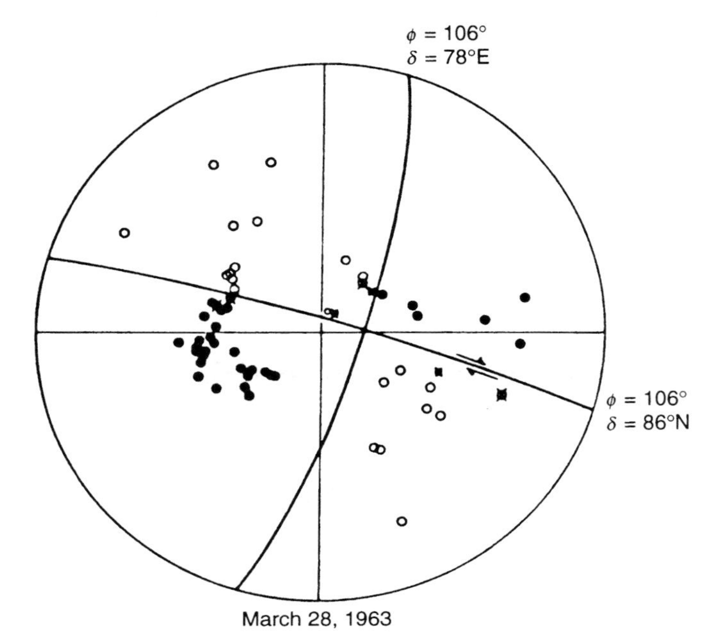

Figure 2.12. Focal mechanism for an earthquake in Iceland. The diagram is an equal-area representation of the lower hemisphere, centred on the focus. The directions of P rays starting from the focus are represented by black dots when the stress before the earthquakes is a compression (first movement towards the recording station), by circles when it is an extension, and by crosses for ambiguous cases. The traces of nodal planes on the hemisphere are also represented (ϕ: azimuth; δ: dip). The arrows show the shear direction in the nodal plane supposed to be close to the fault plane. From Sykes (1967); © AGU.

from the laws of mechanics. But these processes are poorly known, complex, and the computation not achievable. Several source models are therefore tested until the synthetic seismograms are correct reproductions of the actual recorded seismograms. An array of seismographs installed in the area for several months always records some close small earthquakes.

3

Petrography and metamorphism

3.1 MAIN MINERALS IN CRYSTALLINE ROCKS

A *mineral* is defined by a particular chemical composition and a particular crystalline structure, i.e. a particular periodic ordering of its atoms. The chemical composition of a mineral can always be given as a sum of oxides, though these oxide molecules are not individually defined in the crystalline lattice. When they comprise silica SiO_2, the minerals are called *silicates*, even if they are not generally salts of a specific silicic acid.

Silica (the name of a chemical compound) can present itself under eight crystalline forms (*α quartz, β quartz, coesite, stishovite, chalcedony, etc.*). and one amorphous form (*opal*, which along with chalcedony forms flint). These are nine distinct minerals. The two sorts of quartz, α and β, correspond respectively to right- and left-handed Si helices in the crystalline lattice. Although the isolated quartz crystals always are hexagonal prisms, X-ray crystallography shows that the β quartz belongs truly to the hexagonal system, whereas the α quartz (stable below ca. 600°C) is of the rhombohedral system.

In nearly all silicates, the Si have 4 O as nearest neighbours, and the SiO_4 tetrahedra are linked by shared O atoms. This is the case of all the pure-silica minerals quoted above, where the O atoms have no free valency where other elements could fix themselves.

In monoclinic *orthoclase feldspar*, one Si atom out of four is replaced by an atom of Al. This frees up a valency and allows the addition of the alkaline element K in the crystalline lattice. Its formula is Si_3AlO_8K, or $6\,SiO_2\text{-}Al_2\,O_3\text{-}K_2\,O$.

As less oxygen atoms belong to several tetrahedra SiO_4 (and therefore as the ratio Si/O decreases), the free valencies are compensated by more electro-positive elements (Na, K, Ca, Mg, divalent Fe, etc.). In the extreme case where the SiO_4 are totally isolated, they behave like the anions of a fictitious orthosilicic acid SiO_4H_4. This is for example the case of:

- garnets: $(3\,SiO_2\text{-}A_2\,O_3\text{-}3\,B\,O) = (SiO_4)_3\,A_2\,B_3$
- peridots: $(Si\,O_2\text{-}2\,B\,O) = (Si\,O_4)\,B_2$

where A (trivalent) and B (divalent) behave like cations. Only for these orthosilicates (nowadays called *neosilicates*) are the ((SiO_4) grouped in the formula. Let us remark that when they are hydrated, one should not consider the H^+ cations, but the $(OH)^-$ anions.

In a particular crystalline lattice, two atoms can replace each other in all proportions. This provides an *isomorphous series*. For example, the *olivines* are peridots with B = Mg or Fe, and the ratio Mg/Fe can be anything. The end-members of the series are *forsterite* (fo), SiO_4Mg_2, and *fayalite* (fa), SiO_4Fe_2. The formula of olivine is written as $SiO_4(Mg, Fe)_2$. This mineral was once given different names according to the values of the Mg/Fe ratio. It is less pedantic and more precise to say that, for example, most natural olivines have a composition of 90% fo, 10% fa.

The existence of an isomorphous series is more related to the fact that two interchangeable atoms are of similar sizes, than to identical covalencies. Substitutions are very easy for the following couples (with their ionic radii in Ångströms[1]):

Si (0.39)	Mg (0.63)	Na (0.98)	K (1.46)
Al (0.41)	Fe^{++} (0.66)	Ca (1.06)	OH (1.40)

Thus, the *plagioclase feldspars* form a triclinic, isomorphous series, with the formula $(Si, Al)_4\,(Na, Ca)\,O_8$. The end members are:

- albite (ab) = $(SiO_4)_3\,Al\,Na = (1/2)\,(6\,Si\,O_2\text{-}Al_2\,O_3\text{-}Na_2\,O)$
- anorthite (an) = $(SiO_4)_2\,Al_2\,Ca = 2\,Si\,O_2\text{-}Al_2\,O_3\text{-}Ca\,O$

As the electrovalencies must be balanced, it is $(SiNa)^{5+}$ and $(AlCa)^{5+}$ which replace each other. Putting Na/Ca = ab/an = r, any plagioclase may be written $r[Si_2O_{12}Al(SiNa)] + [Si_2O_8Al(AlCa)]$. Consequently:

$$Al/Si = (2 + r)/(2 + 3r) \tag{3.1}$$

Silicates are classified according to the organisation of their SiO_4 tetrahedra:

- **skeleton-forming SiO_4** (*tectosilicates*): quartz, feldspars, and feldspathoids.
- **chain-forming SiO_4**: pyroxenes, orthorhombic (orthopyroxenes) and clinorhombic (clinopyroxenes).
- **strip-forming SiO_4**: formed from juxtaposed rings of SiO_4: amphiboles.
- **sheet-forming SiO_4**: sheets of SiO_4 juxtaposed along an octahedric pattern (*phyllosilicates*): micas, chlorites, serpentine, clays.
- **isolated SiO_4** (*neosilicates*): aluminium silicates, garnets, olivines.

The oxide compositions of the main minerals are given in Table 3.1. A frame covering two columns means that only the sum for the two columns is fixed.

In amphiboles or micas, the ratios are sometimes very different from the simple, stoichiometric ratios, the strips or sheets of a similar crystal having varied compositions.

One can note that, apart from olivines, the minerals with Mg and/or Fe (pyroxenes and amphiboles, biotite, garnet) exhibit dark colours. The rocks in

[1] 1 Å = 10^{-10} m.

Table 3.1. Oxide compositions of the main minerals.

<table>
<tr><th></th><th></th><th>SiO_2</th><th>Al_2O_3</th><th>MgO</th><th>FeO</th><th>Na_2O</th><th>CaO</th><th>K_2O</th><th>H_2O</th><th>Colour</th></tr>
<tr><td>Feldspars</td><td>Orthoclase</td><td>6</td><td>1</td><td></td><td></td><td></td><td></td><td>1</td><td></td><td>White or pink</td></tr>
<tr><td></td><td>Albite</td><td>6</td><td>1</td><td></td><td></td><td>1</td><td></td><td></td><td></td><td></td></tr>
<tr><td></td><td>Anorthite</td><td>2</td><td>1</td><td></td><td></td><td></td><td>1</td><td></td><td></td><td>Whitish</td></tr>
<tr><td>Feldsparthoids</td><td>Leukite</td><td>4</td><td>1</td><td></td><td></td><td></td><td></td><td>1</td><td></td><td>Whitish</td></tr>
<tr><td></td><td>Nepheline</td><td>8</td><td>4</td><td></td><td></td><td>3</td><td>1</td><td></td><td></td><td>Colourless</td></tr>
<tr><td>Orthopyroxenes</td><td>Enstatite</td><td>1</td><td></td><td>1</td><td></td><td></td><td></td><td></td><td></td><td>Yellowish or light green</td></tr>
<tr><td></td><td>Hypersthene[a]</td><td>1</td><td></td><td></td><td>1</td><td></td><td></td><td></td><td></td><td></td></tr>
<tr><td>Clinopyroxenes</td><td>Diopside</td><td>2</td><td></td><td>1</td><td></td><td></td><td></td><td></td><td></td><td>Dark green or black</td></tr>
<tr><td></td><td>Augite</td><td>2</td><td></td><td colspan="2">1</td><td></td><td>1</td><td></td><td></td><td></td></tr>
<tr><td></td><td>to</td><td>3</td><td>1</td><td colspan="2">1</td><td>2</td><td></td><td></td><td></td><td></td></tr>
<tr><td>Amphiboles</td><td>Soapstone</td><td>4</td><td>1</td><td></td><td></td><td></td><td></td><td></td><td>1</td><td>White</td></tr>
<tr><td></td><td>Glaucophane</td><td>8</td><td>1</td><td>3</td><td></td><td>1</td><td></td><td></td><td>1</td><td>Blue</td></tr>
<tr><td></td><td>Hornblende</td><td>6</td><td>3</td><td colspan="2">1</td><td>1</td><td>2</td><td></td><td>1</td><td>Dark green</td></tr>
<tr><td></td><td>to</td><td>7</td><td>1</td><td>5</td><td></td><td></td><td>2</td><td></td><td></td><td></td></tr>
<tr><td>Micas</td><td>Muscovite</td><td>6</td><td>3</td><td></td><td></td><td></td><td></td><td>1</td><td>2</td><td>Colourless</td></tr>
<tr><td></td><td>Biotite</td><td>6</td><td>1[b]</td><td colspan="2">1</td><td></td><td></td><td>1</td><td>2</td><td>Black</td></tr>
<tr><td></td><td>Serpentine</td><td>2</td><td></td><td>3</td><td></td><td></td><td></td><td></td><td>2</td><td>Blue-green</td></tr>
<tr><td>Garnets</td><td>Pyrope</td><td>3</td><td>1</td><td>3</td><td></td><td></td><td></td><td></td><td></td><td>Vivid red</td></tr>
<tr><td></td><td>Almandine</td><td>3</td><td>1</td><td></td><td>3</td><td></td><td></td><td></td><td></td><td>Purple</td></tr>
<tr><td></td><td>Grossularite</td><td>3</td><td>1</td><td></td><td></td><td></td><td>3</td><td></td><td></td><td>Red or grey</td></tr>
<tr><td>Olivines</td><td>Forsterite</td><td>1</td><td></td><td>2</td><td></td><td></td><td></td><td></td><td></td><td>Colour of olive oil</td></tr>
<tr><td></td><td>Fayalite</td><td>1</td><td></td><td></td><td>2</td><td></td><td></td><td></td><td></td><td></td></tr>
</table>

[a] Also called ferrosilite.
[b] Al partly replaced by trivalent Fe.

which they are in abundance are said to be *mafic rocks*. The minerals without Mg or Fe (quartz, feldspars, muscovite) are white or colourless. The rocks in which they are predominant are said to be *felsic rocks*.

Chlorites are strongly hydrated phyllosilicates, forming by alteration of other minerals. The hydration of peridotite, which contains olivine and pyroxenes, produces serpentine. These minerals are green, and the rocks thus altered are often called 'green rocks'.

Amongst non-silicated minerals, one can find:

- *calcite* (rhombohedral) and *aragonite* (orthorhombic), with the same formula: CO_3Ca. With CO_3Mg, calcite forms an isomorphous series (*dolomites*). These minerals are extracted from seawater by the biological activity of corals. They may form thick beds, and even complete mountains (e.g. Jura, Préalpes). But their abundance on the scale of the terrestrial crust should not be over-estimated;
- *anhydrite* (SO_4Ca) and its hydrated form *gypsum* ($SO_4Ca, 2\,H_2O$), more frequent. The transition from the former to the latter is accompanied by an increase in volume and strong heat (when tempering plaster);
- *spinels*, which are minerals crystallized along cubic octahedra, of general formula ($A_2O_3, M.O$) where A is Al, or Fe, or Cr or trivalent Mn, and where M is Mg, Zn, Ni, divalent Fe or Mn. Depending on the nature of A, these will be aluminates, ferrites, chromites or hypo-manganites of M. Thus the *magnetite* (Fe_2O_3, FeO) is an iron ferrite.

3.2 ROCKS – GENERALITIES

Rocks are made of minerals forming grains, visible with the naked eye or aided by a microscope. Two minerals or more can often solidify simultaneously and form a *eutectic* (Figure 3.1). They cannot therefore be detected with the microscope, and can only be reconstituted from chemical analysis. These minerals are not *free*, their presence in the rock is said to be *normative*.

The classification of rocks is much more arduous than the classification of animal or vegetal species, as all intermediates between two well-characterized rocks will exist. Petrographs have therefore coined thousands of rock names, fortunately put aside for the largest part. However, even in the limited dictionary of indispensable rock names, there are names corresponding to their main constitutive minerals (the

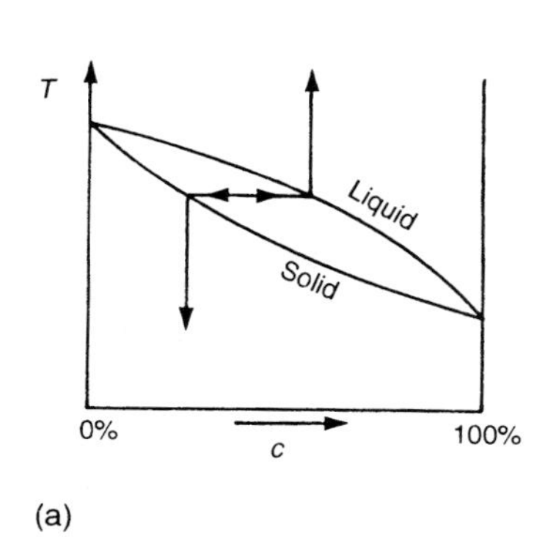

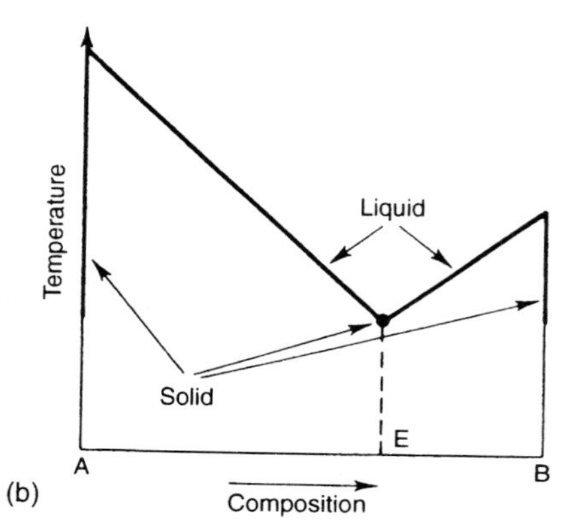

Figure 3.1. Liquid–solid equilibrium, for two chemical compounds which do not react chemically: (a) co-crystallization; (b) precipitation of an eutectic of composition *E*. See Section 3.4 for more details.

others are called *accessory minerals)*, and there are names corresponding to their aspect, or more precisely to their *texture*.

Thus, *shales* are all the rocks with mica flakes nearly parallel, and susceptible of lamination. *Gneisses* are the rocks where quartz and coarse-grained feldspar layers alternate with schistose layers. Very often the whole has a flow structure. *Porphyries* are volcanic lavas where large isolated crystals (*phenocrysts*) are scattered in a matrix without discernible grains, eutectic or glass.

Even when a rock only contains one, or nearly one, mineral, it has a proper name. For example, a rock mainly made of quartz is a quartzite, a rock mainly made of anorthite an anorthosite, and a rock made of olivine a dunite. But beware: amphibolite contains amphibole among its main minerals, but also plagioclases.

Three main classes of rocks can be distinguished: sedimentary rocks, crystalline rocks and metamorphic rocks.

Sedimentary rocks come from the consolidation of loose sediments, simply under the pressure of the overlying terrains (*lithostatic pressure*), or because the pores have been obstructed by a cement, generally calcareous, sometimes siliceous, carried as a solution in water.

Crystalline rocks should rather be called magmatic rocks. They come from the solidification of a *magma* (liquid rock, which may contain solid crystals in suspension). When the magma solidified slowly at depth, the grains are large; the rock is a *plutonite* (from Pluto, god of the underworld). Granite is an example. When the magma solidified quickly by erupting at the surface or on the seafloor, the grains are very fine or invisible, apart sometimes from phenocrysts; the rock is a *volcanic rock*. Basalt is an example. *Obsidian* is the extreme case of a glassy solid, i.e. not crystallized.

Magmas may have formed at depth, at a level where the rock temperature reached the melting point. Being lighter, the magmas have next reached the upper levels. The crystalline rock is then called *intrusive*. But in the case of granite, whose chemical composition corresponds more or less to the average composition of sedimentary terrains, the magma may have formed locally by melting of the existing rocks. What intruded was only the heat necessary to melt the rock. Such a granite is termed an *ultrametamorphic granite*.

Metamorphic rocks are created by mineralogical transformations of sedimentary or crystalline rocks which remained solid. For greater clarity, I exclude from metamorphism the simple modifications of texture, such as the appearance of a schistosity. This latter process is called *dynamo-metamorphism* in geology. It is true that the simple fact of changing the texture also changes the name of the rock.

The source rock, before metamorphism, is called the *protolith*. When this precursor is known, the metamorphic rock may be designated by adding the prefix 'meta' to its name (e.g. a metabasalt). But this is not precise enough, as a rock may be metamorphic in different ways. *Contact metamorphism* concerns only the few metres to hundred of metres around a magma intrusion. *Regional metamorphism* is the only one of interest for geodynamics.

Migmatites are intermediate between crystalline and metamorphic rocks. They exhibit alternating thin layers of metamorphic gneiss and granitic plutonite. Many believe that the granite layers are due to the intrusion of magma from outside, and wrongly call them 'injection gneisses'. They are not always found at a contact with a granitic pluton, and moreover very thin magma intrusions would have solidified over a short distance. For my part, I interpret these as an incomplete ultrametamorphism. When a rock saturated with magma is solidifying, capillary forces must suck the magma toward the solidification contact. More exactly, the liquid phase migrates toward the places which solidification starved from liquid phase. This occurs under the gradient of some thermodynamic potential, due to the superficial energy between the two phases. This is the same process by which ice lenses and ice layers form in a soil easily cracked by frost (a well-known phenomenon, which causes potholes in roads after this ice melts). Another similar case was shown to me in an aluminium-processing plant. When a molten mix of bauxite and cryolite is electrolysed, this magma infiltrates the alumina bricks of the oven, puffs them up and in the long term destroys them.

3.3 MAIN CRYSTALLINE ROCKS

The main series, or *calc-alkalic* series, is made of the following rocks, classified by decreasing *acidity* (silica content). The names of the rocks are different for plutonites and for volcanics. Note that the acidity of petrographs has nothing to do with the acidity of chemists, which is the susceptibility of freeing H^+ ions. This term was borrowed to chemists before the study of crystals with X-rays threw light on the complex world of silicates.

Plutonites:	granite	granodiorite	diorite	gabbro	peridotite
Volcanics:	rhyolite	dacite	andesite	basalt	kimberlite

Granite and rhyolite contain 25 to 40% quartz, 50 to 72% feldspars, mostly alkaline, and 3 to 10% of biotite, sometimes partially replaced by amphibole.

Granodiorite and dacite only contain 10 to 25% quartz (as phenocrystals in dacite), plagioclases and biotite.

Diorite and andesite contain less than 10% quartz, plagioclases and hornblende (which gives its greenish tint to andesite).

Gabbros and basalts contain mainly plagioclases and pyroxenes (augite, hypersthene). The latter give a dark tint to the rock. There is no free olivine when some more quartz remains (free or normative): these are *tholeiitic basalts*, or *tholeiites* as on mid-ocean ridges. Without quartz but with free olivine, these are *olivine basalts* or *alkalic basalts*, as in the Massif Central or the East-African Rift.

Rittman and Kuno pointed out that in the subduction zones of Indonesia and Japan, basalts are tholeiitic ocean-side and progressively become alkalic closer to the continent. This is related to lower water contents and higher pressures, as the oceanic sediments sink along the Wadati-Benioff plane. Laboratory studies (e.g. Green and Ringwood, 1967) showed that these two factors exert the same influence: they increase the melting temperature and yield a higher proportion of olivine. According to Gast (1968), tholeiitic magmas would form around 30 km deep, by melting of 15 to 30% of the source rock, alkalic magmas would form around 60 km deep, with a partial melting of only 5%.

Peridotites are very dense rocks (3.32 at atmospheric pressure), made from olivine and pyroxene, with spinels and garnets as secondary materials. A peridotite nearly exclusively formed from olivine is a *dunite*. If orthopyroxenes are in abundance, this is a *harzburgite*. And if it also contains clinopyroxenes, it is a *lherzolite* (the name comes from Lake Lherz, in Ariège (France)).

Volcanics from the peridotite family are extremely rare. I only mentioned one variety, kimberlite, which is found in the old diamantiferous volcanic chimneys of Kimberley (South Africa). Diamond is stable at 600°C only above 3.2 GPa, i.e. deeper than 100 km, and at 900°C above 4.2 GPa, i.e. deeper than 130 km. These are the minimal depths at which peridotite magma appeared. No recent volcanism is from this type.

Let us mention another series of rocks, with alkali feldspars like the granites, but without any quartz and with abundant K_2O.

Syenites (plutonite) and **trachytes** (volcanics) contain more than 60% feldspars, mostly alkaline, less than 30% biotite or hornblende. Trachytes have a porphyritic aspect, with orthoclase phenocrysts. When they also contain nepheline, these are phonolites.

Monzonites (plutonite) and **trachy-andesites** (volcanics) have roughly equal contents of orthoclase and plagioclase, plus some biotite, hornblende and sometimes pyroxenes.

3.4 MAGMA FORMATION AND INTRUSION

The different minerals which form a rock will melt successively as temperature increases. Rapidly, the molten parts are interconnected and a magma forms with a homogeneous composition. This composition is different from the composition of the solid fraction. Considering only two minerals, the different compositions can be represented by the points of a segment, on the x-axis, and the temperatures can be plotted on the y-axis. The two resulting curves, one for the *solidus* and one for the *liquidus*, show the relative compositions of the liquid and solid parts in equilibrium for a given temperature. Roughly, two cases can occur, depending on whether the two minerals form an isomorphous series (Figure 3.1a) or do not crystallize together (Figure 3.1b). In the latter case, the first magma appearing has the composition of the eutectic.

With three minerals, the rock composition is represented by the points of a triangle. The liquidus and solidus curves become surfaces. Instead of being the summit of the angle drawn by the liquidus, the eutectic is the summit of a trihedron, and always is the lowest temperature achievable by the magma.

Liquidus and solidus depend on the magma water content. A very small water content may decrease the eutectic temperature by several hundreds of degrees. And over 3% water, additional water does not have any influence. In a very hot deep zone, infiltrations of superficial water should be able to trigger the melting. At other

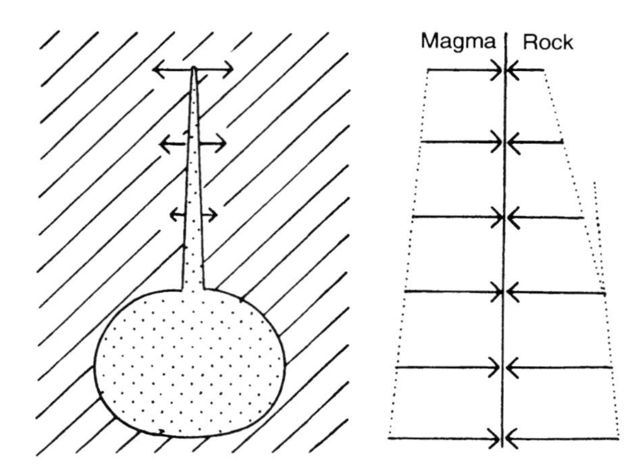

Figure 3.2. In the magma chamber, the hydrostatic pressure is the same as the pressure in the surrounding rock. As magma is less dense, the pressure decreases more slowly with altitude (the effect was exaggerated in this diagram).

times, it can be the dissociation of a hydrated mineral (dehydration melting). A rock which can easily generate magmas is said to be *fertile*. This is the case of the rocks containing a hydrated mineral, such as chlorites. But a first partial melting, dehydrating the rock, may make it sterile.

Let us assume a large magma reservoir has formed (without bothering yet how it formed). This reservoir is at the same pressure than the surrounding rocks. If a crack appears in the roof of the magma chamber (which is likely, melting being accompanied by a volume increase), this crack gets filled with magma (Figure 3.2). The pressure p in the magma decreases with the altitude z, according to the Pascal principle. ρ_m being its density and g the gravity, $dp/dz = -\rho_m g$. It is nearly the same inside the rock, using the rock's density ρ_s. As $\rho_s > \rho_m$, at the top of the crack the magma is overcompressed, which extends the crack upwards as far as the magma does not solidify.

Liquidus and solidus depend on pressure. As it goes upwards and decompresses, the magma temperature may become higher and higher than the eutectic temperature, or lower and solidify. Three possible cases can be outlined:

1. **Outpouring of fluid lava**. Let us consider a basaltic magma, with low silica content. Its temperature is the one of the ternary eutectic albite-enstatite-hypersthene, which decreases with pressure. The magma will therefore remain fluid while rising in the fissure; it will be able to reach the surface, and the lava will pour out. Successive flows of very fluid lavas can form stepped plateaus, called *traps* (*trap* is the Dutch word for stairs). Instead of fissural volcanism, a localized chimney may form at the end, and the lavas will build a strato-volcano. Very rapid cooling during a submarine volcanic eruption produces *pillow-lavas*. Revealed by erosion, the volcanics from a volcanic chimney form a *neck*, and those from a sub-vertical fissure form a *dyke* (Figure 3.3a). Magma

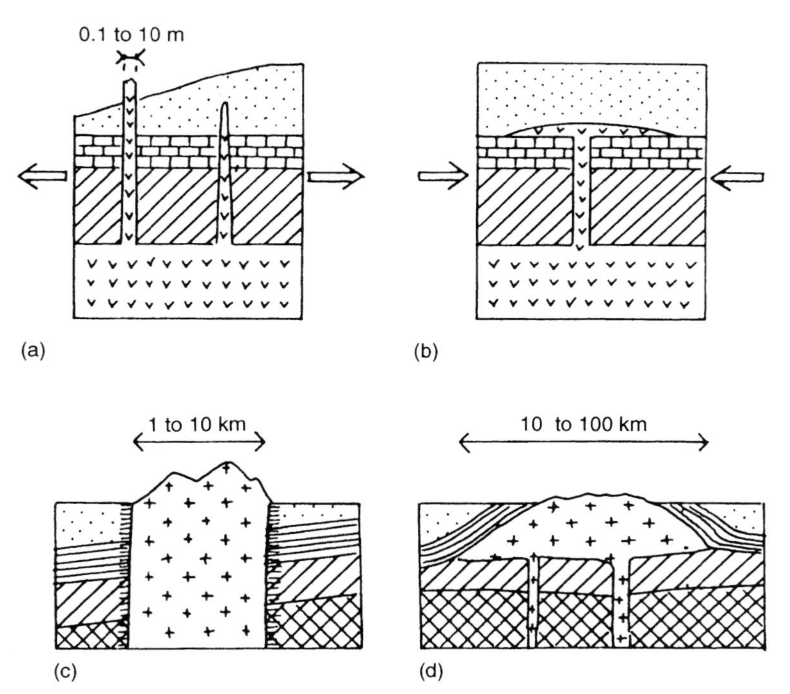

Figure 3.3. (a) Dykes; (b) sill; (c) stock (the fine hashed lines indicate contact metamorphism); (d) laccolith.

can also intrude between two sedimentary layers. If the layer is thin and cooling rapidly, this forms a volcanite *sill* (Figure 3.3b).

2. **Explosive volcanism**. With less acidic lavas (e.g. andesites), the magma can nearly reach the surface without pouring out. The volcano's summit may then explode. Particles of magma are projected into the air. Sometimes, since the adiabatic expansion of gases in the atmosphere makes them cooler and denser; the ascending gas current reverses and there is formation of *a nuée ardente* (*glowing cloud*), a jet of air charged with incandescent pyroclastic rock particles. The dynamics of these jets is the same as an avalanche of powdery snow. Solidified but still very hot, the particles weld together and form *ignimbrites*. In northern Chile, the ignimbrite layers reach several thousand metres in thickness, showing a very long past of explosive volcanism. Under the sea, a volcanic explosion produces a dust of agglomerated glass (*hyaloclastites*).
3. **Formation of a pluton**. A granitic magma, very rich in silica, corresponds to the eutectic quartz-albite-orthoclase, whose temperature increases as pressure decreases. To ascend, its temperature must increase, and therefore part of it must solidify. It will have completely solidified as a *pluton* before reaching the surface. The intrusion will most often be uneven, in layers dislocated by tectonic activity. Only the simplest cases will be discussed here.

An *intrusive pluton*, i.e. a pluton with a clear outline on a detailed geological map, is a *stock* when its area is less than 100 km^2 (Figure 3.3c), a *batholith* when its area is larger. Examples of stocks are the hornblende granite of Flamanville in Normandy, and the Fitz-Roy and Paine massifs in Patagonia.

The granitic magma pocket may migrate upwards in bulk, by three mechanisms which are not mutually exclusive:

- sometimes, if the surrounding terrains are plastic enough, there is massive intrusion (*punching*). Sedimentary layers folded upwards at the contact with the intrusive stock are indicators of this process;
- the walls of the magma chamber may collapse repeatedly, the blocks denser than magma sinking at the bottom (magmatic stopping) (Daly, 1933);
- simultaneously, the roof of the chamber may melt and the floor solidify, heat being continuously transferred from the floor to the roof by thermal convection in the magma. This 'digestion' of the overlaying terrains changes evidently the composition of the magma, and 'contaminates' it. For example, the magma of the Vesuvius eruption had been contaminated by the digestion of dolomitic limestone from the Trias.

In the last two cases, the intrusion occurs without affecting the surrounding terrains, as for ultrametamorphic granite. However, when magma gets inserted between two horizontal sedimentary layers to form a very large plutonite lens (a *laccolith*), the overlying terrains bulge (Figure 3.3d).

The differences between lavas coming from neighbouring sites, or from the same volcano but at different periods, is a factor of complication, as much as of interest, in the study of volcanics.

Easter Island is, for example, situated 500 km away from a very active mid-ocean ridge, with abundance of tholeiitic magmas. But on the island, one finds not only tholeiites, but also trachytes, hawaiites (more acidic and less rich in ferromagnesians than basalt), and even rhyolites as obsidians topping some volcanoes. In the Hawaiian Chain, the first lavas are olivine basalts, which next alternate with andesites and trachytes.

In these islands isolated in the middle of the Pacific Ocean, the emission of andesites or rhyolites cannot be explained by some input of silica by sediments coming from a continent. There must have been differentiation during the solidification of a magma reservoir, the molten remainder becoming more and more acidic or rich in K_2O. A hypothesis often encountered is of a *gravity differentiation*, the first crystals formed falling at the bottom of the magma reservoir. It is, however, difficult to understand how the first crystals could appear inside the magma: the cold coming from outside, they should appear on the walls.

The centre of the Fitz-Roy stock, which the author sampled in 1952, is an amphibolic granodiorite, progressively becoming pure granite at the outskirts. The opposite should be expected, as the centre of the pluton solidifies last. One possibility in the present case could be the contamination by the surrounding schists.

Let us finally remark that hybridation may occur between two magmas of different origins and compositions. During the Miocene, basaltic and rhyolitic lavas flowed out at the Mont-Dore (Massif Central, France). At the end of Pliocene, the lavas were uniformly trachy-andesites; the two magmas should have met and mixed.

3.5 NATURE OF THE OCEANIC CRUST AND UPPER MANTLE

Close to shore, the seafloor is covered by a layer of sediments from continental origin, with widely varying thickness (up to ten kilometres close to the mouths of large rivers). Far from the shores, sedimentation is very small. The shells of mono-cellular marine organisms in marine plankton are dominant: calcareous shells of *Foraminifera*, siliceous shells of cold-water *diatoms* or warm-water *Radiolaria*. Most Foraminifera, like *Globigerina*, live close to the surface. But there exist deep-water (*benthonic*) Foraminifera, like *Cibicidoides* or *Uvegerina*. In the deepest regions, only siliceous shells reach the seafloor, as calcareous shells dissolve at high pressures. Hydrated aluminous silica covers a large portion of the Pacific Ocean seafloor; that is the deep-ocean *red mud*.

Seismology shows that under these sediments, the crust of all major oceanic basins has a very uniform structure. This is called the *oceanic crust*. Below the sediments is a layer of basalt, 0.5 to 2.5 km thick. Below, between the basaltic layer and the Moho, there is a third layer, 3 to 6 km thick, named the *oceanic layer*. The P-wave celerity is ca. 2.2 km/s in the sedimentary layer and 5.2 km/s in the basaltic layer. In the oceanic layer, it increases to 6.7 ± 0.25 km/s. ODP drillings penetrated this layer over 500 m, at the Atlantis II Bank (south of Réunion), in 1987, and over 154 m at the South Pacific mid-ocean ridge in 1994. This layer is made of gabbros, i.e. the plutonic form of basalt.

Basalt and gabbros are coming from the same magma, originating from the mantle at mid-ocean ridges (see Chapter 5 for the details of the process). The thickness of the gabbroic layer increases with the distance from the mid-ocean ridge. It reaches for example 3.9 km on the Mid-Atlantic Ridge, of recent formation, and 4.8 km on the edges of the North Atlantic, where the crust dates back to the Upper Jurassic. It is therefore necessary to admit that the mantle away from mid-ocean ridges has emitted some basaltic magma, which was added below the crust. This process has been suggested by several authors for the continental crust, and named *underplating* (which has nothing to do with lithospheric plates).

The process was thought to be generalized. Recent studies have confirmed the thickening of the crust with age in the North Atlantic (NAT Studying Group, 1985). But they also showed the thickness of the Pacific Ocean crust does not depend on its age (McClain and Atallah, 1986).

The celerity of P_n waves (head waves propagating along the Moho) is relatively uniform around the Earth: 7.9 to 8.2 km/s. The density of the upper mantle is 3.4 g/cm^3 (we shall see in Chapter 17 how it can be estimated). This density is exactly the one of peridotites. *The upper mantle is formed of peridotites*, lherzolite or harzburgite, probably enriched in garnets. The average composition is commonly admitted as being 60% olivine, 30% pyroxenes and 10% garnets. This composition should be maintained down to 410 km, where the celerity of P and S waves increases suddenly. The layer between the Moho and this discontinuity is today called the *upper mantle* (before, the upper mantle designated the layer between the Moho and another discontinuity at 660 km. The layer between 410 and 660 km is now called the *transition layer*).

To account for the exudation of basaltic magmas by the upper mantle, Ringwood (1966) postulated that it was made of a rock unknown at the surface of the Earth. This rock, called *pyrolite*, was dissociating in 3/4 basalt and 1/4 peridotite. Dredges from the mid-oceanic ridges never brought back anything but serpentinized basalts or sometimes peridotites. Despite these facts, this theory endured a long popularity. It is now completely abandoned. Since 1928, Bowen had proved in the laboratory that peridotites generate basaltic magma during partial melting, but a new word has a strong advertising impact.

Chondritic meteorites, or *chondrites*, form two-thirds of the meteorites whose fall has been observed, and have a composition very similar to the composition of peridotites (the other third is made of iron and nickel). Statistics based on meteorites collected on the ground would be highly biased, chondrites being usually unnoticed. Meteorites are admitted to be fragments of one or two planets of sizes similar to the Earth's. The halo of interplanetary dust observed during total eclipses of the Sun is also composed of olivine, as shown by spectrometric measurements (some made in 1972 aboard *Concorde*, following the Moon's shadow during such an eclipse).

Genuine sections through the oceanic crust and upper mantle are known to be outcropping on continents. These are the *ophiolitic complexes*, or *ophiolites*. Their origin was intriguing to geologists (for example, the alpine spillites were attributed to fissure eruptions of unknown age). Joe Cann explained it in 1970. In Section 7.7, will be examined the mechanisms by which in a subduction zone a plate of oceanic crust and uppermost mantle could rise above the continental crust instead of sinking below it.

There are few ophiolites with intact and non-metamorphized rock series: Vourinos in Greece, Troodos in Cyprus, Bay of Islands in Newfoundland, and the Oman mountains. From the top to bottom, one can find:

- deep marine sediments: schists and radiolarites;
- basalts, often chloritized and enriched in sodium;
- gabbros, crossed by basaltic dykes;
- peridotites.

The transition from gabbros to peridotites occurs in less than 100 m, with alternate layers having many chromite veins interspersed.

3.6 METAMORPHIC ROCKS

Metamorphism may consist in simple allotropic transformations of a mineral component, but most often it consists in genuine chemical reactions in a solid state. The rock evolves very slowly toward a new mineralogical composition, more stable at the surrounding pressure and temperature. In both cases, these reactions are reversible. It is therefore possible to wonder why metamorphic rocks formed at depth, under larger pressure and temperature, can be found at the surface. A complete *retrograde metamorphism* should have occurred, bringing the rock back to its original mineralogical state. But this retrograde metamorphism is only obvious after metamorphism at depths larger than 70 km.

The explanation is the existence during metamorphism of a fluid phase in small quantity, which decreased the equilibrium temperature between two mineralogical compositions, and then escaped. Reaction rates then became negligible at the geological time scale, and the metamorphic rock remained in a metastable equilibrium. This fluid often transported dissolved minerals. When the input of outside material is important, petrologists call the process *metasomatism*.

What are these mineralizing fluids, which were mysterious and called *migmas*? By studying the fluid inclusions found in some crystals from metamorphic rocks this point has been cleared up.

- Up to 650°C, it consists in water, generally with 20–25% salts and dissolved carbon dioxide CO_2. Under the Canadian shield, the inclusions are brines with up to 50–60% NaCl.
- Above 650°C, i.e. in the lower crust, there cannot be any water, as its presence lowers the melting point of the rock, inducing magmatism and not metamorphism. The fluid is CO_2, produced by the decomposition of limestone. (Above 500°C, in presence of silica, there is production of CO_2 and wollastonite SiO_3Ca.)

In both cases, at the depths where metamorphism takes place, these are not liquids or gases anymore, but hypercritical fluids. The critical point of CO_2 is at $T = 31°C$ and $P = 7.4$ MPa; the critical point of H_2O is at $T = 374°C$ and $P = 22.8$ MPa. Temperature increases with depth at around 30°C per kilometre, and the lithostatic pressure increases at 3.7 MPa per kilometre.

Deformations are much easier during the atomic readjustments implied by metamorphism. Always present even in the absence of orogeny, the deviatoric (i.e. tectonic) stress yields intense shearing. This explains why all the metamorphic rocks take a schistose or fluidal structure. Regional metamorphism is often called *thermo-dynamo-metamorphism*, but this does not imply that the deformations modify the mineralogical compositions. At most, deformations can only speed up the reaction rates.

Metamorphism always seems to occur during the important mountain-building stages (Alpine cycle, Hercynian cycle, etc.). It can be relatively fast, and it may not occur at the paroxysmal stage. For example, the metamorphism observed on the north side of the Pyrénées is only visible in the terrains of Albian age (mid-Cretaceous, between 104 and 96 Ma Before Present). This metamorphism is older than the Pyrenean Orogeny, which occurred between 50 and 40 Ma BP. In the Vanoise massif, however, metamorphism occurred after the schistosity and thrust sheets produced during the Western Alps Orogeny (Alpine phase *sensu stricto*). The link between the metamorphic rocks observed at the surface and the orogeny can be attributed to two processes:

- the same heating at depth which enabled the orogeny, often caused the metamorphism. Lengthy erosion then exposed the metamorphic rocks (see Section 8.8);
- at other times, metamorphism occurred very deep in a subduction zone. It would have remained unnoticed if some tectonic processes, linked to orogeny, had not *exhumed* the metamorphic rocks, much faster than a simple *denudation* through erosion.

3.7 GEOTHERMO-BAROMETERS

A *facies* is a group of rocks formed in some very well-determined environment. The word of environment is vague on purpose. It is sometimes partially hypothetical, and some authors contest it (for example the flysch and molasse facies of alpine geology, mentioned in Chapter 14). In the case of metamorphic rocks, many laboratory experiments allow us to define the environment as a 'temperature and pressure domain'. The existence of shearing, of a non-hydrostatic stress, does not influence much the mineralogical composition. It can only modify the texture, creating schist or gneiss.

The word facies should be clearly distinguished from *terrain* (often phonetically spelled *terrane* by American scientists). The dictionary of the American Geological Institute defines a terrain as 'a complex group of strata accumulated within a definite geological epoch'. But the word terrane is often used inappropriately, and caution should be exercised.

The facies are named after the metamorphic rocks that are the most frequent, and excluded from the other facies. The granites from Mont Blanc are for example rightly designed as green-schist facies. It only indicates the temperatures and pressures at which these rocks are assumed to have formed. Each facies has its characteristic minerals, primary or secondary. But the absence of a particular characteristic mineral cannot be interpreted as a proof, only as a hint. It should have formed only if the chemical composition of the protolith was appropriate, and this may not have been the case. But the disappearance of a characteristic mineral, with a protolith remaining the same, proves the transition to another facies.

Some minerals only appear in specific temperature intervals (*geothermometers*). Others only appear in specific pressure intervals (*geobarometers*). The simplest case is of a simple allotropic transformation, for example from quartz to coesite, or the transformation of aluminium silicates (SiO_2-Al_2O_3): andalusite and sillimanite, orthorhombics, kyanite (also called disthene), triclinics. The limits between their respective stability domains are shown in Figure 3.4. In this case, the other minerals in the rock do not change.

The list of such allotropic changes is rapidly exhausted, and petrologists had to address real chemical reactions, such as:

albite $\Longleftrightarrow$ jadeite + quartz

lawsonite + 1st garnet $\Longleftrightarrow$ zoisite + coesite + 2nd garnet + H_2O

Jadeite (of formula $4SiO_2$-Al_2O_3-Na_2O) is a green clinopyroxene, looking like jade (which is an amphibole).

Lawsonite has the formula $2\,SiO_2$-Al_2O_3-CaO-H_2O. Zoisite (of formula $6\,SiO_2$-$3\,Al_2O_3$-$4\,CaO$-H_2O) belongs to the epidote family, for which Al can be replaced with trivalent Fe or Mn, Ca with Ce, La, Mg or divalent Mn.

The limits in the temperature/pressure plane depend on the presence of other minerals in the rock, in the common case where atom substitutions are possible, and most often on the amount of water present. One must therefore control the concentration in water and other minerals throughout the whole system, or more exactly control their *activity*. Activity is the thermodynamics parameter, which intervenes in the chemical potential and becomes equal to the concentration when it is very small. For a gas, this is the fugacity. For a hypercritical fluid, activity and fugacity are synonymous.

The subject is very arduous, and reserved to specialists. But the geodynamicist should not ignore their results. Extremely powerful tools of spectral analysis (ion microprobes) have been developed. Together with the immense work of experimental petrologists, they allow us today, in the most favourable cases, to reconstitute the temperatures and pressures experienced by a rock during a continental collision, from the study of only one of its garnets, by analysing it from its centre to its periphery, along with its solid and fluid inclusions.

3.8 FACIES AND METAMORPHIC ROCKS

The stability limits of some minerals are represented in Figure 3.4 along with the domains corresponding to the different metamorphic facies. The positions of some terrains from the Western Alps have also been represented, using the studies of Kerkhove and Desmons (1980) and Schreyer (1995).

Rather than classifying metamorphism by high- and low-pressure, and high- and low-temperature facies, it is more precise to classify them relative to the *geothermal* curve T(P) normal in continents. This curve is plotted on Figure 3.4. As we shall see in Chapter 9, the temperature increases by 30°C/km over the first ten kilometres. The heat to expel is mostly due to the radioactivity of the continental crust, and the geothermal gradient must then decrease at depth. As demonstrated in Section 9.13, a Moho at 40 km should in general be between 700 and 900°C.

First group: temperatures hotter than the average, by ca. 200°C. This is the *Abukuma facies*. By increasing temperatures and pressures:

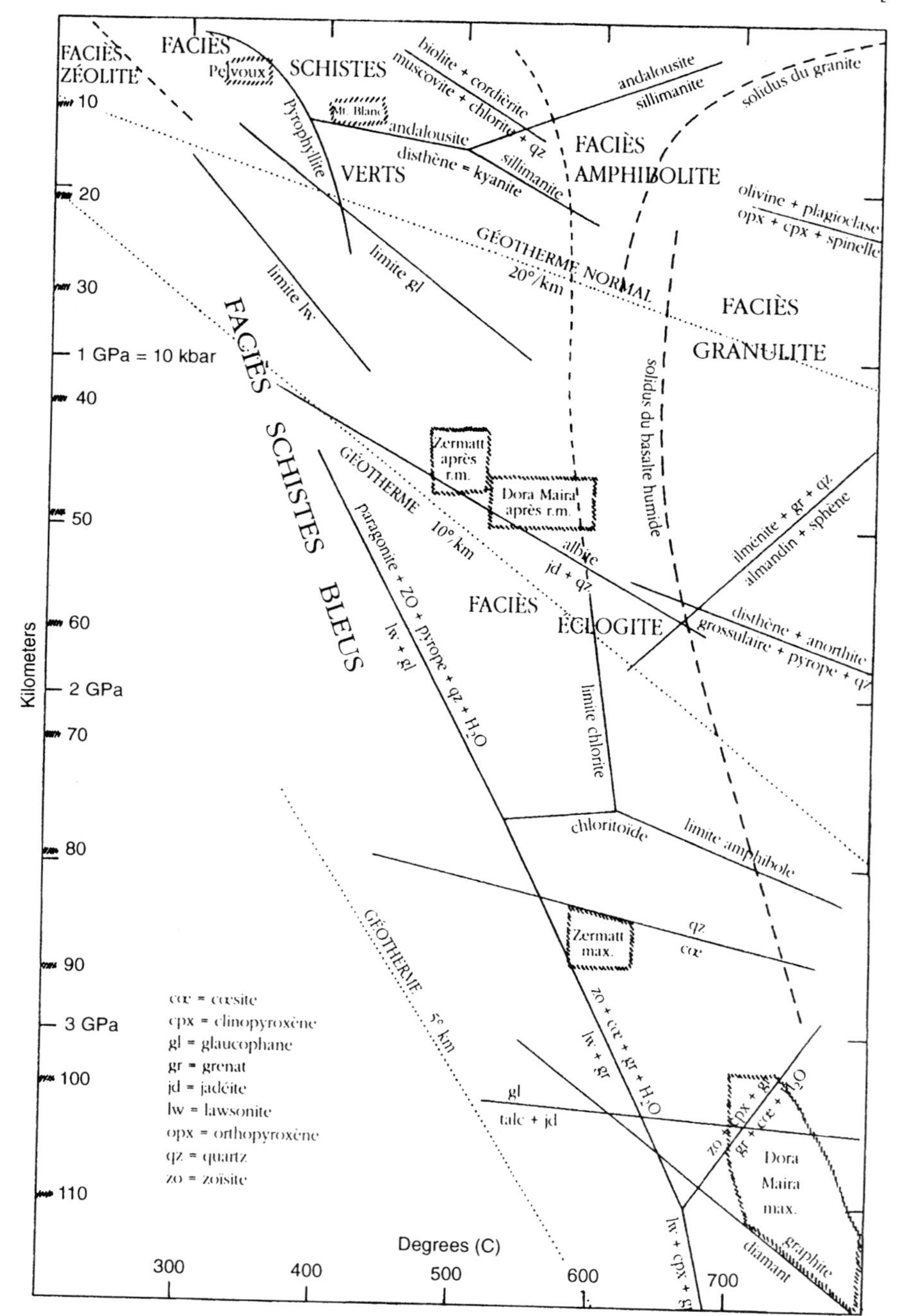

Figure 3.4. Metamorphic equilibrium curves in the temperature/pressure plane. The domains outlined correspond to some minerals from the Dora Maira massif (Piedmont, Italy) or the Zermatt region (max.), and to the general rock facies after retrograde metamorphism (after r.m.).

- *Zeolite facies* ($T < 250°C$, approximately). Zeolites are complex hydrated silicates, most often due to hydrothermal processes.
- *Green schist facies* ($250°C < T < 450°C$, approximately). Green schists owe their name to the amount of chlorites.
- *Amphibolite facies* ($450°C < T < 700°C$, approximately). Amphibolite is a crystalline schist, i.e. the grains are visible with the naked eye, and there is abundant hornblende, alkali feldspar, orthoclase or albite, with or without quartz.

Second group: normal geotherm. This is the *Barrow facies*. We will only refer to:

- *Granulite facies* ($T > 650°C$). Granulite is a gneiss formed mainly of plagioclase and pyroxene (diopside and hypersthene), often enclosing garnets. The name of granulite comes from the German *granat* for garnet, and has nothing to do with granite. Gneiss with dominant hypersthene orthoclase and quartz, i.e. granite with hypersthene instead of biotite, is called a *charnockite*. The granulite facies was formerly called 'charnockite complex'. Note that the temperature of granulite formation is larger than the solidus temperature of wet basaltic magmas.

Third group: pressure larger than the normal geotherm, by around 0.5 GPa, which corresponds to 20 km deeper.

- *Glossy schist facies* (*white schists*, for American scientists). Léon Moret describes the glossy schists from the Western Alps as: 'schistose rock, crystalline, made of alternate thin zones of calcite, quartz and a lining of white mica with ilmenite ($FeTiO_3$) and rutile (TiO_2)'.
- *Blueschist facies* ($0.7\,GPa < P < 1.4\,GPa$), i.e. depths between 25 and 48 km. The blue colour of the schists comes from glaucophane. Although it is not destroyed before depths of around 100 km, the next facies is reached at greater depth than 48 km.
- *Eclogite facies* ($P > 1.4\,GPa$). *Eclogite* is a very dense rock (density of ca. 3.5), made from garnets (generally pyrope) and pyroxenes (generally *omphacite*, a green-looking sodic clinopyroxene). Eclogite is only rarely found at the surface, as nodules in kimberlites and some gneisses. An inclusion of foreign rock is called a xenolith. If the kimberlite magma may well have carried a fragment of foreign rock, in the case of gneiss, retrograde metamorphism has obviously affected the whole rock, while the eclogite remained as a 'metamorphic marker'.

Fourth group: ultra-high-pressure metamorphism. At 600°C, quartz transforms into coesite at 2.7 GPa, i.e. at a depth of 85 km. Chopin (1984) discovered coesite inclusions in metamorphic rocks, undoubtedly of crustal origin, in the crystalline massif of Dora Maira. The concerned terrain covers 40 km^2, and is 1–2 km thick. It was therefore buried deeper than 85 km, then exhumed, and its rocks were subjected to complete retrograde metamorphism. Coesite inclusions were later discovered in garnets and tourmalines from the Zermatt-Saas metamorphosed ophiolites, and in the Precambrian terrains of Mali (see Schreyer (1995) for a review and a discussion.)

Diamond only forms at even higher pressures. Microdiamonds were found in crustal rocks (and not from the mantle like kimberlites). In the Kokchetov massif (Northern Khazakhstan), they form inclusions in garnets and zircons of schists and gneisses with amphibolite facies. In Eastern China, a large area was in the Triassic (200 Ma BP) a collision zone between the Korean craton in the north and the Yangtse craton in the south. There, it is possible to encounter diamond inclusions in the small eclogite lenses found in acid gneiss and marble. This location was therefore the scene of the subduction of some continental crust, followed by its exhumation from depths of at least 120 km.

3.9 NATURE OF THE CONTINENTAL CRUST

Continental crust is any crust with metamorphic rocks and granitic plutons. This corresponds not only to all emerged lands, but also to shallow seas like the North Sea or the Adriatic (*epicontinental seas*), and submarine plateaus rising above the oceanic basins, like some fishing grounds in the North Atlantic. The Earth's surface is 510×10^6 million km^2, and the surface of emerged lands is 149×10^6 km^2. The surface of continental crust reaches approximately one-third of the Earth's surface.

Of these 170×10^6 km^2, two-thirds are *cratons*, also called *shields*, or *platforms*. None of them suffered any orogenic episode in the Phanerozoic (fossiliferous times, the last 600 Ma of the 4,600 Ma of the Earth's history). They are called cratons when metamorphic rocks are outcropping or very close to the surface, platforms when the surface is covered by a thick series of sedimentary beds nearly horizontal.

The orogenic belts, where mountains formed during the Phanerozoic, cover $13 \pm 2\%$ of continental crust. Insular arcs with continental crust (Indonesia–Philippines, Japan, etc.) cover 6.0%. The rest (14.5%) comprises extension zones pervaded with basaltic dykes, for example the Basin and Range province (Nevada, US), or the large West-African Rift, and mainly *passive continental margins* (that is, without subduction zone) and submarine plateaus. The latter are *extended crust*, the extension preceding the splitting of the continent and the appearance of an ocean.

Global statistics were provided recently by Rudwick and Fountain (1995) and Christensen and Mooney (1995). The thickness of the crust, i.e. the depth of the Moho, can vary a lot. The extreme values are 16 km in the very young extended crust of Afar (Djibouti), and 72 km in the orogenic zone of Tibet. The petrographical constitution varies a lot too. But, in average, the crust of cratons and platforms is the weightier, and the average values are close to their mean thickness. This is referred to as the *normal continental crust*.

The **upper crust** is approximately 15-km thick. It is the only one considered by geologists, and penetrated by the deepest drillings (12 km in the Kola Peninsula, above a craton). When geologists talk of 'basement' instead of 'cover', they talk about old metamorphic rocks inside this layer. The granitic basement was once thought to be everywhere at depth, but this proved wrong. Often, we have only large laccoliths fed by underlying granite seams. Examples are found in deep mine tunnels going below the laccolith, or on the valley walls in Gangotri (Himalayas), and in Qônqut (Greenland). The large batholiths, exhumed by the glaciers on the Pacific coasts of Alaska and Patagonia, are only contiguous stocks of very varied ages.

The mean chemical composition of the shallow crust is the one of *granite*, with 70% weight in SiO_2.

In the **middle crust**, between depths of 15 and 25 km, the mean celerity of P waves increases progressively from 6.3 ± 0.3 km/s to 6.7 ± 0.2 km/s. The celerities and densities correspond to the *amphibolite facies*.

The **lower crust**, between 25 km and a Moho at 41.5 ± 5.8 km, presents P-wave celerities increasing from 6.7 to 7.4 km/s. Combined with the xenoliths which come from it, these celerities and densities support the hypothesis that it is formed from *ferromagnesian granulites with garnets*. In a normal continental crust, the Moho is at a temperature below the solidus temperature for these rocks.

Recent large-scale exploration with reflection vibroseismics in orogenic zones showed in the lower crust a very large number of reflectors, not continuous but all parallel to the Moho, as in a superposition of 'plates' 200- to 2,000-m thick. On records, the Moho is often less distinct than these reflectors. Its continuity is not straightforward, so that the beginning of the mantle is inferred from the disappearance of the reflectors. This layered structure is less clear in platforms and cratons, but not all are seismically transparent (Mooney and Meissner, 1992).

Several hypotheses have been brought forward to explain these plates:

(1) They would be different sedimentary layers, very old and metamorphic.
(2) The reflectors would be layers of crushed rock formed when old faults acted, and transformed into rock again (*mylonite*).
(3) The plates would have similar petrographical compositions, but different metamorphic textures.
(4) The reflectors would be sills, formed after intrusions of basaltic magmas from the mantle.
(5) The plates would result from successive underplating of gabbros coming from the mantle.

This would remain speculative if geologists had not discovered the succession of the three continental crust facies in a few outcrops: in Northern Ontario near St. James Bay, in the Ivrea area north from Torino (Italy), etc. (Percival *et al.*, 1992). In all cases, migmatites are found in the granulite facies, which proves that the conditions for ultrametamorphism were reached in the past. This favours explanation (3). Unfortunately, the only outcrops of mantle peridotites following a granulite facies are in the Ivrea region and in Calabria (Italy), so that an old continental Moho cannot be observed. As seismology seems to show that a continental Moho is in fact a transition layer 3- to 5-km thick, the hypotheses (4) and (5) cannot be excluded for the moment.

According to Rudwick and Fountain (1995), the average composition of the whole continental crust would be the following (*in oxide moles*):

SiO_2	63.0%	FeO	5.9%	
Al_2O_3	9.9%	Na_2O	3.3%	Others 1.4%
CaO	7.2%	K_2O	1.3%	
MgO	7.0%	TiO_2	0.6%	

By its acidity, this average corresponds to the limit between granite and granodiorite (prior estimates were favouring granodiorite). Since erosion only affects the upper crust, the average composition of sediments deposited in the oceans is definitely granitic.

A century ago, when nearly nothing was known about the inside of the Earth or about metamorphism, the continental crust was said to be made of *sial*, and the rest (including oceanic floors) of *sima*. These words were introduced by Suess (1871–1901) and are still found in the most dated books. They should be avoided at all costs.

4

Geomagnetism and rock magnetization

4.1 MAGNETIC FIELDS PRODUCED BY MAGNETIC MATERIALS

In the universe, the only two forces working over large distances are gravity (see Chapter 8) and electromagnetism (the other two forces are noticeable only at the atomic scale). The electromagnetic force can be resolved into the electric force and the magnetic force, in different ways according to the reference system used (this is the starting point of the theory of Relativity).

At any point in space, in a given frame of reference, two vectors represent the electric field **E** and the magnetic field **B**. The former exerts on each electric charge q at this point a force: $q\mathbf{E}$. The latter produces on every segment **ds** of a circuit followed by an electric current of intensity I a force: $I\,\mathbf{ds} \wedge \mathbf{B}$ (the Laplace force). The field **E** is the resultant (the vector sum) of electric fields produced in the point considered by any means; *idem* for **B**.

All the following equations will be written in the MKSA system (*Système International*: SI). Its basic units are the metre, the kilogram, the second and the ampere (A). The unit for electric charge is the coulomb (Cb), and the unit for electric field is the volt/m. The unit for the magnetic field is the tesla (T). The Earth's magnetic field at the surface is a fraction of a gauss (G), the old unit from the CGS system, which is still used in places. Variations of this magnetic field are given in nanotesla (nT):

$$1\,\mathrm{G} = 10^{-4}\,\mathrm{T}; \qquad 1\,\mathrm{nT} = 10^{-9}\,\mathrm{T} \tag{4.1}$$

A magnetic field can be produced by magnetized bodies (magnets, permanent or not) or by electric fields. More precisely, both produce at every point a magnetic excitation vector **H**. In magnetic media, a magnetization vector **J** can be defined. And the field **B** is:

$$\mathbf{B} = \mu_0(\mathbf{H} + \mathbf{J}) \qquad \text{with } \mu_0 = 4\pi \times 10^{-7} \tag{4.2}$$

Any volume of magnetic matter small enough can be considered as a *magnetic dipole*, consisting of two fictitious magnetic poles distant of $2a$: a north pole of magnetic mass $+m$ and a south pole of magnetic mass $-m$. The (magnetic) *dipolar moment* is the quantity: $M_2 = 2am$. The dipolar moment vector is the vector directed from the south pole to the north pole, with this intensity. The magnetic excitation produced by a small volume v of magnetic matter depends only on its magnetic dipolar moment vector. So do the forces to which it is submitted when in the field **B**. The moment vector is the sum of all the dipolar moment vectors from its components (even a single electron or nucleus has a dipolar magnetic moment). The magnetization vector is obtained by division by v:

$$\mathbf{J} = \mathbf{M}_2/v \tag{4.3}$$

Two magnetic poles m and m' are repulsive if they have the same sign, and attractive if they have opposite signs. The corresponding force is aligned with the line between these poles. Its expression is identical to the electrostatic force between two charges q and q':

$$F = \frac{mm'}{4\pi r^2} = m'H \tag{4.4}$$

H is the field produced by m at the position of m'. The factor $1/4\pi$ was introduced in the rationalized unit systems so that, when applying the Gauss theorem, the flux coming out of the closed surface is equal to the sum of the magnetic masses inside the surface (see Annex A.1). The *magnetic potential* V can therefore be defined such that:

$$\mathbf{H} = -\mathbf{grad}\ V \tag{4.5}$$

And for an isolated dipole:

$$V = \frac{m}{4\pi r} \tag{4.6}$$

Only elementary dipoles exist, and such an isolated magnetic pole is here only to ease the computation. Consider the dipole consisting of two magnetic masses on the z-axis: $+m$ at $z=a$ and $-m$ at $z=a$. The dipole potential is found by adding the two potentials:

$$\begin{aligned} V &= \frac{m}{4\pi}[x^2+y^2+(z-a)^2]^{-1/2} - \frac{m}{4\pi}[x^2+y^2+(z+a)^2]^{-1/2} \\ &= \frac{m}{4\pi r}\left[\left(1-\frac{2az}{r^2}+\frac{a^2}{r^2}\right)^{-1/2} - \left(1+\frac{2az}{r^2}+\frac{a^2}{r^2}\right)^{-1/2}\right] \end{aligned} \tag{4.7}$$

It may be expressed as a series, since:

$$(1+\varepsilon)^{-1/2} = 1 - \tfrac{1}{2}\varepsilon + \tfrac{3}{8}\varepsilon^2 - \tfrac{5}{16}\varepsilon^3 + \cdots \tag{4.8}$$

The main term is:

$$V_{10} = \frac{M_2 z}{4\pi r^3} = \frac{M_2}{4\pi r^2}\cos\theta \tag{4.9}$$

with r, θ and ϕ denoting the spherical coordinates corresponding to x, y and z.

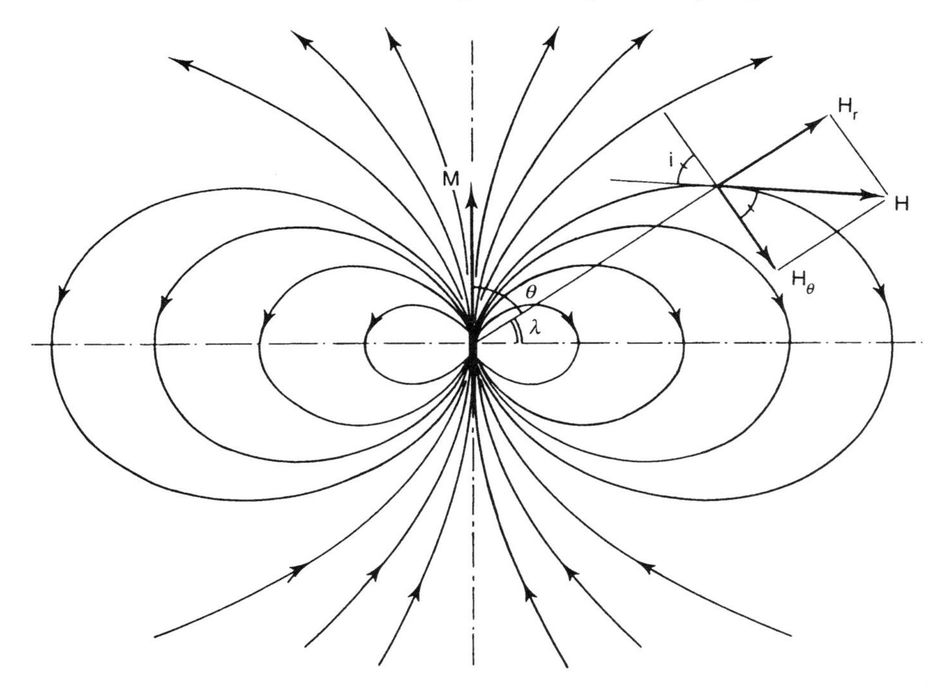

Figure 4.1. Dipolar field. The force lines are in meridian planes and their equation is: $r = C\sin^2\theta$.

The corresponding magnetic field is $\mathbf{B}_{10} = -\mu_0\,\mathbf{grad}\,V_{10}$, and is called a dipolar field. It is the field created by a dipole at a distance $r \gg a$. The equipotential surfaces (V_{10} = constant) being of revolution around Oz, the force lines which are perpendicular to them are in radial planes (Figure 4.1). This field is *axial*, or *zonal*. The radial and perpendicular components of the excitation are:

$$\begin{cases} H_r = -\dfrac{\partial V_{10}}{\partial r} = \dfrac{2M_2\cos\theta}{4\pi r^3} \\ H_\theta = -\dfrac{1}{r}\dfrac{\partial V_{10}}{\partial\theta} = \dfrac{M_2\sin\theta}{4\pi r^3} \end{cases} \tag{4.10}$$

Their ratio only depends on the angle θ:

$$H_r/H_\theta = \tan i = 2\,\mathrm{cotan}\,\theta \tag{4.11}$$

The equation of force lines is obtained by integration of $dr/H_r = r\,d\theta/H_\theta$.

With two magnetic masses $+m$ on the x-axis, at $x = \pm a$, and two masses $-m$ on the z-axis, at $y = \pm a$, (i.e. two identical dipoles head to foot) the main term becomes:

$$V_{22} = \frac{3ma^2}{4\pi r^5}(x^2 - y^2) = \frac{3ma^2}{4\pi r^3}\sin^2\theta\cos 2\varphi \tag{4.12}$$

The factor μ_0 apart, this is the potential of a *sectorial quadrupolar field* (Figure 4.2a).

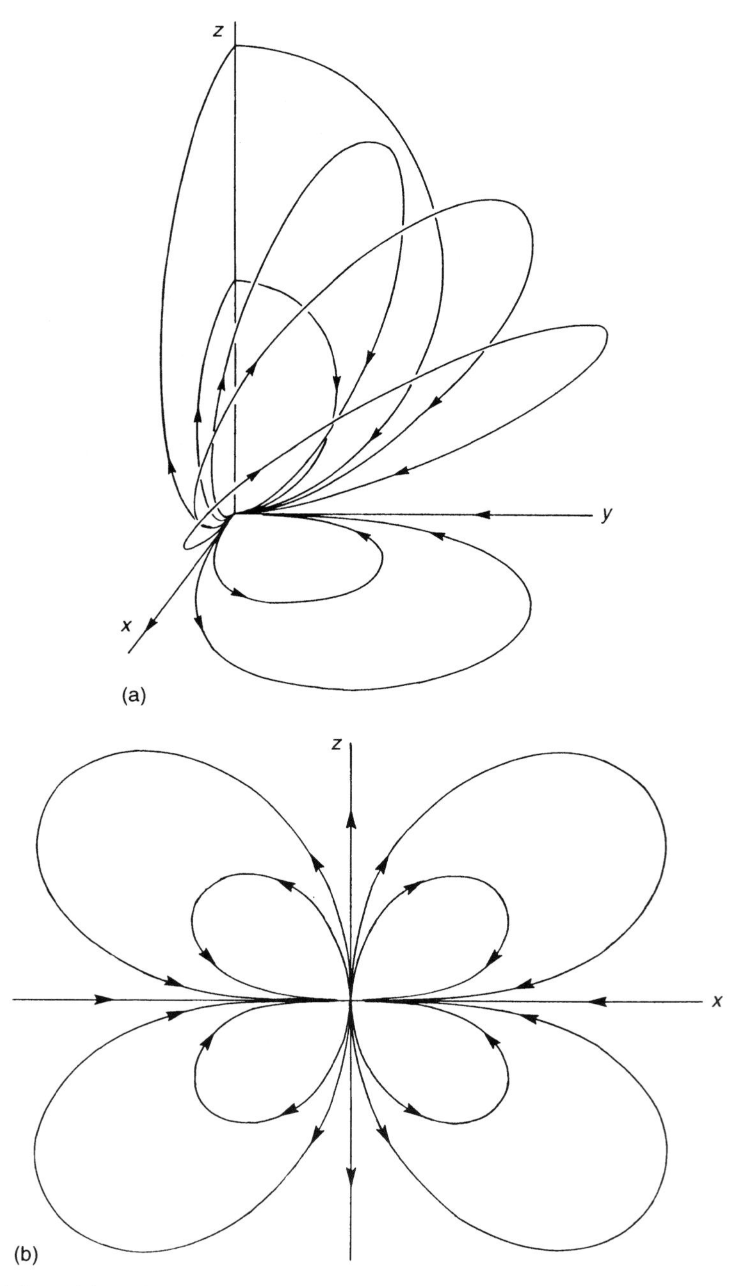

Figure 4.2. (a) Sectorial quadrupolar field. The force lines are the intersections of the revolution surfaces: $r^2 = a^2 \cos^3 \theta$ and the cones: $\cot\text{an}^2 \theta = b \sin 2\varphi$. A few are represented along the perspective $X = y - 0.36x$; $Y = z - 0.48x$. The others can be deduced by symmetry about the three coordinate planes. (b) Axial quadrupolar field. The force lines are in the meridian planes and their equation is: $r^2 \cos \theta \sin^2 \theta = C$.

With two magnetic masses $+m$ on the z-axis at $z = \pm a$ and a mass $-2m$ at the origin (i.e. two identical dipoles end to end, with opposite directions), the main term corresponds to the potential of *an axial quadrupolar field* (Figure 4.2b):

$$V_{20} = \frac{3ma^2}{4\pi r^5}\left(z^2 - \frac{r^2}{3}\right) = \frac{ma^2}{4\pi r^3}(3\cos^2\theta - 1) \tag{4.13}$$

The poles of a magnetic body show themselves only at its surface; inside, they neutralize each other. According to the Gauss theorem, the flux going out of a closed surface inside a magnetic body is zero. The magnetization vector **J** has a conservative flux inside magnetic bodies. However, there is a layer of isolated monopoles at the surface of the magnet. Apply the Gauss theorem to a small surface enclosing an infinitesimal portion of the surface.It is found that the surface layer of monopoles produces, on both sides of its immediate neighbourhood, opposed magnetic excitations $J_n/2$ and $-J_n/2$ (J_n being the component of **J** perpendicular to the surface). **H** being the sum of all excitations, wherever they come from, it follows that the flux of $(\mathbf{H}+\mathbf{J})$ is continuous across the surface. This flux is therefore conservative. So is the flux of $\mathbf{B} = \mu_0(\mathbf{H}+\mathbf{J})$. From the Ostrogradsky theorem (Annex A.1, equation (A.11)), it follows that $\operatorname{div}\mathbf{B} = 0$.

Inside a magnetic body, the excitation created by the monopoles on the surface is called the *demagnetizing excitation*. Indeed, if a body is magnetized through an external magnetic excitation $\mathbf{H_e}$, the resulting vector **J** produces inside the body an excitation $\mathbf{H_d}$ of opposite sign. The matter is therefore only submitted to the excitation $(\mathbf{H_e} + \mathbf{H_d})$, smaller than forecast. This is better understood with Figure 4.3.

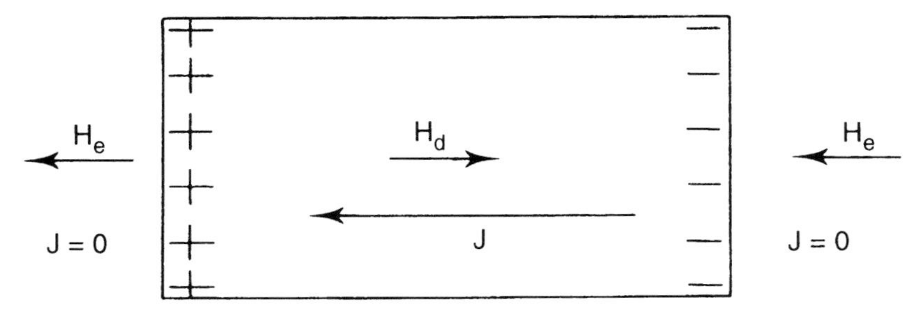

Figure 4.3. Non-compensated magnetic masses (+ and −), demagnetizing excitation H_d and magnetization J for an ideal magnet.

4.2 MAGNETIC FIELDS PRODUCED BY ELECTRIC CURRENTS

Several equivalent laws give, not set out here, the magnetic excitation produced by continuous electric currents. One of these laws expresses the excitation **H** as a curl. Therefore: $\operatorname{div}\mathbf{H} = 0$. *The flux of* **H** *created by currents is conservative*, which was not the case when **H** was created by magnetic substances.

Another law states that **H** remains the same, whether created by a current of intensity I in some circuit (C), or by a very thin *magnetic shell* sealing (C), as far as

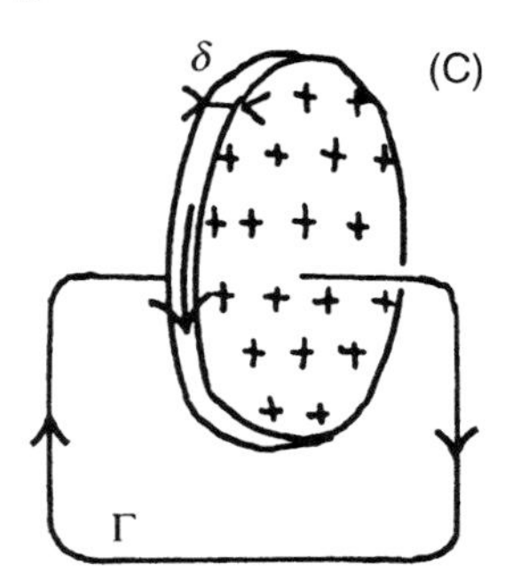

Figure 4.4. Equivalence of an electric circuit and a magnetic shell.

one remains outside (C). A magnetic shell is the fictitious permanent magnet that would be a thin layer of thickness δ, with all north poles on one face and all south poles on the other face. The law states that the dipolar moment of a small portion of the shell, of area dS, must be:

$$dM_2 = (\delta\, dS)J = I\, dS \tag{4.14}$$

Let us calculate the work of **H** along a closed path (Γ) which crosses (C) once (Figure 4.4). Starting from the southern face of the shell, inward, we have the demagnetizing excitation $-J$, for half due to the south face and directed toward it, for half due to the north face and pointing away from it. The work of **H** between the two faces is: $-J\delta = -I$. As the circulation of **H** created by magnetic bodies is 0, the work for the remainder of (Γ), outside the shell, must equal I. With an electric current, not with the equivalent magnetic shell any more, there is no demagnetizing field any more, and the circulation of **H** equals I.

At a distance, it is therefore impossible to know whether a magnetic field comes from a magnetic dipole or from a circular electric current (Figure 4.5a), from a sectorial quadrupole or from a current along two circles forming an 8 (Figure 4.5b), or from an axial quadrupole or two circular currents parallel and with opposite directions (Figure 4.5c).

According to the *Stokes theorem*, the circulation of **H** (whatever its origin) along a circuit (Γ) equals the algebraic sum of the intensities of electric currents crossing (Γ), considered as positive or negative depending on whether they follow the right-handed screw convention or not.

Consider a current I in a coil uniformly wound round a torus, which may be a magnetic core. Following Ampère's theorem, the circulation LH (where L is the length of the circular axis of the coil and n the number of turns) is equal to nI. The circulation along a closed path outside the core equals zero. There is therefore a magnetic excitation inside the core:

$$H = In/L \tag{4.15}$$

This excitation is not observable outside the core. The corresponding magnetic field ($\mu_0\mathbf{H}$ if the core is not magnetic, much more in the case of a transformer whose core

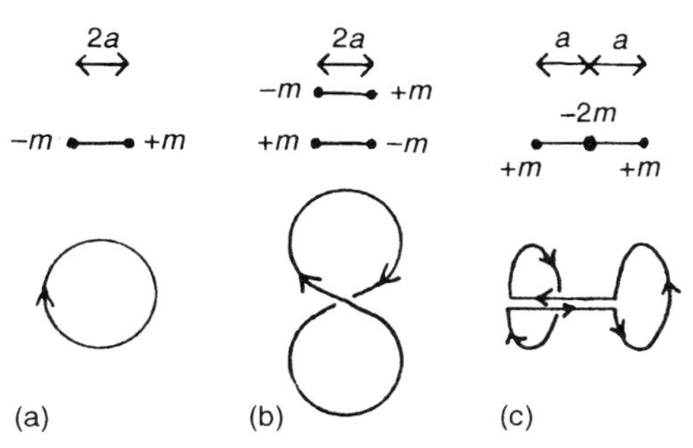

Figure 4.5. (a) dipole; (b) sectorial quadrupole; (c) axial quadrupole.

is made of a soft alloy) is a particular case of toroidal field. The above formula explains why the SI unit for the magnetic excitation is the ampere/metre ($\mathrm{A\,m^{-1}}$).

4.3 MEASURING INSTRUMENTS

Magnetic materials can acquire a strong magnetization **J** for a given excitation **H**. They are used to produce a field $\mathbf{B} = \mu_0(\mathbf{H} + \mathbf{J})$ from a small excitation. The ratio J/H is the *magnetic susceptibility* χ of the material. It is a constant for mild steel and some alloys. It is **B**, and not **H**, which intervenes in the Laplace force (electric motors). The variations of **B** with time create the electromagnetic induction forces. When the flux of **B** crosses a circuit, there appears in the circuit an electromotive induction force (Lenz law):

$$e = \oint \mathbf{E}_i \cdot \mathbf{ds} = -\frac{\partial}{\partial t}\left(\iint \mathbf{B} \cdot \mathbf{n}\, dS\right) \tag{4.16}$$

The electric force e is in Volts, the flux of **B** is expressed in Tesla per square meter, or Weber, and t is in seconds.

The induction electric field E_i adds to the electrostatic field E_e to create the total electric field E. As E_e is a gradient, its circulation is zero, and so is its curl. From the Stokes theorem, the Lenz law follows from the Maxwell–Faraday relationship:

$$\operatorname{curl} \mathbf{E} = -\frac{\partial \mathbf{B}}{\partial t} \tag{4.17}$$

B was first measured with torsion balances (small magnets hung from a wire), by comparison with the Earth's magnetic field. These were *relative measurements*. The very small magnetizations of rocks were measured by the magnetic field that is produced. Then, the reference terrestrial field must be reduced as much as possible. This can be achieved with two flat coils of large identical diameters, with their common axis in the direction of the terrestrial field to cancel. These two coils are crossed by the same current, of intensity and direction suitably chosen (*Helmholtz coils*). The distance between the coils must equal their diameter, so that the field produced is very uniform close to the central point (Figure 4.6).

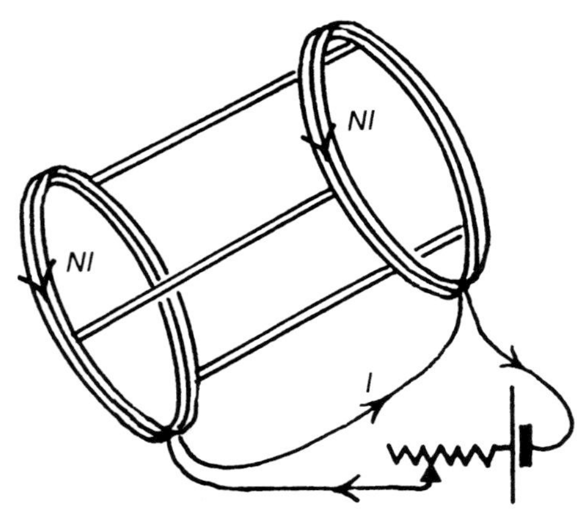

Figure 4.6. Helmholtz coils. The two coils are linked, and can be oriented in any direction.

These rudimentary devices, made by the scientists themselves, were used to discover the laws of magnetism in the nineteenth century, and the laws of rock magnetization in the 1930s (by Émile Thellier). They were superseded by the measurement of alternative electromagnetic forces, which can be amplified as deemed necessary.

Simply by rotating a coil in a magnetic field, it is possible to measure this field; the Lenz law applies in a reference system linked to the coil. This is the principle of *variometers*, which record the terrestrial magnetic field continuously. In *rotating magnetometers*, which measure weak magnetizations, the coil is fixed and the magnetic body (i.e. the rock core) is rotated.

The sensitivity of these instruments has been greatly increased by using coils with a core of mu-metal, an alloy with a very high susceptibility, and a differential technique. These *fluxgates* are used in the modern variometers and rotating magnetometers.

A very high sensitivity to the flux variations is obtained with a 'SQUID' (*superconducting quantum interference device*). This instrument uses a supraconducting circuit, which must be maintained at the temperature of liquid nitrogen, cut by one or several *Josephson junctions*. A Josephson junction is a very thin layer of oxide, insulating, but which the electric current can cross (tunnel effect). This is one of the rare cases where the quantum uncertainty principle is visible at a macroscopic scale. The phase of the current changes importantly while crossing the junction. The smallest variation of the flux of **B** creates a very small electromagnetic induction force in the supraconducting circuit, but changes the phase drastically. The latter can be measured by creating interferences. Squids are now used to measure the very weak magnetism of any sediment or sedimentary rock, whereas before only the magnetism of lavas and baked clays (1,000 times stronger) was accessible.

The direct measurement of the field has also much progressed. The Laplace force is only used in the calibration of ammeters. The *Hall-effect probe* makes use of the small difference in potential (of the order of 1 μV) which appears between the two faces of a very thin conducting layer, crossed by an electric current and submitted to a magnetic field **B**. This probe is simple and of reduced dimensions, but not very accurate.

Around 1955, the first magnetometers using NMR (nuclear magnetic resonance) were a technical revolution. *Proton magnetometers* use the NMR of hydrogen, *magnetic-resonance magnetometers* the NMR of caesium. The resonance frequency ν depends on the magnetic field's intensity B:

$$\nu = \gamma B/(2\pi) \tag{4.18}$$

The constant γ is the gyromagnetic ratio. It is determined with great precision for the proton, much less so for the caesium nucleus. To give an idea, in a field of 0.5 G, v is of the order of 2,000 Hz in a proton magnetometer, 165,000 Hz in the CSF-type atomic resonance magnetometer.

The proton magnetometer enables absolute measurements with an accuracy of 1 nT. The atomic resonance magnetometers are 10 times more accurate, and need to be calibrated with proton magnetometers. Let us reiterate the fact that NMR magnetometers only measure the intensity B of the magnetic field, not its three components B_x, B_y and B_z.

4.4 ORIGIN OF THE REGULAR TERRESTRIAL MAGNETIC FIELD

The terrestrial magnetic field, or geomagnetic field, shows slow variations at the historical scale and rapid fluctuations. Removing the latter, the field corresponds to the *regular field*, originating from inside the Earth. It has a main component, the *regular planetary field* (so called because it is the only one to extend far from the Earth) and *local anomalies*, whose smallest dimensions are 100 km at most.

The regular planetary field has about the same intensity all around the Earth, which shows that its origin is very deep. The mantle and the core are not magnetic (they are too far above the Curie temperature, where magnetic properties disappear). This field can only come from electric fields. The electrical conductivity is of around $10^{-4}\,\Omega^{-1}\,\mathrm{m}^{-1}$ in the upper crust, 10 to 100 times larger in the lower crust, and increases in the mantle transition zone to $100\,\Omega^{-1}\,\mathrm{m}^{-1}$ in the lower mantle. It becomes thousands of times larger in the outer core of liquid iron (a few $10^5\,\Omega^{-1}\,\mathrm{m}^{-1}$). To give a comparison, at ordinary pressure and temperature, the electrical conductivity of iron is $10^7\,\Omega^{-1}\,\mathrm{m}^{-1}$, the conductivity of mercury is $10^8\,\Omega^{-1}\,\mathrm{m}^{-1}$. The only electrical currents of importance are therefore in the liquid core. They would decrease with a time constant of 10,000 years, by ohmic dissipation, if they were not associated to transfers of matter (*convection currents* in the liquid core). Some very small electric currents at the very base of the mantle produce some mechanical coupling, which is used to explain the very small irregularities observed in the rotation of the Earth.

The energy source for this convection will be examined in Section 16.4 (temperature differences like in the Bénard–Boussinesq convection, or density differences related to a variation in chemical composition?). Convection and electrical currents are closely related, as an electric current creates a magnetic field, which exerts a Lorentz force on the conducting material. This can be compared to a dynamo (the mechanical energy of the rotor being transformed into continuous current) whose 'excitation' (electric current producing the magnetic field) is the current produced by the dynamo itself. In such a self-excited dynamo, if the direction of the current reverses without modifications in the rotation of the rotor, the direction of the electric current reverses, the magnetic field reverses, but the Lorentz forces do not change.

The comparison stops here, however. Magnetic fields and convection currents in the liquid core form a much more complicated problem of *magneto-hydrodynamics* (MHD). Let us say only that the average rotation of the core does not follow exactly the rotation of the mantle (which would be the case if the Coriolis force were predominant). At a given point in the core, the ratio between the Lorentz force and the Coriolis force is the *Elsasser number*. According to LeMouël (1976), the two forces would be of about $10^{-4}\,\mathrm{N\,m^{-3}}$, and the Elsasser number would therefore be close to 1. It was assumed to be 0 in the first models of the terrestrial dynamo. The core MHD did not provide certain results before 1995 (see Section 16.4), because the assumed models cannot be tested, and the magnetic field inside the core cannot be deduced from any measurement.

Outside the core, away from the electric currents, the regular field depends on a potential. It can therefore be extended from the surface of the Earth to the surface of the core. Further, the field is no longer the gradient of a potential. The equations of MHD can be divided into two independent systems. The first one yields a field $\mathbf{B} = \mu_0\,\mathbf{curl}(q\mathbf{a})$, where q is a function of the point considered, and $\mathbf{a}$ the unit vector on the radius. This field (*toroidal field*) is everywhere perpendicular to the radius, and therefore not accessible. The other system of equations yields a field, abusively called *poloidal*, whose component radial to the core's surface is accessible, but whose component perpendicular to the radius cannot be accessed either.

The local anomalies of the regular field are measured aboard planes or ships trailing the measuring instrument at a fair distance, to avoid perturbing it. These anomalies are used in *magnetic exploration*. On the continents, they reveal the presence of iron ores (and of other associated minerals), as well as the presence of faults masked by the surface sediments. In the oceans, the local anomalies form parallel stripes; they give the age of the oceanic crust, and the spreading rate of its mid-ocean ridges (see Chapter 5).

4.5 OUTER GEOMAGNETIC FIELD AND MAGNETO-TELLURIC EXPLORATION

The outer geomagnetic field is created by electric currents in the ionosphere, the more or less ionized atmosphere at altitudes between 30 and 500 km. Its variations produce variations of the geomagnetic field. They consist of:

1. *Periodic fluctuations*: daily variations (a few tens of nT), variations related to the Moon and yearly variations (a few nT), due to distortion of the geomagnetic field by the *solar wind* (a continuous emission of particles by the Sun).
2. *Rapid random fluctuations* (a few minutes at most), small (a few nT), due to oscillations in the ionosphere and the magnetosphere. The latter is at altitudes between 2 and 4 terrestrial radii.
3. *'Agitated days' and magnetic storms*: with rapid starts, synchronous all over the Earth. The field may decrease of a few hundreds of nT (which is still not enough to perturb a compass). These storms last several hours, and are linked to the arrival in the magnetosphere of a plasma burst from the Sun. They are accompanied by aurora borealis.

The measures of the inner geomagnetic field are the ones of interest to us, and they should therefore be made:

- during days which are magnetically calm;
- by averaging values over several minutes (when measuring on the ground) or over several orbits (when measuring with satellites), in order to eliminate the pulses.
- with a correction to eliminate periodic fluctuations.

These fluctuations induce in the ground electrical currents, which are called *telluric currents*. Other telluric currents come from human activity: electric railways, noticeable up to 10 km away, high-voltage lines, etc.

All these fluctuations, natural and artificial, with quite diverse periods, are used in *magneto-telluric exploration*. Their origin is external, and the electrical currents produced are not detected below a certain depth, the *penetration depth*. The complex system of equations linking the telluric currents **E** and **B** at different depths, can be simplified by assuming that the variations of **B** are: nearly horizontal (in the xy plane), uniform over a distance large compared to the penetration depth, and are a sine function of time (with a frequency ν). From the resistivities at several depths, a *surface impedance* $Z(\nu)$ can be defined:

$$Z(\nu) = \frac{E_x(\nu)}{H_y(\nu)} = -\frac{E_y(\nu)}{H_x(\nu)} \tag{4.19}$$

Cagniard's method (1953) consists of recording the components $E_x(t)$ and $B_y(t) = \mu_0 H_y(t)$, and performing a spectral analysis (by computing the Fourier transforms: see Annex A.4). This provides $Z(\nu)$, which can be compared to the functions computed for different series of layers with given resistivities.

Magneto-telluric exploration allows the estimation of the conductivity of the underground under a very insulating rock, which insulates from a conducting surface layer. This is not possible with continuous-current electric exploration, where the current stays in the surface layer.

4.6 SPHERICAL HARMONICS OF GEOMAGNETIC POTENTIAL

Any function defined on a sphere can be expanded in a single way as a double series of spherical harmonics (see Annex A.2). Any function $f(r, \theta, \phi)$ in space can therefore be represented by a double series of spherical harmonics whose coefficients are functions of r.

Outside from magnetic matter and electric currents, geomagnetic excitation is the gradient of a potential with a Laplacian of zero, according to Green's theorem. One can demonstrate that this potential is of the form:

$$V = \frac{a}{\mu_0} \sum_{n=1}^{\infty} \sum_{m=0}^{\infty} P_{nm} \left\{ \left[C_{nm} \left(\frac{r}{a} \right)^n + (1 - C_{nm}) \left(\frac{a}{r} \right)^{n+1} \right] g_{nm} \cos m\phi + \left[S_{nm} \left(\frac{r}{a} \right)^n + (1 - S_{nm}) \left(\frac{a}{r} \right)^{n+1} \right] h_{nm} \sin m\phi \right\} \quad (4.20)$$

with a denoting the Earth's radius (6,371 km). The coefficients g_{nm} and h_{nm} are the *Schmidt coefficients*. Note that they are defined here with non-normalized coefficients, and without the factor $(-1)^m$ which was later introduced by mathematicians. There is no term of degree 0 because there are no isolated magnetic poles. The terms in C_{nm} and S_{nm} come from sources outside $r = a$ (electric currents in the ionosphere). For measurements made on magnetically calm days, they are close to zero. This formula allows us to bring the values of fields measured on board satellites back to ground values. They could be extended to the close neighbourhood of the core, if the magnetizations of the terrestrial crust (which create the local anomalies) were well known. But this is not the case.

Local anomalies only create harmonics of high degree (thought to be greater than 14). It is therefore sound to try extending the magnetic field down to the core by using only the harmonics of degree below 14. Unfortunately, the precise measurements performed all over the Earth were made with NMR magnetometers, which measure only the intensity B of μ_0 **grad** V, not its components. All rectangle terms $g_{nm} g_{n'm'}$, $g_{nm} h_{n'm'}$, etc. intervene in its expression, and neglecting the harmonics of higher order introduces large inaccuracies in the computation of low-degree terms.

Measurements on board the six OGO satellites (*Orbital Geophysical Laboratory*) between 1964 and 1969, and aboard the Magsat satellite in 1979/80 were used to determine the Schmidt coefficients to degree 8. By noting t the time since 1 January 1965, in years, the first coefficients are (in nanoteslas):

$$g_{10} = -29{,}877 + 23.0t \qquad g_{20} = -2{,}073 - 15.2t$$
$$g_{11} = -1{,}903 + 10.6t \qquad g_{21} = 3.045 + 3.6t$$
$$h_{11} = 5{,}497 - 21.4t \qquad h_{21} = -2{,}191 - 12.4t$$
$$g_{22} = 1{,}691 + 5.6t$$
$$h_{22} = -309 - 21.8t$$

The harmonics of degree 1 are equivalent to the magnetic field of a dipole at the centre of the Earth, of moment $7.87 \times 10^{22}\ \mathrm{A\,m^2}$, and oriented toward the *south geomagnetic pole*. This should not be confused with the south magnetic pole, the place where the *total* geomagnetic field points to the zenith, and where an explorer always heading toward the south by his compass would arrive. The north magnetic pole, where the total field points to the nadir, is not antipodal to the south magnetic pole. But the *geomagnetic* north pole is always exactly antipodal to the geomagnetic south pole. In 1985, it was located at 78.98°N latitude, 70.9°W longitude. At the Earth's surface, the dipole field is predominant by a factor 10, but this predominance decreases toward the core, because of the factor in $(a/r)^{n+1}$.

The harmonics of degree 2 are equivalent to the field from an axial quadrupole and a sectorial quadrupole, at the centre of the Earth, etc. But these equivalences are purely mathematical. No one thinks nowadays that the dipole field and the non-dipole field have separate origins, which could be separated in models of the terrestrial dynamo. This wrong idea came from measurements that were too limited in space and in time, and led to a belief in the westward drift of the whole non-dipole field.

4.7 MAGNETIZATION PRODUCED BY A MAGNETIC EXCITATION

Electrons, protons and neutrons have their own magnetic moments, measured by their *spin*. The link between an electron and a nucleus produces an additional magnetic moment, the *orbital moment*. This name comes from the old atomic model of Bohr. The electron was thought to be a charged body orbiting around the nucleus, which could be assimilated to a small circular electric current.

Generally, all magnetic moments relative to an atom compensate for each other. But this is not the case for some elements with incomplete inner layers of electrons (Fe, Ni, Co, Mn, rare earths). These atoms have a magnetic moment, and minerals containing them may be magnetized. If they are demagnetized, an excitation will produce a magnetization J. The ratio $J/H = \chi$ is the *magnetic susceptibility*.

Above the *Curie temperature*, all magnetic bodies have a very small susceptibility, well defined and of the order of 10^{-3}. These bodies are called *paramagnetic*. The Curie temperature may be very low, or may not exist, and therefore the material may be paramagnetic down to room temperature. Below the Curie temperature, the material usually has a very strong susceptibility, not very well defined as its magnetization depends on present and on past excitations. The material is then said to be *ferromagnetic*. Iron is the best known example of this behaviour. Nickel, cobalt, several alloys and the magnetic minerals of rocks are also ferromagnetic. The latter may experience a similar state, called *ferrimagnetism*, but the only difference exists at the atomic scale and does not affect their behaviour.

When a ferromagnetic material is subject to a total excitation H oscillating between $-H_m$ and $+H_m$, the magnetization follows with a phase delay. The representative point in the plane (H, J) describes a *hysteresis cycle* (Figure 4.7a). For very large values of H_m, the point describes a limit hysteresis cycle, plus two reversible branches. The excitation necessary to cancel J out is $-H_c$. H_c is the *coercive excitation*. Its value varies a lot and depends on impurities. It reaches

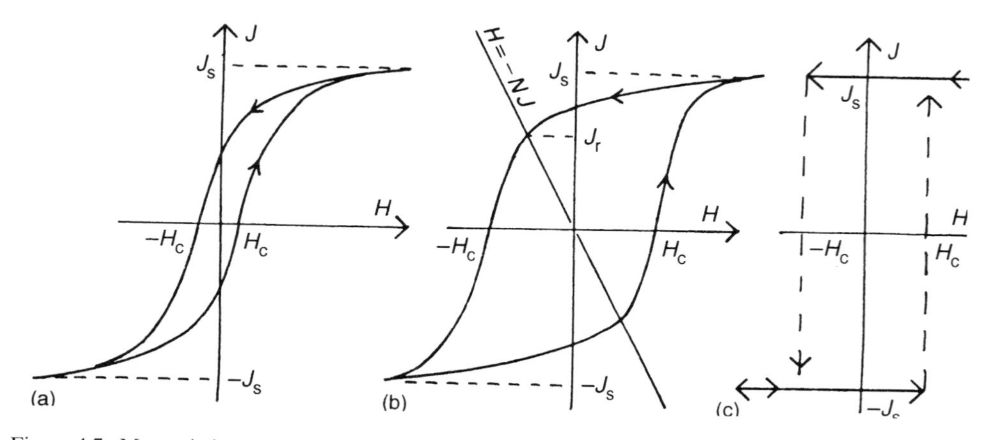

Figure 4.7. Magnetic hysteresis cycles: (a) soft steel; (b) strong steel; (c) perfect magnet.

$2\,\mathrm{A\,m^{-1}}$ for ultra-pure iron, $1{,}700\,\mathrm{A\,m^{-1}}$ for steel with 1% carbon, 40,000 to $100{,}000\,\mathrm{A\,m^{-1}}$ for the special alloys used as magnets. On the reversible branches, the magnetization tends toward a limit value J_S, the *saturation magnetization*.

Now, the total excitation **H** is not equal to the external one $\mathbf{H}_e$: the reverse demagnetizing excitation $\mathbf{H}_d$ due to **J** must be added. Both $\mathbf{H}_d$ and **J** depend on the direction of $\mathbf{H}_e$ relative to the object. If the object is an ellipsoid and $\mathbf{H}_e$ is aligned with one of its three axes, **J**, $\mathbf{H}_d$ and **H** are uniform in the whole object and colinear to $\mathbf{H}_e$. With N constant, the *coefficient of demagnetizing excitation* is:

$$H_d = -NJ; \qquad J = \chi(H_e - NJ) = \chi\frac{H_e}{(1 + N\chi)} \tag{4.21}$$

There are three values of N depending on the axis selected, and their sum equals 1. For a sphere: $N_1 = N_2 = N_3 = 1/3$. For an ellipsoid five times longer than it is wide, the values are $N_1 = 0.0558$, $N_2 = N_3 = 0.4721$. And if the ellipsoid is 10 times longer than it is wide, $N_1 = 0.0203$, $N_2 = N_3 = 0.4899$.

The distance from a point to the axis $H = -NJ$, measured along the H-axis, is $H + H_d = H_e$. The same hysteresis curve gives the magnetization for a given external excitation H_e, and various coefficients N corresponding to different shapes of the magnetic body (Figure 4.7b). The remanent magnetization J_r is the value of J for which $H_e = 0$. To get a large remanent magnetization, J must remain important for $H = 0$, but N must also be the smallest possible. And, if N is fixed, H_c must be important and the hysteresis cycle must be the most rectangular possible (Figure 4.7c). This is the aim in alloys for permanent magnets, and it is achieved naturally in the magnetic inclusions of rocks (but for different reasons).

A ferromagnetic sample can be demagnetized by subjecting it to an oscillating external excitation, of amplitude progressively decreasing down to 0 (this assumes that the geomagnetic field is cancelled with Helmholtz coils). The representative point (H, J) describes progressively smaller hysteresis cycles, centred on the origin.

But if a small constant excitation ΔH_e is added during the process, the result is a strong *anhysteretic magnetization* J_{anh}.

4.8 MICROSCOPIC PROCESSES OF FERROMAGNETISM

The complex laws of ferromagnetism are best understood by studying magnetization processes on the microscopic scale.

The interaction between atoms with a magnetic moment is of a quantum nature, but is correctly rendered by introducing a fictitious 'molecular magnetic excitation'. This interaction aims at making all the moment vectors parallel. It is counterbalanced by thermal agitation, and the alignment of vectors only occurs below a certain temperature, the Curie temperature. Ferromagnetism appears suddenly, the substance taking a *spontaneous magnetization*, which will be seen to equal the saturation magnetization J_S.

If magnetization at the macroscopic scale is much smaller than J_S, it is because a crystal divides into many ferromagnetic domains, also called *Weiss domains*, whose spontaneous magnetizations are oriented differently. These domains are layers a few micrometres to hundreds of micrometres thick. In each layer, the magnetization equals J_S, and its orientation (imposed by a minimum of the *magneto-crystalline energy* stored) has the symmetries of the crystalline structure. In the case of iron, nickel or spinels like magnetite (all with a cubic system), there are three possible directions, all normal to each other and with two orientations. The magnetizations of two neighbouring Weiss domains are at angles of 180° or 90°. The boundary between the two domains (*Bloch wall*) is said to be at 180° or 90°.

The wall is oriented so that the magnetization flux is conserved when crossing it. The walls at 180° are therefore parallel to the magnetizations on each side, the walls at 90° are at 45° of each other. Despite the existence of prismatic domains, the closing of the magnetization flux is imperfect at the crystal boundaries, and the magnetic poles appear.

If the Bloch walls were geometric surfaces without thickness, the exchange energy between two contiguous atoms with magnetic moments oriented differently would be too high. When crossing the wall, the magnetization J_S must turn progressively from one orientation to the next, deviating from the orientations that minimise the magneto-crystalline energy. The latter would be too high if the partition were too thick. Computations show that the Bloch walls are around 0.1 μm thick.

A ferromagnetic body appears demagnetized macroscopically when the magnetizations of all its Weiss domains are cancelling each other. When a magnetic excitation is applied, the domains where magnetization is at acute angle with the excitation grow, at the expense of the domains where the angle is obtuse, by a migration of the Bloch walls. There is no more cancelling, and magnetization gets observable at a macroscopic scale.

When this process cannot work longer (about one-half of the Weiss domain have then disappeared), and excitation increases, the spontaneous magnetization moves away from the direction fixed by the crystalline structure, and aligns itself on the direction of excitation. This is why saturation magnetization equals spontaneous magnetization.

The causes of hysteresis are:

(1) The crystalline structure has some defects: the ones intervening in plastic deformation processes (see Chapter 11), and often microscopic inclusions too. These defects act as anchor points for the Bloch walls, and delay their progression.
(2) Any Weiss domain may disappear when the walls move, but to reappear when the excitation is reversed, it needs a nucleus. This may be a cavity or non-ferromagnetic inclusion large enough (larger than 0.1 μm in the case of iron) (Néel, 1944).

4.9 SINGLE-DOMAIN FERROMAGNETISM OF SMALL INCLUSIONS

The explanations calling on the migration of the Bloch walls are not valid for most inclusions of ferromagnetic materials responsible for the magnetization of rocks, baked clays and sediments. Each inclusion is too small to allow the formation of several Weiss domains. Haematite inclusions in baked clays are for example smaller than 0.15 μm (which is the thickness of a Bloch wall). Magnetite flakes are shorter than 3 μm. These single-domain microcrystals have a rectangular hysteresis cycle, and they behave as perfect permanent magnets of microscopic sizes. For unknown reasons, this permanent magnetic behaviour persists even for larger grains, up to 10 μm, although they contain several Weiss domains (Dunlop, 1995).

The remanent magnetization of some inclusions may reverse spontaneously, the process being activated by thermal agitation. The energy transfers are of the order of kT, T being the absolute temperature and $k = 1.38 \times 10^{-23}\,\mathrm{J\,K^{-1}}$ being the Boltzmann constant. The magnetic energy of an inclusion with volume V, coercive field B_c and magnetization J_s is $VJ_sB_c/2$. A very general law of thermodynamics states that the average magnetization of inclusions must decrease exponentially, and be divided by e after a *relaxation time* Θ:

$$\Theta = A\exp(VJ_sB_c/2kT) = A\exp(VT_C/V_0T) \tag{4.22}$$

T_C is the absolute Curie temperature, and A is a constant.

The relaxation time will be very short if $V < V_0$. Let us estimate V_0. The magnetic excitation inside the inclusion must be lower than the negative coercive excitation, necessary to change J_s:

$$H_e - NJ_s > -H_c \iff H_c > NJ_s - H_e \tag{4.23}$$

The lower limit of H_c may vary from 0 to $J_s/3$. Taking the mean value of $J_s/6$, one finds $V_0 \cong 12kT_c/\mu_0 J_s^2$. V_0 varies from $5 \times 10^{-25}\,\mathrm{m^3}$ for magnetite (the equivalent of a sphere 0.01 μm in diameter) to $3 \times 10^{-20}\,\mathrm{m^3}$ for haematite (the equivalent of a sphere 0.4 μm in diameter).

When V is much smaller than V_0, the relaxation time is very short. The direction of the spontaneous magnetization keeps reversing rapidly, and there is no contribution to remanent magnetization. The inclusion is *super-paramagnetic*. The contribution of such inclusions would have been very small for magnetite, but very important for haematite.

The relaxation time depends on temperature, and, at a given temperature, depends on precise values of H_c and N (varying according to the grains), and on the geomagnetic excitation. This has two important consequences:

1. For inclusions of a fixed volume, only the magnetization acquired below a certain temperature remains (because the relaxation time increases importantly when the temperature decreases). This is the *blocking temperature*. If a sample is heated, away from any outside magnetic excitation, to a temperature T_1, the remanent magnetization is erased from all inclusions up to the size associated to the blocking temperature T_1. The remanent magnetization that is not erased is the magnetization acquired at temperatures larger than T_1. This law was discovered experimentally by Émile Thellier with baked clays, and explained by Louis Néel in 1949.
2. The global remanent magnetization of a sample is a function of the intensity of the geomagnetic field at the time it was gained.

4.10 MAGNETIC MINERALOGY, FERRO- AND ANTI-FERROMAGNETISM

The magnetic properties of rocks come from the inclusions of magnetite, maghemite and haematite. By definition, the global magnetization is the weighted average of the magnetizations from all components of the rock or the sediment.

Magnetite (Fe_2O_3-FeO) is called *ferrimagnetic*, like the other ferrites (Fe_2O_3-CrO or Fe_2O_3-NiO), because the magnetic moments of trivalent and divalent iron are parallel but point in opposite directions (they are *antiparallel*). Their J_S is therefore the difference of the ferromagnetic J_S of the two sub-lattices. Ferrites are very interesting for industrial applications, because they provide perfect permanent micro-magnets, and furthermore they are electric insulators. They are used in magnetic memories.

Maghemite (γ-Fe_2O_3), which has a spinel structure with vacant sites, has magnetic properties close to the magnetite.

When the compensation of the two sub-lattices with antiparallel moments is complete, the material is *antiferromagnetic*. Below the Curie point (then called *Néel point*), the material reacts to magnetic excitations as a paramagnetic body. The ferromagnetic effects independent from the direction of magnetization will still be felt, however (e.g. small volume variations: *magnetostriction*). The French scientist Louis Néel explained ferrimagnetism and antiferromagnetism in the 1940s. But he had to wait for 1970 and the development of magnetic memories to receive the Nobel Prize.

Table 4.1. Magnetic parameters of iron, magnetite and haematite.

	Curie point (°C)	J_S (10^5 A m^{-1})	σ_S (A m^2 kg^{-1})
Iron	770	17.4	227
Magnetite	578	4.8	93
Haematite	675	0.02	0.5

Haematite, or **oligist** (α-Fe_2O_3), rhombohedral, gives its red coloration to bricks and baked clays, lateritic soils, the old Devonian sandstones, etc. This is an imperfect ferromagnetic body, where two Fe sub-systems are not rigorously antiparallel.

The magnetic parameters of magnetite, haematite, and pure iron are presented in Table 4.1. The SI unit for magnetization (magnetic moment per unit volume) is the A m^2/m^3, i.e. the A m^{-1}. This is the same unit as H, as $B = \mu_0(H + J)$ requires. More often the moment per unit mass, called *magnetization coefficient* (σ), is given in A m^2/kg. In the old cgs system, where $B = \mu_0 H + 4\pi J$, J was expressed in gauss (G). As 1 tesla equals 10^4 G, $4\pi J$ (in gauss) equals $10^4 \mu_0 J$ (in A m^{-1}). And therefore J (in A m^{-1}) is equivalent to $10^3 J$ (in gauss).

In magnetite and maghemite, the iron can be substituted by titanium, which decreases the Curie point and the spontaneous magnetization. The end-members of isomorphous series, *ulvospinel* Fe_2TiO_4 and *ilmenite* $FeTiO_3$ respectively, are no longer ferrimagnetic, but only paramagnetic.

4.11 THERMO-REMANENT AND CHEMICAL REMANENT MAGNETIZATION

Ferrimagnetic minerals may appear during the solidification of magma. Thus, volcanic lavas and oceanic crust have inclusions of titano-magnetite. These minerals can also form during chemical reactions at high temperatures. For example, when baking clay in a kiln, the hydrated iron oxides (*limonites*) are dehydrating. *Goethite* (α-FeO_2H) produces haematite, and *lepidocrocite* (γ-FeO_2H) produces maghemite.

In these cases, the ferrimagnetic inclusions appear at a temperature above the Curie point, which they reach when cooling. They acquire a spontaneous magnetization directed in the direction of the outside excitation, i.e. the geomagnetic field of the time. This is *thermo-remanent magnetization* (TRM).

The chemical reaction in the solid state, which produced the ferrimagnetic inclusions observed, may at some other times have occurred at temperatures below the Curie point. Affect by the percolation of seawater containing dissolved oxygen, part of the titano-magnetites created at mid-ocean ridges oxidize into maghemites in a few million years. The resulting spontaneous, remanent magnetization has the same direction as the geomagnetic field, but is now called the *chemical remanent magnetization* (CRM).

TRM and CRM provide clues about the orientation of past geomagnetic fields, at the time of acquisition of the magnetization. Of course, this requires that later tectonic movements did not modify the orientation of the rock. In Chapter 17, the inverse problem will be studied; by assuming the former direction of the geomagnetic field to be more or less known, how can the previous orientation of the rock be deduced from the TRM?

Even better, TRM or CRM can give the intensity of the past geomagnetic field, if magnetic inclusions are single Weiss domains. After measuring the TRM, the sample can be demagnetized and the value of H_e corresponding to this TRM can be determined experimentally.

In all cases, the smallest inclusions must first be demagnetized, because, their blocking temperature being too low, they may have been re-magnetized later. This is achieved through moderate heating of the samples, protected from any magnetic field, to a temperature smaller than the temperature of TRM or CRM, but larger than the temperatures undergone later by the rock. The maghemite sometimes needs to be removed, as it may have formed at some indeterminate time by oxidation of magnetite. This is accomplished by keeping the samples above 300°C for some time (the maghemite transforms into haematite) and cooling away from any magnetic excitation. These steps are called the *magnetic cleaning* of the sample.

4.12 ARCHAEOMAGNETISM

The angle between the geomagnetic field and the horizontal, in a given location and at a given time, is called the *magnetic inclination* (i). The angle between the geomagnetic field and the meridian is the *magnetic declination* (δ). Both have been measured in London since the sixteenth century, in Paris since 1660. It is possible to go back further in time by using well dated bricks or Roman tiles. The TRM was acquired during cooling, the brick generally lying in the kiln on one of its two longest sides. The curved tiles were generally arranged vertically (the high number of measurements allows the identification of erroneous cases, where the bricks or tiles were not in the expected position). This gives the inclination, but not the declination. With kiln hearths, or clay baked below prehistoric fireplaces, it is possible to know both.

Knowing i and δ, it is possible to compute the geographic coordinates of a *virtual geomagnetic pole*. This is the position that the north magnetic pole would occupy if the geomagnetic field had been a perfect dipole field. This virtual pole is located on the large Earth circle, cutting the meridian of the observation point with an angle δ, and at an angular distance θ given by equation (4.11).

E. Thellier determined the magnetic inclinations in Paris for the last 2,000 years (Figure 4.8). The inclination varied between 55° and 73° (and therefore θ varied between 51.5° and 31.5°), without any periodicity.

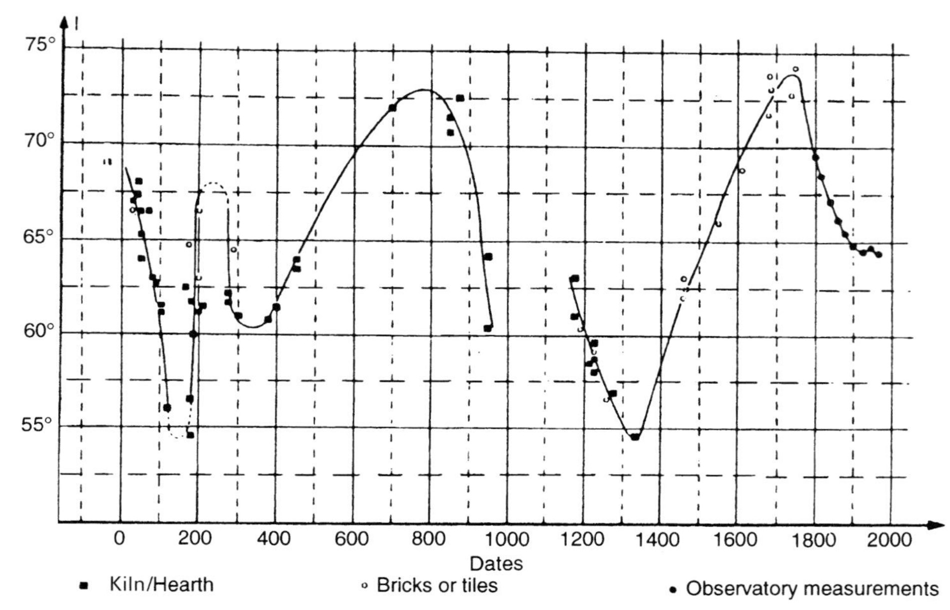

Figure 4.8. Variations of the magnetic inclination in Paris for the last 2,000 years (from Thellier (1971); © Gallimard).

The virtual geomagnetic pole for Bulgaria has moved with time (Figure 4.9). Its path is completely random, but the virtual pole never moved more than 20° from the geographic north pole.

4.13 GEOMAGNETIC FIELD REVERSALS

Lava flows showing a TRM with a direction opposite to the present geomagnetic field have been known since 1905, but for a long time, physicists did not believe the measurements. Until the works of Elsasser and Bullard were published in the 1950s, the origin of the field was subject to controversy. In 1951, Néel suggested an explanation, which proved valid for a Japanese lava. During cooling, a first mineral acquires a normal TRM. In the immediate neighbourhood of these inclusions, the magnetic excitation they cause is stronger than the geomagnetic excitation, and of opposite direction. Another mineral with a lower Curie point then acquires a reverse TRM, which finally becomes more important. This requires a particular combination of magnetic parameters, when more and more reversals were discovered. In 1955, Néel himself did not believe his explanation was the right one. But it was only in 1958 that an experiment performed by Rikitake changed the mind of physicists. Two dynamos were coupled, the current provided by one feeding the electric magnet of the other; the direction of the two currents reversed with time in a random and chaotic way. This was very far from the self-sustained terrestrial

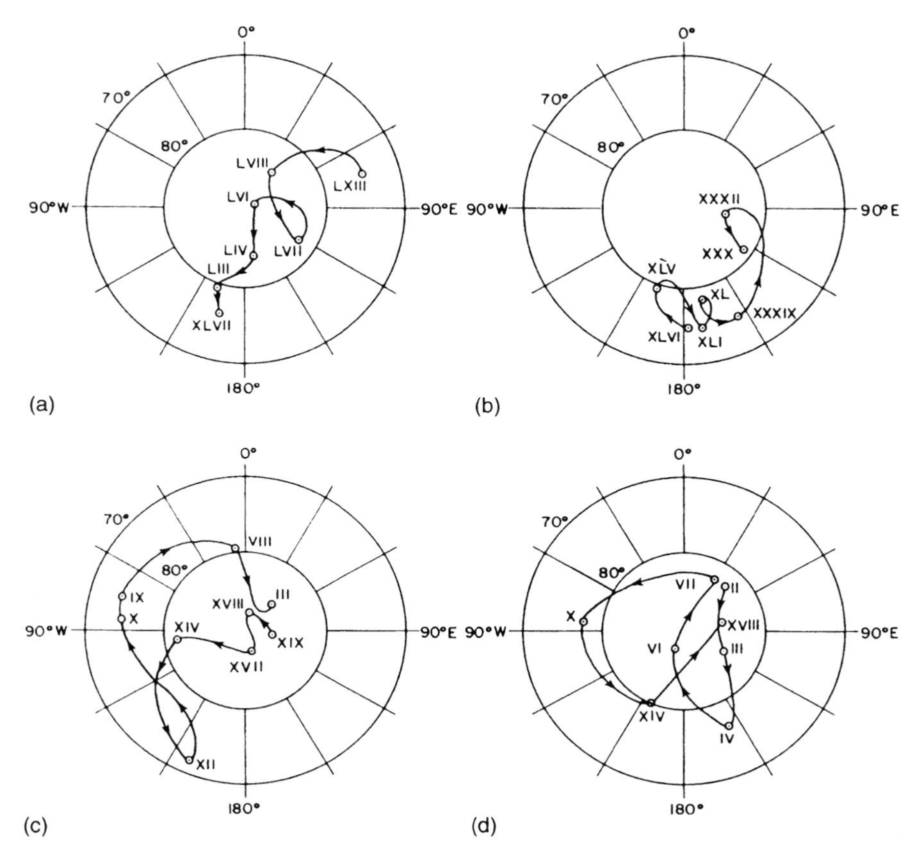

Figure 4.9. Virtual geomagnetic pole for Bulgaria. The Roman numerals correspond to the century (AD or BC): (a) 6,300–4,600 BC; (b) 4,600–2,900 BC; (c) 1,900–200 BC; (d) 100–800 AD. From Kovacheva (1980); © EGS.

dynamo, but it illustrates the unexpected behaviour of some systems of non-linear equations (which had been forgotten since their study by Poincaré at the beginning of the century).

In 1960–1966, Cox, Doell and Dalrymple in the United States, McDougall and Chamalum in Australia, studied systematically the TRM of lavas from the last Ma. Dating by ^{40}Kr/^{40}Ar radioactivity was not very precise then, and therefore they analysed the chronological series of lavas emitted in the same places. They demonstrated that *the virtual geomagnetic poles rarely moved more than 25° away from the geographic pole, but that polarity reversed several times, synchronously, all over the Earth.*

A fine-scale study of a reversal is possible by extracting juxtaposed horizontal cores in a cliff made from a horizontal layer of sedimentary rock (e.g. Valet and Laj, 1981). During the reversal, which lasts around 4,000 years, the intensity of the geomagnetic field decreases to one-fifth of its normal value. The virtual geomagnetic pole moves from the proximity of one geographic pole to the other, but the results obtained in different locations around the Earth are inconsistent. In fact, the dipole component of the geomagnetic field changes sign by going through zero, so that for 4,000 years the main component of the field is a quadrupolar field, and the virtual pole loses any meaning.

In 1969, Cox published a chronology of reversals for the last 4.5 Ma. It shows *epochs*, lasting around a million years, during which the field keeps the same polarity, except during shorter *events*, of opposite polarity. The whole does not present any periodicity. Epochs and events started to receive names, but this was halted because of their sheer number.

In the next chapter, we will see how the local anomalies of the geomagnetic fields in the oceans allowed the extension of this chronology to 160 Ma, to the beginning of the Upper Jurassic. This led to a division in *chrons*, including an epoch with normal dominant polarity (by convention, the current polarity is considered as normal) and the next epoch with reversed dominant polarity. Half-chrons are noted n (normal polarity) or r (reverse polarity), and short events within a chron .1n, 2n,... or .1r, 2r,...

Exceptionally, the first chron (C1) only corresponds to the current normal epoch of *Brunhes*. The C2 chron includes the reverse period of *Matuyama* (0.78 Ma to 1.77 Ma) and the normal period of Gauss (1.77 Ma to 3.58 Ma). The next periods are presented in the table at the end of the book. Minor revisions have been added since (Cande and Kent, 1995; Wei, 1995). The Cretaceous–Tertiary boundary is very well delimited in palaeontology (complete extinction of ammonites and dinosaurs, appearance of nummulites and placental mammals); it occurred 65.0 Ma ago, and corresponds to the limit between chrons C29 and C30.

We shall come back to polarity reversals in Chapter 16. Let us only say that there were a hundred of them in the last 25 Ma, but very few during long periods, called *superchrons* (C34n superchron, between 83 and 118 Ma, or between 260 and 315 Ma, with a polarity mostly reversed).

4.14 REMANENT MAGNETIZATION OF SEDIMENTS

The remanent magnetization of sediments is of ca. $10^{-5}\,\mathrm{A\,m^2\,kg^{-1}}$, a thousand times less than for lavas. Its study was only possible with modern magnetometers, which can reach $10^{-7}\,\mathrm{A\,m^2\,kg^{-1}}$.

This magnetization mainly comes from magnetite, and not only from the fact that sedimentation occurred in an anoxic, reducing environment. There exist bacteria, which transform limonite into magnetite. Some think they use these magnets to orient themselves, as more evolved animals do (bees, tuna fish, pigeons). Chitons, common marine molluscs without shells, may have magnetite in their calcareous blades because they feed on these bacteria.

Far from the coasts, the oceanic sedimentation is of around 1 mm per century. The Brunhes–Matuyama transition, 780,000 years ago, can therefore be found approximately 8 m deep. This is an invaluable marker during oceanic drillings (as worms and other animals occupy the 10 cm near the surface, this layer is highly perturbed, and the variables are somewhat smoothed over 10,000 years). Drills used up to now did not have compasses, and the earlier declinations cannot be extracted from the seafloor. But the inclination can be extracted, because sediment compaction is not significant on the first 20 metres. The *inclination shallowing* produced by sediment compaction is, however, important when the sediments are on land. But the declination can then be measured. In both cases, the field intensities can be determined.

The magnetization of particles already magnetized, which orient themselves during sedimentation as so many tiny compasses, adds itself to the chemical or biochemical remanent magnetization at low temperature. This is *depositional remanent magnetization*.

The remanent magnetization of sediments is very small, and was acquired at low temperature. Its precision is therefore limited by its partial modifications with time. As the geomagnetic field changes, some grains re-magnetize themselves in conformity with the new field. This is *viscous remanent magnetization*, which cannot be counteracted.

Measuring magnetic susceptibility all along a sediment core gives precious details about the variations of sedimentation with time, and therefore about the past environment (see Verosub and Roberts (1995) and the references therein). The routine measurements consist in:

- The apparent reversible susceptibility χ'. The sample is submitted to an alternative excitation of amplitude ΔH_e. The resulting magnetization oscillation has the amplitude ΔJ, and $\chi' = \Delta J/\Delta H_e$. The true reversible susceptibility is $\chi = \Delta J/\Delta H$, and from equation (4.21), we have:

$$\chi' = \frac{\chi}{1 + N\chi} \tag{4.24}$$

- An apparent anhysteretic susceptibility $\chi'_{anh} = J_{anh}/\Delta H_e$, where ΔH_e is the small excitation constant during demagnetization. In the case of normal ferromagnetic bodies with several domains, Néel showed in 1943 that it was linked to the true anhysteretic susceptibility χ_{anh} by the relation:

$$\chi'_{anh} = \frac{1}{N}[1 - e^{-N\chi_{anh}}] \tag{4.25}$$

Only the susceptibilities of super-paramagnetic or ferromagnetic inclusions are important, ranging between 100 and 1000. At these values, the two apparent susceptibilities are nearly equal to $1/N$, and the meaning of their ratio, as used by some scientists, is not clear.

5

Mid-ocean ridges and hot spots

5.1 GEOPHYSICAL EXPLORATION OF THE OCEANS

The study of the deep seafloor was long limited to bathymetry. For example, in the 1950s, the depths of the Pacific Ocean were determined from the Soviet oceanographic ship *Vitiaz* and from the American navy ship *Cape Johnson* (commanded by Harry Hess, geologist by background). Already by this time, the magnetic field at the surface of the oceans was systematically measured. The impetus came from Maurice Ewing, the director of the Lamont–Doherty laboratory (Palisades, New York). A result was, between 1960 and 1967, the unexpected discovery of seafloor spreading at mid-ocean ridges, and the elaboration of plate theory. From 1967, similar measurements were performed every year in the Indian Ocean from the ship chartered by TAAF (*Terres Australes et Antarctiques Françaises*, the French Austral and Antarctic Territories Administration).

The considerable stir of plate theory started an ambitious programme, dedicated to the drilling of marine sediments in deep water. From 1968 to 1975, the five main oceanographic institutions in the US and the University of Washington in Seattle were associated in the *JOIDES* programme (*Joint Oceanographic Institutions Deep Earth Sounding*). They were followed by the *DSDP* (*Deep Sea Drilling Project*) with a specially designed ship, the *Glomar Challenger*. In May 1978, this ship had been through 60 scientific cruises (*legs*), each two months long. 105 km of cores had been extracted from 461 different sites. The current drilling programme is named *ODP* (*Ocean Drilling Program*), and uses the American ship *Joides Resolution*. Its support is international, with an important French participation.

5.2 MID-OCEAN RIDGES

Abyssal plains form a large part of the oceanic basins, with a slope smaller than 1/1000 and 5 to 6 km deep. Only 1 to 2 mm of clay are deposited per century.

Limestone (in particular from marine plankton) does not deposit as it dissolves under high pressures.

Isolated *seamounts* may rise above the abyssal plains. They are of volcanic origin, and rise from 0.5 to 4 km above the seafloor. Much larger features are the *rises*, *plateaus* and *mid-ocean ridges*.

The rises are large elongated domes of very gentle slopes, parallel to the deep oceanic trenches close to the continents (see Chapter 13) or in the middle of the ocean. Mid-ocean rises often support a series of aligned volcanoes. The volcanic chain of the Hawaiian Islands is a perfect example.

Mid-ocean ridges do not have any continental equivalent. Using averaged dimensions, when the values range widely, the crossing of a ridge would look as follows (Carbotte and McDonald, 1994). The seafloor remains horizontal over 1,000 km, but there are 50-m scarps every kilometre, all corresponding to *normal* faults. These thousand parallel steps are ascending or descending, but the altitude is usually rising closer to the axis of the ridge. There, the basaltic crust lays bare, whereas far from the axis the fault scarps end up buried under the sediments. (These scarps are in average 10 km long. This is why, on the old bathymetric charts made from vertical sonar soundings, the ridges were looking like series of elongated convex hills. Their true morphology was only revealed with sidescan sonars).

The French word for ridges is *dorsale*, and was already used by French oceanographers long before the Russian or American scientists studied them. It is more specific than the English 'ridge' or the Russian 'khrebet', which can designate a proper ridge as well as a series of isolated volcanoes, or even the crest of a mountain. 'Ridge' comes from the same Germanic root than the German word 'Rücken' (back).

Two detailed examples of ridge topography are shown in Figures 5.1 and 5.2. At the axis of most ridges, there is a ditch (*rift*, or *graben*), approximately 10-km wide, sometimes more. It is the locus of volcanic processes and of frequent shallow earthquakes. Focal mechanisms show an extension perpendicular to the rift.

Here again, the terminology needs to be specified. *Rift* is a word of Scandinavian origin, and generally designates a crack or a gap due to a separation. This is the proper word in this case, as there was transverse extension and separation of the two sides. *Graben* is a German word, with the same root as 'grave', and designates a large ditch of tectonic origin.

Fracture zones are more or less perpendicular to mid-ocean ridges. They cut the ridge into segments and move its axis laterally, usually always in the same direction, so that the general direction of the ridge at a 5,000-km scale is always oblique to the direction of its segments.

The topography of a fracture zone is steep. The deepest point of the Atlantic Ocean lies in such a zone, near the Puerto Rico Trench (Romanche Gap, −7,558 m). The difference between the two sides of a fracture zone may reach several kilometres. The part of the fracture zone between the two rift axes is a vertical fault with uniform strike-slip, from one end to the other. This is a *transform fault*. Transform faults are the locus of swarms of very shallow earthquakes (Figure 5.3). Their focal mechanisms correspond perfectly to strike-slip movement along a vertical fault.

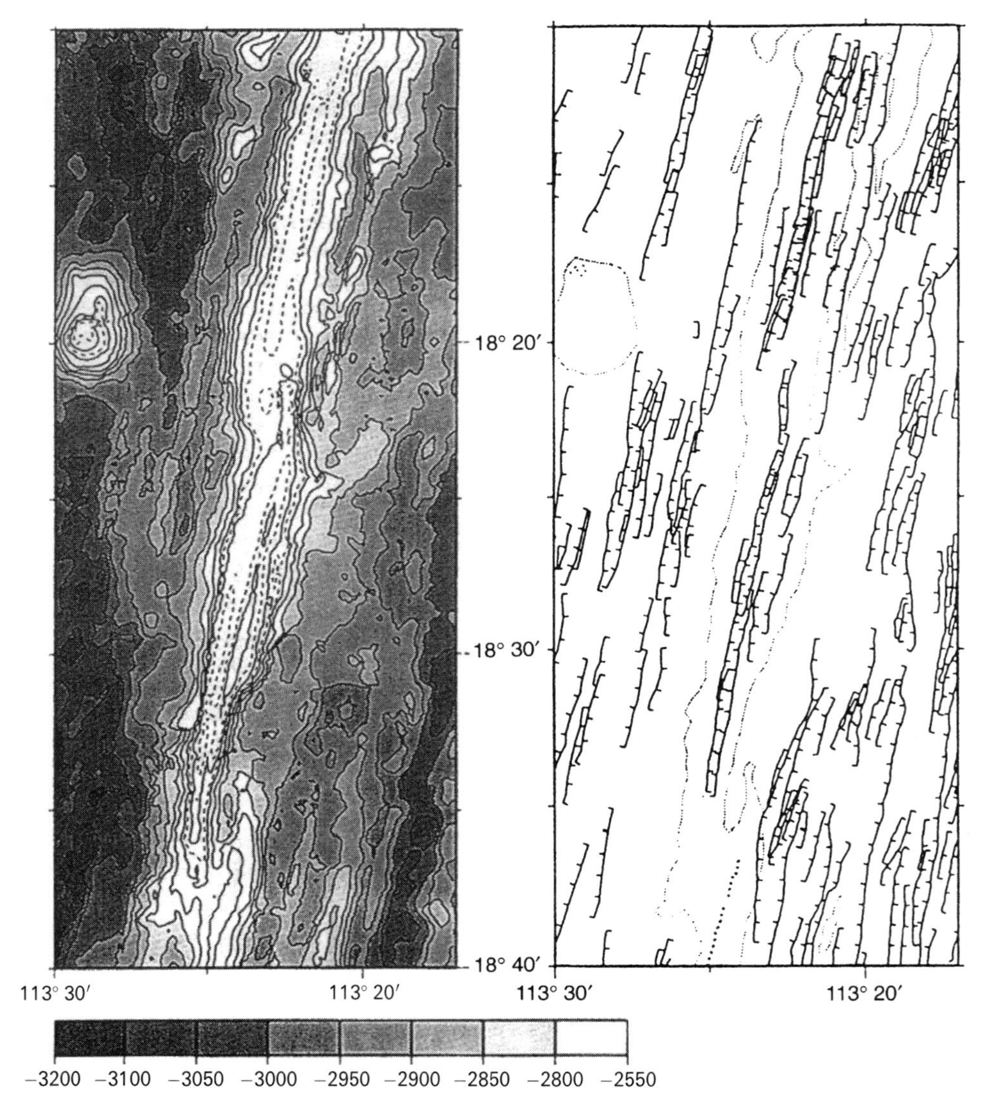

Figure 5.1. East-Pacific Ridge. The scale can be deduced, knowing that 10′ in latitude correspond to 18.5 km. Left: topography with contour levels every 50 m. Right: normal fault scarps, detected with a sidescan sonar. There is a central graben at the ridge axis (the grey tones cover zones shallower than 2,800 m). Elsewhere, there may be horsts as well as grabens. From Carbotte and McDonald (1994); © AGU.

The main active ridges are on the one hand a series of mid-ocean ridges surrounding the Antarctic continent at great distances, which cross the *triple junctions* (intersection of three ridges) of Bouvet, Rodriguez and the Southeast Pacific; on the other hand three north-going ridges leaving these intersections:

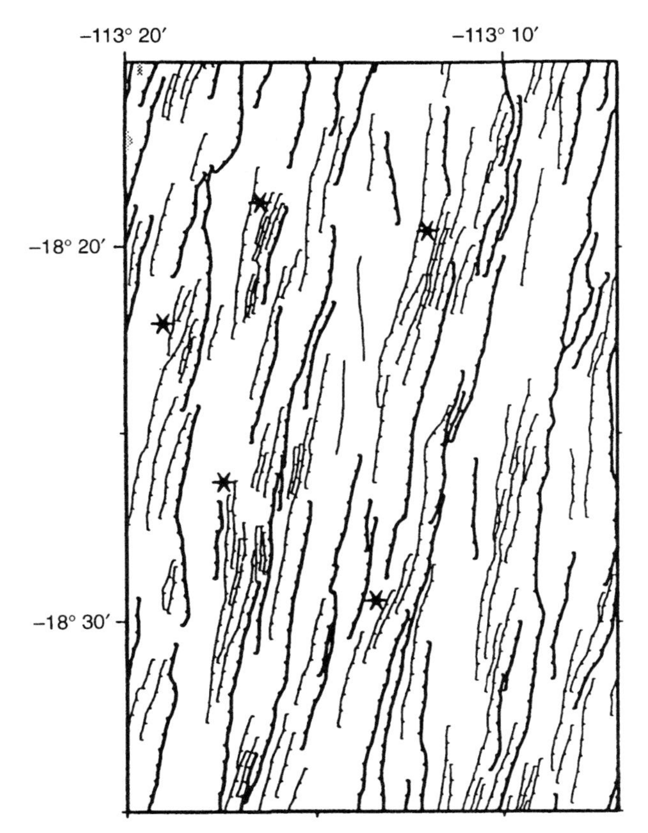

Figure 5.2. Extension of the rightmost map of Figure 5.1. The stars indicate the groups of short parallel faults, 100 to 250 m apart. From Carbotte and McDonald (1994); © AGU.

- The Mid-Atlantic Ridge extends from a triple junction close to Bouvet Island, and lies at the same distance from the Old and New Worlds. Around the Equator, long transform faults move the ridge westwards (Figure 5.3). The ridge next moves toward Iceland (*Reykjanes Ridge*), crosses it, passes between Spitsbergen and Greenland, and disappears progressively in the Arctic Basin (*Nansen Ridge*) (Figure 5.4). On the other side of the Arctic Basin, in the Canada abyssal plain, there is an inactive ridge, perpendicular to the shores of the Beaufort Sea (the Canadian Basin is supposed to have opened 50 to 40 Ma ago). Two parallel ranges, the *Mendeleev Khrebet* (*Alpha Ridge* for US scientists) and the *Lomonosov Khrebet* (*Beta Ridge*) delimit an isolated central abyssal plain, the Makarov Basin. These ranges consist of folded sedimentary rocks, and must therefore be of continental origin.
- The Mid-Indian Ridge starts from the triple junction close to Rodriguez Island, 1,000 km east of the Mascarene Islands (Mauritius and Réunion). Through the Carlsberg Ridge, it reaches Northeast Arabia and extends into the Red Sea.

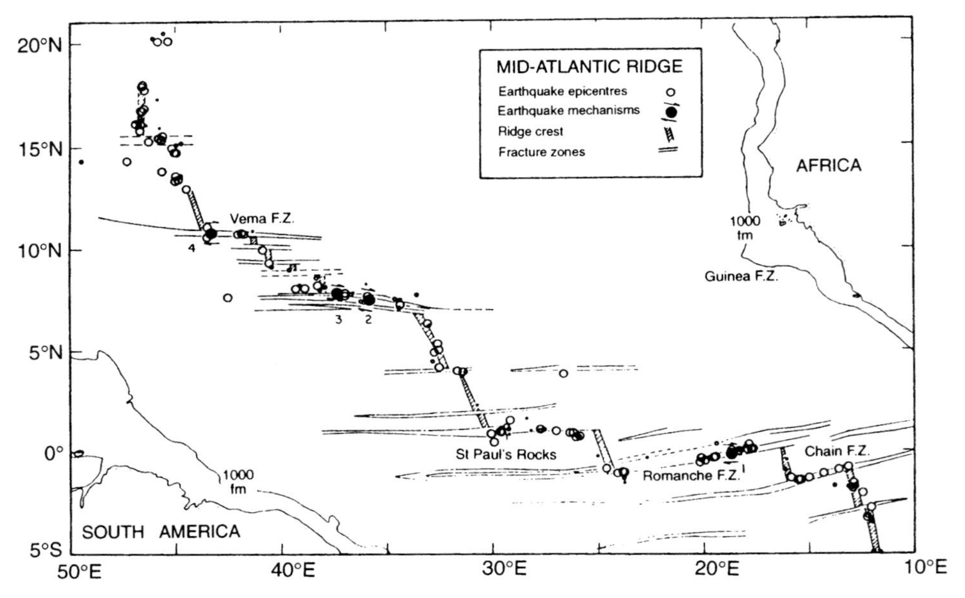

Figure 5.3. Earthquake epicentres in the axial graben and on the transform faults of the Mid-Atlantic Ridge, between 1955 and 1965. The focal mechanisms in transform faults show dextral strike-slip movements. From Sykes (1967); © AGU.

- The East-Pacific Ridge leaves from a triple junction close to latitude 35°S, sometimes called Juan Fernandez triple junction although it lies 2,500 km west of these islands. This ridge goes northward 500 km west of Easter Island and ends in the Gulf of California. There, the transform faults are much more developed than the ridge segments.

5.3 MAGNETIC ANOMALY STRIPES

Mapping the local anomalies of the magnetic field in the oceans requires a very precise navigation. The first survey was performed in 1955–1960, using a ship offshore from California, and benefiting from many radio-location stations. The second survey, airborne, was performed in 1966 on Reykjanes Ridge, using the US base of Keflavik (Iceland) and its radars. Nowadays, precise positioning is available anywhere around the Earth with GPS (*Global Positioning System*), which uses the constellation of artificial satellites Navstar. A database regrouping magnetic anomalies all over the Arctic, the North Atlantic and the surrounding lands, was compiled by the Canadian Geological Survey, following an international cooperation. In 1996, this database contains more than 40 million measurement points, one every 3 kilometres approximately.

By filling in black the zones of positive anomalies, the maps show *stripes*, more or less regular and continuous (Figure 5.5). The word lineation is sometimes used, but

Figure 5.4. Physiography of the Arctic Basin. The bold lines show the limit of the continental platform. From EOS (1980); © AGU.

should be reserved to magnetic exploration, its original domain. Morley, Vine and Matthews correctly explained in 1963 the stripes shown off California. The local anomalies of the present-day geomagnetic field are coming from TRM of the oceanic crust. When it formed, the oceanic crust recorded the field of the time. The successive black and white stripes correspond to successive positive and negative polarities of the geomagnetic fields. This is similar to the recording of information on a magnetic tape, using successive 0s and 1s. The anomaly stripes must therefore be parallel to the mid-ocean ridge from which they come, and nearly symmetric around the ridge. This was exactly what the mapping of anomaly stripes on each side of the Reykjanes Ridge confirmed. It was possible to find there the magnetic field reversals of the last

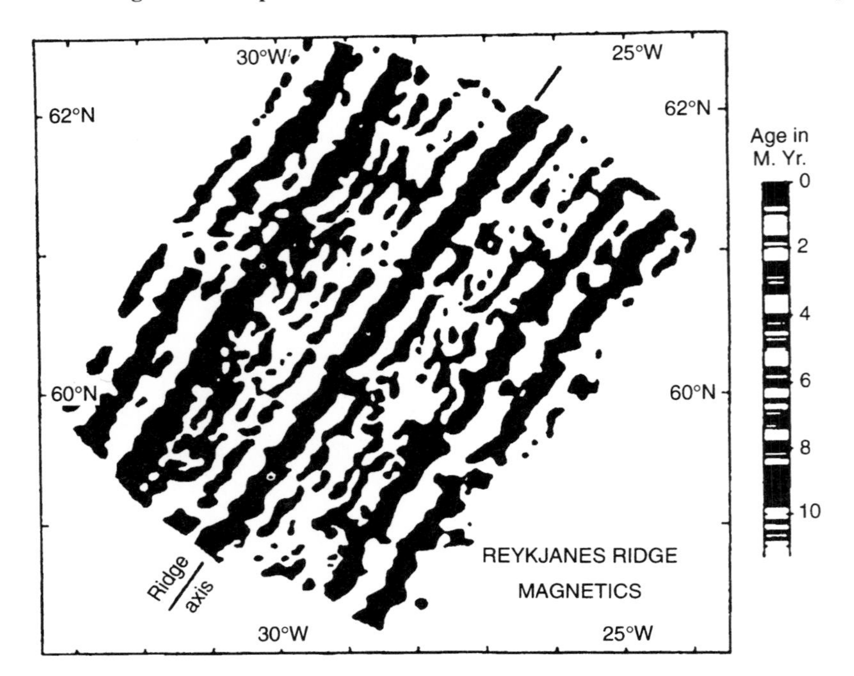

Figure 5.5. Magnetic anomaly stripes over Reykjanes Ridge. The positive anomalies are filled in black. From Vine (1966); © Amer. Assoc. Adv. Sci.

10 Ma, as they had been determined at the time by the TRM of lavas of known ages (Figure 5.6).

In Figure 5.6, the spreading rate was assumed as 20 km/Ma[1], the magnetization J as 0.030 G for the central block, 0.012 G elsewhere, and the mean thickness of the magnetized layer was assumed to be 400 m. In a normal and constant magnetic field, the topography alone would generate the fluctuations represented in the upper curve. The computed anomaly is represented by the curve just above the cross-section. The observed anomaly (middle curve) is a bit smoother, because of contamination effects. The position of the actual peaks sometimes differs slightly from the computed peaks, because of irregularities in the spreading rate.

The idea of *seafloor spreading* was already in the air. Oceanographers had noticed the progressive increase in sediments away from mid-ocean ridges. Isolated volcanic islands had been dated, and found to be older as they were further from a ridge. And there was the close fit between coastlines and geology of the two sides of the South Atlantic, well recognized since the days of Wegener. The magnetic anomaly stripes were not only providing an irrefutable proof of the spreading; they could also measure the average speed on a million-year scale. For Reykjanes Ridge, the

[1] I recommend that the speeds averaged over million years which are inferred from magnetic stripes be given in km/Ma, whereas the ones measured using spatial geodesy be noted in mm/a. One cannot affirm *a priori* that the two spreading rates are identical.

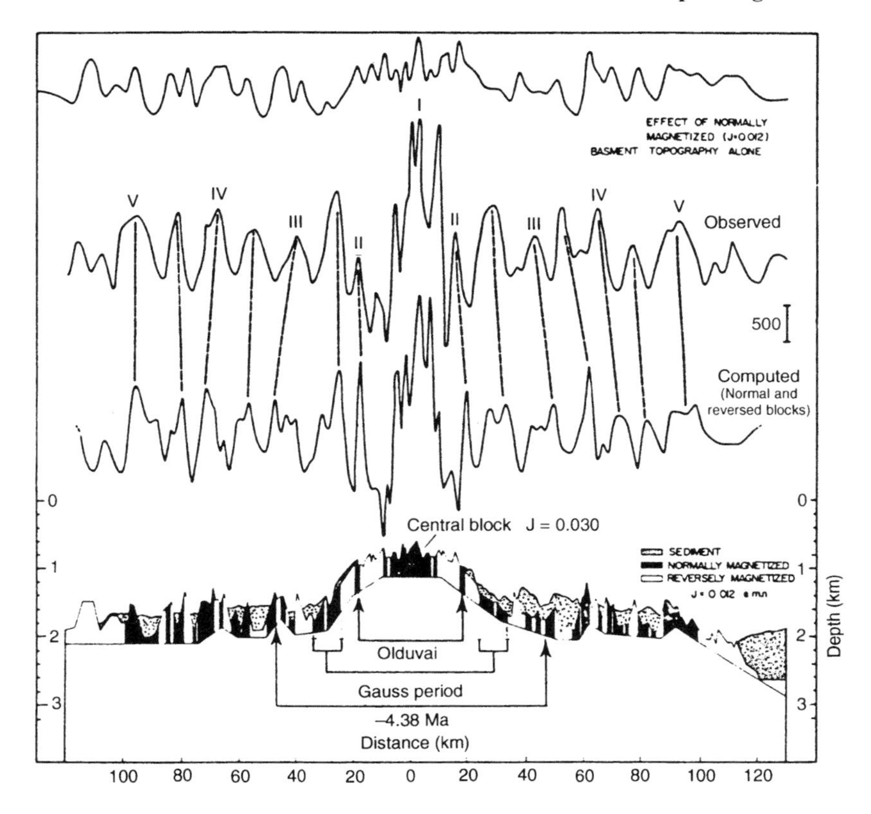

Figure 5.6. Interpretation of magnetic anomaly stripes around the Reykjanes Ridge. From Talwani *et al.* (1971); © AGU.

spreading rate is 20 km/Ma. Its maximum value is 162 km/Ma for the East-Pacific Ridge.

Gabbro cores extracted from the oceanic layer showed that the TRM of this layer could produce up to half of the local magnetic anomaly observed at the surface. When the spreading rate is low, or when gabbros are rapidly cooled by hydrothermal circulation, this TRM is in phase with the TRM of the basaltic layer above. Otherwise, the acquisition of TRM by the gabbros would be out of phase, which would decrease the signal. This process adds to the slow alteration of titano-magnetites, with acquisition of a CRM. It makes the magnetic anomaly stripes fuzzier than the zebra patterns shown in some popular textbooks.

5.4 TRANSFORM FAULTS AND SPREADING DIRECTION

The strike-slip movement along a transform fault progresses at the same speed than the spreading of the neighbouring ridge (Figure 5.7a). The offset introduced equals

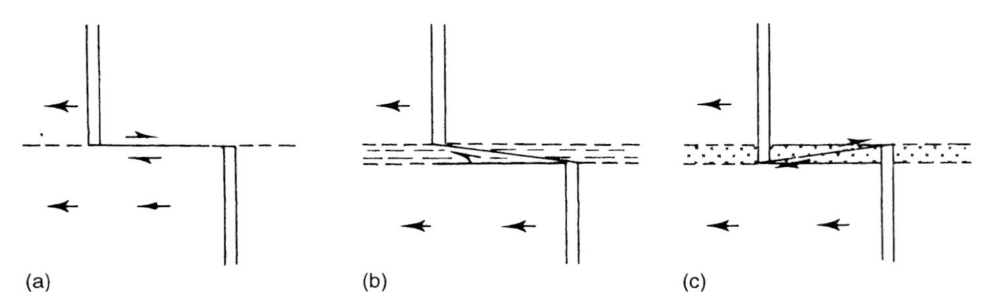

Figure 5.7. Transform fault. The rightmost plate is assumed fixed: (a) ideal case; (b) ditch formation; (c) ridge formation.

the length of the transform fault, and stays constant all along the fracture zone, a 'scar' of the old transform fault. This made possible the recognition of some fracture zones, even if the original transform fault had disappeared. It is the case with the very long parallel fracture zones, oriented WSW–ENE, between longitudes 120°W and 160°W in the Eastern Pacific. From North to South, these are the Surveyor, Mendocino, Pioneer, Murray, Molokai, Clarion, Clipperton, Galapagos, Marquesas and Australs fracture zones.

The magnetic anomaly stripes were used to recognize mid-ocean ridges, active or not, in all oceans, even when they were very faint. But two important corrections were needed:

1. The spreading rates are not always equal on each side of the ridge. For example, in the SE Pacific, the distance between similar anomalies is larger on the west side than on the east side, with a 3 : 2 ratio. It is therefore not recommended to publish spreading half-rates, as was done at the beginning.
2. The transform faults are only rarely exactly perpendicular to the ridge axes. These faults have the direction of the velocity of one side of the ridge relatively to the other. This velocity is therefore not perpendicular to the axial graben, in most cases. In many cases (e.g. for 70% of the Mid-Atlantic Ridge north of 15°20′N), the angle lies between 80° and 45°. When this angle changes with time, the two sides of the transform fault may be compressed, forming a ridge (Figure 5.7c), or separate from each other and create a ditch (Figure 5.7b). This was the case with the Romanche gap, in the Equatorial Atlantic (−7,758 m).

5.5 THE MODEL OF LITHOSPHERIC PLATES (SHELLS)

Magnetic anomaly stripes cover all oceans. Assuming constant spreading rates, the ones determined for the last 10 Ma, the whole seafloor is younger than 160 Ma, 1/30th of the Earth's age. There is no reason for this crust formation process not to have existed at mid-ocean ridges in the early Earth. This requires that the amount of oceanic crust formed at ridges (2.9 km^3/year) disappears in subduction zones. Proofs that this is the case will be provided in Chapter 7. This circulation of matter, at the surface and then in the mantle, is in good agreement with the idea of a thermal convection as the main driver of the movement (see Chapter 18).

The whole surface of the Earth can be divided by a net of ridge axes, transform faults and subduction zones. The domains so delimited (about 20) were first called *plates*, but I shall also call them *shells*, which is a more exact term for fragments of a sphere.

A second important step was taken when LePichon (1968) showed that the relative velocities of the seven largest shells, measured along the mid-ocean ridges, were compatible (considering the measurement accuracy, and assuming the shells are rigid and keep their shapes). This was demonstrated again, later, using the velocities of 14 plates and with a greater accuracy (see Chapter 6).

The question was to determine how deep the Earth's superficial layer could be considered as rigid. The deep rocks indeed have to deform, as otherwise no convection would be possible. Long ago, geodesy and gravimetry had shown that the weight of terrains above a deep enough horizontal surface (above the *compensation level*) was everywhere the same. This was true, the surface being a plain, a high plateau or a mountain range (in this case, considering the average altitude). This *isostasy*, with *regional compensation* of the relief, can be explained by the presence below the superficial layer, which can only deform elastically (the *lithosphere*), of a medium which can deform viscously (the *asthenosphere*). In fact, a particular body can only be considered elastic or viscous with an associated time scale. The asthenosphere defined with gravimetry behaves as a viscous fluid on a 1,000-year time scale, but on the second or minute time scale, it lets the seismic waves propagate as in a perfectly elastic body.

In plate theory, the lithosphere is the layer bound to the surface and moving at the same horizontal velocity. But a small fluidity, too small to make the velocities different, may be enough for the response to a load to be non-elastic. It is therefore necessary to distinguish between the *rigid lithosphere of plate theory* and the *elastic lithosphere of gravimetry*. These are two different models, the latter dealing with vertical movements on a 1,000-year time scale, the former dealing with horizontal movements on a 1-Ma time scale. In Chapter 13, it will be reported that the elastic lithosphere of gravimetry is much thinner than the rigid lithosphere of plate theory (do not forget that these are models, not genuine objects). In general, both encompass the whole crust and part of the upper mantle, the *mantle lithosphere*. And it will be stressed that in both models, permanent deformations or fractures might occur above some stress thresholds. For example, the folding of the lithosphere in a subduction zone shows a *plastic* behaviour.

The composition of the upper mantle is rather uniform, and therefore viscous behaviour is mainly controlled by temperature. The lithosphere–asthenosphere boundary is on first approximation an isotherm surface. *A cooling plate thickens*, some of the asthenosphere transforming into lithosphere.

In plate theory, the upper limit of the asthenosphere below the oceans can be fixed quite precisely to the depth where the celerity of S waves stops increasing, and falls

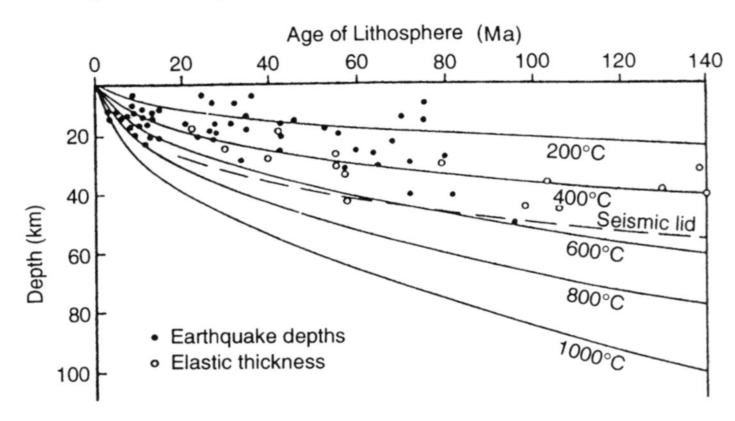

Figure 5.8. Focal depths of intra-plate earthquakes as a function of the age of the oceanic plate (black dots). The white dots represent the base of the elastic lithosphere, deduced from the gravity anomaly (see Section 14.1-2). The isotherms were determined using the model of the thermal plate (see Section 9.5). From Anderson (1995); © AGU.

from around 4.50 to 4.25 km/s. $v_S = 4.50$ km/s is only reached again at a depth of 220 km. This layer is known as the *Low Velocity Zone (LVZ)*. The whole LVZ must indeed be at the solidus temperature (which increases with pressure. Thus within the LVZ temperature increases with depth.) The LVZ should contain liquid micro-inclusions, exactly like the ice in glaciers from temperate regions (Lliboutry, 1969). According to Anderson and Spitzler (1970), 1% magma is enough to account for the decrease noticed in v_S (in temperate glaciers, the liquid content is of the same magnitude).

The asthenosphere–lithosphere boundary should then be around the 1,280°C isotherm (according to plate theory). The limit of the elastic lithosphere (according to gravimetry) is around 600°C only. Most intra-plate earthquakes are coming from the elastic lithosphere (Figure 5.8). But they are very few compared to those originating at plate boundaries.

LVZ and asthenosphere should not be identified, in order to find a lower limit to the latter. The LVZ is an objective fact, whereas the lower limit of the asthenosphere depends on the model chosen to account for the convection currents in the mantle. We shall see that the convection in the lower mantle is 10 times slower than above. In the models where it is neglected, the asthenosphere ends at the 660-km discontinuity, and the lower mantle becomes a rigid *mesosphere*. In the more refined models where the circulation in the lower mantle is accounted for, the asthenosphere extends down to the core.

5.6 LIMITS OF TRADITIONAL PLATE THEORY

Having exposed the bases of plate theory, it is important to mention its limits as well, anticipating the results of the next chapter (relative plate velocities) and Chapter 14 (tectonic mechanisms).

Plates are not always separated by very narrow strips, which are reduced to a line on maps of the World. This is only the case for mid-ocean ridges, and transform faults in the middle of the oceans. On a mid-ocean ridge, the boundary corresponds to an axial strip 8- to 2-km wide, subject to extension (active normal faults, possible dyke intrusions). Only the middle 1–2 km are clearly neovolcanic (lava flowing out of fissures or small volcanic cones). The oceanic transform faults are even closer to a line, 0.5- to 2-km wide most often. But the boundary zones at continent margins, in insular arcs or in the middle of continents are a few hundred kilometres wide, and sometimes a few thousand kilometres. An extreme case is the boundary between the Indian plate and the Eurasian plate, encompassing the whole of Central Asia between the Ganges plain and the Siberian plain, from the Himalayan foothills to the graben of Lake Baikal. This is explained by the very thick, but very irregular, continental lithosphere.

These boundary zones present multiple faults, grabens, and large thrusts and overthrusts, as well as zones of diffuse compression, extension and shear, and folds. They include large, rigid continental blocks, which are often called plates, although they are not bounded by a continuous line of ridges, transform faults and subduction zones (see Figure 15.3, for instance). It is better not to call these continental blocks plates or micro-plates, to keep the strong and precise definition of these words.

Genuine *micro-plates* exist in oceans, where the lithosphere is thin. They are 100 to 1,000 times smaller than the large plates considered up to here. Micro-plates bounded by ridge axes and transform faults are the *Easter Island microplate* (so called because it is close to this island, even if it does not support it), of area 400 × 400 km, and the *Juan Fernandez microplate* (very far from these islands), of area 350 × 350 km, at the triple junction of the Pacific, Nazca and Antarctic plates. Examples of microplates located between a ridge axis and a subduction zone are the *Bismarck microplate* (900 × 400 km) and the *Solomon microplate* (600 × 250 km), NE of New Guinea.

As more and more data on the relative movements of lithospheric plates accumulated, it became necessary to divide some plates from the initial theory (e.g. the American plate, recognizing the autonomous existence of the Caribbean plate). The Indo-Australian plate was divided into an Indian plate and an Australian plate, separated by a diffuse seismic zone, with many faults. The oriental part of the African plate (African Horn and Western Indian Ocean) is also distinct from the rest of the plate. Another example is Alaska and the rest of the North-American plate.

These results are summed up in the sketch published by Gordon and Stein (1992) (Figure 5.9). These non-rigid boundaries between plates cover a good fifth of the Earth. It is easy to understand why the geologists interested in these regions, long declared that the lithospheric plates discovered by oceanographers were of no interest to them. But the proofs of the model of large lithospheric plates covering the Earth were so strong, that the first article showing the limits of the model was published in a major journal of geophysics in 1992 only.

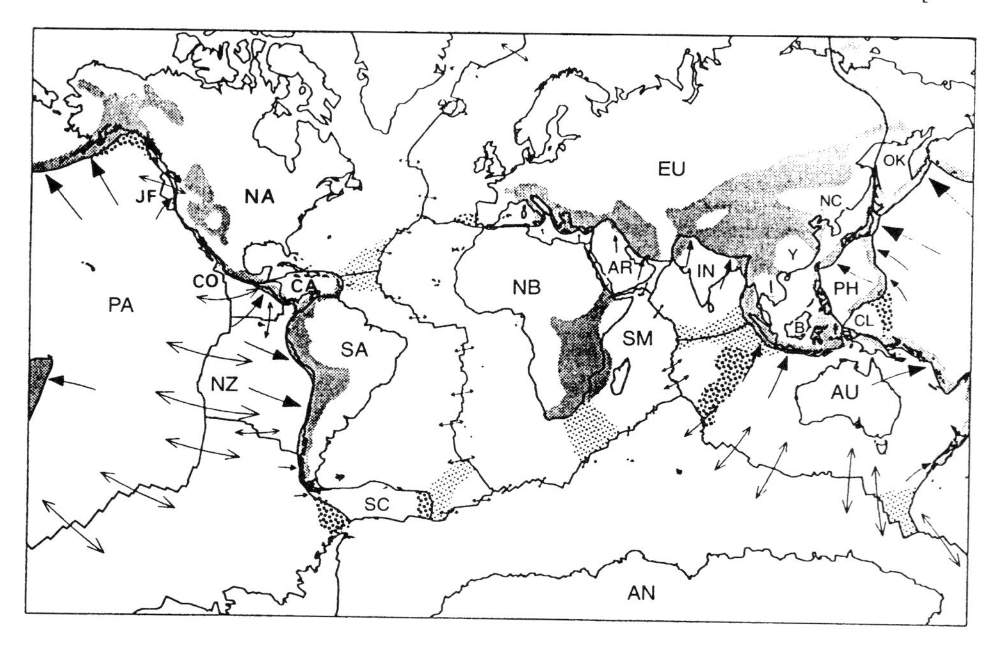

Figure 5.9. Non-rigid boundaries between plates. The fine points below the oceans correspond to deformations deduced from the relative movements of the neighbouring plates. These generally are seismic regions. The coarser points correspond to other seismic submarine regions, with no direct proof of the deformation. The arrows show the relative displacements between neighbouring plates, in 25 Ma, at current spreading rates. From Gordon and Stein (1992); © Amer. Assoc. Adv. Sci.

5.7 MID-OCEAN RIDGE PROCESSES

The inner structure of mid-ocean ridges is now well established. Ocean-bottom seismographs, television cameras, sidescan sonars and many drillings have contributed to this knowledge. For example, in 1994, an international expedition with an important French participation drilled the axial valley of the Mid-Atlantic Ridge, 15 times down to 70 m (at 23°20′N and 23°30′N) and 17 times down to 80–125 m (at 26°08′N), on the location of an important hydrothermal site. The most studied ridges are the East-Pacific Ridge, where the spreading rate is larger than 100 km/Ma ('fast-spreading ridge'), and the Mid-Atlantic Ridge, where the spreading rate is less than 30 km/Ma ('slow-spreading ridge').

The fast-spreading East-Pacific Ridge does not have an axial valley, and sometimes shows the opposite, a *horst*. But it is as valid to say that there are two parallel grabens side by side. Always normal, the faults are statistically shorter than in slow-spreading mid-ocean ridges (15% with lengths greater than 10 km, compared to 25%), and closer (the distance between faults is less than 5 km, compared to 10 km). Earthquakes related to the movement of these faults over a strip 10 to 20 km wide never have magnitudes greater than 3 on a fast-spreading ridge. Their magnitudes are often greater than 5 on slow-spreading ridges. The spreading of a 'fast' ridge therefore occurs with closer and smaller events than on 'slow' ridges.

Magma pockets were found below the axial valley, 1.5 to 3.5 km deep, always during studies of the East-Pacific Ridge (e.g. Wilcock *et al.*, 1995), but not always on the Mid-Atlantic Ridge. Depending on their location, these magma pockets are 2 to 8 km wide, and ca. 1 km wide. They are in the gabbroic layer, above the Moho. But this Moho is a transition zone 500-m thick. This transition zone can be easily explained by analogy with frost-heave in soils. A small portion of the basaltic magma from the mantle below the ridge must not reach the axial pocket, but solidify as a horizontal layer or lens, on the 'freezing front'. As the plate moves away from the ridge axis and cools down, this 'freezing front' descends by steps, creating successive gabbro layers.

The oceanic crust is 6- to 7-km thick in the middle of the ridge segments between two transform faults, but decreases at the extremities. Close to transform faults, the crust is only 2.5-km thick. The location of these faults is therefore linked to the places where the magma is in decreased quantities. A mid-ocean ridge could therefore be seen as the product of a series of magma pockets, one in the middle of each ridge segment: permanent pockets for fast-spreading ridges, temporary pockets for slow-spreading ridges.

This magma comes from a larger zone of the upper mantle, at least 50 km deep in its centre, where the melting of peridotites is important enough to enable communication between the magma veins and magma lenses at grain boundaries, and the flowing of magma to the above pocket. In old models (e.g. Talwani *et al.*, Figure 5.10), this zone was called the 'crust–mantle mix' and was at least 1,000 km wide. This model is not valid for low magma input at transform faults. Recent models are therefore considering a zone approximately 100 km wide at the base, thinning upwards to the thickness of the magma chamber. The melting is supposed to be very important, and this zone is often called the *mush zone*. This seems to me to be going a bit far. Judging from soils or firn (old, dense snow), a liquid content of 5% is largely enough for a communicating network. Considering the uncertainty, I suggest that this zone should be called the *magmagenetic zone*.

Important hydrothermal activity occurs in the axial valley of mid-ocean ridges. Seawater infiltrated in basalt is heated up, dissolving many secondary minerals (in particular metallic sulphurs), and gets out at temperatures of ca. 300°C. Cooling SO_4Ca precipitates and forms cones 1 to 15 m high, and precipitating sulphurs blacken the hot water. These are the *black smokers*. Intense animal life is present around them, contrasting with the near-total absence of life in the abyssal depths. Its metabolism is not based on the chemical reactions of oxygen, but of sulphur, which fascinates the biologists interested in the origins of life.

5.8 EVOLUTION OF MID-OCEAN RIDGES WITH TIME

The axial valley of mid-ocean ridges is not continuous. It can be interrupted, and another ridge axis takes over, a few kilometres to the left or right. The two co-exist

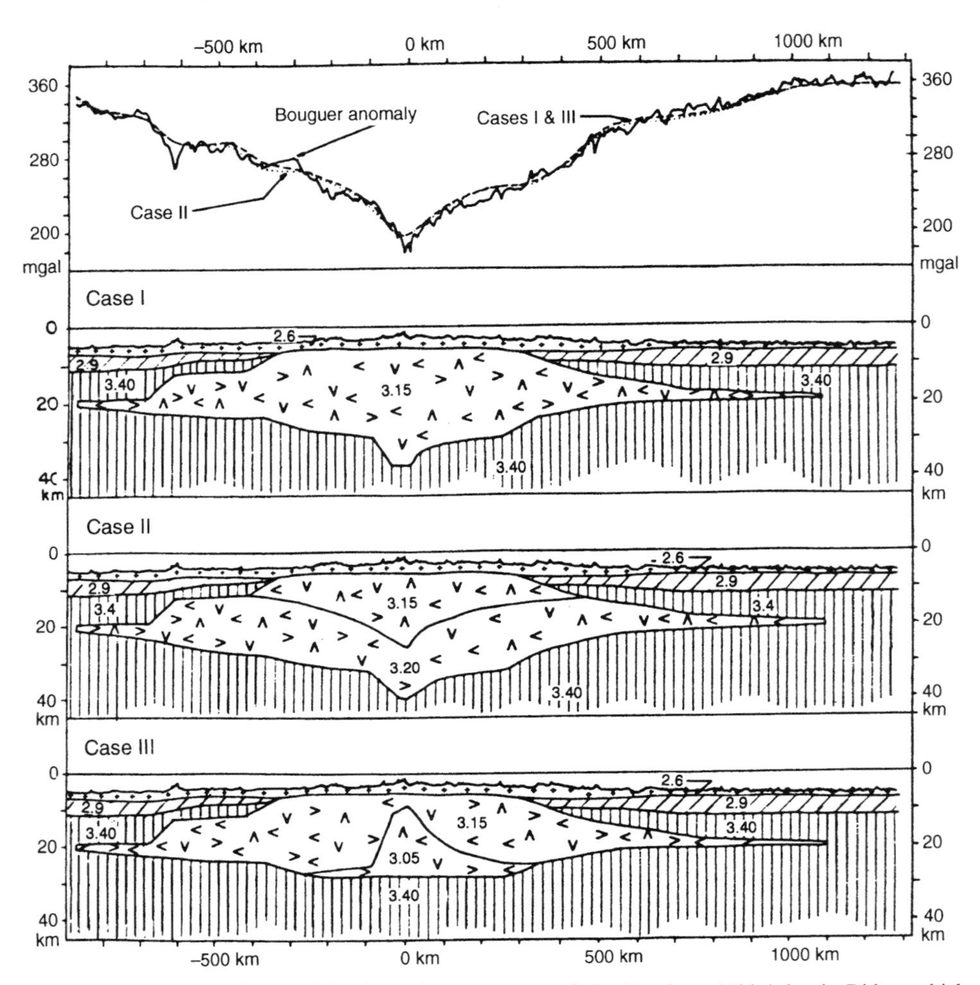

Figure 5.10. Three possible models of the deep structure of the Northern Mid-Atlantic Ridge, which explain both the gravimetry data (top diagram: Bouguer anomaly), and the seismic measurements. Densities are shown in the layers. Vertical exaggeration: ×10. From Talwani *et al.* (1965); © AGU.

along a few kilometres. These are *overlapping spreading centres* (OSC). The whole overlap is in the narrow ductile boundary, so that no transform fault is necessary. The network of parallel faults shows that there is sometimes rotation of a small block of oceanic crust (Figure 5.11). In this particular example, the magnetic anomalies suggest the overlapping spreading centre migrated by ca. 20 km between 0.9 and 0.7 Ma.

The study of magnetic anomaly stripes revealed that mid-ocean ridges can migrate, appear, or become inactive.

An example of migration (Hey *et al.*, 1980) is the mid-ocean ridge between the Nazca and Cocos plates (both near the East-Pacific Ridge). This ridge moved over

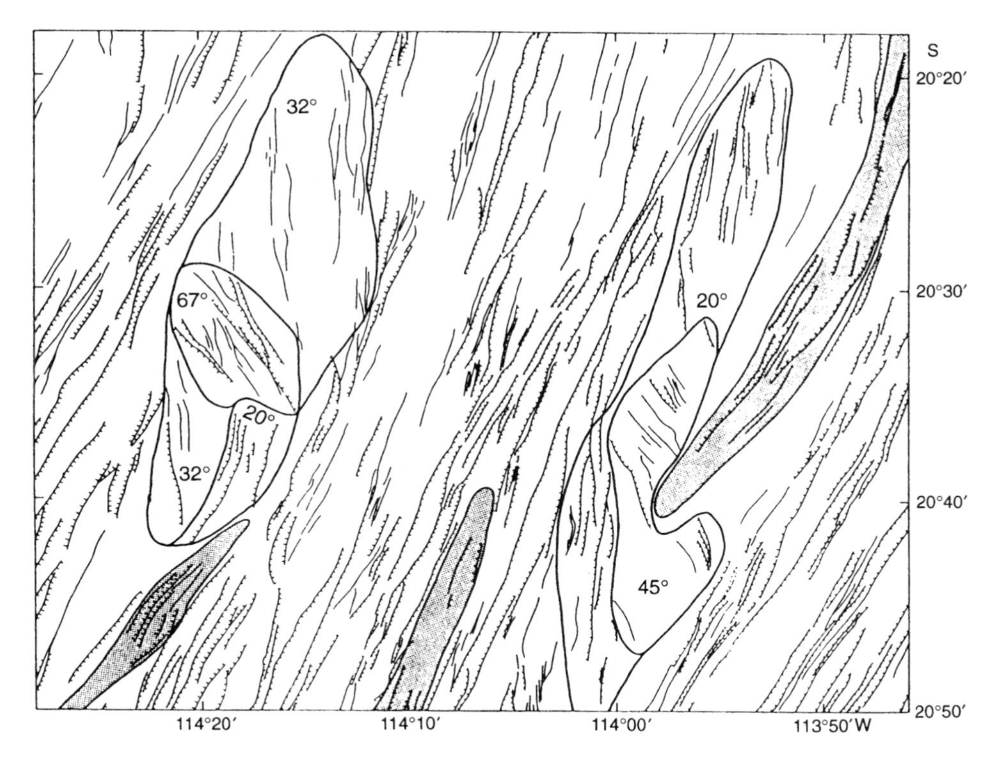

Figure 5.11. Tectonic map of an OSC in the East-Pacific Rise. Small blocks were submitted to small anti-clockwise rotations (continuous outlines). The smallest one only rotated by 67°, and is 16 km long. From Perram et al. (1993); © AGU.

100 km in the last 3 Ma, by six successive jumps of amplitudes increasing from 6 to 26 km.

Some mid-ocean ridges were active during a few Ma only. East of the Gulf of California (where the East-Pacific Ridge enters), between the Murray and Molokai Fracture Zones, the sequence of magnetic anomaly stripes jumps from 1,300 km to 1,900 km (at anomaly 13) (Figure 5.12). In between, there are two sequences of anomalies 8 to 13, symmetrical around an extinct mid-ocean ridge, which was therefore active between chrons 13 (−35 Ma) and 8 (−27 Ma). In the centre of the Pacific Ocean, another secondary ridge, 500-km long, would have been active between 126 Ma and 121 Ma (Tamaki and Joshima, 1979). In Chapter 7, we talk about ridges which appear in back-arc basins and are active between 5 to 20 Ma.

Other ridges can migrate. For example, the magnetic anomalies around the East-Indian Ridge only go to anomaly 17 (Figure 5.13). This ridge therefore appeared during chron 17, 40 Ma ago. This was when the Australian continent and the southern underwater Tasmania plateau detached from Antarctica. Northwards, the ages decrease from chron 33 to chron 17, the latter being close to the Sunda subduction zone. An active mid-ocean ridge therefore existed in the North. As the

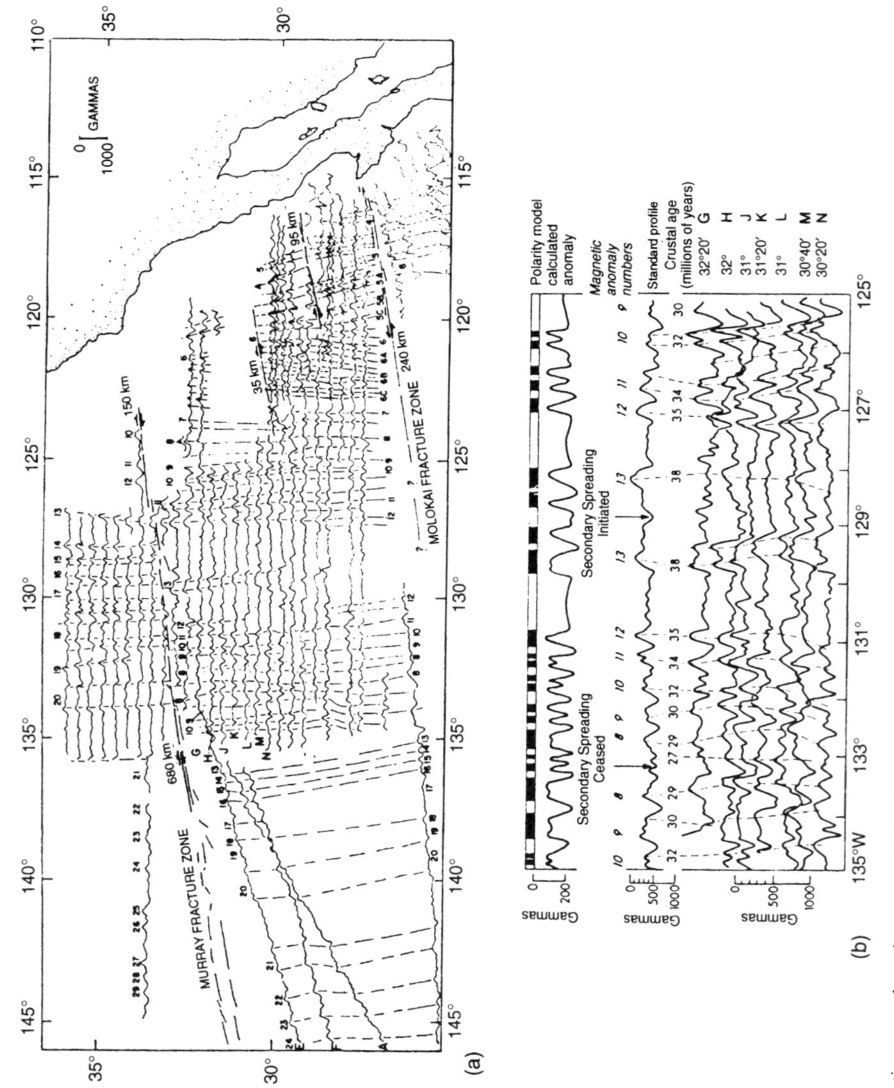

Figure 5.12. Magnetic anomaly stripes west of the Gulf of California: (a) ship-collected profiles, with the general anomaly subtracted; (b) anomalies 9 to 13, showing the existence of a short-lived mid-ocean ridge (the calculated anomaly was assuming a magnetic layer 2-km thick, and $J = 0.050$ G). From Malahoff and Handschumacher (1971); © AGU.

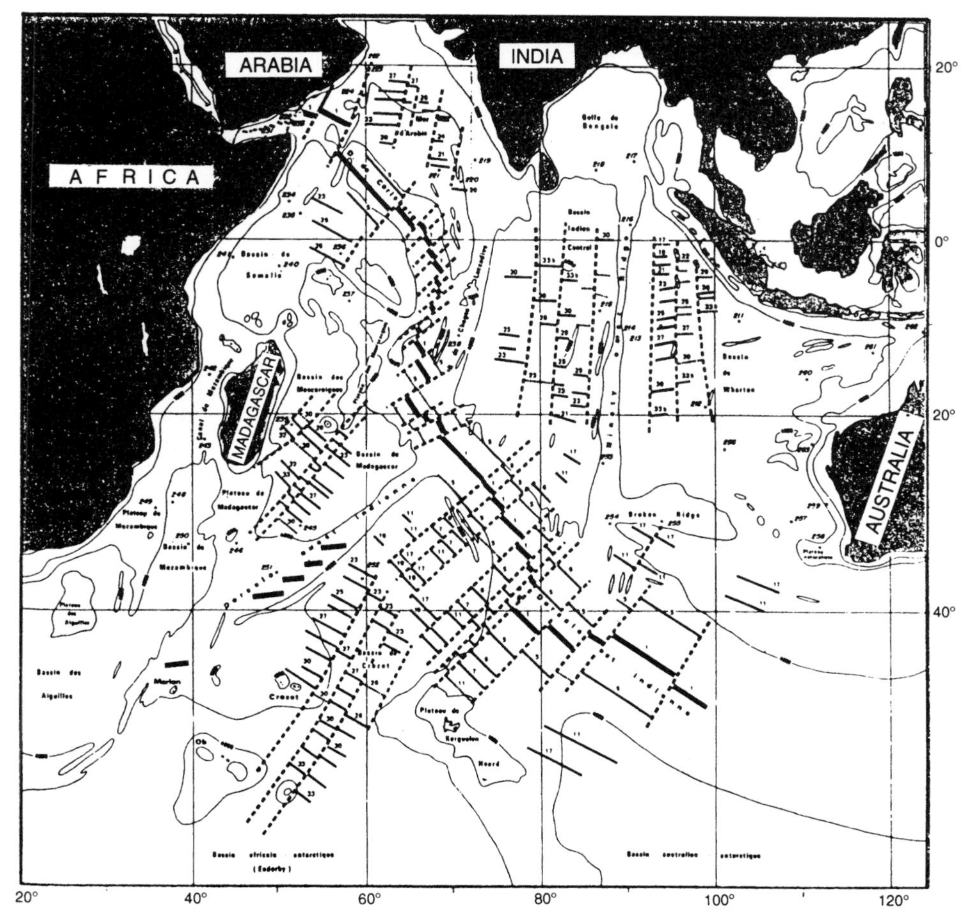

Figure 5.13. Fracture zones (large dots) and most noticeable magnetic anomalies in the Indian Ocean. From Schlich (1975).

subduction was destroying more crust than was formed at this ridge, the ridge moved closer and closer to the subduction zone, which 'neutralized' it. Subduction presupposes a descending mantle flow, whereas ridge activity presupposes an ascending mantle flow. The two cannot coexist in the same place, and only the strongest vertical flow remains.

5.9 PASSIVE CONTINENTAL MARGINS

The exploration of passive continental margins provides information about the first stages of the appearance of a ridge in the middle of a continent. The mantle processes

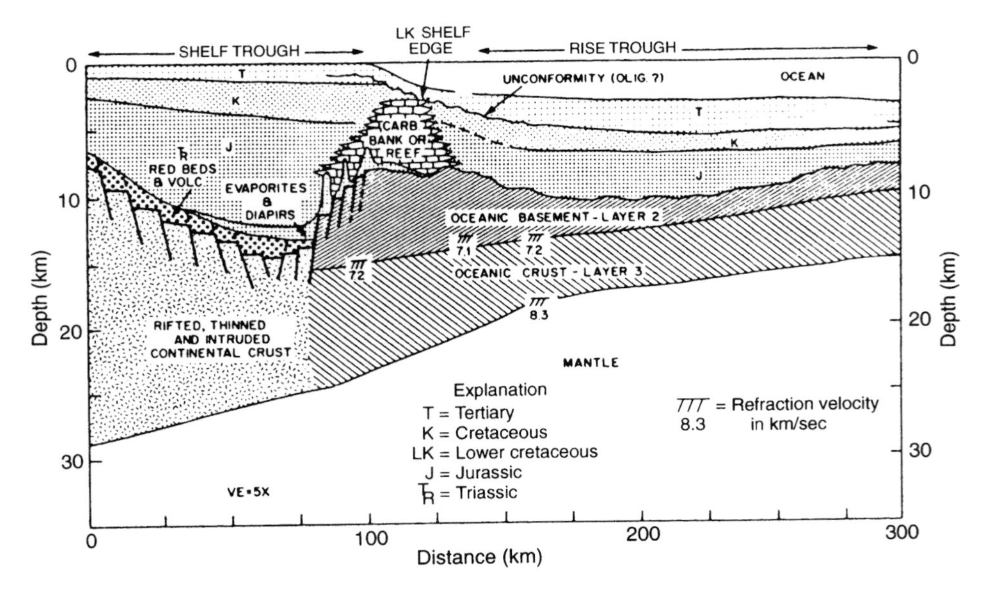

Figure 5.14. Cross-section of the stable continental margin east of the United States. From Folger (in Talwani *et al.* (1980); © AGU).

that may have caused the fission of a continent will be discussed later, and only the proven facts will be presented here.

The best-studied passive continental margin is the eastern coast of the USA, because of oil exploration interests (Figure 5.14). Stretching often thinned the continental crust (estimates of the stretching range from 20 to 600 km, depending on the areas). Basalt intrusions produced dykes. Thinner and heavier, its surface went down. Conversely, the first oceanic crust, formed by development of a continental rift (see Section 14.11), was much thicker than normal. This produced a depression of the continental platform (*shelf trough*, *marginal rift basin*).

Corals grew on the outer perimeter of the marginal rift basin (this was possible because of the latitude at the time; it is not a general occurrence). Successive invasions of the depression by the sea, followed by drying, led to the formation of thick layers of evaporites (mainly rock salt). Sediments from the neighbouring continent then infilled the depression. Rock salt being lighter than the overlaying sediments, it sometimes migrated toward the surface, creating *salt diapirs*, which are often oil traps.

Some continental margins were extremely enriched in basalts, by underplating and large surface flows (*flood basalts*). The Parana basalts, near the Iguaçu Falls, extend all the way to the coast. The rest of the flood basalts is on the East, on the African coast of Namibia (Etendeka basalts). The basalts from the East coast of Greenland (in front of Iceland) continue with the basalts from the Voring plateau, offshore Norway, and Hatton Bank, north of Ireland. These submarine plateaus correspond to crusts between a normal continental crust and a normal oceanic crust.

These intermediate crusts will be described in Section 15.7, in the case of the Western Mediterranean, where they are abundant.

Even in passive margins without any apparent volcanism, intrusive basalts show that, during continental fission, the upper mantle below the crust was at the solidus temperature, and the lithospheric plate was locally reduced to the crust. Depending on the authors, the temperature excess varies between 100° and 250° above the normal geotherm. The difference between the two cases resides in the high volume of magma extruded in the latter. Such events, with a possible duration of a million years and without current example, weaken the hypothesis of uniformitarian theory.

5.10 INTRA-PLATE VOLCANISM AND HOT SPOTS

Submarine volcanoes away from mid-ocean ridges are many, and mostly extinct. The Geosat and ERS-1 satellites determined the surface of the geoid with a decimetric precision, and revealed twice the number of volcanoes previously estimated. A seamount 2,000-m high produces a 2-m elevation of the geoid. The discovery of seamounts was the unexpected result of a space programme conceived with a different goal, the improvement of spatial geodesy.

Other volcanoes form islands or atolls (see Section 9.8). We only consider here the linear chains of volcanoes, aligned on a rise, such as the Hawaiian chain. These volcanoes are only a part of the intra-plate volcanoes, but they are particularly interesting. They are older at one extremity of the chain, suggesting that there is a unique source at depth, in the mantle, over which the lithospheric plate moved. This deep source is called a *hot spot*. When the plate has moved enough, the *plume* of hot material has deviated too much and returns back to its original trajectory, giving birth to a new volcano.

The velocities of hot spots relative to each other are much smaller (by one order of magnitude) than the relative velocities of plates. This can only be explained if their origin is in a lower mantle less moved by convection currents than the upper mantle. This hypothesis seems to be confirmed by the geochemical analyses of lavas (see Section 17.10). All hot spots can be grouped in two large antipodal regions (Figure 5.15). The first one covers the whole Southern Pacific and extends to Hawaii, the hot spot that has emitted by far the most lava. The second region extends from Iceland to the islands Bouvet and Kerguelen, surrounding Africa (excluding the Mediterranean). I will call them 'Pacific group' and 'African group', respectively. These two groups happen to correspond to very-long-wavelength geoid anomalies, probably due to a warmer lower mantle. This gives indications of a very slow convection in the lower mantle.

All scientists agree on the major oceanic hot spots, the ones emitting more than 1 km^3 of volcanics per century. But the identification becomes problematic for hot spots with low activity (less volcanoes, and all submerged) and for hot spots in the middle of continents. Only the Yellowstone hot spot is recognized by everybody,

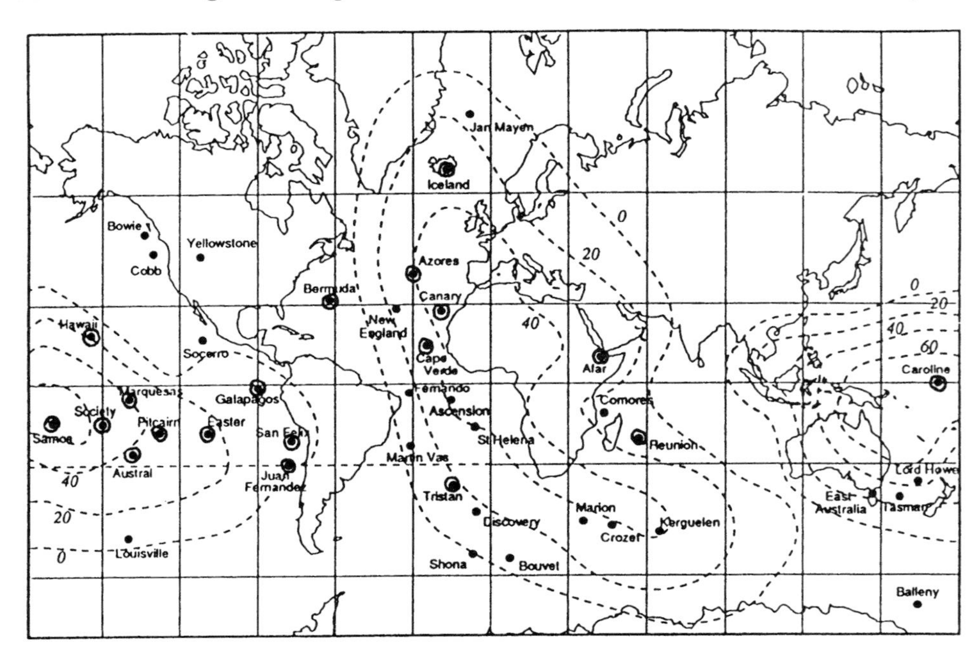

Figure 5.15. Terrestrial hot spots (Duncan and Richards, (1991), © AGU). The hot spots emitting the most magmas are circled. The contour levels show the geoid anomalies (in metres above the reference ellipsoid). These positive anomalies are correlated with slower seismic velocities in the lower mantle. The location of most hot spots in these regions seems to confirm the hypothesis that they originate in the lower mantle.

whereas the Hoggar and Eiffel hot spots are still debated. The number of identified hot spots varies therefore considerably, depending on the author: 19 for Morgan (1972), 66 for Wilson (1973), 42 for Crough and Jurdy (1980), 117 for Vogt (1981) and 38 for Sleep (1990). Of course, the higher the number, the more often the author included intra-plate volcanism not coming from a deep hot spot, and the more his list can be used to affirm that the distances between hot spots change with time. In recent years, precise dating by O'Connor and Duncan (1990) could establish the identity of many controversial hot spots, and reconstitute their history (Figure 5.16). The list of the 40 hot spots recognized by Duncan and Richards (1991) is given in Table 5.1.

The trajectories of some hot spots on the surface of the Earth cross mid-ocean ridges, proving that the deep hot spot is decoupled from the lithospheric plates moving above. This is the case of the Azores hot spot, currently at the boundary of the Eurasian and African plates (in this place, the relative movement of the plates is negligible). In the Cretaceous era, this hot spot was producing submarine volcanoes SE from Newfoundland, on the American plate. Similarly, the Réunion hot spot, on the African plate, produced the Mascarene Plateau on this plate, but was producing the Chagos-Maldive Rise on the Indian plate, more than 30 Ma ago.

The high number of hot spots at the axis of mid-ocean ridges or in the neighbourhood cannot be attributed to chance. Intuitively, the intense but

Table 5.1. Hot spots active today or in the recent past. From Duncan and Richards (1991); © AGU.

Plate above the hot spot	Hot spot	Resulting volcanic chain
Pacific Group		
Australian	Bass Strait	Volcanoes of Eastern Australia
	Tasmania Basin	Chain 150–350 km E of Australia
	Lord Howe Island	Chain to Chesterfield Island
Antarctica	Balleny Island	Along Eastern Tasmania plateau
		Eastern Carolinas
Pacific	60 km SE of Bowie Seamount	Kodiak-Bowie Seamount chain
	Hawaii	Hawaiian Islands and Midway
	Fatu-Hiva (Marquesas)	Marquesas Islands
	Pitcairn (Gambier Islands)	SW Tuamotu Islands
	Tahiti	Society Islands
	McDonald Seamount	Austral and Cook Islands
	Rose Atoll (Eastern Samoa)	Samoa and Wallis Islands
	Burton Seamount	Louisville Ridge
Juan de Fuca	Cobb Seamount	
Rivera	Socorro Island	
Cocos-Nazca	Galapagos Islands	Cocos Ridge on Cocos
		Malpelo Ridge on Nazca
Nazca	Easter Island	Sala y Gomez Rise
	San Felix Island	
	Juan Fernandez Island	
North America	Yellowstone	
African Group		
North America	Bermudas	
South America	Fernando de Noronha Islands	
	Martin Vaz Island	
North America–Eurasia	Jan Mayen Island	
	Iceland	Iceland–Greenland and Iceland–Faeroe rises
Eurasia–Africa	Azores Islands	Newfoundland seamounts
Africa	Canary Islands	
	Great Meteor Seamount	New England seamounts
	Cape Verde Islands	
	Ascension Island	
	St. Helena Island	Cameroon Rise
	Tristan da Cunha Island	Walvis Ridge on African plate
		Rio Grande Rise on South American plate
	Discovery Seamount	Rise WSW of The Cape
America–South Africa	Shona Seamount	Meteor Rise (SW of The Cape)
America–South Africa–Antarctica	Bouvet Island	
Africa–Antarctica	Marion Island	Plateau South of Madagascar
Antarctica	Crozet Island	
	Kerguelen Islands	Kerguelen Plateau
Africa	Réunion Island	Chagos-Maldive Rise
	Comoros Islands	Seychelles Islands
	Afar Territory	

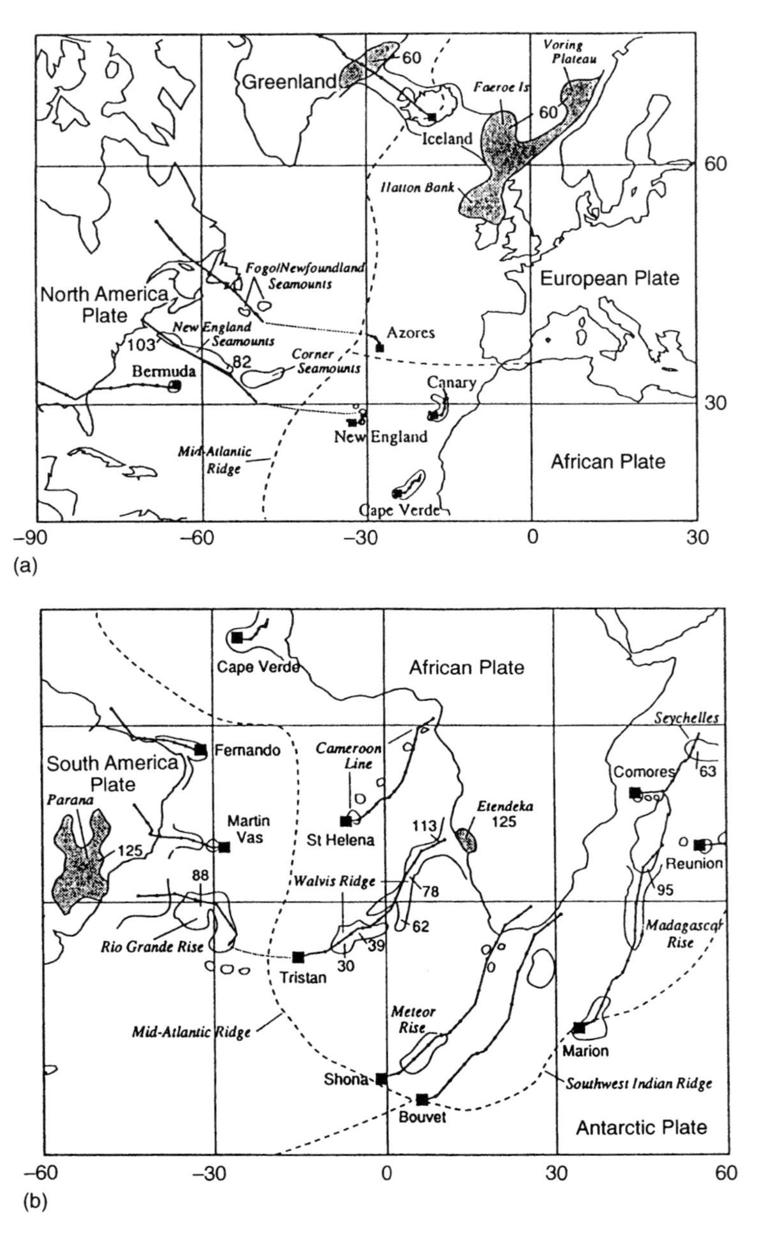

Figure 5.16. Traces of hot spots on the lithospheric plates. The points along the traces correspond to 10-Ma increments. These traces were obtained by assuming the hot spots were fixed, and reconstituting the former positions of plates from the magnetic anomalies, with a single global adjustment between traces and volcanic chains. The good fit for all volcanic chains at the same time, the agreement of volcano ages measured by radiometry (ages in Ma, on the figure) with the ages calculated, concur to prove that the relative velocities of hot spots are smaller than 5 km/Ma. The grey patches correspond to the basalt floods just before the emergence of a hot spot. From Duncan and Richards (1991); © AGU.

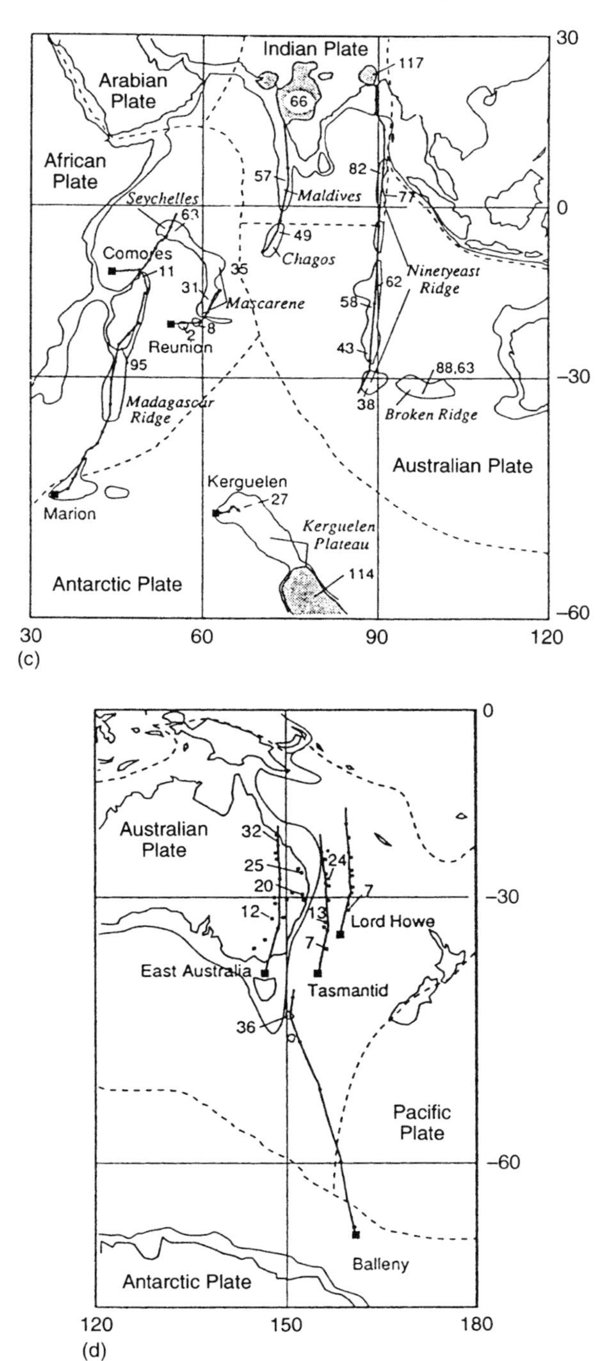

Figure 5.16 (*continued*) Parts (c) and (d).

localized magmato-genetic zone of a hot spot can 'attract' the long magmato-genetic zone of a developing ridge. An extreme case is the opening of the North Atlantic, which progressed from South to North from −160 Ma. At the beginning, the opening propagated northwards, separating Greenland from Baffin Island. But, with the emergence of the intense Iceland hot spot at −60 Ma, the Baffin Sea ridge became inactive, and a new mid-ocean ridge pointing NE (Reykjanes Ridge) joined the Northern Mid-Atlantic Ridge and Iceland.

When a hot spot is situated on a mid-oceanic ridge, a volcanic chain appears simultaneously on both plates. But this cannot last, as a ridge is rarely fixed relative to the lower mantle and the hot spots (as the African and Antarctic plates are completely surrounded by ridges, their areas increase with time; therefore, the global mid-ocean ridge system cannot be fixed). For example, the Tristan da Cunha hotspot in the South Atlantic produced simultaneously the Walvis ridge on the African plate, and the Rio Grande rise on the South American plate. But, between −70 Ma and −50 Ma, successive jumps of the Mid-Atlantic Ridge moved it 400 km west of Tristan da Cunha. This hotspot is currently inactive.

According to Duncan and Richards, the formation of a hot spot would have often been associated with very extensive flood basalts, which exist on some passive continental margins. This important phenomenon is not explained by the current theory of hot spots, explained in Section 19.10.

6

Movement of lithospheric shells

6.1 KINEMATICS OF A RIGID FIGURE IN THE PLANE OR ON THE SPHERE

This chapter will only consider kinematics, which is a matter of geometry, adding the time variable. The relative velocities of lithospheric shells can only be determined with an accuracy of a few per cent. Therefore, to calculate them, the Earth may be considered as a perfect sphere, since the equatorial and polar radii differ only by 1/300th. A lithospheric shell and the centre of the Earth form a rigid figure, i.e. a figure that cannot be deformed, which has a fixed point. The movements are easier to imagine if this fixed point is infinitely far, and the shells become plates moving in a plane. We shall therefore look first at the movements in the plane. Demonstrations and results can be easily extended to the case of the sphere.

Consider first the movement of a rigid plane figure (hatched, in Figure 6.1), relative to a rigid plane containing it (in this example, the paper sheet). A moves to A′, M to M′. The axes of AA′ and MM′ are intersecting in P. The triangles PAM and PA′M′ are equal (three equal sides). Consequently, the angles MPM′ and APA′ are equal. Any point M in the figure has rotated by the same angle θ about P.

If θ tends toward 0, P becomes the *instantaneous centre of rotation* I. For a continuous movement of the figure, I describes a curve on the sheet, called the *base*. Relative to the mobile figure, it describes another curve, called the *caster*. The caster turns, relative to the base around I, and during the next infinitesimal rotation around I′, which is both on the base and on the caster (Figure 6.2). The caster must therefore be tangent to the base in I, and roll without slipping.

For two distant positions of the mobile structure, the centre of the finite rotation which makes them coincide is usually neither on the base, nor on the caster (cf. the example in Figure 6.3).

The product of two *finite* rotations (P_1, θ_1) and (P_2, θ_2) is a rotation of angle $(\theta_1 + \theta_2)$ about a centre, which depends on the order of the rotations: P_{12} or P_{21}, symmetrical around P_1P_2 (Figure 5.4). A rotation (P_1, θ_1) brings P_{12} to P_{21}, and a rotation (P_2, θ_2) brings it back to its starting position. This point did not move, and

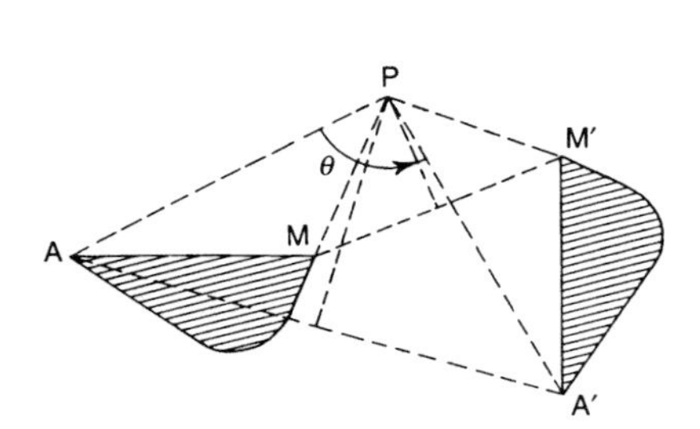

Figure 6.1. Rotation pole P and rotation angle θ.

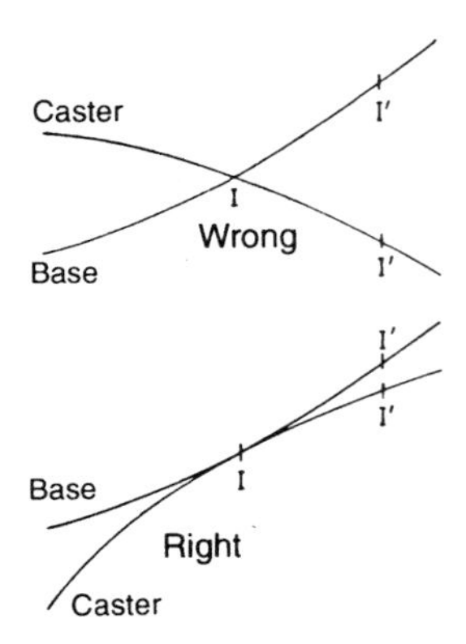

Figure 6.2. The caster rolls on the base without slipping.

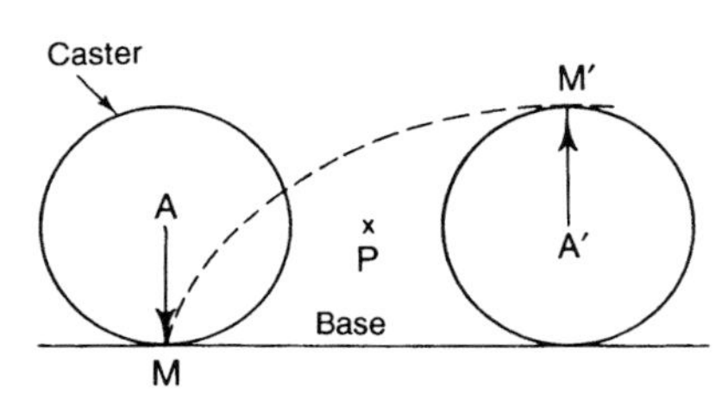

Figure 6.3. The pole P of the finite rotation from AM to A′M′ may be neither on the base, nor on the caster, geometric locations of the instantaneous rotation pole in the 'fixed' plate and in the 'mobile' plate.

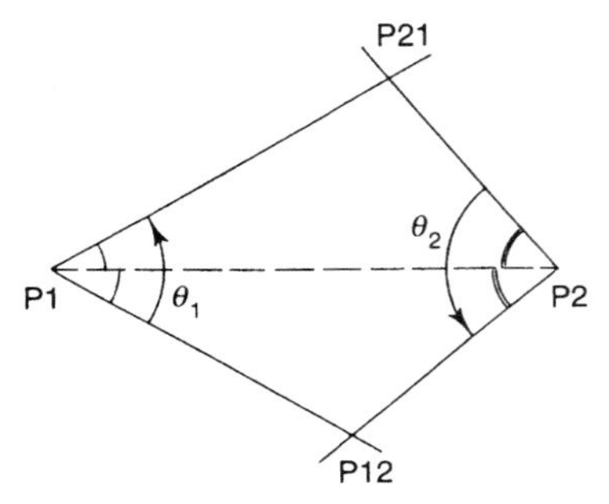

Figure 6.4. The product of two finite rotations is not commutative (see text).

thus is the centre of the resulting rotation when the order is 1 and then 2. The same reasoning may be made with P_{21}, and the order 2 then 1.

When the rotations of angles θ_1 and θ_2 become infinitesimal, P_{12} and P_{21} tend toward the same point I of P_1P_2, for which: $P_1I \times \theta_1 = P_2I \times \theta_2$. The product of infinitesimal rotations does not depend on the order in which they are performed. This is the same with *instantaneous angular velocities*, defined by a centre I and an angular speed $\omega = d\theta/dt$.

In the case of a spherical shell, the instantaneous rotation occurs about a diameter of the sphere, which can be defined by one of the points where the diameter emerges from the sphere. This point is chosen so that, for an observer outside the sphere, further on the diameter, the rotation is anti-clockwise. If a shell A is considered fixed, the infinitesimal movement of a shell B provides an instantaneous rotation axis and a point P on the sphere (the *Eulerian pole* of B relative to A). If B is considered as fixed, and A as mobile, the same relative movement provides the same rotation axis, but the Eulerian pole is now at the antipode of P.

Consider now three shells A, B and C. The instantaneous angular velocity of B relative to A can be represented by a vector originating at the centre of the Earth, directed toward P, of intensity equal to the corresponding rotation, the vector $\boldsymbol{\omega}_{B/A}$. In the same way, it is possible to define $\boldsymbol{\omega}_{C/B}$ and $\boldsymbol{\omega}_{A/C} = -\boldsymbol{\omega}_{C/A}$. We have the following vectorial equations:

$$\begin{cases} \boldsymbol{\omega}_{C/A} = \boldsymbol{\omega}_{C/B} + \boldsymbol{\omega}_{B/A} \\ \boldsymbol{\omega}_{C/B} + \boldsymbol{\omega}_{B/A} + \boldsymbol{\omega}_{A/C} = 0 \end{cases} \tag{6.1}$$

In the case of the very slow relative movements of lithospheric shells that are considered in plate tectonics, the 'instantaneous' angular velocities correspond to average velocities over several million years. The Eulerian pole of a plate relative to another is not fixed on a scale of 100 Ma. It describes a base on the plate supposed fixed, a caster on the plate supposed mobile. Plotted on the Earth, the two curves, bound to their respective plates, roll on each other without slipping. To reconstruct the position of the mobile plate more than 10 Ma ago, it is necessary to determine the finite rotation pole and the finite rotation angle which moved it from its older position to the current one. This finite rotation pole is generally none of the Eulerian poles, past or present. The interest of the following equations is therefore limited to the study of 'current' relative plate movements.

6.2 EQUATIONS OF 'INSTANTANEOUS' PLATE KINEMATICS

Once it has been precisely determined which plate is considered fixed, the instantaneous rotation velocity vectors can be defined in two ways:

- by giving the components in a Cartesian frame of reference (x, y, z), defined in Figure 6.5. They allow the rapid determination of the relative movement of any two plates, by calculating the difference between two instantaneous rotation vectors (i.e. the difference between their three components).
- by giving the geographic coordinates of an Eulerian pole P: north latitude λ_P, between $-90°$ and $+90°$, and east longitude φ_P, between $-180°$ and $+180°$, as well as the instantaneous rotation velocity $\omega > 0$. These three values will be termed the 'Eulerian coordinates'.

Transformation from Eulerian coordinates to Cartesian coordinates

$$\begin{cases} \omega_x = \omega \cos \lambda_P \cos \varphi_P \\ \omega_y = \omega \cos \lambda_P \sin \varphi_P \\ \omega_z = \omega \sin \lambda_P \end{cases} \tag{6.2}$$

Transformation from Cartesian coordinates to Eulerian coordinates

$$\begin{cases} \omega = \sqrt{\omega_x^2 + \omega_y^2 + \omega_z^2} \\ \lambda_P = \arcsin(\omega_z/\omega) \\ \text{if } \omega_x > 0, \varphi_P = \arctan(\omega_y/\omega_x) \\ \text{if } \omega_x < 0, \varphi_P = \arctan(\omega_y/\omega_x) \pm 180^\circ \end{cases} \tag{6.3}$$

Formulas using Cartesian components of plate velocities

Let M be a point on the Earth, with the geographical coordinates (λ, φ). **R** is the vector OM, **m** is the unit vector in M pointing northward along the meridian and **p** is the unit vector pointing eastward along the parallel. The Cartesian components of these three vectors are:

$$\begin{cases} \mathbf{R} = \begin{bmatrix} x \\ y \\ z \end{bmatrix} = \begin{bmatrix} R\cos\lambda\cos\varphi \\ R\cos\lambda\sin\varphi \\ R\sin\lambda \end{bmatrix} \\ \mathbf{m} = \begin{bmatrix} -\sin\lambda\cos\varphi \\ -\sin\lambda\sin\varphi \\ \cos\lambda \end{bmatrix} \\ \mathbf{p} = \begin{bmatrix} -\sin\varphi \\ \cos\varphi \\ 0 \end{bmatrix} \end{cases} \tag{6.4}$$

With ω in radians per time unit, the speed vector in M is:

$$\mathbf{v} = \boldsymbol{\omega} \wedge \mathbf{R} = \begin{bmatrix} \omega_y z - \omega_z y \\ \omega_z x - \omega_x z \\ \omega_x y - \omega_y x \end{bmatrix} = \begin{bmatrix} R(\omega_y \sin\lambda - \omega_z \cos\lambda \sin\varphi) \\ R(\omega_z \cos\lambda\cos\varphi - \omega_x \sin\lambda) \\ R\cos\lambda(\omega_x \sin\varphi - \omega_y \cos\varphi) \end{bmatrix} \tag{6.5}$$

Therefore:

$$\begin{cases} v_\lambda = \mathbf{v}\cdot\mathbf{m} = R(\omega_x \sin\varphi - \omega_y \cos\varphi) \\ v_\varphi = \mathbf{v}\cdot\mathbf{p} = R(\omega_z \cos\lambda - (\omega_x \cos\varphi + \omega_y \sin\varphi)\sin\lambda) \end{cases} \tag{6.6}$$

Angular velocities are generally given in degrees/Ma. From the first definition of the metre (approximate), $R = (20{,}000/\pi)$ km. The values in km/Ma are therefore:

$$\begin{cases} v_\lambda = \dfrac{1{,}000}{9}(\omega_x \sin\varphi - \omega_y \cos\varphi) \\ v_\varphi = \dfrac{1{,}000}{9}(\omega_z \cos\lambda - (\omega_x \cos\varphi + \omega_y \sin\varphi)\sin\lambda) \end{cases} \tag{6.7}$$

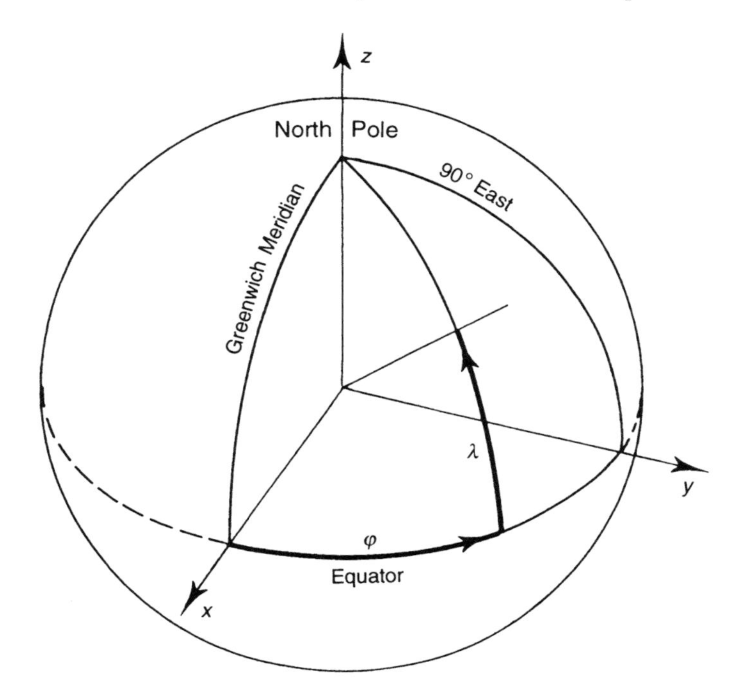

Figure 6.5. Cartesian and spherical coordinates.

Formulas using Eulerian coordinates of plate velocities

The relative velocity at point M, in km/Ma, can be calculated without using the Cartesian components, with the following equations:

$$\begin{cases} v_\lambda = \dfrac{1.000}{9}\omega \cos\lambda_P \sin(\varphi - \varphi_P) \\ v_\varphi = \dfrac{1.000}{9}\omega[\sin\lambda_P \cos\lambda - \cos\lambda_P \sin\lambda \cos(\varphi - \varphi_P)] \end{cases} \tag{6.8}$$

Azimuths are counted positively clockwise, starting from the north. The speed v and its azimuth A are linked to v_λ and v_φ by:

$$\begin{cases} v_\lambda \cos A + v_\varphi \sin A = 0 \\ v_\lambda \sin A - v_\varphi \cos A = 0 \end{cases} \tag{6.9}$$

If M is on a plate boundary with azimuth A_l, and this plate is assumed to be fixed in the preceding calculations, the normal velocity v_n and the tangential velocity v_t at the limit with the neighbouring plate are:

$$\begin{cases} v_n = v_\varphi \cos A_l - v_\lambda \sin A_l \\ v_t = v_\varphi \sin A_l + v_\lambda \cos A_l \end{cases} \tag{6.10}$$

6.3 DETERMINATION OF THE RELATIVE ANGULAR VELOCITIES

The normal velocity v_n is obtained by dividing the distance between two magnetic anomaly stripes around a mid-ocean ridge by their age (as the rotation velocities must be 'instantaneous', this age must be a few Ma only). At a transform fault, $v_n = 0$. In both cases, one obtains a linear relation between ω_x, ω_y and ω_z.

The velocity cannot be measured directly at subduction zones. The focal mechanisms of shallow earthquakes are instead used, and are supposed to give the orientation of the relative movement between the overlaying plate and the subducted plate. But the measurement error can reach 25%, and these measures have little weight. They are only used because they are in large quantity, and that a more trustworthy average orientation can be computed. Considering N plates on the Earth, one of which is supposed fixed, $3(N-1)$ angular velocity components ω_i need to be determined from a much higher number of linear equations, all more or less imprecise. The most probable estimate is provided by the traditional least-squares method, each type of data being given a certain weight (quite subjectively).

This method is based on the very likely assumption that the accuracy errors are independent from each other and follow a Gaussian probability law. Each *estimate* of a variable is given by a linear expression of the data, the *estimator*. The data coefficients in the estimator are worth knowing, as they highlight the data influencing the results, and therefore the areas where more and better data need to be acquired.

In 1978, Minster and Jordan published a model of angular velocities with 11 plates (instead of the 7 originally considered by Le Pichon). They used 110 spreading rates, 78 transform fault azimuths and 142 focal mechanisms to determine the 30 variables. In 1990, DeMets et al. published the more precise NUVEL-1 model, with three more plates (Australia, Philippines and Juan de Fuca). They amended it in 1994 (NUVEL-1A model) to account for more precise dating of the first magnetic anomaly stripes.

Oceanic sediment cores indeed provide information, not only about the inversions of the geomagnetic field, but also about climatic variations induced by perturbations of the Earth's orbit by other planets. This gives an *astronomical chronology* of inversions for the last millions of years, more precise than the chronology based on radioactivity. As some still write to the contrary, it is worth making clear that these astronomical perturbations do not have stable periods over more than a few Ma. They are chaotic, and astronomers must calculate them by going back through time without stopping, and then rely on the periods found. The accuracy of these celestial mechanics computations decreases when the time span increases. This explains why the older inversions are not dated by this way (see Section 17.1 for details).

NUVEL-1A is obtained from NUVEL-1 by multiplying all the velocities by 0.9562. They are reproduced in Table 6.1, with two modifications. First, the angles are not expressed in radians anymore, but in degrees. Second, the Eurasian plate was supposed to be fixed, and not the Pacific plate. The reason is that the present book mainly deals with the regions around the Eurasian plate: Mediterranean, Alps, Central Asia and Japan.

Table 6.1. Angular velocities relative to the Eurasian plate, in degree/Ma (model NUVEL-1A)

Plate	ω_x	ω_y	ω_z
Africa	0.10727	−0.04032	0.04410
N. America	0.07100	−0.06898	−0.18940
S. America	−0.00323	0.05046	−0.23049
Antarctica	0.00920	0.03977	0.03167
Arabia	0.43922	0.10738	0.20665
Australia	0.50535	0.43084	0.17932
Caribbean	0.04602	−0.05670	−0.09007
Cocos	−0.54107	−1.10066	0.44536
Eurasia	0	0	0
India	0.43840	0.13954	0.20841
Juan de Fuca	0.34280	0.59646	−0.46097
Nazca	−0.03156	−0.35419	0.36992
Pacific	−0.03030	0.41457	−0.75189
Philippines	0.652	−0.316	−0.752

Let us note, however, that the first relative velocities obtained with spatial geodesy (Section 6.9) seem to question the astronomical chronology. Therefore NUVEL-1 may still be used.

6.4 EXTENDING THE PALAEOMAGNETIC CHRONOLOGY

The age of magnetic anomaly stripes far from mid-ocean ridges was first estimated by assuming that spreading rates had remained the same since the Cretaceous superchron. Established with TRM of lava dated using ^{40}K radioactivity, the palaeomagnetic chronology of the last few million years was extended to chron 33, around −80 Ma (Upper Cretaceous). This chronology was not very accurate, for two reasons:

- There is no reason why the spreading rates should have been constant. This would even be surprising, as important changes in plate geometry occurred in the last 80 Ma.
- The palaeomagnetic chronology of the last Ma was not very accurate (the half-life of 1,340 Ma adopted for ^{40}K was too long). Astronomical chronology is now preferred.

Deep-sea oceanic drillings allow studying the microfossils in the oldest sediments, in contact with the basaltic layer. Palaeomagnetic chronology can now be related to *stratigraphic chronology*, grounded on palaeontology.

Many age determinations with radioactive methods have established with a very good accuracy the age of several markers of stratigraphic chronology. The most recent chronology of the Mesozoic Era (or Secondary Era) was published by

Table 6.2. Main periods of the Mesozoic Era.

Period	Age (in Ma)
	65.0 ± 0.1
Cretaceous	
	144.2 ± 2.6
Jurassic	
	205.7 ± 4.0
Trias	
	248.2 ± 4.8

Gradstein *et al.* (1994) and is reproduced at the end of the book. It gives the ages of the 31 stages of the Secondary. For scientists who do not need to perform detailed stratigraphic studies, knowing the names of these periods is now completely useless. The limits of the three main periods making up the Mesozoic are given in Table 6.2.

To reconstruct the relative positions of continents at the time of a specific chron, the continents need to be moved until the two magnetic anomalies of this chron on each side of the ridge coincide. This can be achieved with desk computers and plotting machines, finding by trial and error the finite rotation poles.

One result is that in the Lower Jurassic, around −180 Ma, all the continents were merged into a super-continent, called *Pangaea*. This had already been suggested by Wegener, but he believed that this Pangaea existed since the formation of continental crust, at the beginnings of the Earth. In fact, Pangaea was only one of the relative disposition of continents, and had a very short life. The complete history of continent drift will only be examined at the end of the book (Chapters 15 and 16). Let us only mention that the break-up of Pangaea started with the opening of the North Atlantic in the Jurassic. The Caribbean Sea was born in the Upper Jurassic, the South Atlantic and Indian Oceans opened in the lower Cretaceous. At the same time as Africa and India, Australia was already starting to separate from Antarctica at −130 Ma (contrary to what was believed around 1980). But the separation only accelerated after −45 Ma.

6.5 NO-NET-ROTATION FRAME OF REFERENCE AND NNR ABSOLUTE VELOCITIES

For fixed plate limits and angular velocities, it is possible to look for the frame of reference which minimizes the following integral, extended to the whole area of the Earth:

$$\iint_{Earth} v^2 \, dS = \iint_{Earth} [(\omega_y z - \omega_z y)^2 + (\omega_z x - \omega_x z)^2 + (\omega_x y - \omega_y x)^2] \, dS \tag{6.11}$$

This frame of reference is called a *no-net-rotation system* (NNR), and its instantaneous angular velocities are the *NNR absolute velocities*.

Let a, b, c be the Cartesian components of the NNR absolute velocity of the plate, assumed fixed, ω_x, ω_y and ω_z being provided. The instantaneous absolute NNR rotation velocity of any plate has the following Cartesian components:

$$\Omega_x = \omega_x + a, \qquad \Omega_y = \omega_y + b, \qquad \Omega_z = \omega_z + c \tag{6.12}$$

To minimize the integral in (6.11), the three partial derivatives in a, b, c must equal zero. One finds:

$$\iint_{Earth} [\Omega_x(y^2 + z^2) - \Omega_y xy - \Omega_z xz]\, dS = 0 \tag{6.13}$$

and two similar equations that are obtained by cyclic permutation of x, y and z. The reader can check that the same conditions are obtained by imposing:

$$\iint_{Earth} \mathbf{R} \wedge \mathbf{V}\, dS = 0 \tag{6.14}$$

where $\mathbf{V} = \boldsymbol{\Omega} \wedge \mathbf{R}$ is the absolute NNR velocity.

It is possible to prove that any vector field defined in any point M of the surface of a sphere, and tangent to M at the surface (which is the case of the velocities V), can be written using two scalar functions P and T of coordinates:

$$\mathbf{V} = \mathbf{grad}\, P + \mathbf{curl}(T\mathbf{R}) \tag{6.15}$$

The first term is the *poloidal* velocity, the second term is the *toroidal* velocity. For rigid plates, the poloidal velocity equals zero on a shell, is very high at the boundary between shells, and even infinite if the boundary strips are reduced to simple lines. Conversely, the toroidal velocity jumps at these boundaries.

P and T can be expressed as spherical harmonics. The ones of degree 0 are arbitrary constants without interest. I showed that the ones of degree 1 were zero when using the NNR system (Lliboutry, 1987, pp. 338–340). This justifies the name of 'no-net-rotation system'.

To calculate a, b, c, adopt the terrestrial radius as unit of distance. Symmetry considerations show that:

$$\begin{cases} \displaystyle\iint_{Earth} (y^2 + z^2)\, dS = \iint_{Earth} (z^2 + x^2)\, dS = \iint_{Earth} (x^2 + y^2)\, dS = \frac{8\pi}{3} \\ \displaystyle\iint_{Earth} yz\, dS = \iint_{Earth} xz\, dS = \iint_{Earth} xy\, dS = 0 \end{cases} \tag{6.16}$$

Therefore, using equation (6.13), we can write:

$$\frac{8\pi}{3} a + \iint_{Earth} [\omega_x(1 - x^2) - \omega_y xy - \omega_z xz]\, dS = 0 \tag{6.17}$$

This allows the calculation of a. Be aware that the ω_i are different on each shell. The formulas obtained by cyclic permutation yield b and c. The values obtained numerically were so far quite imprecise, because the relative angular velocities were so, and because some large boundary areas (Central Asia, Mediterranean, China Sea, etc.) were replaced by simple lines with zero thickness.

I introduced the NNR velocities in 1972 by starting from a very rough model of convection inside the Earth: a lithosphere with the same thickness over the oceans and the continents, convection only in the lithosphere and the upper mantle, upper mantle without lateral viscosity variations, lower mantle assimilated to a rigid mesosphere on which the system of reference is fixed. As the moment of all forces applied to the mesosphere must equal zero, the condition of equation (6.14) can be proved. The system of reference becomes the *no-net-torque reference frame*. Let us remark that, if the hypotheses of this model are not verified, the definitions of the NNR system as minimizing (6.11) or verifying (6.14) remain equivalent and valid. But the NNR system is no longer fixed to a rigid lower mantle.

The first calculations of a, b and c were made by the author, using relative angular velocities published by Chase in 1972 (Lliboutry, 1974). This was performed again by Minster and Jordan (1978), with their new determination of relative velocities (model AMO-2). This was performed once more by Argus and Gordon (1991) with NUVEL-1 (model NNR-NUVEL-1). The difference of the absolute velocities from this model and AMO-2 reach 15 km/Ma. When multiplying them by 0.9562 to pass from palaeomagnetic chronology to astronomical chronology, the absolute NNR rotation speed of the Eurasian plate becomes (model NNR-NUVEL-1A):

$$a = -0.0555; \qquad b = -0.1347; \qquad c = 0.1773 \qquad \text{(in degrees/Ma)}$$

Some absolute velocities in the NNR-NUVEL-1A model are shown in Figure 6.6. Their quadratic averages on the different plates are (in km/Ma):

Juan de Fuca:	19	Whole Earth:	39	Arabia:	44
North America:	19			India:	56
Caribbean:	9			Australia:	63
South America:	11			Philippines:	40
Antarctica:	15			Pacific:	64
Eurasia:	24			Cocos:	74
Africa:	29			Nazca:	76

The Arabian, Indian and Australian plates are more or less interdependent. Otherwise, the absolute velocities in the most oceanic hemisphere (right column) are larger than the absolute velocities in the most continental hemisphere (left column).

6.6 HOT-SPOT FRAME OF REFERENCE AND HS ABSOLUTE VELOCITIES

It is acknowledged today that the relative velocities between hot spots are small compared to the relative velocities between plates. Inside the Pacific or the Africa

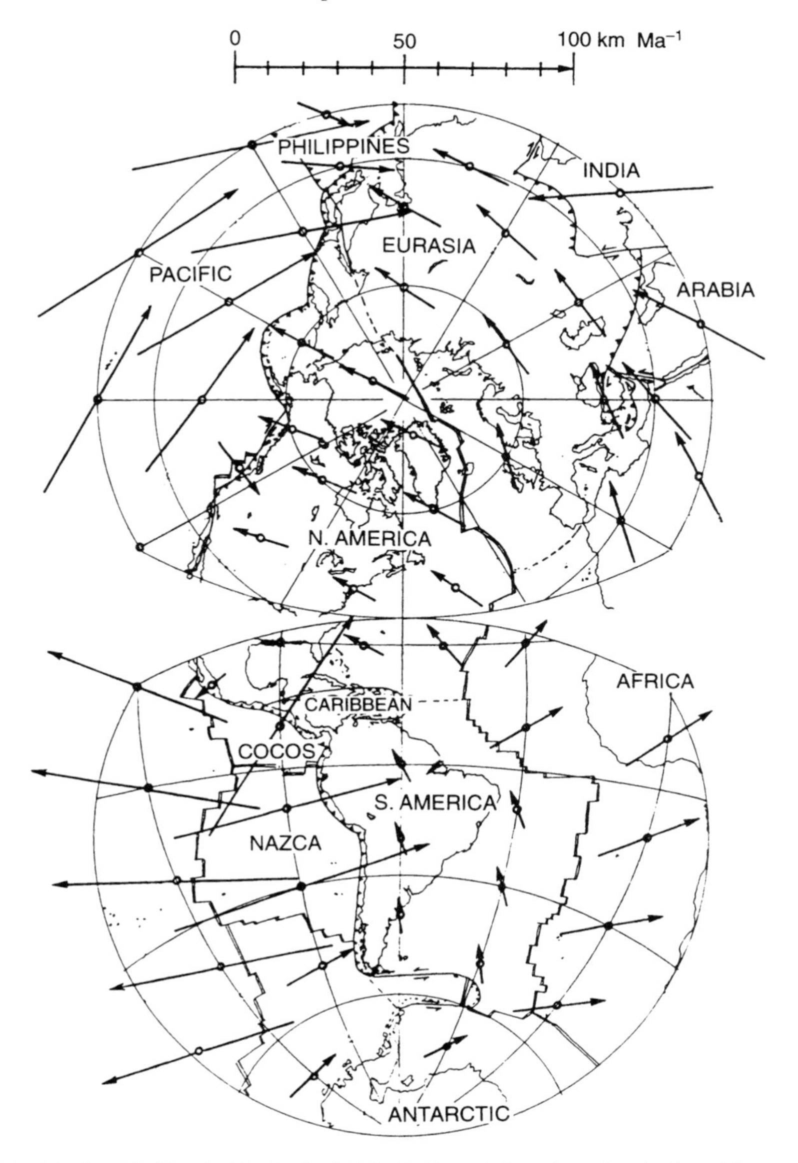

Figure 6.6. Absolute NNR velocities in the NUVEL-1A model, at the points indicated by small circles. Stereographic projection, which conserves the angles, has been used.

group, they are at most of a few km/Ma, and between the two groups of 5–10 km/Ma. It is therefore possible to define a hot-spot system of reference (*HS*), linked to the lower mantle. The velocities of the plates in this system are called the *HS absolute velocities*. They are not known as accurately as the relative velocities, but it is sufficient for geodynamical studies at the scale of the Earth.

To determine the HS frame, Minster and Jordan (1978) used the migration of volcanism (velocity and azimuth) for the hot spots of Hawaii, Marquesas, Society, and Australs, and the azimuths only for the hot spots of Juan de Fuca, Yellowstone and Galapagos (which is on two plates: Cocos and Nazca). Five plates were concerned, and their absolute velocities had to be all determined simultaneously. Furthermore, the accuracy was ±20% for the estimated migration velocities, and ±15° for the azimuths. To establish their model HS2-NUVEL-1, Gripp and Gordon (1990) used the relative rotation velocities of the NUVEL-1 model, but kept the very imprecise volcanic migration data from Minster and Jordan (1978).

It seems more reasonable to determine the HS frame of reference with the migration of volcanism only on the Pacific plate. Volcanic chains there are very long, and their azimuths can be determined to ±1°. We will use the following data:

Hot spot	End of the chain	Middle coordinates	Azimuth of the chain
Hawaii	Miluoki Seamount	26.0°N; −171.9°E	−66.4°
McDonald Seamount	Palmerston Atoll	−18.4°N; −155.1°E	−67.5°
Pitcairn	Duke of Gloucester Island	−22.3°N; −137.5°E	−70.0°

In fact, Miluoki Seamount, 3,520 km from Hawaii, is not the end of the volcanic chain, but a bend in the chain. The azimuth of the chain then becomes −10° in the Emperor Seamounts, which extend 2,080 km to the Western end of the Aleutians, in a subduction zone.

With the values of the above table, the compatibility between the three linear equations in $\Omega_x, \Omega_y, \Omega_z$ is better than 1%. The imprecision comes from the speed of volcanism migration. In 1970–1971, datations of 10 volcanoes along the 500 km between Hawaii and Kauai yielded a migration velocity of 148 ± 14 km/Ma during the last 3.4 Ma. By extrapolation, the age of the chain bend would then be 24 Ma. But in 1973, lavas were dredged from the Koko Seamount, in the Emperor chain and 3,000 km from the bend. They were dated to 46.4 ± 1.1 Ma. As the oldest volcano in the chain is around 70 Ma old, the migration speed of volcanism on the Emperor chain would have been only 88 km/Ma, and the age of the bend would be 42–44 Ma. The results of Jurdy (1979), set out in the next section, show that the actual age of the bend is 36 Ma, which gives an average velocity of 95 km/Ma between the bend and Hawaii.

Depending whether the velocity adopted is 148 km/Ma or 95 km/Ma (as in Argus and Gordon, 1990), the absolute HS velocity of the Pacific plate will be different. Compared, the imprecision on the relative inter-plate velocities is not important.

The HS frame of reference is therefore ill-defined, and in the three cases it differs significantly from the NNR frame, which rotates eastward relative to the hot spots. I suggest that it is due to the pushing by the lower mantle of subducted plates that reach it, in the West Pacific.

Table 6.3. Absolute angular velocities of the Eurasian plate, in degrees/Ma.

HS-NUVEL-1	a	b	c
Gripp and Gordon, 1990	0.034	0.054	−0.063
Present computation ($v = 148$ km/Ma)	−0.130	0.145	−0.695
Idem ($v = 95$ km/Ma)	−0.072	−0.062	−0.165
NNR-NUVEL-1	−0.058	−0.141	0.185

6.7 EVOLUTION OF THE PACIFIC PLATE

Following other workers, Jurdy (1979) reconstructed the earlier boundaries and the relative displacements of the Pacific plates from magnetic anomaly stripes. He went back in time to superchron 34-Cn, in the Cretaceous (reconstructions from elsewhere will be presented with continental drift in Chapter 15).

Some 80 Ma ago, the Pacific plate was less important than it is now. It was bounded in the north by the *Kula* plate, now disappeared in the Aleutian subduction zone, and in the east by the *Farallon* plate. Reconstructions differ, depending on whether Antarctica is assumed to have always formed a single plate, or whether East and West Antarctica lay on two different plates. We are now sure that the latter hypothesis is the correct one. The two Antarctic plates became closer 59 to 36 Ma ago. This coincided with a change in the trajectory of the Pacific plate relative to

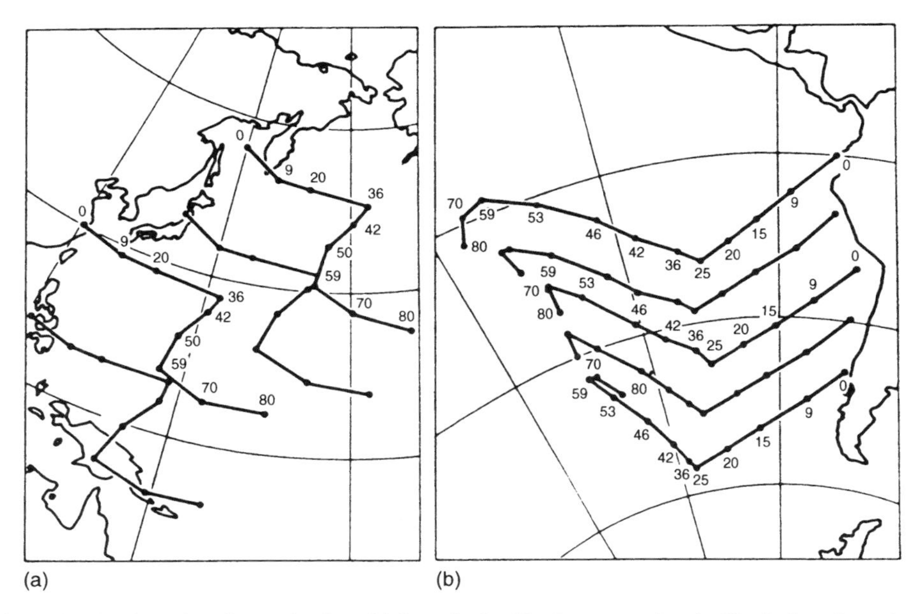

Figure 6.7. Trajectories of several points: (a) from the Pacific plate, assuming the North-American plate is fixed; (b) from the Farallon plate, assuming the South-American plate is fixed. The dates are in Ma Before Present (BP). It has been assumed that Antarctica formed two distinct plates, between 59 and 36 Ma BP, according to the model of Molnar *et al.*, 1975. From Jurdy (1979); © AGU.

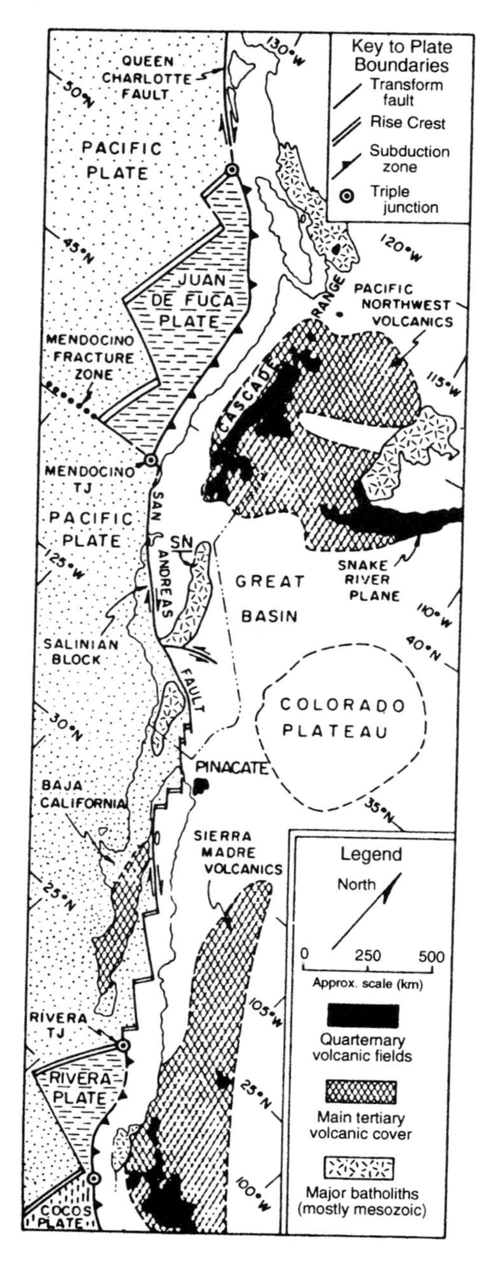

Figure 6.8. Schematic map of the plates on the US Pacific coast. The map uses a Mercator projection with a pole at 53N, 53W. According to Morgan (1968), this is the Eulerian pole of the rotation of the American plate about the Pacific plate. With this projection, the transform faults between the plates (San Andreas and Queen Charlotte) should become parallel lines. This is only partially true, because of the plastic deformations in the boundary area. The dashed line corresponds to the State of California. SN: Sierra Nevada. From Dickinson and Snyder (1979); © AGU.

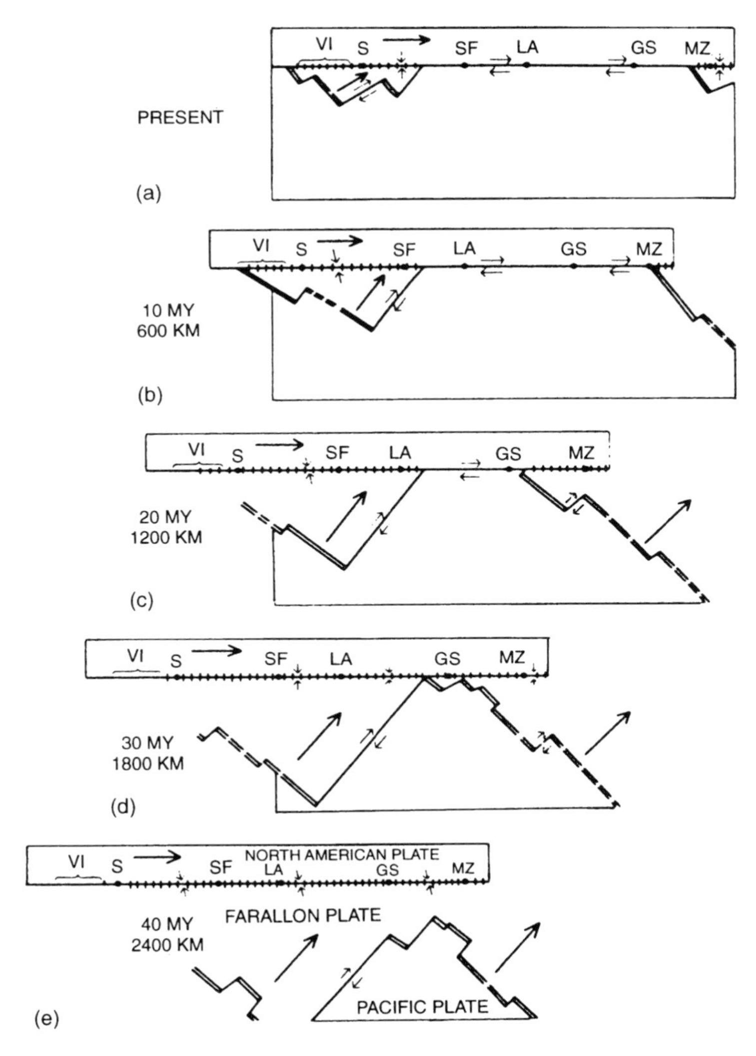

Figure 6.9. Subduction of the early Farallon plate (schematic). VI: Vancouver Island; S: Seattle; SF: San Francisco; LA: Los Angeles; GS: Guaymas (in the Gulf of California); MZ: Mazatlan. From Atwater (1970); © Geological Society of America.

Asia, and an accelerated subduction of the Farallon plate below the South American plate (Figure 6.7).

About 30 Ma ago, the East Pacific Ridge met a subduction zone which extended along the US Pacific Coast. The northern end of the Farallon plate, isolated, became the present *Juan de Fuca* plate. Extension and subduction neutralized each other, and only one transform fault subsisted between the Pacific plate and the American plate. It is the *San Andreas fault* (see Figures 6.8 and 6.9).

Farther south, the same conditions separated the *Rivera* plate from the Farallon plate (the name of Rivera is a tribute paid by Californian oceanographers to the Mexican painter). The south part of the Farallon plate divided 25 Ma ago into the Cocos and Nazca plates, with the emergence of the E–W Galapagos ridge.

6.8 SPATIAL GEODESY TECHNIQUES

The very precise localization of points on the surface of the Earth and the precise determination of the surface with altitude 0, the *geoid*, form the discipline of *geodesy*. Details of all its aspects can be found in my book (Lliboutry, 1992), and I will therefore keep this section short.

On land, traditional geodesy consists in determining, separately, the position of many geodesic points making triangular lattices of 30–40 km aside, and then their altitudes by precision levelling. With modern laser tools, distances of 30–40 km can be measured with an accuracy of 1 cm. Over a decade, this method made it possible to determine the deformation of tectonically active regions, such as California or Japan. This is a very lengthy technique, demanding many qualified personnel in the field. It is no longer used to create new geodetic networks, or even to check the previous networks and find movements of the ground. Instead, *spatial geodesy* is used.

Spatial geodesy uses bodies outside the Earth: artificial satellites or even *quasars*, stellar objects very far from the Earth and emitting a very powerful radioelectric 'noise'. There are four problems to solve, highly imbricated:

1. The determination of the geoid, which brings about the determination of the gravity field outside the Earth, to which artificial satellites are submitted.
2. The very precise location of the observation points, which will provide their relative velocities when renewing the measurements at intervals of several years.
3. The local variations of the vertical at the observation points. These variations have two causes: (a) the variation of the instantaneous rotation axis of the Earth in a Galilean system (the one in which the laws of mechanics are valid); (b) the variation of this rotation axis relative to the Earth, i.e. the displacement of the geographic North Pole. The terrestrial frame of reference needs to be precisely determined, as lithospheric shells move with respect to each other.
4. The perturbations of the Earth's rotation velocity.

The third point will be examined in Chapter 16, and the fourth point will be briefly mentioned in Section 8.13.

Laser ranging

The first technique used consisted in tracking artificial satellites from a network of ground stations, such as the one near Grasse (France). Their radial speeds were measured relative to the station by Doppler effect, emitting laser beams toward on board reflectors. This is known as *satellite laser ranging* (SLR). It uses specific geodetic satellites, heavy, placed on high orbits (around 5,000 km), and provided with a drag-compensation device to avoid perturbations of their orbits. The LAGEOS geodetic satellite is a sphere of 411 kg, with 426 reflectors uniformly distributed over its surface.

The orbit parameters, the geoid and the location of the stations are computed simultaneously. Laser telemetry provided geoid models like the GEM-T model of Lerch *et al.* (1994), from NASA (*National American Space Agency*), or the GRIM-4 model of Balmino *et al.* at GRGS (*Groupe de Recherches en Géodésie Spatiale*, Toulouse, France). Adding gravity measurements from land, Rapp and Pavlis (1990) from the Ohio State University in Columbus obtained the geoid model OSU-89, which is used in the GPS localization system discussed below.

DORIS

The French system DORIS (*Doppler Orbitography and Radiopositioning Integrated by Satellite*) is an improvement on laser telemetry. It uses 49 ground stations, distributed on nine different lithospheric plates. The stations emit upwards, and the satellite itself measures the radial speeds. These measurements are stored on board and retransmitted when it flies over Toulouse.

The orbit is known with an excellent precision, which explains why a DORIS receptor was installed on the remote sensing satellites SPOT (French), ERS-1 (European) and on the oceanographic satellite Topex-Poseidon (French–US). The latter measures the average heights of the ocean with radar altimetry, with a footprint of a few kilometres in diameter. The main source of error is the imprecision of the orbit. With DORIS, the ocean heights are known within a few centimetres. DORIS was used to improve the geoid model GRIM-4. It contributes to the determination of the variations in the Earth's instantaneous rotation axis. And it can be used to determine the relative velocities between emitting stations.

VLBI

The second spatial geodesy technique is *Very Large Base Interferometry* (VLBI). The measurements of electromagnetic noise of an extra-galactic quasar are recorded by different radio-telescopes (Figure 6.10). The phase differences between these measurements equal $(B \cos \alpha)/c$, but α varies continuously. The measures are therefore performed simultaneously for 24 hours. This necessitates the synchronization of atomic clocks in the different observatories to 10^{-8} s accuracy (which is already in itself a special programme, using the Lasso satellite).

VLBI provides continuous measures of the angular variations in the terrestrial rotation axis. However, the *Service International du Mouvement Polaire* (International Service of Polar Movement) has used for a long time the variations of the angle between the Earth's rotation axis and the vertical, at 5 *ad hoc* observatories all located at latitude 39°08′N (see Section 15.2). The main advantage of VLBI has

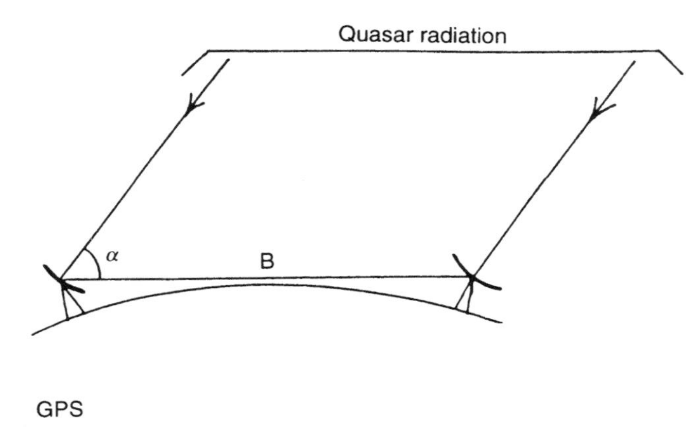

Figure 6.10. Principle of Very Large Base Interferometry (VLBI).

therefore mainly been the measurement of distance variations between points distributed on the Earth, with accuracies of a few millimetres.

GPS

A third technique is the *Global Positioning System* (GPS). This system is an improvement of the Transit system. The user receives modulated decimetric radio waves transmitted by a constellation of satellites, which allows him to calculate their radial speeds by Doppler effect. This is made with two frequencies, to allow removing the perturbation from the ionosphere. Each satellite also sends the user its orbit parameters and its location on the orbit as a function of time, i.e. its ephemeris, updated continuously. The calculator of the receiver holds in memory the geoid parameters, performs the necessary computations, and the user can read, in real time, the geographical coordinates of his own location.

GPS uses the constellation of the 24 Navstar satellites, launched between 1985 and 1995. There are six circular orbits at 20,190-km altitude (they are described in 12 sidereal hours), and four satellites per orbit. This means that several satellites can always be seen from anywhere on Earth. The satellite ephemerides are computed from SLR data obtained at five terrestrial stations, and are updated three times a day.

The system was originally conceived for the exclusive needs of the military, and was meant to be classified, but the economic interest of its larger use was too important. GPS is therefore accessible to all, but with data voluntarily degraded (C/A code), providing a resolution of only 25 m in the localization of a point, and 1 m for the distance between two points. The undegraded system (P-code) gives a precision of 0.1 m for localization, and a few millimetres for distances. The commercial success was immediate, and the fall in the cost and weight of the GPS receiver phenomenal, one explaining the other. In 1984, two people were necessary to carry a receiver, which cost $150,000. At the end of 1995, the receiver can fit in a pocket and costs only \$200. All vehicles might soon be equipped with GPS. Forecasts for the year 2000 are of a turnover of \$5 billion for the GPS industry.

GPS is already used in Earth Sciences for positioning the sampling sites, the location of mobile stations, etc. In a few years, the confidentiality of P-code, which is less and less justified, will probably disappear. A multitude of geodynamic studies should follow.

6.9 RELATIVE MOVEMENTS OF LITHOSPHERIC SHELLS ON THE DECENNIAL SCALE

Relative movements of the lithospheric shells, obtained by satellite laser ranging or VLBI, have been published since 1985. But it was necessary to wait another few years to get more values, with a better precision. Some references are: Smith *et al.* (1990) for laser telemetry (10-year displacements); Fallon and Dillinger (1992) for VLBI (5–8 year displacements); and Soudarin and Cazenave (1995) for DORIS (3-year displacements).

A main result is that *no really significative difference was found between the relative plate velocities measured over a few years, and the average velocities over a few million years*, deduced from magnetic anomaly stripes. The NUVEL-1 palaeomagnetic chronology seems a better match than the NUVEL-1A astronomical chronology.

The number of stations is still too small to determine only with spatial geodesy the relative rotations of every large plate. Some stations must even be excluded because they are on ridges (DORIS stations of Reykjavik and Djibouti), or in boundary areas like California. At Fairbanks (Alaska), the DORIS and VLBI measurements do not agree. So far, the only sure results are the rotations of the North American and Africa plates relative to the Pacific plate.

Figure 6.11 shows the Pacific/North America Eulerian pole, obtained by VLBI, and compared with the Eulerian poles from successive models based on palaeomagnetism. The agreement with NUVEL-1 is nearly perfect (and therefore also with NUVEL-1A). The rotation is $\omega = 0.783°/\text{Ma}$ with VLBI, $0.806°/\text{Ma}$ with NUVEL-1 and $0.771°/\text{Ma}$ with NUVEL-1A.

The Eulerian components of the rotation Africa/Pacific are:

DORIS	$\lambda_p = 56.0 \pm 4.0°$	$\varphi_p = -68.0 \pm 8.0°$	$\omega = 0.981 \pm 0.051$
NUVEL-1	$\lambda_p = 59.2°$	$\phi_p = -73.2°$	$\omega = 0.969$
NUVEL-1A	$\lambda_p = 59.2°$	$\phi_p = -73.2°$	$\omega = 0.935$

On six other plates, there is only one reliable DORIS station, which is not enough to determine the relative rotation of the whole plate. There are many SLR stations in Europe, and two in Australia, but Smith *et al.* (1990) did not publish the rotations of these plates, only the variations of distance between the stations.

Soudarin and Cazenave (1995) used 3 years of measurement of the DORIS systems on board the satellites Spot-2 and Topex-Poseidon, to study the movements of 14 sites over eight different plates. (An 'instantaneous' position is in fact obtained

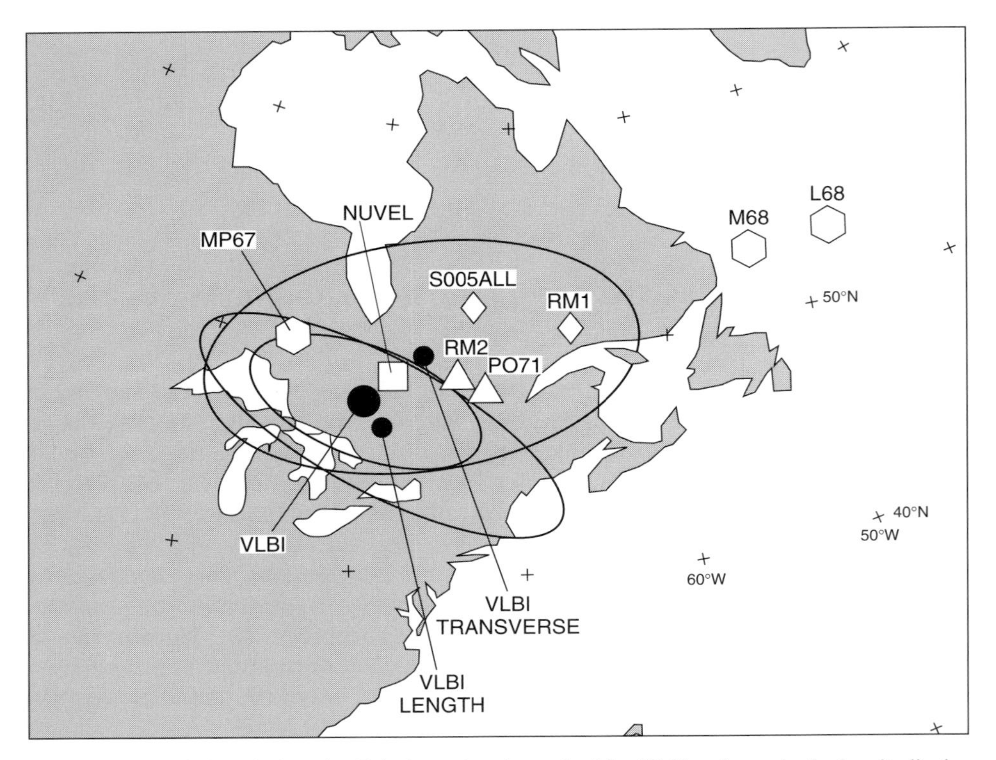

Figure 6.11. Pacific/North America Eulerian poles, determined by VLBI, using only the longitudinal or transverse displacements (small black dots), or using everything (large black dot). The Eulerian pole has 95% chances of being inside the ellipses centred on these points. The other symbols show the Eulerian poles obtained by several authors, using magnetic anomaly stripes and transform fault directions: MP 67 (McKenzie and Parker, 1967); M68 (Morgan, 1968); L68 (Le Pichon, 1968); S005ALL (Chase, 1972); RM1 (Minster *et al.*, 1974); P071 (Chase, 1978); RM2 (Minster and Jordan, 1978); NUVEL-1 (DeMets *et al.*, 1990). From Angus and Gordon (1990); © AGU.

from the measurements during 3 consecutive months). Except for Fairbanks, which is in the centre of Alaska, and for Easter Island, the match with NUVEL-1 is excellent (2% difference). It is less good with NUVEL-1A, the geodetic velocities being $5 \pm 2\%$ faster.

According to these authors, Arequipa would be fixed on the South American plate. And Easter Island would migrate toward the East Pacific Rise, 200 km west, with a velocity of 32 mm/year relative to the Nazca plate on which it is supposedly located. However, considering that Easter Island is fixed and on the Nazca plate, Smith et al. (1990) find consistent results from 10 years of measurements (Figure 6.12). According to them, Arequipa moves relative to South America.

Three years do not seem enough time to get reliable velocities. The subduction proceeds mainly by jumps during earthquakes. Several hundreds of kilometres are needed for the plate's elasticity to dampen these jumps, and Arequipa or Easter Island are too close to active faults. Measures stacked over at least 10 years may

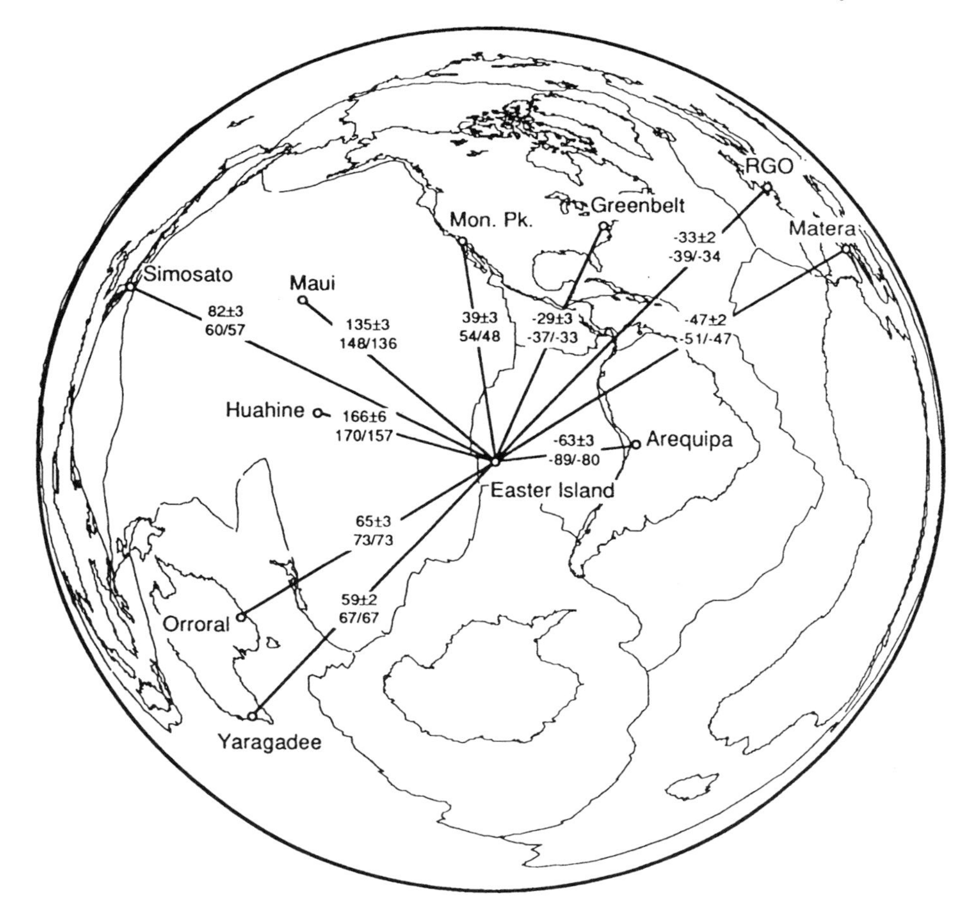

Figure 6.12. Equal-area projection centred on Easter Island, with relative velocities in mm/year (i.e. in km/Ma). The upper numbers correspond to laser telemetry measurements over 10 years, the lower numbers to values deduced from the magnetic anomaly stripes. On the left, AMO-2 model of Minster and Jordan (1978) and on the right, NUVEL-1 model of DeMets *et al.* (1990). From Smith *et al.* (1990); © AGU.

well be necessary to smooth out the effects of seismicity and make the velocities uniform.

6.10 MOVEMENTS IN THE BOUNDARY ZONES

Spatial geodesy confirmed that the relative movement between the North American and Pacific plates in California was not only a simple strike-slip movement along the San Andreas transform fault, or its combination with the movements of a few parallel faults. This strike-slip movement is accompanied by the deformation of a large area encompassing California and the neighbouring states of Nevada and Arizona. Diffuse shear accounts for half the strike-slip forecast by plate theory (56 mm/year). We shall come back to this in Chapter 12, as it is primordial when trying to predict the Californian earthquakes.

Similarly, the strike-slip movement between the Australian and Pacific plates along the Alpine Fault in New Zealand does not occur only along this transform fault and its immediate neighbourhood. Only 2/3 of the strike-slip movement forecast by plate theory is accounted for (this result was obtained by resuming the old geodetic network with the GPS).

If the SLR results are exact, Arequipa, south of Peru, moves eastwards of ca. 17 mm/year relative to the South American plate. This is despite the fact that this station is 200 km from the Peru Trench (where the Nazca plate subducts at a velocity of 80 km/Ma). East of this town, there are active reverse faults and a compression of the whole region is probable.

Still according to SLR geodesy, Simosato, on the coast of the Kii Peninsula (south of Osaka, Japan) is in the same situation. Offshore, the Philippines plate is subducted under the Eurasian plate at the speed of 57 km/Ma. Simosata would move west of it at a speed of 25 mm/year. There is no trace of such compression in the Kii Peninsula. Control with the Japanese geodetic network shows that the whole south of Japan, as a rigid block, must be moving (the north of Japan moving even faster than the west).

A bit farther south, Shanghai would move relative to the Eurasian plate with a velocity of 8 mm/year oriented ESE (Molnar and Gipson, 1996). Shanghai is on the NE edge of the southern China continental block. It is pushed slightly eastward by the compression due to the impact of India against Eurasia. The idea was put forward by Tapponnier and Molnar in 1975 and accepted by everybody, but, until now, this push was thought to be much faster.

7

Subduction zones and island arcs

7.1 OCEANIC TRENCHES AND ISLAND ARCS OF THE PACIFIC

The descent of a lithospheric shell into the mantle is called *subduction*. Evidence of this process include: (1) many earthquakes with foci in a layer approximately plane and dipping obliquely into the mantle; (2) a very small dampening of seismic waves (a very high Q factor) in this layer.

We will treat later subductions that might exist inside a continent. Only the lithospheres with oceanic crust are normally subducting. But they may subduct under another oceanic lithosphere as well as under a continent.

Two processes always accompany subduction:

- A deep oceanic trench, where subduction begins. It goes down several kilometres deeper than the floor of abyssal plains, but is partially filled with sediments from the neighbouring continent. An extreme case is the Java Trench, which extends along Malaysia, where it is filled with sediments from the Ganges and Brahmaputra.
- A strong volcanism on the overthrusting plate, a few hundred kilometres away from the trench.

Apart from the Antilles, Southern Antilles (Scotia Arc) and Eastern Mediterranean, all current subduction zones are at the periphery of the Pacific Ocean or in the NE Indian Ocean. The geography of Pacific trenches will therefore be explained first (see map on Figure 7.1).

The two sides of the Pacific Ocean are very contrasted. On the East side, subduction takes place below a continent with a mostly linear coast (South or Central America), bordered with parallel cordilleras and very high plateaus. On the West side, subduction takes place below strings of islands arranged in arc circles, sometimes as simple volcanoes produced by the subduction itself (Aleutian Islands, Kuriles, Mariannas, etc.), sometimes as fragments detached from the continent (Japan, Philippines, etc.).

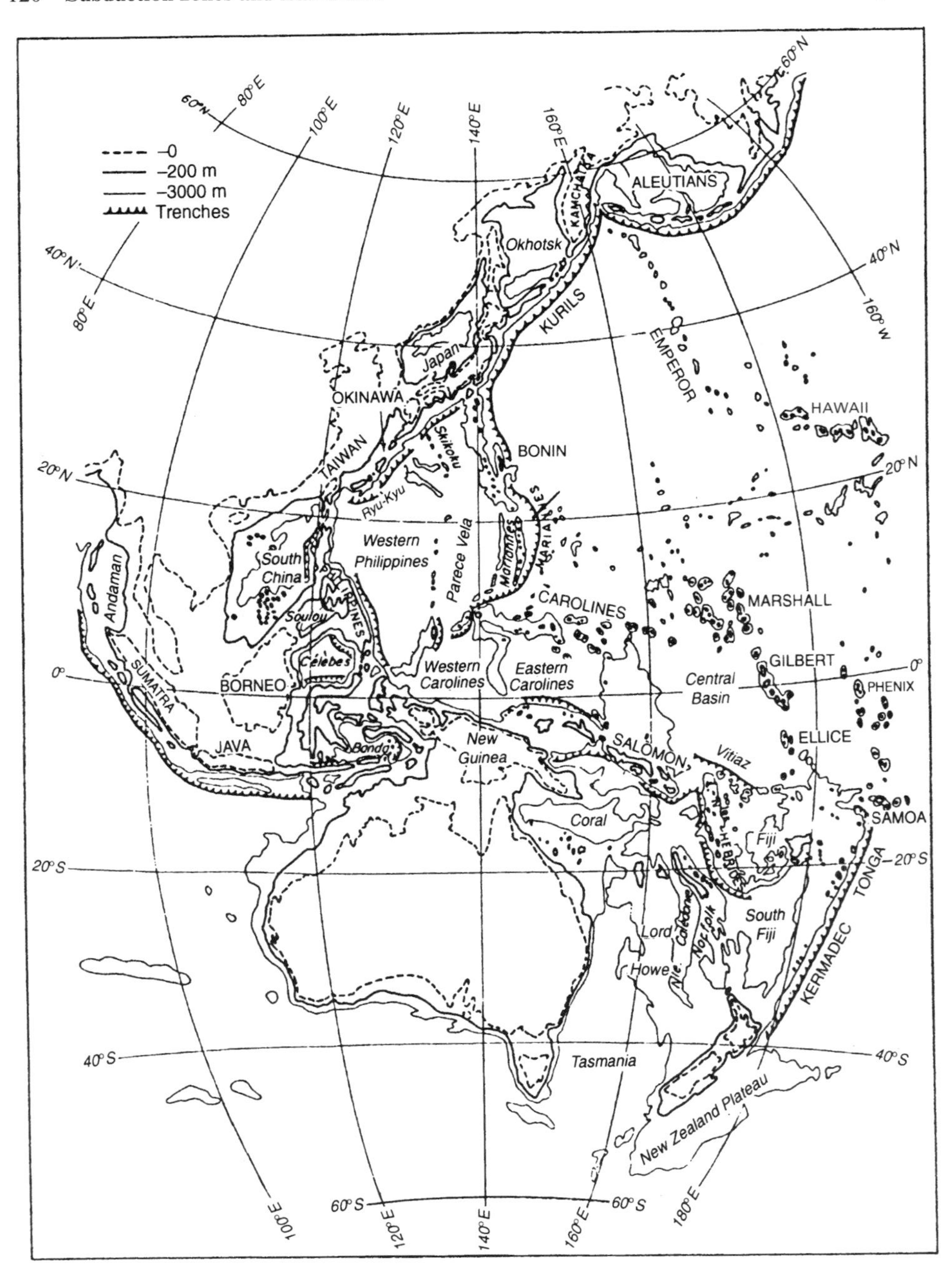

Figure 7.1. Basins (in italics) and subduction zones of the Western Pacific. For subduction zones, the tickmarks are pointing toward the overthrusting plate. Bold lines (−200 m contour) approximately delimit the continental platforms and thin lines (−3,000 m contour) the basins. The map uses an azimuthal equal-area projection.

The island arc may be more or less pronounced (some can be sub-linear, as the Izu-Bonin or Tonga-Kermadec islands). But if there is an arc, *its concavity always points out towards the overthrusting plate*. On this plate, inside the arc, there always is a deep basin with oceanic crust, the *back-arc basin*. Back-arc basins form a large part of the South China Sea, most of the Japan Sea, but only a portion of the southern parts of the Okhotsk Sea (Kuriles Basin) and the Bering Sea (Aleutian Basin). From the continent to the back-arc basin, the crust thins out, gets injected with basalt sills, and progressively transforms into oceanic crust. It shows thick sediments, often with oil, i.e. all the characteristics of a passive continental margin.[1]

Subductions in the SE Pacific roughly show a transverse compression, a push of the subducted plate, whereas subductions in the Western Pacific show an extension, a past or present traction, as if the subduction zone were attracting (or had attracted) the continent. We are now only setting the problem, and the extension processes will be studied in detail later in the book. But these details must not obscure the difference between Eastern and Western Pacific. The SE Pacific Ridge is the region of the Earth where the most oceanic lithosphere is produced, enough to feed subduction on both sides. And the SW Pacific is the region where the most disappears, especially if the Indo-Australian subduction is considered. World-scale general convection must therefore be involved in the asymmetry of subduction zones around the Pacific.

The list of trenches around the Pacific Ocean is given below, with the subduction velocity as predicted in the middle of the trench by the NUVEL-1A model (Table 7.1). The reader may be surprised that the North-American plate, which encompasses the eastern edge of Siberia, actually extends to the middle of Japan. This was, however, confirmed by spatial geodesy from the Kashima station, east of Tokyo.

Trenches as deep exist north and east of Australia; the Tonga Trench reaches down to $-10,882$ m, the Kermadec Trench to $-10,047$m. As subduction occurs below a series of small plates, it cannot be predicted.

7.2 WADATI–BENIOFF SURFACES

Shallow earthquakes occur close to oceanic trenches, on both sides. Focal mechanisms generally show a sub-horizontal traction perpendicular to the trench. The plate moves away from the horizontal to dip into the mantle, and this flexure requires a torque acting on each side. There is indeed traction in the upper layer, as well as compression in the lower layer. If the former breaks up, this is because the confining lithostatic pressure is lower (see Chapter 12).

Lower, from around 30 to 60 km deep, focal mechanisms show a friction between the subducted and overthrust plates. Sliding is therefore accompanied, at least in part, by abrupt sliding at each earthquake. The faults visible at the surface and active

[1] Western geophysicists do not use this term, because they mostly work with very small-scale maps, where the subduction seems very close. One must realize that the North and South China Seas, from Malaysia to Japan, have the same dimensions as the Mediterranean Sea.

Table 7.1. Trenches around the Pacific Ocean, with subduction rates and azimuths

	v	A
Nazca Plate/South America Plate		
Peruvian Trench (deepest point: −8,066 m)	79 km/Ma	82°
Cocos/Caribbean		
Central America Trench (−6,662 m)	106 km/Ma	22°
Pacific/North America		
Aleutian Trench (−7,822 m)	72 km/Ma	−42°
Kurils Trench (−10,542 m)	81 km/Ma	−58°
Japan Trench (−8,412 m)	83 km/Ma	−64°
Pacific/Philippines		
Izu-Bonin Trench (−9,985 m)	58 km/Ma	−72°
Mariannas Trench (−11,022 m)	38 km/Ma	−44°
Carolinas Trench (−8,527 m)	19 km/Ma	−57°
Philippines/Eurasia		
Ryu-Kiu Trench (−7,507 m)	57 km/Ma	−52°
Philippines Trench (−10,497 m)	80 km/Ma	−63°
Australia/Eurasia		
Java Trench (−7,450 m)	73 km/Ma	19°

during these earthquakes are, however, subvertical. It is necessary to admit that edges form on the sides of the overthrusting plate, which the subducting plate tries to drag down (Figure 7.2).

Even further away from the trench, the seismic foci cluster along an inclined layer no more than 20-km thick, which can reach down to the 660-km discontinuity. This is the *Wadati–Benioff surface*. The contour levels of such surfaces are shown in Figures 7.3 and 7.4.

This lithospheric plate is colder than the surrounding mantle, as demonstrated by the good transmission of seismic waves (high Q) when, coming from a deep source,

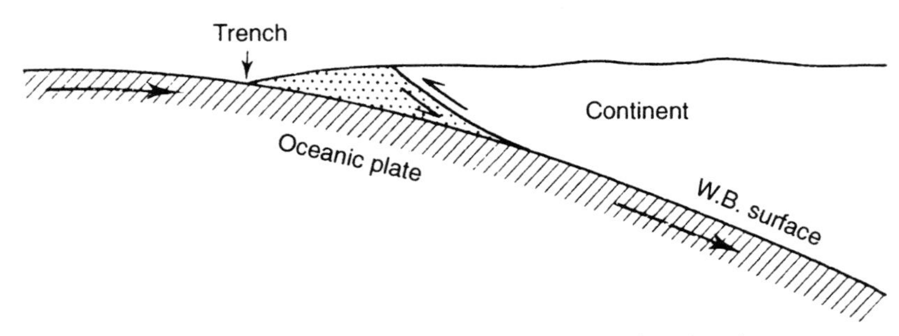

Figure 7.2. The subducting plate may drag an edge of the overthrusting plate down.

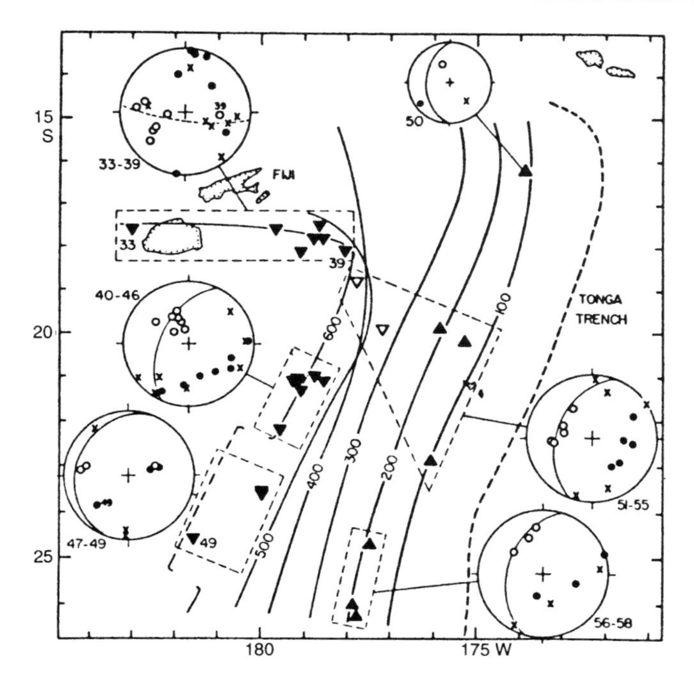

Figure 7.3. Contour representation of the Wadati–Benioff surface (with dashed lines when hypothetical), and focal mechanisms in the Tonga Islands subduction zone (SW Pacific). The levels are in km, the black triangles show earthquake sources. Focal mechanisms are shown in the lower hemisphere, in equal-area projection, with the trace of the Wadati–Benioff surface, the directions of maximal compression (small circles), the directions of minimal compression and tectonic traction (black dots) and the directions of the main intermediate stress (crosses). From Isacks and Molnar (1971); © AGU.

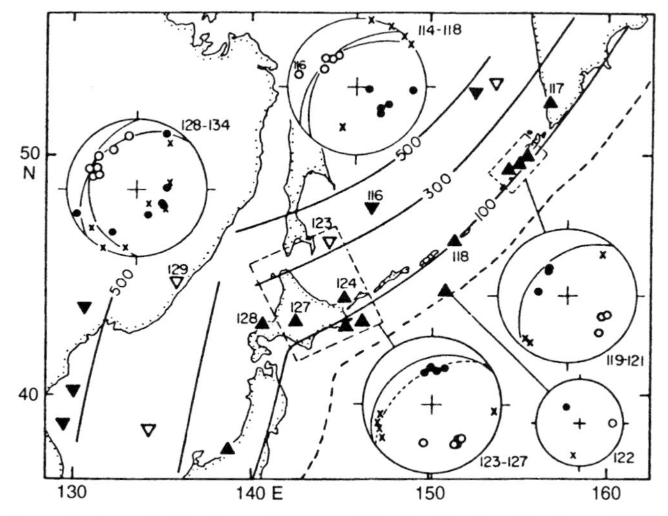

Figure 7.4. Focal mechanisms in Northern Japan and in the Kuriles island arc. The symbols and the original reference are as in Figure 7.3.

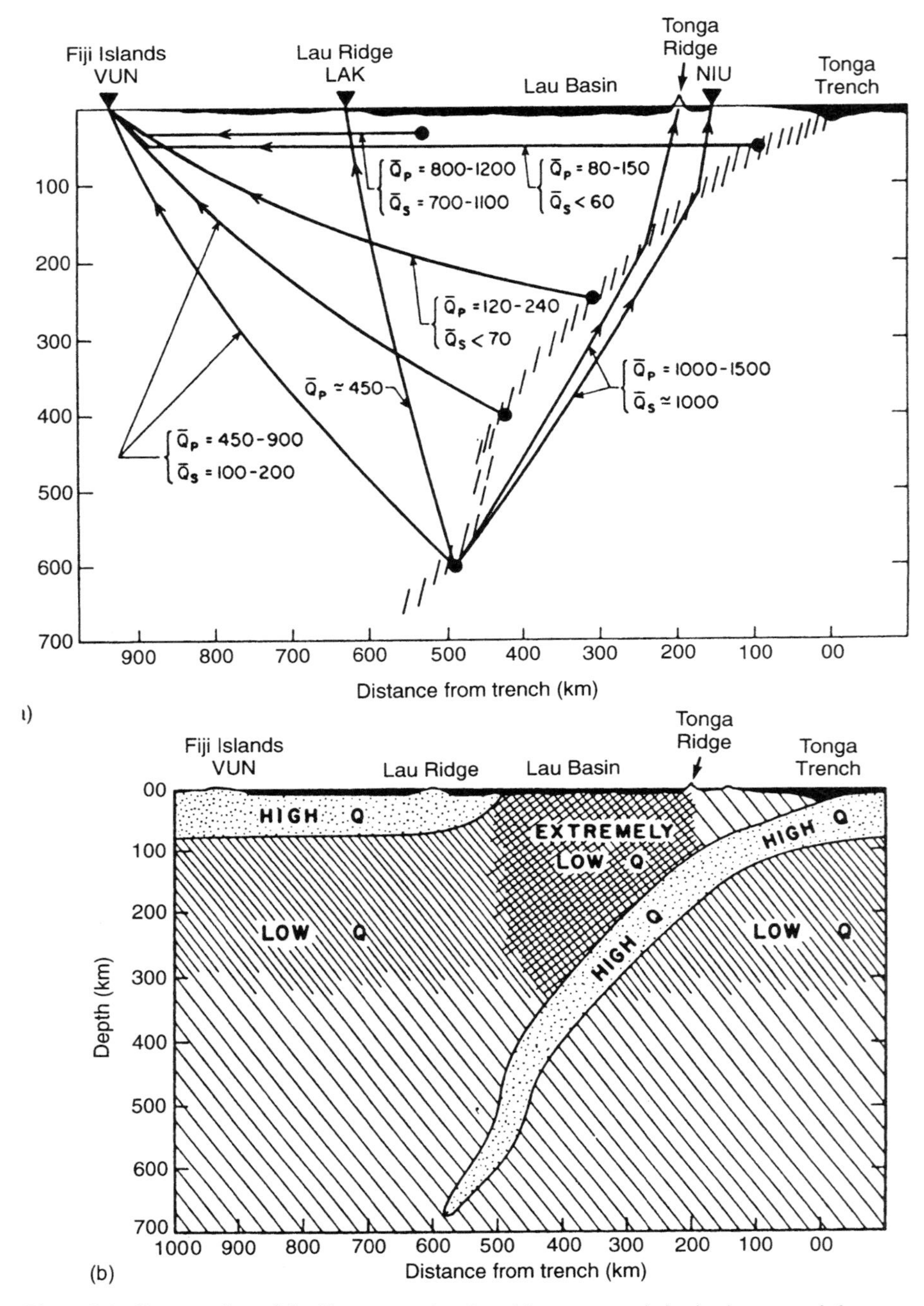

Figure 7.5. Cross-section of the Tonga arc, showing: (a) some recorded seismic rays and the corresponding average Q, from Barazangi and Isacks (1971); (b) the zones associated to a good transmission of seismic waves (high Q) and the zones associated to some absorption (low Q), from Isacks and Molnar (1971); © AGU.

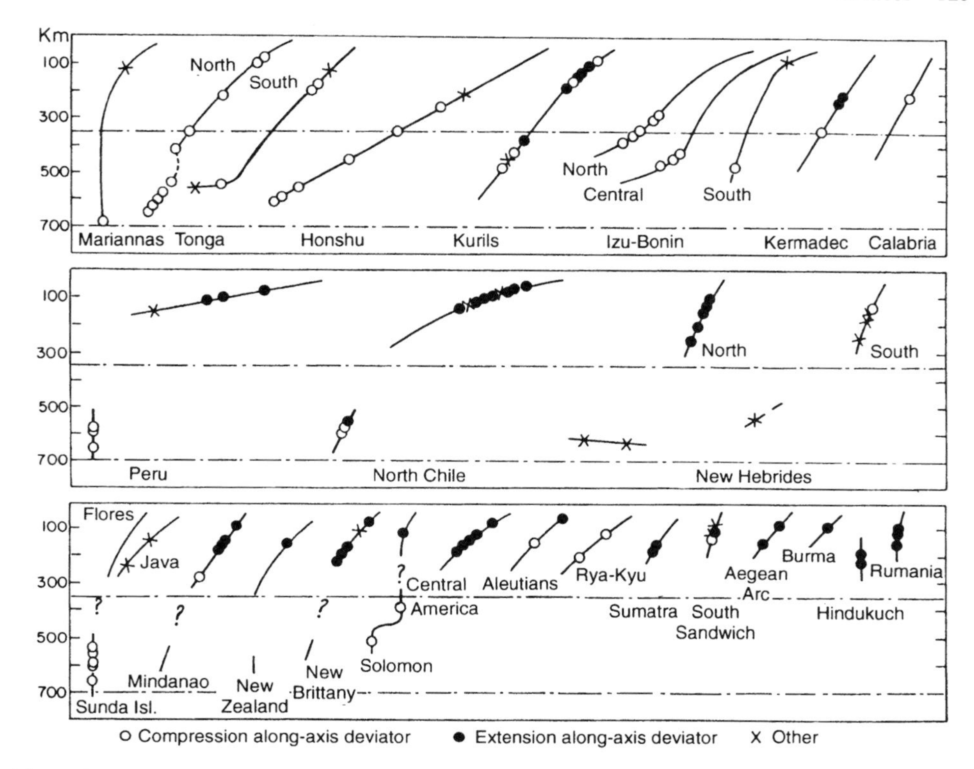

Figure 7.6. Summary of the focal mechanisms determined before 1971, for several subduction zones. The lines indicate longitudinal sections of the Wadati–Benioff surfaces. From Isacks and Molnar (1971); © AGU.

the trajectory to a station is in a Wadati–Benioff surface. Figure 7.5 shows an example taken from Barazangi and Isacks (1971), about the Tonga Islands subduction (small islands east of the Fiji Islands),

Focal mechanisms generally show traction of the dipping plate in the direction of the subduction, down to 200 km deep, and always compression along this direction, beyond. (This is deviatoric stress, not accounting for the isotropic pressure of hydrostatic type which is superposed.) This is represented in the classic diagram of Isacks and Molnar (1971) (Figure 7.6). It must therefore be admitted that the subducting plate, driven by convection, either enters a denser medium, or stops being driven by the surrounding mantle, which instead resists its dipping. To affirm that the longest subducted plates 'pull' more the upper part of the plate runs counter to the facts.

Several interesting cases are worthy of mention:

- The **Nazca Plate**, subducting under Peru, remains subhorizontal, close to contact with the lithosphere of the South American continent, over more than 500 km (which corresponds to 6 Ma of subduction). Earthquake sources then

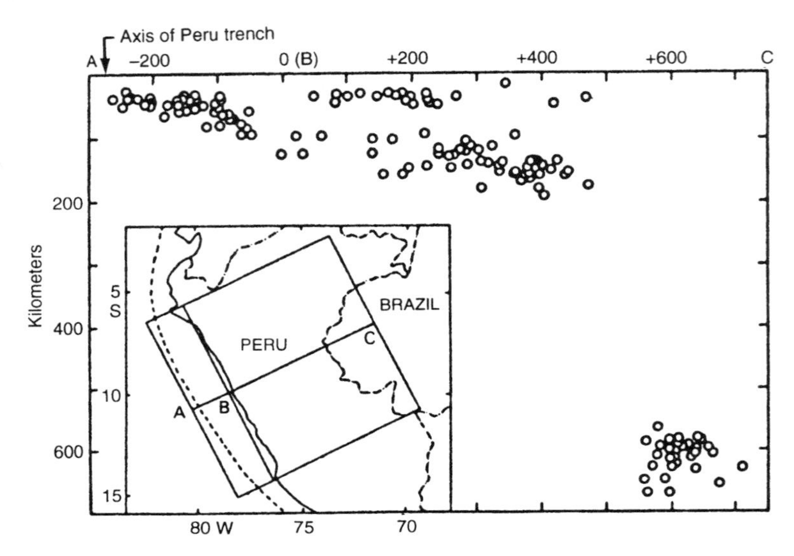

Figure 7.7. Earthquakes in Peru (cross-section). From Isacks and Molnar (1971); © AGU.

disappear. Some are found again 600 km deep, but no further horizontally, showing the dipping became subvertical (Figure 7.7). The absence of earthquakes between 200 and 500 km depths cannot be interpreted as a temporary halt in the subduction between −13 Ma and −6 Ma; there are no traces of such an event in the magnetic anomaly stripes of the Nazca plate. It must therefore be admitted that the Nazca plate crosses this zone without being submitted to strong deviatoric stresses.

- The **New Hebrides** subduction, toward the east of the Australian plate below the Fiji plateau is even more curious. This time, the folding of the subducting plate becoming subvertical is well marked with earthquake sources. There are none between 300 and 600 km deep. Further east, deep sources are evidence of a lithosphere 'floating' over the 660-km discontinuity. It seems to have been dragged along at the bottom of the upper mantle by a back-current, which will feed the East Pacific Rise. In this case and in the previous one, the occurrence of deep earthquakes in a lithosphere which is neither pulled, nor pushed, nor bent is an intriguing fact, which will be examined in Section 12.12.
- Subduction below the **Aleutian Islands** has been very well studied. Once the friction zone between the subducting Pacific plate and the overthrusting North American plate has been crossed, the Wadati–Benioff surface divides into two surfaces more or less parallel, approximately 30 km apart (Figure 7.8). The same happens under the Kuriles Islands and under Japan. The upper surface is in compression, along the dipping direction, and the lower surface is in extension. In the Kuriles arc, more precise localization of the sources show that the distance between the 2 surfaces decreases progressively from 50 to 25 km as the plate goes down (Kao and Chen, 1995).

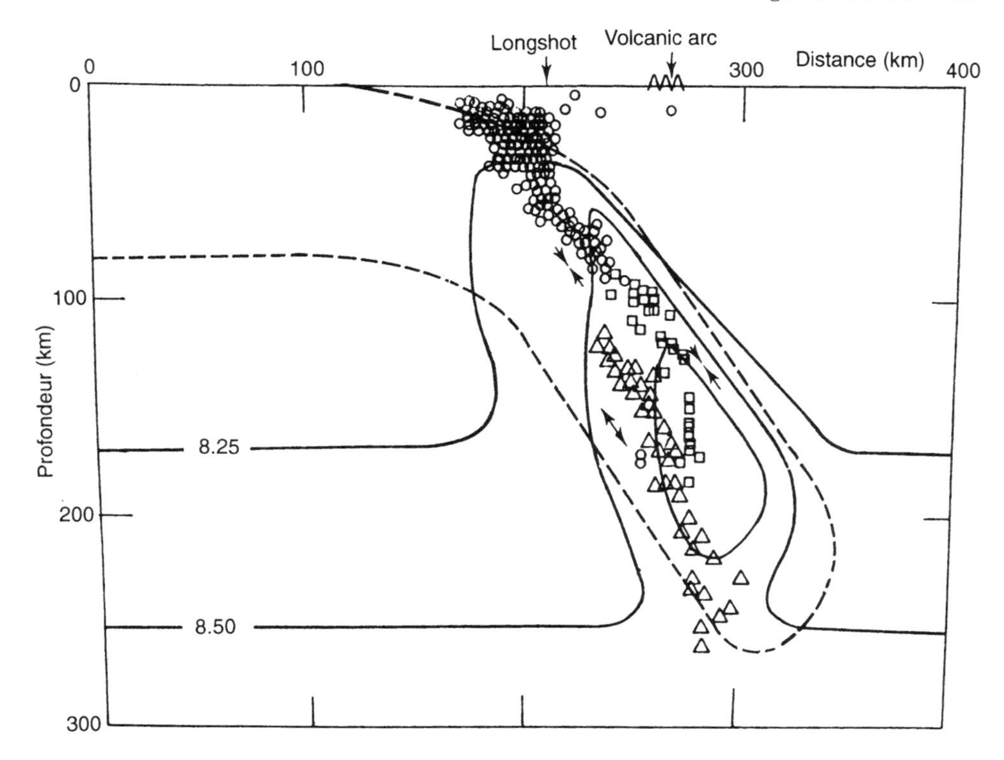

Figure 7.8. Earthquake sources below the Aleutian arc, projected on a vertical plane perpendicular to the arc. The contour levels correspond to the P velocities, according to a thermal model of Engdahl et al. (1977), and the dashed line shows the probable contour of the subducted plate. The triangles indicate extension in the direction of the plate, the squares indicate compression, and the circles another type of focal mechanism. The unbending of the plate is evident. Adapted from Engdahl and Scholtz (1977).

This division of the surface in two is easy to explain. The subducted plate must curve and bend to go down. As the lithosphere is not as elastic as rubber, this must be a permanent, plastic deformation. The plate would remain curled if there were no later unbending, induced by currents in the mantle. This time, the upper part of the plate is in compression, the lower part is in traction. The 5–25 km distance between the surfaces where earthquakes occur, because traction and compression are maxima, is the thickness of the *elastic* lithosphere. It thins out as the subducting plate warms up. Deeper still, the plate is unbent, but a compression of the whole plate remains and produces the earthquakes.

7.3 VOLCANISM AND GEOTHERMAL FLUX

Volcanism related to subduction seems, at first, easy to understand. The subducted oceanic crust melts partially, as it descends into a hotter mantle. The sediments of continental origin on the top produce andesitic magmas, very abundant around the Pacific Ocean, whereas the basalts produce basaltic magmas. But this simplistic explanation comes up against a difficulty: the heat source necessary to obtain magma is not evident. If it is the hot mantle of the LVZ below the plate, why did it wait for subduction to act? And the edge between the subducting plate and the overthrusting plate is isolated, and cannot indefinitely give up calories.

Between 30 and 60 km deep, the friction of plates (Sugimura and Uyeda, 1970) can dissipate some heat. But this does not seem enough to me, and the formation of magma persists well below 60 km. The main factor that was forgotten is *water*, whose presence decreases the solidus temperature. It is therefore possible to have simultaneous decrease of the temperature and, with the resulting calories, partial melting. The subducting oceanic crust is filled with water. In the case of the sedimentary layer, part of the water must be used for transition to the amphibolite facies. Amphiboles contain 2.2 to 2.7 weight per cent of water in their structure. They can carry it down to 80 km deep, where pressure decomposes them. The rock melts by dehydration of this mineral. This is not the only case. According to Poli and Schmitt (1995), lawsonite (with 11 weight per cent of water) and phengite (4%) can carry 1% of water down to 200 km deep.

Whereas the geothermal flux is low near the trench (less than 40 mW/m^2, which is the average value below the oceans), it is higher than 100 mW/m^2 in back-arc basins. This is caused by a local roll of *mechanical* convection, whose only engine is the drag by the subducting plate (Uyeda, 1977). Even if the asthenosphere upwelling velocity below the back-arc basin is low, it carries magmas from the subducted plate, and therefore latent heat. Heat was taken from the plate, and provided to the crust of the back-arc basin when magma solidified (whether there was intrusion of dykes or underplating). The dampening of seismic waves shows an extremely low Q in the whole area between the two plates, an indication of the presence of a liquid phase. This is particularly visible below the Lau back-arc basin, currently in extension (see Figure 7.5).

7.4 EXTENSION PHASES OF BACK-ARC BASINS

In the 1970s, the mechanism described above was thought to function permanently, during the whole duration of subduction. In the 1980s, numerical models were set up a good dozen times, as this mechanism is relatively easy to model, with 2 space dimensions and no time dimension (a 2-D problem, and not a 4-D one).

However, between 1976 and 1979, the age of 10 back-arc basins had been determined, and a synthesis was published by Jurdy (1979). The magnetic anomalies of these basins are very difficult to interpret, because of the very slow spreading rate, making possible an important hydrothermal alteration of lavas, and because of the short length of the spreading ridges. The map of Jurdy, reproduced in Figure 7.9, showed a few mistakes. For example, ODP drillings established that the opening of the Sea of Japan did not take place between 60 and Ma ago, but between 23 and 14 Ma ago. However, even if the identification of chrons is subject to errors,

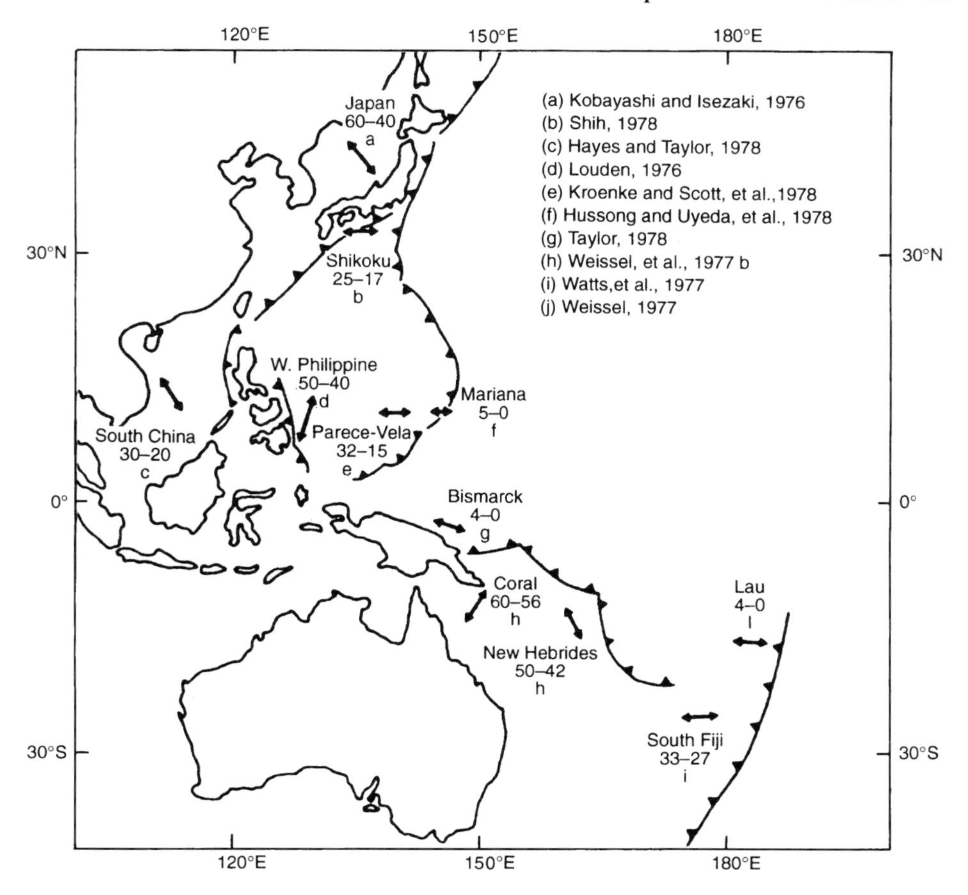

Figure 7.9. Opening of the Western Pacific marginal basins, established from magnetic anomalies. The numbers indicate the time of spreading, in Ma BP, and the arrows indicate the spreading direction. From Jurdy (1979); © AGU.

the single fact that magnetic anomaly stripes are in finite number is sufficient to show that spreading was not permanent. No one doubts this fact now. The chrons identified in two recent studies are shown in Figure 7.10.

Each back-arc basin opened in a limited time, ranging from 4 to 17 Ma, and therefore short compared to the duration of subduction. The opening periods of the basins can be grouped in 3 'extension phases':

- *A phase of early extension, from 60 Ma to 40 Ma BP.* Examples are the Coral Sea (60–56 Ma), and the Philippines Sea SW of the plate with the same name (50–40 Ma).
- *A phase of more recent extension, from 33 Ma to 14 Ma BP.* Examples are the Fiji Basin (33–27 Ma), the Parece Vela Basin (32–15 Ma) in the southern part of the Philippines plate, the Shikoku Basin in the northern part of this plate (25–

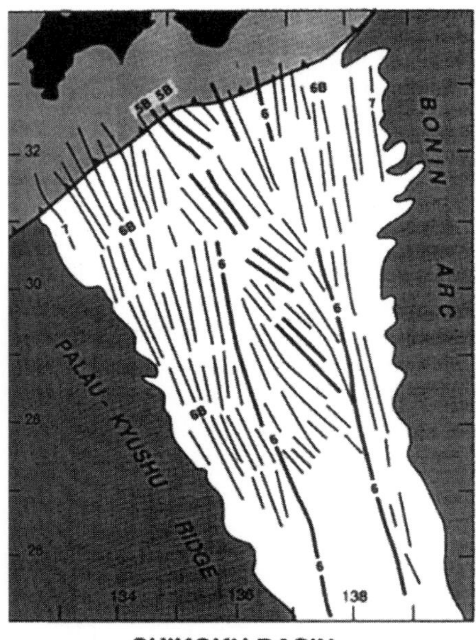

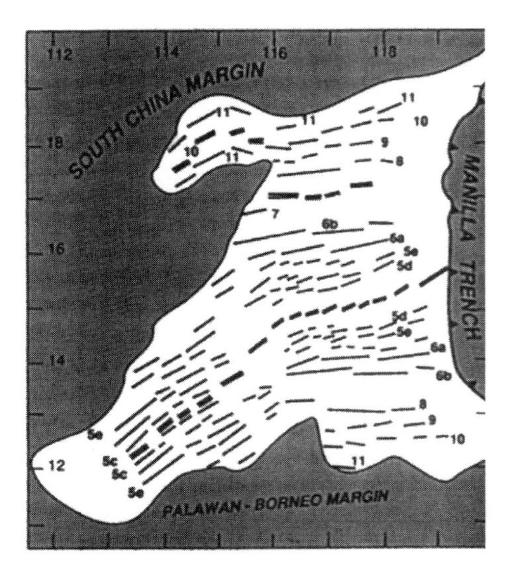

Figure 7.10. Magnetic anomaly stripes in two back-arc basins, with the corresponding chrons. From Jolivet et al. (1994); © AGU.

15 Ma), the South China Sea (30–16.5 Ma), the Japan Sea and the Kuriles Basin (23–14 Ma for both). Let us remember that, according to the first spatial geodesy results, Japan actually moves closer to Asia.

- *A present phase, since 5 Ma BP*. Examples are the basins of Lau (4–0 Ma), the Bismarck Islands (4–0 Ma), and the Mariannas (5–0 Ma).

These basin extensions during finite periods remain a mystery. Why does the spreading of the back-arc basin stop, while rapid subduction continues until now?

We shall come back to this later, but it should be noted that the many studies of the Mediterranean converge to show the opening of basins with similar durations. Western Mediterranean basins opened at a time synchronous with the second phase in the Pacific. In particular, Corsica and Sardinia separated from the Maures and the Esterel, and rotated anti-clockwise to occupy their present positions between −23 Ma and −19 Ma (see Sections 14.17 and 15.6 for more details).

7.5 ACCRETION PRISMS AND FORE-ARC TROUGHS

On the outer side, subduction is most often accompanied by the building of an *accretion prism*, formed by the piling up of all the sediments from the oceanic plate, which are not carried at depth. The overthrusting plate acts like a huge mud-scraper, pushing back all the unconsolidated sediments, and even some submarine volcanoes. This mechanism is called *off-scraping*.

Most of the sediments come from the neighbouring continent. This explains the absence of accretion prisms in the middle of the Pacific Ocean, in the Mariannas arc or in the Tonga–Kermadec arc. There are no accretion prisms either along Peru or northern Chile, but for a different reason. Not only are the sediments are dragged down by the subduction, but the edge of the overthrusting continent is eroded and dragged as well; this is *ablative subduction*.

The most developed young accretion prism is in front of the Sunda Islands. Offshore from Sumatra, it forms a string of islands (Simaulue, Nias, Mentawei) and extends as shoals off Java, Bali, Lombok and Sumbawa (the situation is more complicated in the east, and we will come back to it later). In Western Indonesia, we can find from the SW to the NE:

- The trench, which appeared with the subduction 28 Ma ago (Oligocene, in the middle of the Cenozoic).
- The accretion prism, with a high seismicity.
- An elongated basin, parallel to the trench, with a thick layer of recent sediments, neither folded nor perturbed: the Mentewei Trough.
- The Sunda Islands, made of folded Cenozoic rocks and Pleistocene volcanoes. Volcanism is intense and explosive. In 1883, the explosion of the Krakatau volcano, between Sumatra and Java, covered the sea with a layer of pumice so thick that a French ship in the vicinity could not approach the scene. The explosion of the Agung volcano, in Bali, in 1963, scattered ash throughout the atmosphere, which can be found in Antarctic ice cores.
- An old depression filled with sedimentary rocks, not very folded: the Burma Trench.
- The Shan plateau, folded in the Mesozoic, which extends to Indochina, Malaysia and the Gulf of Siam.

The presence of the trough is due to the fact that the trench did not appear exactly along the continent, but off shore, leaving a fragment of oceanic crust attached to the overthrusting plate. Seismic exploration revealed this oceanic crust below a thick layer of sediments.

Accretion prisms consist of a mixing of all sorts of sedimentary rocks, of both marine and continental origins, and volcanics. One example is an old accretion prism near San Francisco, dating back to the time where the Farallon plate was subducting below California. This terrain, called *Franciscan*, comprises (in complete disorder), the following:

- Dark sandstones with very angular grains and water cement, with volcanic elements (*greywackes*). Black schist layers are interspersed.
- Thin layers of red *cherts*, with abundant radiolarites.
- Pillow basalts, altered, green.

- Metamorphic rocks from the blue-schist facies, demonstrating some low-temperature burying in the subduction zone.

In other mixings, for example in Mentawei, it is possible to find ophiolitic flakes.

In the light of these studies, European geologists were able to understand the nature of the mixes, called *Wildflysch* in the Alps and *argille scagliose* in the Apennines. They are old accretion prisms.

7.6 ABLATIVE SUBDUCTION

Investigating the subduction of the Nazca plate below Peru, Rutland (1971) came to the conclusion that the edge of the continent had at several times been torn away and carried down by subduction. This explains the following facts:

1. Orogenesis, plutonism and volcanism have (roughly) migrated from west to east during the successive phases.
2. The Gondwanides, mountain chains formed in the Carboniferous, are cut by the coastline at an angle of 45°, which seems to indicate that the continent extended further west.
3. This coastline is everywhere a (Pliocene?) fault scarp, and the terrains at the surface are from the Jurassic or earlier.
4. The oceanic trench contains only young sediments, neither folded nor perturbed.
5. There is an intense calc-alkalic volcanism along the coast, which must be fed somehow.

This idea did not provoke much interest until it was formulated again in 1992, presented as brand-new and with the name of *ablative subduction*.

This year saw the study of the accretion prism at the junction of the South American plate (overthrusting) and the Nazca and Antarctic plates (subducted), near the Taitao peninsula. Drillings showed that, during successive periods lasting 2 to 3 Ma each, there was either accretion, or ablative subduction along ca. 40 km, or nothing at all. (In the Taitao Peninsula itself, ophiolites 3–5 Ma old are found.)

7.7 OPHIOLITE OVERTHRUST SHEETS

Ophiolites were noted as quite frequent in accretion prisms. At other places, the ophiolitic assemblage, a fragment of the oceanic crust, overlays the continental crust. This peculiar overthrust is termed *obduction*.

This term was introduced by Coleman, who proposed that an upper layer of the oceanic plate would split off when the plate bends and disappears below the continent (Figure 7.11a). Such a mechanism is difficult to explain, as the continental margin is always very fractured and fragile, and not likely to act as a knife peeling an orange.

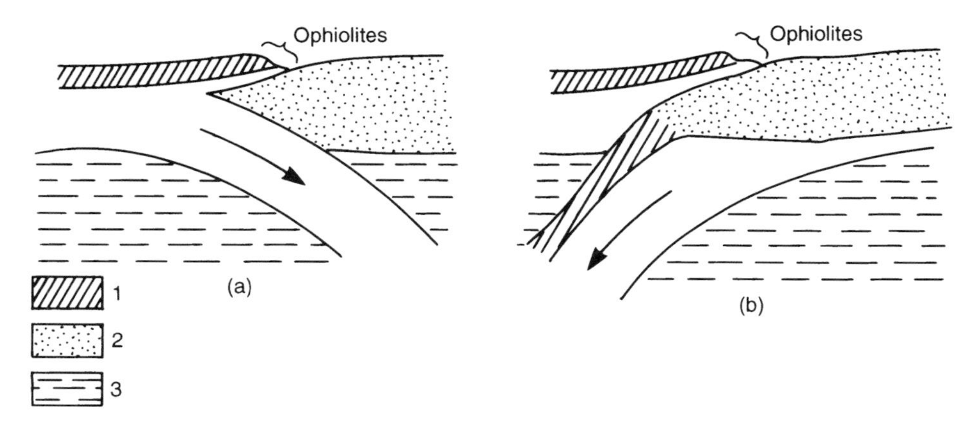

Figure 7.11. Two mechanisms are possible to explain the obduction of ophiolitic complexes: (a) splitting of the oceanic plate; (b) very limited subduction of the continental margin, more probable. 1: Oceanic crust; 2: continental crust, 3: asthenosphere.

According to Davies, the oceanic crust would not be overthrust, pushed by subduction on the edge of the continent, but the opposite would occur: underthrusting of the edge of the continent, under the oceanic crust. Even if continental crust is lighter than oceanic crust, a limited subduction remains possible, especially for a thinned crust. With this hypothesis, the direction of subduction must be reversed (Figure 7.11b). This is how Allègre (1980) explains the obduction of ophiolites in New Guinea and New Caledonia (Figure 7.12).

Ophiolite belts are visible in the middle of the present continents, and show an ocean was there, which has disappeared through subduction. The current consensus is that ophiolite belts are the 'scar' of a suture line between continents or continental blocks, which were separated. But their setting is not always as simple as in the simple cases of Figure 7.11.

The continental collision of India against Asia is the most stupendous of the last 100 Ma. The suture is outlined by an ophiolite belt along the Tibetan valleys of Indus and Zangbo (upper part of the Brahmaputra). North of the suture, the trans-Himalayan chain (where there is volcanism) and the high Tibetan plateau are reminiscent of the Andes and the high Bolivian plateau. It is therefore thought that the subduction was oriented northwards. Between the ophiolite belt and the Himalayas (more recent), there are sedimentary terrains from the Mesozoic of marine origin, deposited on the old Indian continental shelf, and Upper Cretaceous *Wildflysch*. This is impossible to explain with the mechanism of Figure 7.11a. The Asian continent would not have been able to push back and mix the sediments deposited on a passive continental margin. Further west, in Ladakh, this situation has been interpreted by assuming the existence of an island arc off the Asian continent. The mixings would be the old accretion prism, and the ophiolites further north would come from the oceanic crust of the old back-arc basin. But the problem remains for the north of the Himalayas.

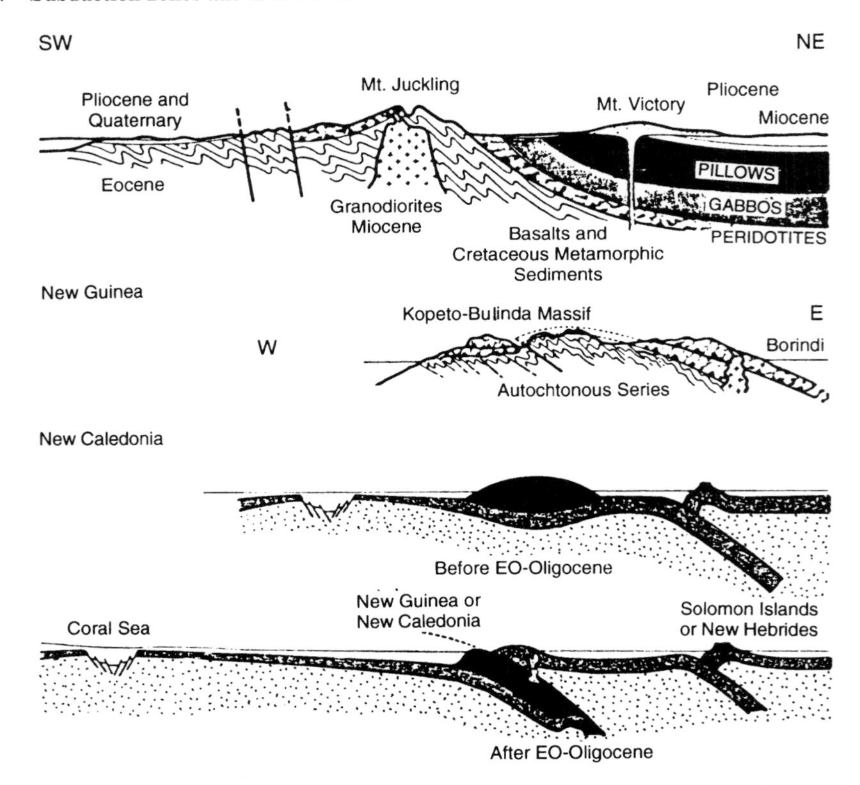

Figure 7.12. Two examples of ophiolitic complexes, in New Guinea (from Davies) and in New Caledonia (from Guillou). Bottom: possible geodynamic evolution. (From Allègre (1980); © Pour la Science SARL 1980.)

7.8 SUBDUCTION OF OCEANIC BACK-ARC BASINS BELOW CONTINENTAL ISLAND ARCS

The apparition of back-arc basins with oceanic crust separated Japan and the Philippines from Asia, but this stopped 14 Ma and 16.5 Ma ago, respectively. These basins are now closing, with subductions toward the East, under the islands with continental crust. Let us detail this, going from North to South.

Japan

A transversal zone of young sedimentary rocks, overlain in places by large volcanic edifices (e.g. Fuji-Yama), over oceanic crust, cuts the main island (Hondo) in the middle (west of Tokyo). This is the *Fossa Magna*, which separates Northern Japan, on the North American plate, from Southern Japan, on the Eurasian plate. The Eulerian pole of these two plates is in Eastern Siberia (between Yakutsk and Oimiakon, according to the NUVEL-1 model). The two plates separate along the Arctic ridge, which diminishes and disappears close to the mouth of the Lena.

Symmetrical to the Eulerian pole, the two plates get closer, and eastward subduction appears in the Tatar Strait (between the continent and the Sakhalin Islands), and in the Sea of Japan. This is marked by earthquake sources in a Wadati–Benioff plane, but there is no distinct trench.

Philippines

The Asian continent extends southward with shallow epicontinental seas (Gulf of Siam, Java Sea) down to Java, with the large island of Borneo on its eastern side. The Philippines are islands 1,000 to 1,500 km further East. They have an old pre-Jurassic continental crust, with many granodiorite or diorite plutons. As in Borneo, the mountains culminate at more than 2,000 m. Between the Philippines and the Asian continent, there are 3 deep oceanic basins (from North to South, deeper than 4,000 m to deeper than 5,000 m). These basins are the South China Sea, the Sulu Sea, and the Celebes (Sulawesi) Sea. Thin continental slices linking Borneo to the Philippines separate them: the long island of Palawan and the Sulu archipelago.

Relative to Asia, the Philippines plate moves WNW at approximately 90 km/Ma. The subductions enabling this are different, north and south of the Philippines. In the north, the China Sea is subducted eastward below Luzon Island, in the Manila Trench. The Wadati–Benioff plane goes as deep as 200 km. The subduction of the Philippine plate west of Luzon is barely visible. In the south, however, the subduction of the Philippine plate is noticeable down to 700 km deep as a Wadati–Benioff surface. This does not preclude the eastward subduction of the seafloor in the Sulu and Celebes seas, below the Southern Philippines and the extending Sangihe Rise. The associated Wadati–Benioff plane is visible down to 150 km deep.

Furthermore, it seems that there was in the past a N–S extension of the Celebes basin, which rotated from 90° the long peninsula of Northern Celebes, giving the island its crooked shape. South of the basin, there are now a trench and a subduction below Celebes.

Moluccas

The Molucca Sea extends between the Sangihe Rise and the Northern Moluccas, whose main island is Halmahera. It measures 400 km from north to south, and only 200 km from east to west. It barely reaches depths of 2,000 m, and it is divided in its middle by a rise emerging at the Talaud Islands. This is supposed to result from an *arching* of the plate, due to a strong transverse compression between two divergent weak subductions (see Section 14.5).

7.9 INITIATION OF SUBDUCTION

Subduction may appear in the middle of an ocean (e.g. the subduction of the Pacific plate below the Philippine plate), or, more frequently, close to a continent or continental fragment (e.g. the above case). This cannot be understood with the traditional lithospheric plates at least 60-km thick. A plate should sink by at least 60 km in a dense asthenosphere before being able to slide below the other plate. Even being cold and overloaded with sediments as it goes down, this contradicts the isostasy principle. Therefore, some scientists argued that subduction zones persist since the early formation of the terrestrial crust, without new ones forming, whereas Uyeda (1977) suggested that they form out of transform faults, where the plate reduces to a thin oceanic crust.

The overthrusting plate must end as a wedge for subduction to be possible. How did this wedge form? To detach a whole corner of the plate, the force should be larger than 80 MPa, as computed by McKenzie (1969).

It is therefore necessary to admit that the edge of the overthrusting plate is not completely rigid, a notion we saw in the study of passive margins with thinned, stretched crust. We shall also see, in Chapter 14, that the lower continental crust is quite ductile, more deformable than the mantle lithosphere. Would the deformation occur at this level, and would there be underthrusting of portions of the mantle lithosphere? The problem remains.

7.10 SUBDUCTION OF THE AUSTRALIAN CONTINENT BELOW THE BANDA SEA OCEANIC PLATE

To end this chapter about subduction, a particular and complex case will be examined: the Banda Sea, in Eastern Indonesia, between the Western Moluccas and the small Sunda Islands (Figure 7.13).

This sea is deeper than 4,000 m in most areas, 1,200-km long from east to west and 500-km wide from north to south. The oceanic plate below appears at first glance to be an extension of the Eurasian plate. Now, the Australian plate moves NNE with a velocity of 76 km/Ma relative to the Eurasian plate, and supports the Australian continent. If the emerged parts of Australia are still far, the Sahul continental shelf, which prolongs it, is reaching the vicinity of the small Sunda Islands. This is the reverse of the traditional situation found further west, where the Indian Ocean is subducted below a continent at the Java Trench. The Australian continent starts being subducted below an island arc before an oceanic basin.

The trench is therefore wide and shallow, so much so that it is more often called the Timor Trough than the Timor Trench. The accretion prisms are the uplifted islands of Sawu, Timor, Leti, Babar and Tanimbar. Below the mixings, one finds calcareous massifs, crystalline and eruptive, created up to the Upper Miocene (Lemoine, 1959). The trough is irregular, widening to 200 km in the small Sawu Sea. And the volcanic arc (islands of Flores, Alor, Wetar, etc.) emerges less and less, as it built over a deep seafloor (Figure 7.14).

Northeast of the Banda Sea, the situation is symmetrical, with a Wadati–Benioff plane dipping southward this time. The Seram trough/trench mirrors the Timor trough/trench, the islands Buru and Seram corresponding to the islands between Timor and Tanimbar. But the volcanic arc is reduced to the small islet of Banda.

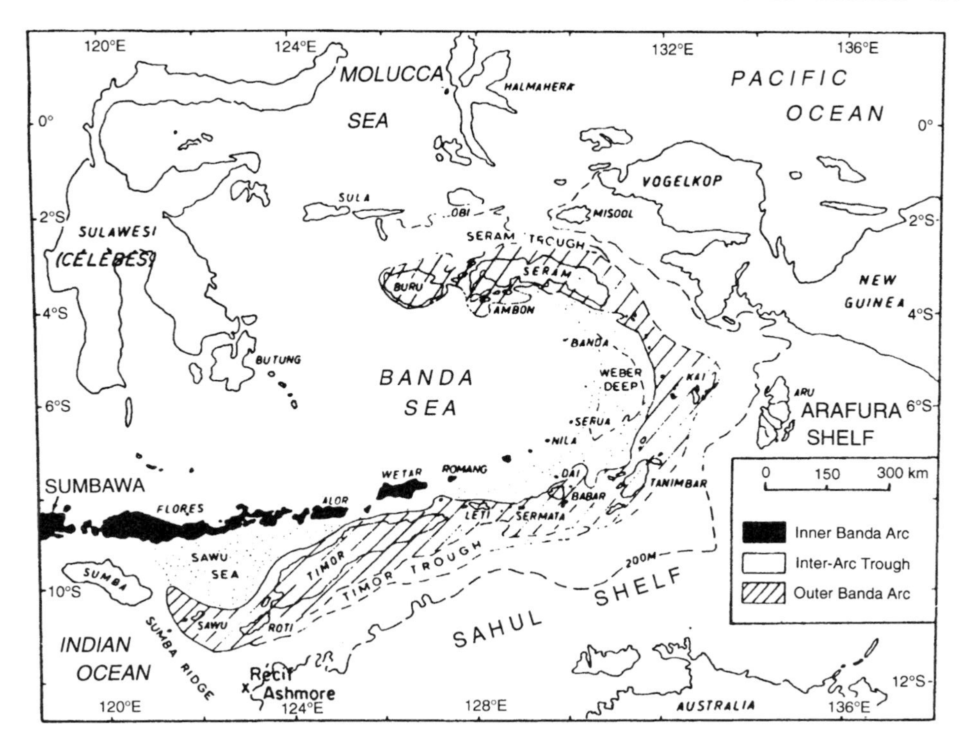

Figure 7.13. Map of the Banda arc (Eastern Indonesia). From Audrey-Charles *et al.* (1980); © BRGM.

A small continental block, formed by the western end of New Guinea replaces the Australian continent. But this block does not move south, relative to Eurasia. Consequently, the NNE movement of Australia is absorbed partly in the Timor subduction, partly in the Seram one. This implies that the Banda Sea is not bound to Eurasia, but moves NNE, more slowly than Australia (at 20 mm/a, according to recent GPS measurements). This is still not enough, as a small subduction in the south, under the island of Alor, starts to be visible (Figure 7.15).

Let us explain why the small continental block of the western tip of New Guinea has no velocity along the north–south direction. This block is made of the peninsulas of Doberai (called *Vogelkop*, bird's head, by Dutch cartographers) and Bomberai. It is assumed that, in the past, it rotated 90° clockwise, but we will only look at the current situation. An E–W sinistral fault separates this block from Australia, and another E–W sinistral fault (the Sorong fault) separates it from the Pacific plate. In this area, the Philippines and Pacific plates have nearly the same movement (their Eulerian pole is nearby), and they cannot be distinguished anymore. Relative to Eurasia, they move WNW with a velocity of 86–90 km/Ma. The Vogelkop continental block therefore moves WNW, with a speed approximately half smaller.

But this is not all. From the above description, the two subduction zones below the Banda Sea (S and NE) should join, making an acute angle. This is not the case, as

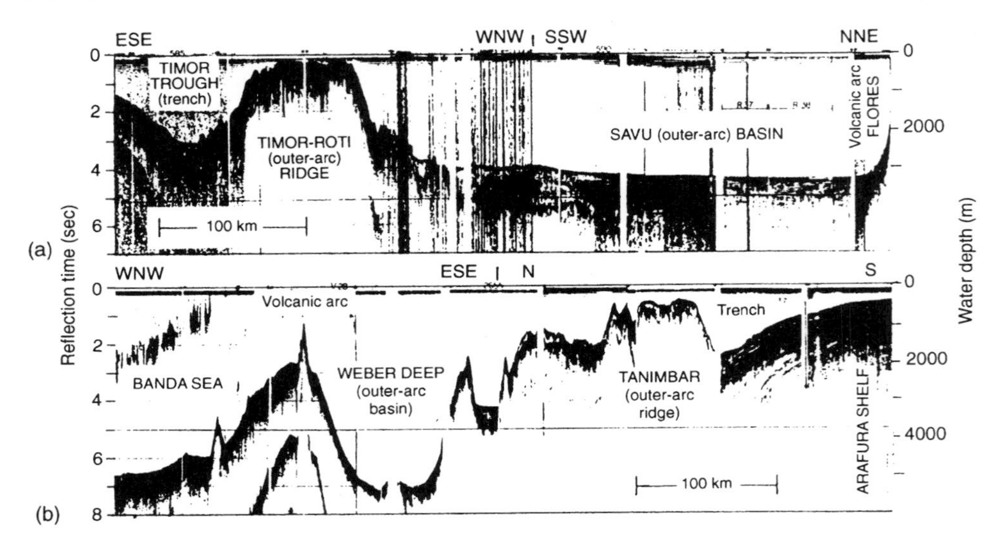

Figure 7.14. Reflection-seismics profiles across the Banda arc, collected by the Lamont–Doherty laboratory. (a) Timor region. Left, the Australian continental platform goes into the Timor trough, but the subduction there is stopped. There is no trough in the Savu basin. (b) Tanimbar region. Right, the Australian continental platform is subducted. There is no trench in the Weber Deep. From Hamilton (1977); © AGU.

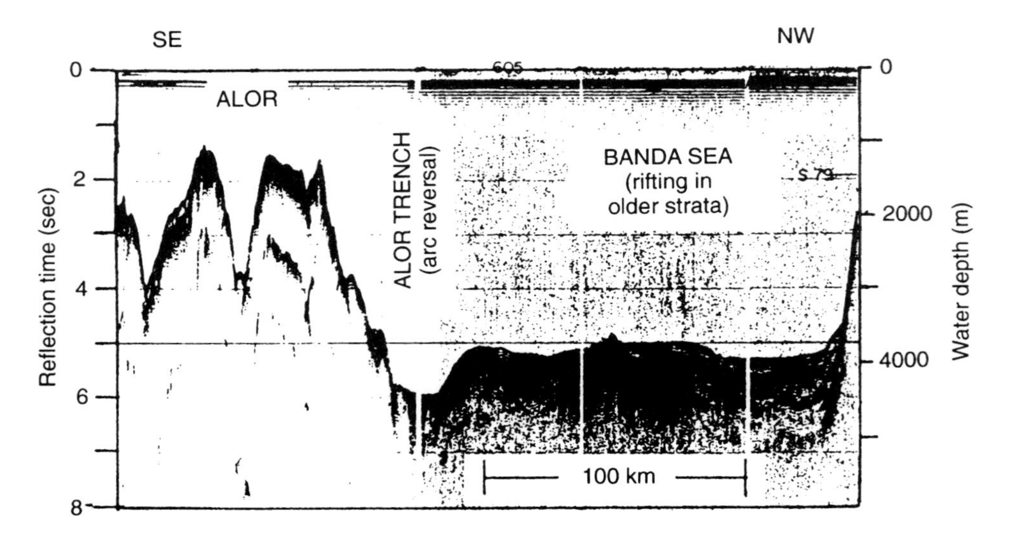

Figure 7.15. Reflection-seismics profile across the Banda arc, north of Timor. A subduction starts in the south, below Alor. From Hamilton (1977); © AGU.

they join as a superb half-circle. What should be the trough reaches down to 7,440 m, the normal depth of a large subduction trench (and therefore it is called neither trench, nor trough: this is the Weber Deep).

The only explanation I see is that the small plate of the Banda Sea is deforming. Compressed N–S, it stretches eastwards. It abuts against the Arafura continental platform, linking Australia and New Guinea, and must start being subducted below.

According to my analysis, the northward movement of the Australian plate is absorbed in the Banda Sea with three subductions and one global compression. This a particularly instructive example of a 'boundary area'.

8

Earth rotation, gravimetry, and isostasy

8.1 EARTH ROTATION AND THE CORIOLIS FORCE

In this chapter, the relative plate movements will be neglected. The Earth thus becomes a frame of reference defined unambiguously. In a Galilean system, linked to the stars, and for which the laws of mechanics are valid, the Earth rotates about an axis going through its gravity centre, the *world's axis*. One rotation takes a *sidereal day* of 86,164 seconds (and not 86,400 seconds, the average solar day; a year has 1 more sidereal day than the number of solar days). The angular rotation velocity (in rad/s) equals $\Omega = 2\pi/86{,}164$, and the distance to the rotation axis is $r\sin\theta$. In a Galilean frame of reference, any point fixed relative to the Earth has an axipetal acceleration (i.e. directed toward the world's axis following the shortest distance, which is the perpendicular), of value $\Omega^2(r\sin\theta)$.

Let V be a solid body, and Δ an axis going through its gravity centre, with direction cosines α, β, γ (Figure 8.1). If V rotates around Δ with the angular velocity ω, any point of V with the mass dm has a momentum equal to $(MH)\,\omega\,dm$. The moment of this momentum about Δ (the *angular momentum*) is a vector along Δ, of intensity $(MH)^2\omega\,dm$. The rotational kinetic energy is $\frac{1}{2}(MH)^2\,\omega^2\,dm$.

These quantities are additive, and V has an angular momentum $I_\Delta\omega$ and a rotational kinetic energy equal to $I_\Delta\omega^2/2$, with:

$$\begin{cases} I_\Delta = \iiint_V (MH)^2\,dm = I_x\alpha^2 + I_y\beta^2 + I_z\gamma^2 - 2P_{yz}\beta\gamma - 2P_{zx}\gamma\alpha - 2P_{xy}\alpha\beta \\ I_x = \iiint_V (y^2+z^2)\,dm, \text{ etc.,} \qquad P_{yz} = \iiint_V yz\,dm, \text{ etc.} \end{cases} \tag{8.1}$$

It may be demonstrated that the coordinate axes can always be chosen so that P_{yz}, P_{zx} and P_{xy} equal 0. The axes x, y and z are then the *principal inertia axes* of V, and $I_x = A, I_y = B$, and $I_z = C$ are the *principal inertia moments* of V.

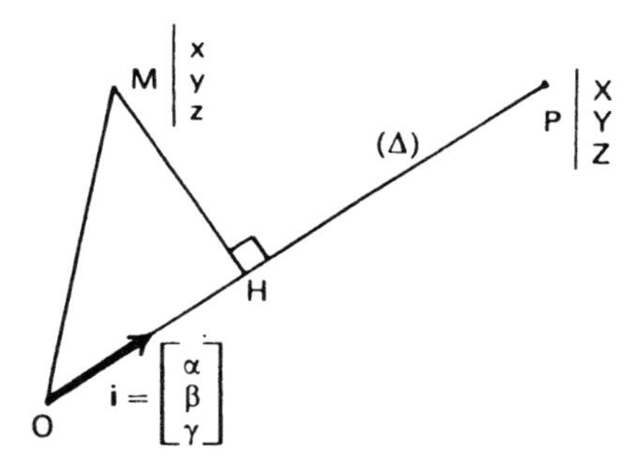

Figure 8.1. Calculation of the kinetic moment and the kinetic energy for rotation.

It can also be demonstrated that, in the absence of outside forces, the axis of rotation of the solid body is stable when very close to the principal axis corresponding to the highest moment of inertia (Lliboutry, 1992, pages 93–100).

This is indeed the case with the Earth. It is more or less an ellipsoid, slightly flattened at the poles. Its moment of inertia about the axis of the poles, i.e. C, is 0.5% larger than the 2 other moments of inertia, about the equatorial diameters. (Because of the axial symmetry, any equatorial diameter may be considered as a principal axis, and $A = B$). More precisely, $C - A = 2.63 \times 10^{35}\,\mathrm{kg\,m^2}$, whereas $A = B = 540 \times 10^{35}\,\mathrm{kg\,m^2}$, approximately. In Chapter 17, the variations with time of the orientation of the principal inertia moment, relative to the Earth (*polar drift*) will be examined.

The rotation of the Earth must be accounted for when applying the laws of mechanics with speeds and accelerations relative to the Earth. Let $\boldsymbol{i}, \boldsymbol{j}, \boldsymbol{k}$ be the unit vectors along the 3 coordinate axes x, y, z linked to the Earth. In a Galilean system of reference, and noting $\boldsymbol{\Omega}$ the terrestrial rotation vector, we have:

$$\left\{\begin{array}{l} \dfrac{d\mathbf{i}}{dt} = \boldsymbol{\Omega} \wedge \mathbf{i} \\ \dfrac{d\mathbf{j}}{dt} = \boldsymbol{\Omega} \wedge \mathbf{j} \\ \dfrac{d\mathbf{k}}{dt} = \boldsymbol{\Omega} \wedge \mathbf{k} \end{array}\right. \tag{8.2}$$

Let M be a point with the coordinates x, y, z:

$$\mathbf{OM} = x\mathbf{i} + y\mathbf{j} + z\mathbf{k} \tag{8.3}$$

A terrestrial observer, considering $\boldsymbol{i}, \boldsymbol{j}, \boldsymbol{k}$ as fixed, will measure the relative speed of M:

$$\mathbf{v_r} = \frac{dx}{dt}\mathbf{i} + \frac{dy}{dt}\mathbf{j} + \frac{dz}{dt}\mathbf{k} \tag{8.4}$$

In a Galilean system, the absolute speed of M is:

$$\mathbf{v}_a = \frac{d\mathbf{OM}}{dt} = \frac{dx}{dt}\mathbf{i} + \frac{dy}{dt}\mathbf{j} + \frac{dz}{dt}\mathbf{k} + x\frac{d\mathbf{i}}{dt} + y\frac{d\mathbf{j}}{dt} + z\frac{d\mathbf{k}}{dt}$$
$$= \mathbf{v}_r + \Omega \wedge \mathbf{OM} \tag{8.5}$$

This is the sum of the relative speed and of the speed of the reference frame $\boldsymbol{v}_e = \Omega \wedge \mathbf{OM}$. Its time derivation provides the absolute acceleration:

$$\gamma_a = \frac{d\mathbf{v}_a}{dt} = \left[\frac{d^2x}{dt^2}\mathbf{i} + \frac{d^2y}{dt^2}\mathbf{j} + \frac{d^2z}{dt^2}\mathbf{k}\right] + \left[x\frac{d^2\mathbf{i}}{dt^2} + y\frac{d^2\mathbf{j}}{dt^2} + z\frac{d^2\mathbf{k}}{dt^2}\right]$$
$$+ 2\left[\frac{dx}{dt}\frac{d\mathbf{i}}{dt} + \frac{dy}{dt}\frac{d\mathbf{j}}{dt} + \frac{dz}{dt}\frac{d\mathbf{k}}{dt}\right]$$
$$= \gamma_r + \gamma_e + 2\Omega \wedge \mathbf{v}_r \tag{8.6}$$

A mass m is subjected to the gravity forces, resulting in $m\mathbf{g}_a$. Its acceleration is $\gamma_a = \mathbf{g}_a$, as the same mass intervenes in the gravitation and in the law of inertia (all this is valid in non-relativist mechanics; for speeds close to the speed of light, the law of inertia is no longer valid, nor the addition of speeds as we wrote it). In a terrestrial system of reference, one must write:

$$\begin{cases} m\gamma_r = m(\mathbf{g}_a - \gamma_e) + \mathbf{F}_C \\ \mathbf{F}_C = -2m\Omega \wedge \mathbf{v}_r \end{cases} \tag{8.7}$$

γ_e must therefore be subtracted from the gravitational field, and a force $\mathbf{F}_C$ must be added to the acting forces. $\mathbf{F}_C$ is the Coriolis force, and depends on the relative speed.

At point M, $\boldsymbol{\Omega}$ can be split up in one component along the meridian $\Omega \cos \lambda$, and another component along the horizontal, $\Omega \sin \lambda$. If $\mathbf{v}_r$ is horizontal, the former produces a vertical Coriolis force, upwards if $\mathbf{v}_r$ is directed eastward, and the latter produces a horizontal Coriolis force, directed to the right relative to $\mathbf{v}_r$ in the northern hemisphere and to the left in the southern hemisphere.

The Coriolis force is essential in the dynamics of the atmosphere and the oceans. For example, around a barometric depression, there exists at high altitude a *geostrophic wind*, anti-clockwise in the northern hemisphere. Its speed is such that the Coriolis force balances the barometric gradient (which requires considerable speeds in the tropical regions where $\sin \lambda$ is small). The geodynamic processes with which we are dealing have relative speeds smaller than 10^{-8} m/s, and the Coriolis forces are therefore completely negligible.

8.2 NEWTONIAN POTENTIAL IN A GALILEAN FRAME OF REFERENCE

The movements of artificial satellites are analysed in a Galilean frame. The acting forces will therefore only be the forces related to Newtonian attraction (and the very small drag from the atmosphere, if they are on low orbits). The attraction from planets other than the Earth is negligible, the attractions from the Moon and the Sun are easily calculated, and the forces resulting from tide-induced modifications can be computed from the empirical formulae which predict the tides. There remains the gravity field $\mathbf{g}_a$ in the Galilean 'absolute' frame. It is deduced from a potential U, the *Newtonian geopotential*:

$$g_a = \mathbf{grad}\, U; \qquad U = G \iiint_{Earth} \frac{dm}{r} \tag{8.8}$$

The summation is extended to all Earth masses; r is their respective distance to the point considered, G is the gravitation constant:

$$G = (6.673 \pm 0.003) \times 10^{-11}\,\mathrm{N\,m^2\,kg^{-2}} \tag{8.9}$$

As the attraction law is in r^{-2}, the Gauss theorem is applicable (see Annex A.1). The flow of $\mathbf{g}_a$ coming out of S, closed, equals $-4\pi G$ times the sum of the masses inside S.

U can be represented as a series of spherical harmonics, whose coefficients are functions of r. This corresponds to a result already seen when representing the geomagnetic potential (Section 4.6). We use the *geocentric latitude*, defined in Figure 8.2. The following equation is written for the Newtonian geopotential, M being the mass of the Earth, a its mean equatorial radius (defined more precisely later in the text), θ the geocentric co-latitude, ϕ the longitude, $P_{nm}(\cos\theta)$ the associated Legendre polynomials and functions:

$$U = \frac{GM}{r}\left[\begin{array}{l} 1 - \displaystyle\sum_{n=2}^{\infty}\left(\frac{a}{r}\right)^n J_n P_{n0} \\ + \displaystyle\sum_{n=2}^{\infty}\sum_{m=1}^{n}\left(\frac{a}{r}\right)^n P_{nm}(C_{nm}\cos m\phi + S_{nm}\sin m\phi) \end{array}\right] \tag{8.10}$$

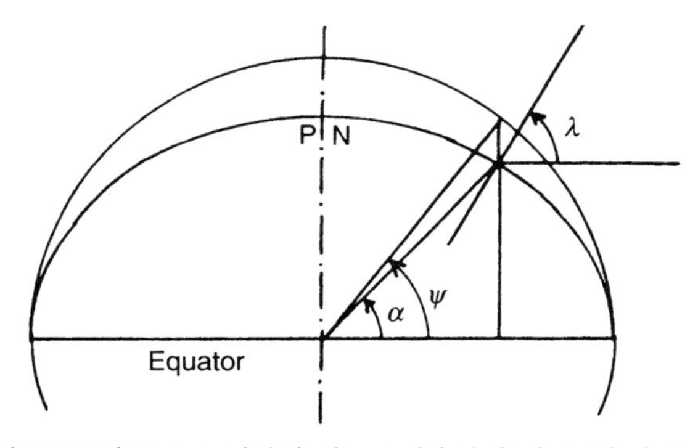

Figure 8.2. Difference between the geocentric latitude α and the latitude λ. The flattening of the Earth has been exaggerated 60 times.

US scientists use a+ instead of a− in the first sum, and use fully normalized P_{nm}, which I note $\bar{P}_{nm}$. They verify:

$$\iint_{Sphere} \bar{P}_{nm}^2 \, dS = 4\pi; \qquad \bar{P}_{nm} = \sqrt{4\pi} \, N_{nm} P_{nm} \tag{8.11}$$

Thus, $\bar{P}_{20} = \sqrt{5}\, P_{20}$. The corresponding coefficient is $\bar{J}_2 = -J_2/\sqrt{5}$.

The main term of the equation, Gm/r, is the geopotential in the *Kepler approximation*, assuming that the Earth is perfectly spherical. $\mathbf{g}_a$ must then be everywhere pointing toward the centre of the Earth, and its intensity must only depend on the distance r to the centre. Applying the Gauss theorem to a sphere of radius r encompassing the Earth, one finds:

$$g_a = -\frac{GM}{r^2} \tag{8.12}$$

as if all the mass of the Earth was concentrated at its centre.

The value of G can only be measured from the minute attraction between two masses, in a laboratory, and therefore it is only known with a precision of 1 : 2000 Conversely, the value of GM is perfectly established, with the extreme accuracy available from the observation of artificial satellites:

$$GM = 3.986004405 \times 10^{14} \, \mathrm{m}^3 \, \mathrm{s}^{-2} \tag{8.13}$$

This yields the mass of the Earth ($M = 5.973 \times 10^{24}$ kg) and its mean density (5.515), which is only reached at the base of the lower mantle. The mean crustal density is 2.7, the density of the upper mantle is 3.2, but the outer liquid core has a density of around 10, and the inner core of approximately 14. The intensity of the gravitation therefore increases by 10% from the surface to the mantle–core boundary, and then decreases rapidly to 0 at the centre of the Earth. If the density had been uniform, it would have decreased linearly from the surface to the centre of the Earth.

Putting the origin at the centre of gravity, it is possible to demonstrate that the expression of U does not contain any first-degree harmonics. The following term:

$$-\frac{Gma^2}{r^3} J_2 \left(\tfrac{3}{2}\cos^2\theta - \tfrac{1}{2}\right) \tag{8.14}$$

is negative, and quite large because of the flattening of the Earth. More precisely, it can be demonstrated that it depends on the difference between the principal inertia moments:

$$J_2 = \frac{C - A}{Ma^2} = 1.08263 \times 10^{-3} \tag{8.15}$$

The other coefficients J_n, C_{nm} and S_{nm} are of the order of 10^{-6}. The current status of their determination will be explained later in the text.

8.3 GRAVITY POTENTIAL AND GEOID

The gravity measured in gravimetry, in a terrestrial frame of reference, is $\mathbf{g} = \mathbf{g}_a - \gamma_e$. It is possible to consider $-\gamma_e$, which goes away from the rotation axis, as the gradient of an *axifugal* potential:

$$\mathbf{g} = \mathbf{grad}\, U_r; \qquad U_r = U + \frac{\Omega^2}{2}(r \sin\theta)^2 \tag{8.16}$$

By definition, the vertical is the direction of $\mathbf{g}$. The surfaces U_r = constant are always perpendicular to the gradient, and are therefore horizontal. The one coinciding with the average sea level is the *geoid*. It is defined if the altitude N of the geoid is defined everywhere above a *reference ellipsoid*, chosen so as to minimize the quadratic mean of N. There was an international agreement in 1980 to use a flattened ellipsoid with the following dimensions:

$$\left| \begin{array}{l} a = b = 6,378,137m \\ \textit{flattening}: \left(\dfrac{a-c}{a}\right) = \dfrac{1}{298.257} \implies c = 6,356,752m \end{array} \right. \tag{8.17}$$

The mean quadratic value of N is thus reduced to ca. 30 m. From these values, it is possible to calculate the quarter meridian of the reference ellipsoid. It was meant to be 10^7 m when the metre was selected, and it is in fact 1.000205×10^7 m. The radius of the sphere with the same volume as the reference ellipsoid is $(a\ b\ c)^{1/3} = 6,371,001$ m.

The geocentric latitude $\alpha = 90^\circ - \theta$ differs from the usual latitude λ of maps, which is the complementary angle of the vertical and the Earth's axis. The equation of the meridian of the ellipsoid being $x = a\cos\psi$, $z = c\sin\psi$, it follows that:

$$\tan\alpha = \frac{c}{a}\tan\psi = \frac{c^2}{a^2}\tan\lambda \tag{8.18}$$

Writing $\lambda - \alpha = \Delta\lambda$ (in radians), we get:

$$\begin{cases} \tan(\lambda - \Delta\lambda) = \tan\lambda - \dfrac{\Delta\lambda}{\cos^2\lambda} = \dfrac{c^2}{a^2}\tan\lambda \\ \Delta\lambda = \dfrac{a^2 - c^2}{2a^2}\sin 2\lambda \text{ (in radians)} \\ = 23.014 \sin 2\lambda \text{ (in sexagesimal minutes)} \end{cases} \tag{8.19}$$

According to a theorem by Stokes, an ellipsoid with the dimensions of the reference ellipsoid, a mass M (and therefore a Newtonian potential U tending toward GM/r^2 infinitely far) and a surface where U is constant, shows at its surface a gravity γ, perfectly defined whatever the inner mass distribution. γ can be calculated if one finds a mass distribution, however unrealistic, which makes the ellipsoid an equipotential. Somigliana found the following distribution: a uniform density inside the surface, and a negative surface mass. The gravity thus obtained will be used as a reference, and is expressed as:

$$\begin{aligned}\gamma &= 978{,}032.7 + 5{,}278.9 \sin^2 \lambda + 23.5 \sin^4 \lambda \\ &= 980{,}681 - 2{,}651 \cos 2\lambda + 3 \cos 4\lambda \text{ (in mgal)}\end{aligned} \tag{8.20}$$

($1\text{ mgal} = 10^{-5}\text{ m.s}^{-2}$).

Gravity increases with latitude, for 2/3 because of the decrease in the vertical component of the centrifugal force, and for 1/3 because of the decrease in r.

After correcting to the altitude of the reference ellipsoid, the difference $(g - \gamma)$ is the *free-air gravity anomaly*. It would be almost the same if the reference level was the geoid, but its calculation is immediate with the reference ellipsoid (see Section 8.6). This difference can be represented by a double series of spherical harmonics g_{nm}. The same can be performed with the altitude Z of the geoid relative to the ellipsoid. Let Z_{nm} be the spherical harmonics of this *geoid undulation*. It can be demonstrated that:

$$\frac{g_{nm}}{g_0} = (n - 1)\frac{Z_{nm}}{a} \tag{8.21}$$

g_0 is the approximate value of the gravity $g_0 = GM/a^2 = 979{,}835\text{ mgal}$.

When n increases, which corresponds to shorter and shorter wavelengths, the Z_{nm} decrease faster than the g_{nm}. *The geoid undulations are a smoothed expression of free-air anomalies*. The latter are interesting for underground exploration, but the geoid undulations provide insights about geodynamical processes at the regional scale and even, keeping only the harmonics of degree less than 7, about general mantle convection.

8.4 GEOID DETERMINATION

A 'model' of geoid was published in 1966 with only the harmonics of degree up to 7. It had been obtained mainly with measurements from the American military geodesy satellite Geosat. Geosat data and the improvements on the model were then classified. The Cold War was still on, and the data might be used for the guiding of ballistic missiles. In 1994, Geosat data north of latitude 30°S were still confidential.

The coefficients of spherical harmonics were determined up to the 50th degree, with an accuracy of 2×10^{-9}, through the continuous work of 3 civilian groups:

- The NASA Goddard Space Flight Center (GSFC). In 1990, it published the model GEM-T2, obtained with 31 artificial satellites (and using 4 techniques other than laser telemetry). In 1994, the same group was publishing the model GEM-T3, obtained with radar altimetry measurements over the oceans, from the satellites Geos-3, Seasat and Geosat, as well as ground gravimetry measurements. Finally, the model GEM-3A adds data from the DORIS system on board Spot-2, which provides better coverage of some regions (Nerem *et al.*, 1994; Lerch *et al.*, 1994).

- The CNES *Groupe de Recherches en Géodésie Spatiale* (GRGS), in Toulouse (France), associated with the *Deutsches Geodätisches Forschungsinstitut* (DGFI). The last model published is called GRIM–4C2.
- The space research centre of Texas University, which published the model TEG–2.

These last two models use observations from satellites, from DORIS, from satellite altimetry and from ground measurements. Their precision is comparable to the GEM-3A model, and the difference with it is smaller than the inaccuracy.

To cater for the needs of the GPS system, the Department of Geodetic Sciences and Surveying from the University of Ohio established the model OSU–91A, from the model GEM–T2 and from gravity data from measurements or models of the underground. This model provides the spherical harmonics up to degree and order 360. It reaches geoid undulations with a half-wavelength of 50 km, but its accuracy varies around the Earth.

A third stage in international trust and collaboration has been reached with the joint exploration by GSFC, GRGS, and DGFI of altimetric data from the oceanographic satellite Topex-Poseidon (a joint NASA/ESA programme). The orbit of this satellite is known with a precision of 2 cm, and the geoid can be determined over the oceans with a precision of 10 cm. The model OSU–91A was improved by determining precisely the 5037 coefficients of spherical harmonics up to degree 70 (included): this became the model JGM–2. With this model, the inaccuracy over g is 1–3 mgal over the oceans, and 4–8 mgal over the continents. The short-term goal is the correct determination of all 130,317 coefficients up to degree 360.

The best map of free-air anomalies published to data uses the model JGM–2 for the first 70 degrees, OSU–91A for the others (Nerem *et al.*, 1994). The following facts can be noted:

- *Subduction zones*. In all subduction zones, there is a positive anomaly of around +10 mgal over the outside arching, a large negative anomaly around −80 mgal at the trench, and +50–80 mgal spread over the island arc and the back-arc basin, or in the adjacent mountains (Andes, Himalayas).
- *Hot spots*. Hot spots are characterized by a moderate positive anomaly. But it extends over a vast area in Iceland, in the South Pacific, and mostly in Kerguelen.
- *Mid-ocean ridges*. The Northern Mid-Atlantic Ridge and the Iceland hot spot create a very large and strong positive anomaly. The Southern Mid-Atlantic Ridge and the Indian Ocean ridge create a narrow positive anomaly, curiously mirrored west by a narrow negative anomaly. This leads to believe that their feeding is not symmetrical. *The East-Pacific Rise does not appear at all in the map of anomalies*, even if it is the most active spreading ridge on Earth.
- *Passive margins*. Passive continental margins are underlined by narrow anomalies, generally positive, sometimes negative (e.g. along the southern coast of Australia, where it is exceptionally wide).
- *Regions covered with ice 10,000 years ago*. A negative anomaly over Eastern

Canada and the Baltic Sea is a sign that the isostatic bouncing after the melting of ice has not ended yet.

- *Unexplained negative anomalies*. Finally, there exist large and important negative anomalies of unknown origins: in the Atlantic near the Bahamas and off Guyana; in all Eastern Mediterranean (contrasting with a strong positive anomaly in Turkey); in the centre of Zaire; in the Xinjiang Basin (Chinese Turkestan); in SE India.

8.5 ALTITUDE OF A PLACE

The *orthometric altitude* of a place is its elevation above the geoid. It is provided by the non-debased GPS system with accuracies of a few centimetres. But GPS only uses an imperfect model of the geoid (OSU–89B, and later OSU–91A) smoothed at a 50-km scale. It should therefore be mentioned 'GPS altitude', and the date of the measurement should also be given.

Conversely, the *ellipsoidal geometric altitude* is the distance to the reference ellipsoid. Here again the reference ellipsoid changed with time. Each national institute had its own. The determination of geometric altitudes was the aim, and they still figure on maps. For example, Mount Everest is 8,887 m high according to geodetic measurements from the ground, in 1954. But according to measurements of 1993, its GPS altitude is 8,846 m.

To determine geometric altitudes with ground measurements, there were two problems:

- The choice of altitude zero. In France, this was selected as the mean level of the Mediterranean at the tidal gauge of Marseilles. Before spatial geodesy, the reference ellipsoid was always passing through this point.
- The geodetic levelling between two points A and B consists in measuring a high number of successive differences in altitude ($h_1, h_2, \ldots, h_n$) between levelling rods. If their sum is taken as the difference in altitude between A and B, it depends on the path chosen. The work of gravity between A and B is independent of the path chosen; $\Delta U_r = g_1 h_1 + g_2 h_2 + \cdots + g_n h_n$ (g_i are the successive values of the gravity). Therefore, the difference in the *geopotential altitude* $\Delta U_r / g_0$ was selected, the value of g_0 being fixed once and for all.

8.6 GRAVIMETRY

Absolute gravimetry

For a long time, the absolute measurement of g with a very high accuracy was only achievable in a few metrology institutes such as the *Bureau des Poids et Mesures* in France. The free fall of a body, or even better the ascent and descent of a body thrown upwards, are measured by interferometry. This technique yields measurements of g at a few microgals.

Portable absolute gravimeters are now available. An International Absolute Gravity Base-stations Network (IABGN) is now being set up. The goal is to measure g with an accuracy better than 0.01 mgal in 36 stations scattered around the Earth. They will be used as starting points for the relative regional measurements, and will prove useful to measure the variations of g with time, mainly due to variations in meteorological and oceanic circulation.

Relative gravimetry

Measuring the difference in g between two points is much easier. This is made with *gravimeters*, sorts of extremely sensitive spring scales. The best have accuracies of 0.01 mgal. An excellent levelling is required, as a difference of 1 meter is enough to induce a variation of 0.30 mgal. It is also necessary to correct for the tides, as g varies by ± 0.16 mgal along the day, and the Earth's tide changes the altitude. Minute variations of the springs, lengthening with time, can never be entirely corrected. To reach the best precision, it is necessary to come back every 2 hours to a fixed point, in order to compute this *gravimeter drift* and correct it. Gravimeters of a same model may be of different qualities, and they improve with age (like violins).

Gravimetric exploration can be performed from planes or helicopters. The vertical acceleration of the aircraft can be simultaneously measured with laser altimetry, or by the phase of the 19-cm carrier electromagnetic wave of the GPS system. The final precision reaches 2 mgal, but, as measures need to be smoothed over 2.5 minutes, the resolution is only 10–20 km.

The errors induced by accelerations of the aircraft are suppressed with the measure of the gradient of g. Called *gradiometry*, this technique was developed by the US Army. It does not hold its promises, and therefore the European Space Agency never adopted the successive projects of spaceborne gradiometers.

Bouguer anomaly

Gravimetric exploration consists in mapping the local anomalies of g, and in interpreting them by comparing with the anomalies produced by models of the underground which are geologically plausible. Its small cost means it is used in oil exploration before any seismic exploration (the exploration drillings, very costly, are only used as a last resort).

The anomaly used is the Bouguer anomaly, the difference between the measured value of g and its 'normal' value (considering the altitude and topography). It is obtained by starting from γ (function of the latitude), and performing three corrections:

- *The free-air correction.* The 'free-air' gravity is obtained by no accounting for the Newtonian attraction by ground above the reference ellipsoid. If the

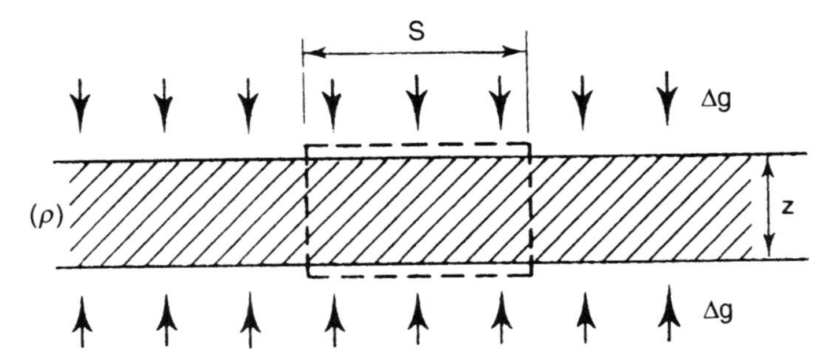

Figure 8.3. Plateau correction, as used in gravimetry.

ellipsoidal altitude is z and the terrestrial radius is R, measuring γ in milligals and z in metres, we have:

$$\gamma\left(\frac{R}{R+z}\right)^2 = \gamma\left(1 - \frac{2z}{R}\right) = \gamma - 0.308z \tag{8.22}$$

- *The correction from the attraction by the plateau.* The attraction from terrains above the ellipsoid can be easily calculated if they form a horizontal plateau, very wide compared to its thickness, and with a uniform density ρ. Close to the plateau and far from its edges, the attraction must be vertical and uniform, and of opposite direction depending whether we consider the neighbourhood above or below the plateau (Figure 8.3). The Gauss theorem gives its value:

  $$2\pi \times G\rho z = 0.0419\, dz \tag{8.23}$$

  where d is the density relative to water: $d = \rho/(1,000\,\text{kg/m}^3)$. This correction is proportional to z, like the previous one, and the two corrections are performed together. This is the *plateau correction.*
- *The topographic (or terrain) correction,* when the ground is not flat and horizontal, but presents strong slopes. It can be a mountain rising above a plateau or a valley below, but the 'normal' gravity must always be *decreased* from a quantity T (Figure 8.4). It is estimated by placing over the topographic map a transparent sheet with a grid, and estimating the average altitude in each grid cell (unless the regional topography has already been digitalized). If $-\gamma_T$ is this variation due to the topography, the *Bouguer anomaly* is:

  $$g - [\gamma - (0.308 - 0.0419\, d)z + \gamma_T] \tag{8.24}$$

8.7 ISOSTASY

When regional maps of the Bouguer anomaly were drawn with the hope that they would provide details of geodynamic interest, they looked like the smoothed negative of the topography. To avoid strong anomalies, the plateau correction had not to be performed, only the free-air one.

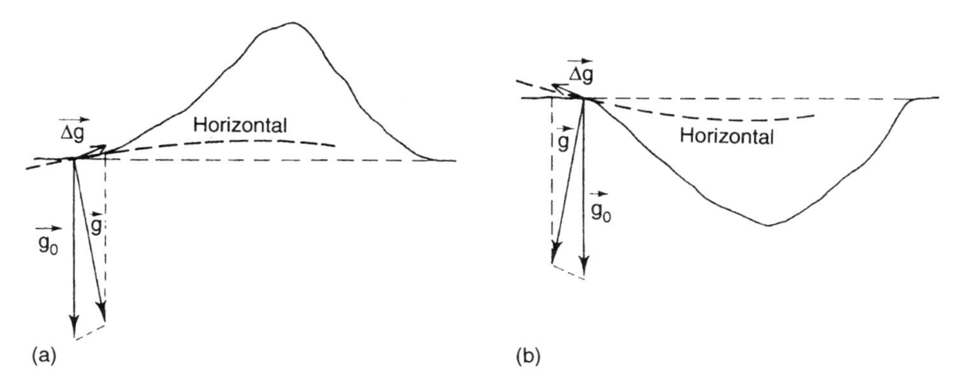

Figure 8.4. The attraction from a neighbouring mountain (a) or valley (b) makes the vertical deviate, and in both cases decreases the intensity of the gravity.

The immediate conclusion is that the mass of terrains above a certain level, the *compensation level*, is everywhere the same, whatever the mean elevation of the region. It will therefore produce the same Newtonian attraction at the surface. Below the compensation level, there are no noticeable variations of the density.

This will be the case if, below a certain level (*which can be the compensation level or a level below*), the rocks behave like a fluid in hydrostatic equilibrium. This is the *asthenosphere*, and the rigid layer above is the *lithosphere*. The lithosphere 'floats' on the asthenosphere. This equilibrium is called *isostasy*.

This phenomenon was discovered in 1854 by the Indian Geodetic Survey, led by Everest. Close to the Himalayas, the deviation from the vertical was only a third of what the topography predicted. Two explanations were suggested straight away, corresponding to the two explanations of orogenesis competing at the time:

- Orogenesis would be caused by the intrusion of lighter granitic plutons. (In 1930, Haarmann made the theory more specific, by admitting the existence of a layer between sial and sima, the basaltic *salsima*, which would dissociate into sial and sima when buried at the bottom of a geosyncline (see Section 14.2).) This leads to the *Pratt hypothesis*: the terrains above the compensation level are lighter when they are mountains. We know today that this is not true. Nevertheless the Pratt hypothesis holds for oceanic lithosphere, an increase in density resulting from the slow cooling since its formation at mid-ocean ridges.
- Orogenesis is caused by a transverse contraction, which leads to the *Airy hypothesis*: the lighter upper layer of the Earth has the same density everywhere, but is thicker where there is a mountain (below a mountain, there is a light *root*). The compensation level then goes through the lowest point of this root. This opinion was the correct one: orogenesis was indeed caused by a transverse contraction. Its assumed origin (the cooling of the Earth) was wrong, but the idea persisted until the discovery of seafloor spreading.

Let us calculate the thickness of the root in the Airy hypothesis. Let ρ_u, ρ_l and ρ_m be the respective densities of the upper crust (forming a mountain of mean thickness h), of the lower crust (forming a root of mean thickness H), and of the mantle. The variation of gravity with depth is negligible and therefore the isostasy condition is:

$$\begin{cases} \rho_c h + \rho_i H = \rho_m H \\ H = \dfrac{\rho_c}{\rho_m - \rho_i} h \approx 6.5h \end{cases} \tag{8.25}$$

In the example of the Tibetan plateau (of altitude $h = 4.4$ km), the isostasy condition requires that $H = 28.6$ km. The normal thickness of the crust is 40 km, and it must therefore be $40 + 4.4 + 28.6 = 73$ km thick below Tibet. This is indeed (more or less) the case.

Conversely, isostasy requires that the Moho rises below a depression. This is sometimes called an *antiroot*.

In 1914, Barrell introduced the words of asthenosphere and lithosphere. He insisted on the fact that compensation should be only *regional*, the lithosphere behaving like an elastic plate. Jeffreys (1952) and Vening-Meinesz developed this model to perform an *isostatic correction* and detect variations from the isostasy condition. But, as we shall explain in Chapter 13, the flexural rigidity of the lithosphere used in this calculation is highly variable with space and cannot be predicted. It can only be determined in one place by already assuming isostasy.

Deviations from the isostasy condition are therefore induced from spatial geodesy and the geoid. The regionalization is automatically performed with the spherical harmonics up to degree 50 only. It is found that isostasy is verified only to 90%, the principal cause of deviation being the convection currents in the mantle.

8.8 ISOSTATIC EQUILIBRIUM AND ARCHIMEDES THRUST

To avoid misunderstandings, let us go back to the calculation of hydrostatic equilibrium. The model used is rough and very simple:

(1) The density of the upper mantle is assumed constant, nearly the same in the mantle lithosphere and in the asthenosphere.
(2) The regionalization is not accounted for, and we simply look at the equilibrium of vertical prisms. Their horizontal extension is assumed to be sufficient for their vertical displacements to be considered independent.

With these approximations, the four models of Figure 8.5 lead to the same ratio H/h. Model (a) corresponds to the predominant idea when the Moho was discovered: the lithosphere would be the crust and the asthenosphere would be the mantle. The distinction is clearly made in the other three models. In (b), the lithosphere is assumed to have a normal thickness below the mountain. In (c), it was admitted that the mountain resulted from a transverse compression of a whole normal lithosphere. And in (d), the lithosphere was reduced to the crust before the transverse

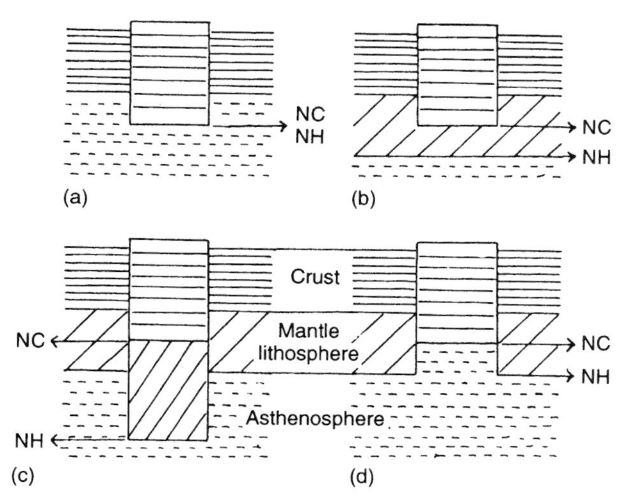

Figure 8.5. Approximate calculation of isostasy. NC = compensation level; NH = level where the pressure is hydrostatic.

compression. This last model is the more realistic one. It explains why, while the whole plate was compressed transversally, there was only one region deforming, the one where the mountain appeared. The horizontal forces in the lithosphere were balancing each other, and the overpressure (relative to the hydrostatic pressure) is inversely proportional to the thickness.

In both models (b) and (c), a fluid in hydrostatic equilibrium exists everywhere only below the level marked NH, below the compensation level NC. Between NH and NC, the stresses in the mantle lithosphere are not hydrostatic. But the *lithostatic pressure* is the same, and is the only one used in the calculation of the weights of prisms. As we assumed that the mantle lithosphere and the asthenosphere have the same density, using NC or NH for the lower limit of the prisms does not change the result. It is the same as with model (a).

Isostasy requires the existence of roots below the mountains, but it should not be said that 'the root exerts an Archimedes thrust'. The Archimedes thrust does not result from the mass forces over all the particles of a body. It results from all pressure forces over the *surface* of a body, when it is entirely immersed in fluids in equilibrium (one of these fluids can be the atmosphere). It is therefore illicit to talk about Archimedes forces over the different parts of the body.

Figure 8.6 shows where this mistake can lead. The quarter of a drum able to rotate about its axis is immersed in a denser fluid, inside a container. If this quarter of the body were submitted to an Archimedes thrust, the drum would rotate and provide work without absorbing energy, realizing the famous 'perpetual motion'. In fact, the Archimedes theorem does not apply, because the fluids around the drum are not in hydrostatic equilibrium. The forces on the drum all go through its axis and cannot rotate it.

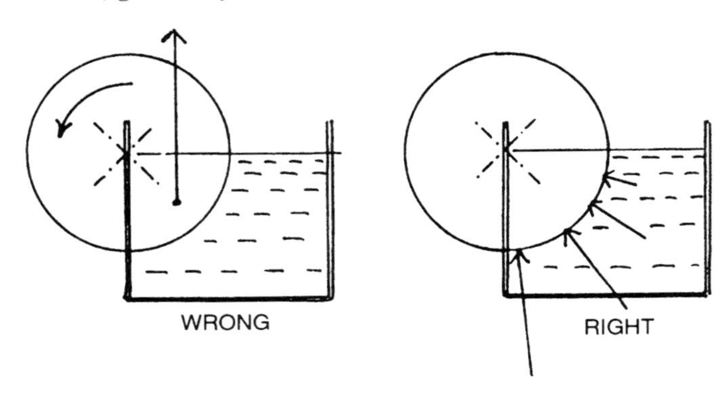

Figure 8.6. The Archimedes thrust is not some kind of 'anti-gravity'. It results from the pressures applied to the surface of the body, completely surrounded by fluids in hydrostatic equilibrium.

It is therefore absurd to think (as some do) that the oceanic crust in a plate going downward can undergo *delamination* (i.e. block separation), only because it is less dense than the surrounding asthenosphere.

But in the case of a continental crust, as we shall see in Chapter 14, its lower portion shows some fluidity and allows the transmission of hydrostatic pressures. If this lower crust communicates freely with asthenosphere, the mantle lithosphere is completely surrounded with fluid and the Archimedes theorem applies. This cold upper mantle could undergo delamination and sink in a warmer asthenosphere (less dense). Lenardic and Kaula (1994) recently used this process, to explain that there has been no mantle lithosphere below Tibet for 10 Ma. But this model is far from being accepted by everybody (see Section 14.10).

8.9 VERTICAL MOVEMENTS DUE TO EROSION OR SEDIMENTATION

Let us consider again the rough model of an upper mantle with constant density and with no regionalization. We may use it to estimate the vertical movements caused by erosion or sedimentation, which restore isostasy. As the four models of Figure 8.5 are equivalent, model (a) will be used.

Let h be the initial average height of a mountain, e the average erosion it undergoes, Δh the resulting uplift. The height of the mountain has become $h - e + \Delta h$, the thickness of the root has become $H - \Delta h$. We call ρ_u, ρ_l and ρ_m the respective densities of the upper crust, lower crust and mantle. Hydrostatic equilibrium before and after erosion can be written as:

$$\begin{cases} \rho_u h + \rho_1 H = \rho_m H \\ \rho_u(h - e + \Delta h) + \rho_1(H - \Delta h) = \rho_m(H - \Delta h) \end{cases} \tag{8.26}$$

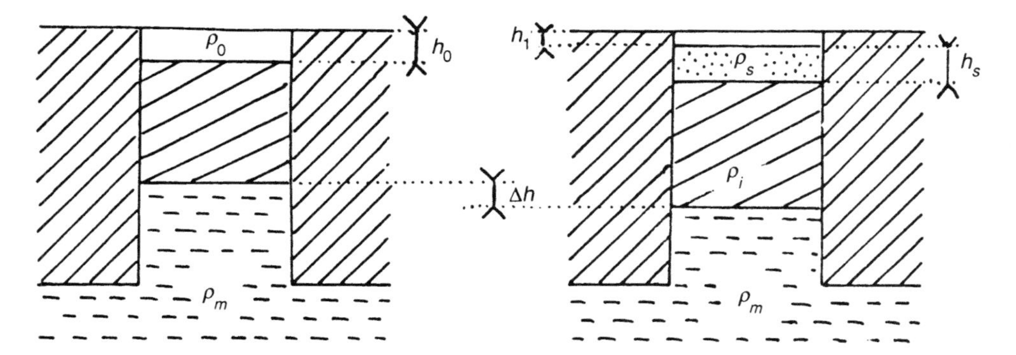

Figure 8.7. Isostatic sinking under the weight of sediments.

Subtracting one equation from the other yields:

$$\Delta h = \left(\frac{\rho_u}{\rho_u + (\rho_m - \rho_l)} \right) e = (0.85 \pm 0.07)e \tag{8.27}$$

Erosion will stop when the root will have disappeared, i.e. for $\Delta h = H$. Because of their values above, this condition reads:

$$e = \left(1 + \frac{\rho_u}{\rho_m - \rho_{li}} \right) h = (7.0 \pm 0.5)h \tag{8.28}$$

The height of the mountain will have changed to $h - e + H = 0$. *Erosion will have removed around 7 times the volume of the mountain. Rocks metamorphosed at depths of 6 times the average height of the mountain will be brought up to the surface.*

Let us now look at sedimentation in an epicontinental sea, of depth h_0 before the sedimentation, and h_1 after a sediment layer of thickness h_S has been deposited (Figure 8.7). The subsidence of the lithosphere has been $\Delta h = h_S + h_1 - h_0$. Let $\rho_0 = 1$ and $\rho_1 = 2.2$ be the respective densities of seawater and the sediments. Isostasy before and after sedimentation means that:

$$\rho_0 h_0 + \rho_m \Delta h = \rho_0 h_1 + \rho_S h_S \tag{8.29}$$

By elimination of $(h_1 - h_0)$, we get:

$$\Delta h = \frac{\rho_S - \rho_0}{\rho_m - \rho_0} h_S \approx 0.52 h_S \tag{8.30}$$

Sedimentation will stop when $h_1 = 0$. Then $h_S = \Delta h + h_0$, and:

$$h_S = \frac{\rho_m - \rho_0}{\rho_m - \rho_S} h_0 \approx 2.1 h_0 \tag{8.31}$$

The isostatic sinking of the continental platform under the weight of sediments can only double the thickness of the sediment layer. *The thick sediment layers deposited in shallow seas* (e.g. in the Parisian Basin in the Jurassic and Cretaceous periods) *require another geodynamic process of prolonged subsidence*. They are too localized on the Earth to be explained by a continuous rise of the sea level (see Section 9.14).

8.10 TIME NECESSARY FOR THE ISOSTATIC ADJUSTMENT

The asthenosphere only shows a relative fluidity, and the setting of isostatic equilibrium takes thousands of years. Erosion or sedimentation are slow enough that this did not need to be accounted for in the previous section. But the situation is very different for the melting of an ice cap, which can be likened to erosion. The ice cap thickness can decrease by 1 m/year, and even 10 times more at its edges.

The discovery of isostasy explained the continuous uplifting of the Baltic coasts, observed for centuries. The town of Stockholm (Sweden) was founded in 1260 on a small island which controlled a long inlet. This is now the Lake Mälar, and ships from the Baltic Sea must pass through two locks before reaching it. This uplift is coming from the melting of the ice cap which covered the whole of Finland and Scandinavia (*Fennoscandia*), the Baltic Sea and further during the last glaciation. As the uplift is not finished, the gravimetric deficit is $\Delta g = -20$ to -30 mgal in Fennoscandia. In Hudson Bay, the same process of *glacio-isostasy* yields a $\Delta g \approx -50$ mgal.

The glacio-isostatic rebounds of Fennoscandia and Eastern Canada provide the most details about mantle viscosity. This complex and difficult subject will be dealt with in Chapter 18.

8.11 EARTH–MOON ANGULAR MOMENTUM

The following theorem is assumed to be common knowledge. In a Galilean frame of reference, the moment vector resulting from the forces applied to a solid, relative to its centre of gravity, equals the time derivative of its angular momentum vector. Especially, if a solid is not submitted to any outside force, its angular momentum is constant. This theorem is also valid for a system of solids not in contact, when the only forces between them are the Newtonian attraction forces. We will apply this to the Earth–Moon system. The lunar orbit can be very well approximated by a perfectly circular orbit exactly in the equatorial plane of the Earth.

In a Galilean frame of reference, the Earth-Moon system rotates in 27.496 sidereal days about its centre of mass. The attraction force between the Earth and the Moon balances the centrifugal force due to the rotation. The ratio between the mass of the Moon (m) and the mass of the Earth (M) is $\mu = 0.012300$. The distance between the centre of the Moon and the centre of the Earth (O) is $D = 384,000$ km. The centre of mass G of the system is at the distance $e = D\mu/(1+\mu) = 4,666$ km of O, on the side of the Moon. The trajectory of the Earth will therefore undulate slightly about its elliptical orbit (in 1 lunar month, the Earth travels 76 million km in its orbit around the Sun). These undulations, and even the curvature of the Earth's orbit, will be neglected in the following rough calculation.

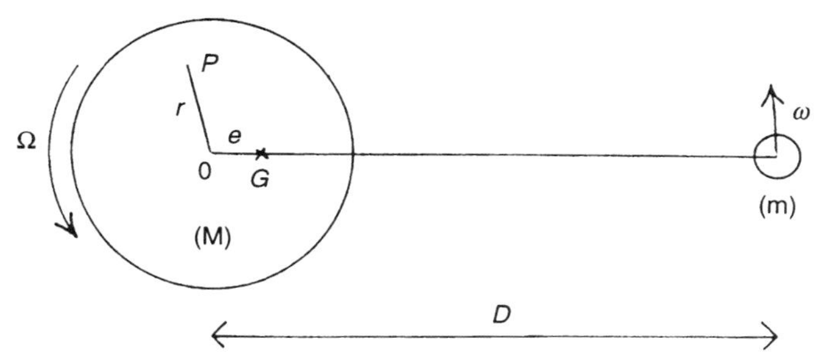

Figure 8.8. Earth–Moon system (not to scale).

Let us calculate the angular momentum of the Earth relative to the z-axis passing through G and perpendicular to the equatorial plane (Figure 8.8). The Earth rotates at the speed Ω around a parallel axis passing through O (the axis of the world), which rotates with a speed ω around the z-axis. With a suitable time origin, the Galilean, Cartesian coordinates of a point P, with the mass dM, are:

$$\begin{cases} x = e\cos\omega t + r\cos(\Omega t + \varphi) \\ y = e\sin\omega t + r\sin(\Omega t + \varphi) \\ z = \text{constant} \end{cases} \tag{8.32}$$

r is the distance from P to the axis of the world. The components of the angular momentum ($\mathbf{GP} \wedge \mathrm{d}\mathbf{GP}/\mathrm{dt}$) are:

$$\begin{cases} -z\dfrac{dy}{dt}dM \\ z\dfrac{dx}{dt}dM \\ \left(x\dfrac{dy}{dt} - y\dfrac{dx}{dt}\right)dM = [e^2\omega + r^2\Omega + er(\omega + \Omega)\cos(\Omega t + \varphi - \omega t)]dM \end{cases} \tag{8.33}$$

When performing the summation for all points of the Earth, the components along x and y become zero, because of the symmetry around the equatorial plane. The term with φ also becomes zero, because of the axial symmetry. The angular momentum of the Earth is therefore:

$$M_E = e^2\omega M + \Omega\int_{Earth} r^2 dM = \omega D^2\frac{\mu m}{(1+\mu)^2} + C\Omega \tag{8.34}$$

The angular momentum of the Moon is computed in the same way, replacing e with $D - e$, which equals $D/(1+\mu)$, replacing M with m, C by the Moon's moment of inertia I, and Ω with ω. The Moon must indeed rotate as fast as it rotates around the Earth, to always present the same face. We get the equation:

$$M_M = \frac{m\omega D^2}{(1+\mu)^2} + I\omega \tag{8.35}$$

An upper limit of I is the value it would have if the density were homogeneous, the same from the surface to the centre: $(2/5)\,m\,R_M^2$, where R_M is the radius of the Moon. As $R_M/D = 0.00453$, this term can be neglected. The angular momentum of the Earth–Moon system is therefore:

$$M = M_E + M_M = \frac{m\omega D^2}{1+\mu} + C\Omega \tag{8.36}$$

To prepare for later, the current values of Ω, D and ω are indexed 0. The current ratio of the two terms in the angular momentum is:

$$\left.\begin{aligned} C &= 5.4 \times 10^{37}\,\mathrm{kg\,m^2} \\ m &= 7.35 \times 10^{22}\,\mathrm{kg} \\ D_0 &= 3.84 \times 10^{8}\,\mathrm{m} \\ \frac{\Omega_0}{\omega_0} &= 27.396 \end{aligned}\right\} \Longrightarrow \alpha = \frac{m\,D_0^2\,\omega_0}{(1+\mu)C\Omega_0} = 7.24 \tag{8.37}$$

From these values, it can be inferred that 86.8% of the angular momentum of the Earth–Moon system comes from the orbital momentum of the Moon, 12.1% comes from the rotation of the Earth, and 1.1% from the rotation of the centre of the Earth round the centre of mass of the system.

8.12 SLOWING DOWN OF THE EARTH'S ROTATION AND INCREASING DISTANCE TO THE MOON

For each portion of the Earth, the attraction by the Moon differs slightly, in its direction and in its intensity, from what this portion would undergo at the centre of the Earth. This difference is the *lunar tidal force*. It points toward the Moon and is maximal (around 0.11 mgal) at the point of the Earth's surface closest to the Moon. It points in the other direction and has the same value at the furthest point.

The same goes for the Earth–Sun system, and there is a solar tidal force. The mass of the Sun is 27 million times larger than the mass of the Moon. But the Sun is 390 times further from the Earth, and the tidal force varies with the cube of the distance. The solar tidal force is therefore 2.2 times smaller than the lunar tidal force.

Moon–Sun tidal forces create an elastic deformation of the whole Earth, the *terrestrial tide*, of the order of 0.5 m, along with the tides in the oceans. Their amplitude is about 1 m in the ocean, but can be strongly amplified close to the coasts. There are normally 2 tides per lunar day of 24 hours 50 minutes. Oceanic tides are more complex, and can only be predicted using empirical formulae.

Apart from the corrections it brings about in gravimetry, the tide is interesting in this book only because it slowed down the rotation of the Earth on the geological time-scale. The mechanism is shown in Figure 8.9, where the axes are Galilean and the viewpoint is above the North Pole. The Earth rotates around itself in the anti-clockwise direction with the angular speed Ω, and the Moon rotates around the

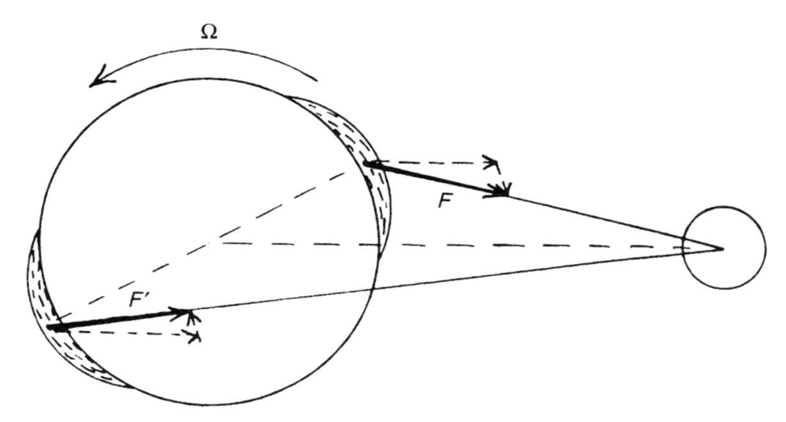

Figure 8.9. Torque (F, F′) exerted by the Moon on the tide bulges, because they are carried a few degrees eastward by the rotation of the Earth. This effect was grossly exaggerated in the figure. This torque slows down the rotation of the Earth.

Earth, also in the anti-clockwise direction, with the angular speed ω. Because of frictions, the bulges induced by the lunar tides are not exactly on the Earth–Moon axis, but carried a few degrees eastward. The attraction the Moon exerts on them creates a clockwise torque, which slows down the Earth's rotation.

At first glance, the modified Earth–Moon attraction would seem to increase the speed of the Moon. In fact, it modifies slightly its orbit and drives it away from the Earth, which has the opposite effect (because of Kepler's Third Law: $\omega^2 D^3 = \omega_0^2 D_0^3$). Some angular momentum is therefore transferred from the Earth to the Moon.

Let us look at the slowing of the Earth's rotation, neglecting the small amount of angular momentum transferred from the Sun to the Earth by the solar tide. Define:

$$\begin{cases} \left(\dfrac{\omega_0}{\omega}\right)^{\frac{2}{3}} = \dfrac{D}{D_0} = x^2 \\ \dfrac{\Omega}{\Omega_0} = y \\ \left.\dfrac{dy}{dt}\right|_{t=0} = -f \end{cases} \tag{8.38}$$

The angular momentum becomes $M = \Omega_0 C(\alpha x + y)$ and its conservation reads:

$$\alpha \frac{dx}{dt} + \frac{dy}{dt} = 0 \tag{8.39}$$

As x and y equal 1 for $t = 0$, we deduce that:

$$\begin{cases} \alpha(x-1) = 1 - y \\ \left.\dfrac{dD}{dt}\right|_0 = \left(2x\dfrac{dx}{dt}\right)_0 D_0 = 2D_0\dfrac{f}{\alpha} \end{cases} \tag{8.40}$$

How the Earth's rotation slowed down on the geological time-scale can be estimated with the periodicity in coral growth or in sedimentation. Corals grow one layer each day, and its thickness depends on the light, which has a yearly periodicity (the duration of the year cannot have varied much). The number of days per year at 4 different geologic periods have been published by Stoyko (1970). From these data, $f = 0.37\,\mathrm{Ga}^{-1}$ between 440 Ma and 270 Ma. The value at 270 Ma, a fifth value at 72 Ma and the current value concur to yield $f = 0.19\,\mathrm{Ga}^{-1}$, which seems wrong.

$$1\,\mathrm{Ga} = 1 \text{ giga-year} = 1 \text{ billion years} = 1 \text{ eon}$$

Another method uses coastal marine sediments. Tidal currents bring thin sand layers, alternating with thin mud layers. After lithification, they form *tidalites*. A bimonthly lunar periodicity is superimposed to the diurnal or semi-diurnal periodicity, as there are strong tides, called *spring tides* at the full Moon and new Moon (when the lunar and solar tidal forces are in phase), and small tides, called *neap tides*, at the first and last quarters, when these forces are in quadrature. The number of days for each moon can therefore be counted, and, owing to a yearly periodicity of the sedimentation, the number of moons per year (Ω/ω) can also be measured. The latter is usually more reliable. In practice, the Fourier analysis of the distances from all layers to a reference layer yields peaks corresponding to the different periods: year, lunar month and half-month, day and half-day (Sonett *et al.*, 1996).

From one formation in Utah, aged of 900 ± 100 Ma, and one formation in Indiana, aged 305 ± 5 Ma, these authors find that $f = 0.35\,\mathrm{Ga}^{-1}$.

To estimate f in a more distant past, it would also be possible to use the fact that the Earth–Moon system formed around 4.2 Ga ago, and that $x = (D/D_0)^{1/2} \approx 0$ (see Section 8.13). It is indeed possible to write a second equation in $x(t)$ and $y(t)$, expressing the dissipation of energy by the terrestrial and oceanic tides. The lunar tidal force is proportional to D^{-3}. The solar tidal force is 2.2 times smaller, and independent from D. The work of the two is proportional to Ω. And it must equal the dissipation of kinetic energy. Neglecting the kinetic energy of the rotation of the Moon around itself, the kinetic energy of the Earth–Moon system is:

$$E = \tfrac{1}{2}m\left(\frac{D}{1+\mu}\right)^2\omega^2 + \tfrac{1}{2}M\left(\frac{\mu D}{1+\mu}\right)^2\omega^2 + \tfrac{1}{2}C\Omega^2 = \frac{1}{2}\left(\frac{mD^2\omega^2}{1+\mu} + C\Omega^2\right) \tag{8.41}$$

It can expressed as a function of x only, because $y = 1 + \alpha - \alpha x$.

$$E = \tfrac{1}{2}C\Omega_0[\alpha\omega_0 x^{-2} + \Omega_0(1 + \alpha - \alpha x)^2] \tag{8.42}$$

The work of the tidal forces, $-dE/dt$, is proportional to:

$$\Omega\left[\left(\frac{D_0}{D}\right)^3 + \frac{1}{2.2}\right] = \Omega_0(1 + \alpha - \alpha x)\left(\frac{1}{x^6} + \frac{1}{2.2}\right) \tag{8.43}$$

And, knowing that for $x = 1$, $dx/dt = f/\alpha$, we get the equation:

$$\frac{x^3 + (\Omega_0/\omega_0)(1 + \alpha - \alpha x)x^6}{(1 + \alpha - \alpha x)(x^6 + 2.2)}dx = \left(1 + \frac{\Omega_0}{\omega_0}\right)\frac{f}{3.2\alpha}dt \tag{8.44}$$

Variable x increased from 0 to 1 while t was increasing from -4.2 Ga to 0. A simple numerical quadrature yields $f = 0.29\,\text{Ga}^{-1}$. A value smaller than the current value should not be a surprise, because the continental crust only appeared progressively (Chapter 17). According to geochemists, it only reached its current volume 1 Ga ago. Now, a factor was omitted from the previous calculation: for given values of Ω and D, the energy dissipated by the tides depends on the area of epicontinental seas. It also depends on their latitudes and shapes, but continental drift must smooth this factor over 100 Ma. A more realistic model would express the last equation with f as an increasing function of t.

In any case, numerous impacts by planetary bodies must have modified the energy and the kinetic moment of the Earth–Moon system up to 3.2 Ga ago. The previous calculation becomes very inaccurate before this period. Admitting that $f = 0.35\,\text{Ga}^{-1}$ since, 3.2 Ga ago, we had $y - 1 = 1.12$, and therefore $x^2 = D/D_0 = 0.715$. The Moon was 43 terrestrial radii away (compared to 60 now), the day was 14.5-hour long (using current hours), and the tides were 2.74 times stronger in amplitude.

8.13 NON-TIDAL VARIATIONS IN THE ANGULAR VELOCITY

The lengthening of the day during the last 2,700 years can be deduced from the dates and hours at which eclipses were observed in antiquity. They occurred several hours sooner in the day than predicted using the current duration of the day. It provides a relative slowing $f = 2.1 \times 10^{-10}\,\text{year}^{-1}$, i.e. only 6/10 of its 'geological' value.

There has therefore been a non-tidal acceleration in the last thousand years. This is due to the glacio-isostatic uplift of the Arctic regions, which required a transfer of matter inside the Earth toward regions closer to its rotation axis. As the inertia moment C decreased, the rotation velocity increased. We will come back to this effect in Chapter 18, to estimate the viscosity of the mantle.

At the decade scale, the rate at which the Moon moves away is measured with laser telemetry, using reflectors placed during the Apollo missions. It is $dD/dt = 3.82 \pm 0.07$ cm/year. From that, it is deduced that $f = 0.36\,\text{Ga}^{-1}$, i.e. more or less its 'geological' value. One conclusion is that the effect of the glacio-isostatic rebound is no longer perceptible, or that it is masked by an inverse effect. For the last century, glaciers and polar ice caps have decreased in volume, the melting water being immediately redistributed all over the Earth, and before any isostatic readjustment the inertia moment C increases.

The accuracy of modern atomic clocks makes possible the measurement of the variations in the Earth's rotation from one week to the next. Sudden increases or decreases of the angular velocity have been observed, with relative values up to 3×10^{-9}. They mask any variation at the scale of the decade or the millennium or on the geological time-scale. These events could be associated with transitory magneto-hydrodynamic processes in the liquid core, or with modifications of the mass distribution following large earthquakes.

8.14 EARTH AND MOON FORMATION

Here is the occasion to talk briefly of the current thinking about the formation of the Earth–Moon system and its age (even if this will be used in the last chapters of this book), as the reasoning uses the conservation of the angular momentum.

Isotopic geochemistry shows that the Solar System formed around 4.7–4.5 Ga ago. Mercury, Venus, the Earth, the Moon, Mars and the asteroids come from a dust disk which was then circling around the Sun. Newtonian attraction agglomerated this dust into planetary bodies of increasing sizes.

The oldest minerals found on Earth are zircons aged from 4.3 to 4.0 Ga. The Moon was sampled during the Apollo missions, and most *mares* (large dark basaltic plains, called seas before) date back 3.8 to 3.2 Ga., whereas the highlands (70% of its surface) formed 4.2 Ga ago. The craters visible on the surface are traces of impacts by numerous planetary bodies up to 3.2 Ga. The most violent may have created the *mares*. Their frequency then dropped, without falling to zero (the Copernicus crater is 850 Ma old, and Tycho is 100 Ma old). The Earth must have received similar impacts, but the traces from the oldest have disappeared. The largest impact craters known on Earth (140 km in diameter) are also the oldest: Vedrefort in South Africa and Sudbury in Ontario (Canada), respectively 1.97 Ga and 1.84 Ga old.

The Moon is not much smaller than Mercury, and therefore increases strongly the angular momentum of the Earth–Moon system. The two smallest satellites of Mars were without doubt captured, but the formation of the Earth–Moon system is much more difficult to understand. 4 scenarios were proposed successively.

1. *The Moon was torn away from the Earth.* Even if the whole angular momentum is taken by a single planet, a bit larger than the Earth, this is still very far from the counteracting of gravity by the centrifugal force. A tremendous tidal force would have been necessary, which could only result from the proximity of Jupiter. But it does not account for the creation of this large angular momentum.
2. *The Moon was captured by the Earth after their formation.* This was even dated to 1.8 Ga by some authors, as orogenesis seemed intense at this time. For a capture, a large portion of the Moon's kinetic energy must be dissipated by the tidal effect from the Earth, and this seems impossible.
3. *The Moon and the Earth formed simultaneously, very close to each other*, by agglomeration of small planetary bodies. The co-accretion theory does not explain the high value of the angular momentum, nor the composition difference between the Earth and the Moon. The latter does not have an iron core, or a very small one, and the samples brought back do not have siderophile elements (i.e. dissolving in liquid iron).
4. Therefore, the current consensus is that *a proto-Earth was impacted around 4.2 Ga by a large planetary body*, comparable to Mercury. Part of the silicates melted or vaporized, the iron from the incident body joined the terrestrial core, and the system acquired a strong angular momentum from the side impact.

9

Terrestrial heat

9.1 HEAT DIFFUSION

The heat flux is the quantity of heat crossing a unit surface per unit time. At any point it equals $\boldsymbol{\varphi} \cdot \boldsymbol{n}$, with $\boldsymbol{n}$ denoting the unit normal to the crossed surface and $\boldsymbol{\varphi}$ denoting the *heat flux vector*. In a thermally isotropic medium, the diffusion of heat follows the *Fourier law*:

$$\boldsymbol{\varphi} = -K \, \mathbf{grad} \, T \tag{9.1}$$

T is the temperature of the medium, K its thermal *conductivity*. In a thermally anisotropic medium, $\boldsymbol{\varphi}$ and $\mathbf{grad}\, T$ have not in general the same direction, but the Cartesian components of $\boldsymbol{\varphi}$ are linear functions of the Cartesian components of $\mathbf{grad}\, T$. Therefore, for at least 3 directions perpendicular to each other, equation (9.1) holds, with different K each time. For instance, in phyllosilicates or in schists, K is 5 to 6 times smaller perpendicular to the schistosity than along the schistosity plane. In the following, thermally anisotropic media will be excluded.

Let us consider a motionless medium, without heat sources or heat sinks (in particular, there are no phase changes). Its density is ρ, its heat capacity per unit mass C. Following Ostrogradski's theorem (Annex A.1), the conservation of heat implies that:

$$\mathrm{div}(K \, \mathbf{grad} \, T) = \rho C \frac{\partial T}{\partial t} \tag{9.2}$$

If K and ρC are independent from temperature, this equation can be written:

$$K \nabla^2 T = \frac{\partial T}{\partial t}, \qquad \kappa = \frac{K}{\rho C} \tag{9.3}$$

The ratio κ is the thermal *diffusivity*. Its SI unit is the $\mathrm{m^2\,s^{-1}}$, but depending on the applications it may be better to use the year or Ma for the time unit (1 year $= 3.1556 \times 10^7$ s; 1 Ma $= 10^6$ years). Here are a few approximate values:

	$K(W\,m^{-1}\,K^{-1})$	$\rho(Mg\,m^{-3})$	$C(kJ\,K^{-1}\,kg^{-1})$	$\kappa(m^2\,a^{-1})$
Granite	2.1	2.6	0.96	26.5
Basalt, gneiss	2.5	2.8	1.01	27.9
Upper mantle	3.1	3.3	1.17	25.3

We aim at solving equation (9.3) in a domain with *fixed boundaries*, part or all of the limits can be infinitely far (as in the case of a half-space limited by a plane). Any function obeying equation (9.3) is a *general solution*. The particular solution to look for must also obey *boundary conditions* and, when it depends on time, *initial conditions* at t = 0.

As (9.3) is linear in T and with constant coefficients, if the general solution is multiplied by a constant, or if two general solutions are added, we still have a general solution. This is often useful to separate the problem into two or more easier problems (of course, boundary conditions and initial conditions must be multiplied by the constant or added, as the general solutions). In particular, it is possible to look separately for the temperature in a steady regime and the additional transitory perturbation.

In *steady regime* ($\partial T/\partial t = 0$), the heat diffusion equation (9.3) reduces to the Laplace equation $\nabla^2 T = 0$. This is the same equation as for electrostatic or gravity potentials. The problem has one single solution if the temperatures are given on all the limits of the domain, or the temperatures on one portion and the perpendicular component of **grad** T on the remainder. This is equivalent to defining the heat flow crossing the boundary. It is also possible to give only the heat flows, and the temperatures will be determined close to an additive constant.

An example of analytical solution (i.e. expressed with known and tabulated functions, infinitely derivable except in a few single points) is given further, for its teaching value. But numerical computations are generally used. Any computation centre has programs for solving the Laplace equation, whatever the shape of the domain and the boundary conditions.

For *time-dependent temperatures*, analytical solutions are known only if there is a single spatial variable. Here, this will be the altitude z. Analytical solutions could be deduced for problems with spherical symmetry, as the equation (9.3) could be rewritten using the radius r:

$$\kappa \frac{\partial^2 (rT)}{\partial r^2} = \frac{\partial (rT)}{\partial t} \tag{9.4}$$

Note that if $T(x, y, z, t)$ is a general solution of equation (9.3), its partial derivatives relative to one of the 4 variables are also general solutions.

In this case, the numerical solutions consist in computing T at successive time steps from $t = 0$, with regular time increments. Each time, the temperature is computed at all points of the space domain (the *nodes* of the *grid*). This is simple if there is only one space variable, but not obvious, as any error should not get amplified at each successive time step. Now, the partial derivatives are replaced by

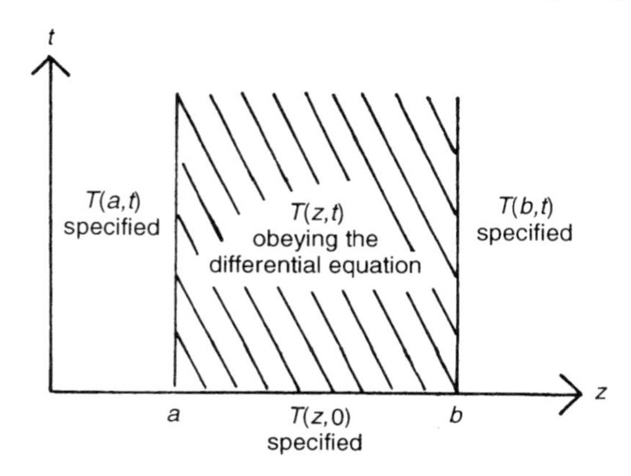

Figure 9.1. Integration domain for equation (9.2), and boundary conditions providing a single solution.

finite differences, not infinitesimal, and this introduces a *truncating error*. The numbers handled by the computer have limited number of decimals, and this introduces a *rounding error*. The reader will find in many books and articles details of how to write programs, including in Chapter 2 of my book *Very slow flows of solids* (Lliboutry, 1987).

With the single spatial variable z, the integration domain is represented on Figure 9.1. $T(z,0)$ is fixed for $a < z < b$; so are $T(a,t)$ and $T(b,t)$ for $t > 0$. When this leads to 2 different values for $T(a,0)$ or $T(b,0)$, a singular point exists, where the numerical computation cannot start. An analytical solution is necessary in the neighbourhood of this singular point.

9.2 MEASURING THE GEOTHERMAL FLUX

At an underground depth not reached by the seasonal fluctuations, there is a permanent upward heat flux of a few tens of mW/m^2, the *geothermal flux*. To determine it, it is necessary to measure the temperature gradient (*geothermal gradient*), of a few hundredths of a degree per metre, and the thermal conductivity of the medium. This is achieved by drilling, waiting for the heat produced by the drilling to dissipate, measuring the temperature differences in the hole and measuring the conductivity in the laboratory. The measurements are simple, but there are many sources of error. The main one comes from the perturbation induced by the circulation of underground water.

On continents, deep drillings are made to reach below the aquifer. The air does not refrigerate them, because they are filled with the drilling mud. On the seafloor and close to mid-ocean ridges, the intense hydrothermal circulation irregularly decreases the geothermal gradient. Part of the geothermal flow is dissipated by the submarine hot water springs. Their temperature and output should be measured, but this has only been done for a few black smokers and cannot be generalized.

Measurements need to be made in non-stratified terrains, or with a sub-horizontal stratification. If layers with a high inclination are present, the geothermal flux will concentrate in the most conductive ones.

The surface must be flat and horizontal. On a convex ground, the fluxes will be too small and on a concave ground, the flows will be too high. The ground is indeed a near-isothermal surface, and the vector $\varphi = -K\,\mathbf{grad}\,T$ is almost perpendicular to it. Its force lines diverge toward a summit, converge toward a hollow. The section of a force tube increases with altitude in the first case, decreases in the second. The flux of φ being conserved in a force tube, its intensity φ is inversely proportional to this section.

In a two-dimensional case, this modification of the flux can be estimated by tracing isotherms and orthogonal force lines, which respect the flux conservation of **grad** T. An analytical solution valid for a specific profile group is given below, more to show the traps to avoid in the modelling, than for the results.

9.3 INFLUENCE OF THE TOPOGRAPHY: CASE WITH AN ANALYTICAL SOLUTION

For a plane problem, where y does not intervene, the function $\ln(x^2 + z^2)$ is a traditional solution of the Laplace equation. The derivative relative to $z, 2z/(x^2 + z^2)$ is another general solution. The next solution is reached by changing the origin of z. Let z be the altitude relative to the altitude for which $x \to +\infty, \gamma$ the geothermal gradient for $z \to -\infty$, and λ the temperature gradient in the atmosphere.

The temperature in the atmosphere is:

$$T = T_0 - \lambda z \tag{9.5}$$

The temperature in the ground is:

$$T = T_0 - \gamma z + (\gamma - \lambda)\frac{h(H-h)(H-z)}{x^2 + (H-z)^2} \tag{9.6}$$

The surface profile is:

$$z = \frac{h(H-h)(H-z)}{x^2 + (H-z)^2} \tag{9.7}$$

The temperature of the atmosphere and the ground are indeed equal in each point of this profile. But beware that for $x = 0, z = H$ must not be in the ground or on the surface, as the temperature gradient would be infinite there. This point must be definitely above. But, when $H < 2h$, the cubic curve used as a profile intersects the line $x = 0$ only in a real point, precisely for $z = H$.

It is therefore necessary that $H > 2h$. The curve is then intersected by $x = 0$ in 3 points: $z = h,\ H - h$ and H. Apart from a curve which may represent the transverse profile of a symmetrical rise of height h, the cubic curve contains a closed curve passing through the 2 other points (Figure 9.2a). One should not be

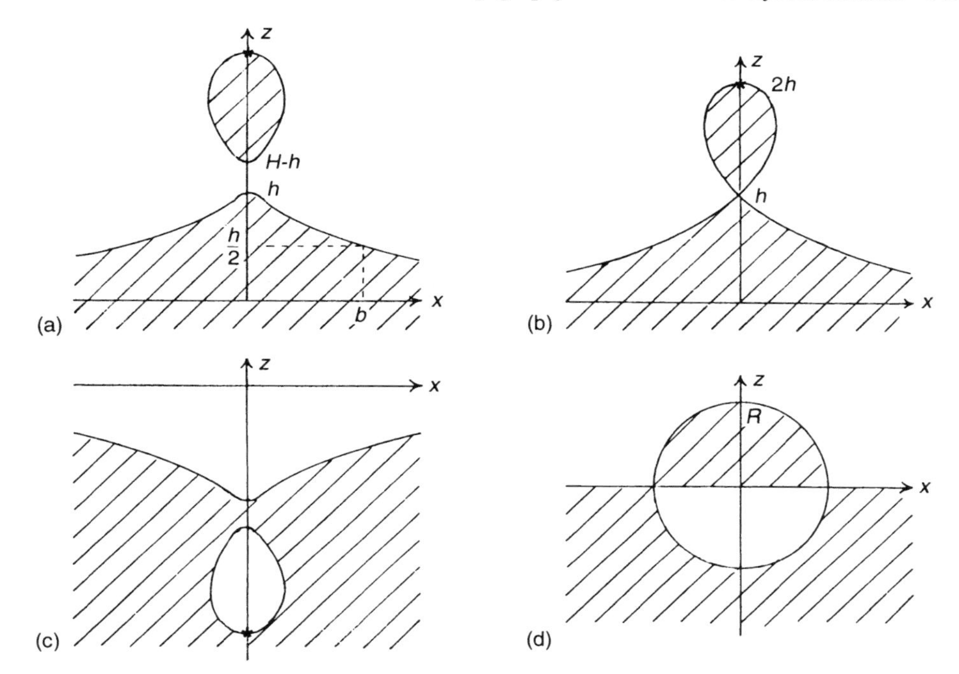

Figure 9.2. Influence of topography on the geothermal flux, in a case where there is an analytical solution (see the text). The asterisk indicates a singular point for the geothermal flux.

worried by this parasite terrain floating in the atmosphere which is introduced by the analytical solution.

The parameter H can be deduced from the half-height width $2b$ of the rise ($z = h/2$):

$$H = h + \sqrt{b^2 + h^2/4} \tag{9.8}$$

The geothermal gradient at the summit of the rise is:

$$-\left.\frac{\partial T}{\partial z}\right|_{0,h} = \gamma\left(1 - \frac{h}{H - h}\right) + \lambda\frac{h}{H - h} \tag{9.9}$$

In the limit case where $H = 2h$, the rise becomes a ridge with 45°-slopes at the top (Figure 9.2b). The geothermal gradient is the same as in the air. All the excess geothermal heat has been dissipated on the sides.

The same cubic curve, inverted with h and H negative, could not represent the transverse profile of a valley, as its loop would correspond to an underground tunnel below the valley (Figure 9.2c). But there is the following solution.

The temperature in the atmosphere is:

$$T = T_0 - \lambda z \tag{9.10}$$

The temperature in the ground is:

$$T = T_0 - \gamma z + (\gamma - \lambda)\frac{R^2 z}{x^2 + z^2} \tag{9.11}$$

The surface profile is:

$$z(x^2 + z^2 - R^2) = 0 \tag{9.12}$$

The cross-section of the valley is semi-circular and the rest of the degenerate cubic curve does not modify the problem (Figure 9.2d). At the centre of the valley, the geothermal gradient is nearly doubled:

$$-\left.\frac{\partial T}{\partial z}\right|_{0,-R} = 2\gamma - \lambda \tag{9.13}$$

9.4 TEMPERATURES IN A MOVING MEDIUM

Let us now write the most general heat equation, with a thermally isotropic medium as the only condition. New terms appear because:

(1) There is production of a heat quantity Q per unit volume and unit time, due to the deformation of the medium and/or its radioactivity.
(2) Any material point has a velocity $\mathbf{v}$ (and therefore, in general, the medium is deforming).

The heat balance is written for a small *fixed* unit volume. The medium transports a heat quantity ρCT per unit volume. This *heat advection flux* must be added to the diffusion flux:

$$\mathrm{div}(K\,\mathbf{grad}\,T - \rho CT\mathbf{v}) + Q = \rho C\frac{\partial T}{\partial t} \tag{9.14}$$

If ρC is a constant, and if $\mathrm{div}\,v = 0$ (deformation with no volume change):

$$\mathrm{div}(\rho CT\mathbf{v}) = \rho C\mathbf{v}\cdot\mathbf{grad}\,T \tag{9.15}$$

If K is constant too, the heat equation becomes:

$$\kappa\nabla^2 T + \frac{Q}{\rho C} = \frac{\partial T}{\partial t} + \mathbf{v}\cdot\mathbf{grad}\,T \tag{9.16}$$

The right hand side is the *material derivative* of T, i.e. the time derivative following the material point in its movement. If $\mathbf{v}$ is uniform and a steady state is dealt with, the same equation can be translated in two ways:

- *Eulerian viewpoint*. The coordinate axes are fixed, and the medium moves with the velocity $\mathbf{v}$. Then, in a steady state, $\partial T/\partial t = 0$.
- *Lagrangian viewpoint*. The coordinate axes are moving with the medium, and $\mathbf{v} = 0$, but in a steady state, $\partial T/\partial t \neq 0$.

9.5 SOLUTIONS OF THE FORM $T = f(z) \times g(\kappa t)$

The solution of $\partial^2 T/\partial z^2 = \partial T/\partial(\kappa t)$ is found by using the common mathematical method of looking for a solution of the form $f(z) \times g(\kappa t)$. Adding dashes to designate the ordinary derivatives, this means that we must have:

$$\frac{f''}{f} = \frac{2f'f''}{2ff'} = \frac{\frac{d}{dz}(f'^2)}{\frac{d}{dz}(f^2)} = \frac{\frac{dg}{dt}}{g(\kappa t)} \tag{9.17}$$

The common value can only be a constant, say a^2. We get the solution:

$$T = C\exp(a^2\kappa t \pm az) \tag{9.18}$$

a can be chosen as negative or complex, which introduces sine functions. Two cases are interesting: a^2 purely imaginary, and a purely imaginary.

Penetration of a periodic temperature oscillation

When the surface temperature is a periodic function of time with a period Θ, it can be expressed as a Fourier series. Using the pulsatance $\omega = 2\pi/\Theta$, we have:

$$T(0,t) = T_0 + \sum_{n=1}^{\infty}(A_n\cos n\omega t + B_n \sin n\omega t) \tag{9.19}$$

Let the z-axis point downward, with its origin at the surface. When z is very large, T tends toward $T(\infty, t) = T_0 + Gz$. The solution corresponds to $a^2 = \pm in\omega/\kappa$, and therefore $a = \pm(1+i)\sqrt{n\omega/(2\kappa)}$:

$$T = T_0 + Gz + \sum_{n=1}^{\infty}\exp\left(-\sqrt{\frac{n\omega}{2\kappa}}z\right) \times \left[A_n\cos\left(n\omega t - \sqrt{\frac{n\omega}{2\kappa}}z\right) + B_n\sin\left(n\omega t - \sqrt{\frac{n\omega}{2\kappa}}z\right)\right] \tag{9.20}$$

Each harmonic of the surface perturbation diffuses downward with a constant phase velocity:

$$c_n = \sqrt{2\kappa n\omega} = 2\sqrt{\pi\kappa n/\Theta} \tag{9.21}$$

which decreases exponentially. At a given time, it is in phase opposition with the surface for:

$$\sqrt{\frac{n\omega}{2\kappa}}z = \pi \iff z = \sqrt{\frac{\pi\kappa\Theta}{n}} \tag{9.22}$$

Its amplitude is then divided by $e^{\pi} \approx 23$. I shall call this depth the 'penetration depth'. It varies as $1/\sqrt{n}$. The fundamental harmonic ($n = 1$) penetrates the deepest.

The penetration depth in sound compact rock is half a metre for the diurnal oscillation, around 9 m for the yearly oscillation. These values are less in a porous soil, more insulating. For a damp soil, if there is periodical freezing of the surface, the problem is completely different and much more complex, as the condition on the freezing front (whose position varies with time) must be obeyed: this is a *Stefan problem*.

Applying the above solution to the successive ice ages, with $\Theta = 100,000$ years, a penetration depth of 2.8 km is found. This shows that it is possible to look for the past temperatures averaged over a century, using temperature profiles collected in oil drillings from dry regions, where the variations of underground water flows should not have modified the signal.

Model of the 'thermal plate'

In the model of the 'thermal plate', the lithosphere and asthenosphere are assumed to move away from the spreading centre, along the x-axis, with a constant velocity v and down to a fixed depth $z = H$. $Q = 0$ is assumed everywhere. $T = 0$ at the upper boundary $z = 0$. The temperature at z=H, and also for $x = 0$, remains constant and equals T_m. Defining:

$$T = T_m z/H + T_1(x, z) \tag{9.23}$$

the boundary conditions for $z = 0$ and for $z = H$ become $T_1 = 0$.

The increase in temperature T_1 can therefore be expressed with the series:

$$T_1 = \sum_{n=1}^{\infty} F_n(x) \cdot \sin \frac{n\pi z}{H} \tag{9.24}$$

The equation (9.16) can be written with Eulerian coordinates:

$$\kappa \left(\frac{\partial^2 T}{\partial x^2} + \frac{\partial^2 T}{\partial z^2} \right) = v \frac{\partial T}{\partial x} \tag{9.25}$$

This comes down to a differential equation in x only:

$$\frac{d^2 F_n}{dx^2} - \frac{v}{\kappa} \frac{dF_n}{dx} - \frac{n^2 \pi^2}{H^2} F_n = 0 \tag{9.26}$$

As F_n must remain finite when x becomes infinite, only one of its two solutions is acceptable. Defining $Pe = vH/(2\kappa)$ (the *Péclet number*), the solution is:

$$F_n(x) = C_n \exp\left[-\left(\sqrt{Pe^2 + n^2\pi^2} - Pe \right) x/H \right] \tag{9.27}$$

It must obey the condition for $x = 0$:

$$T_1(0, z) = T_m \left(1 - \frac{z}{H} \right) = \sum_{n=1}^{\infty} C_n \sin \frac{n\pi z}{H} \tag{9.28}$$

Only the term in C_n remains when the two sides are multiplied by $\sin(n\pi z/H)$ and integrated over one period. As T_1 is an odd function of z:

$$\int_0^H T_m\left(1-\frac{z}{H}\right)\sin\frac{n\pi z}{H}\,dz = C_n\int_0^H \sin^2\frac{n\pi z}{H}\,dz \tag{9.29}$$

One finds:

$$C_n = \frac{2}{n\pi}T_m \tag{9.30}$$

The surface geothermal gradient is:

$$\gamma_0 = \left.\frac{\partial T}{\partial z}\right|_{z=0} = \frac{T_m}{H}\left[1+2\sum_{n=1}^{\infty}\exp\left(-\beta\frac{x}{H}\right)\right]$$

$$\beta = \sqrt{Pe^2+n^2\pi^2}-Pe \tag{9.31}$$

Note that it reaches infinity for $x=0$. For $x>0$, the first term ($n=1$) rapidly becomes the only one. And as Pe varies between 17 and 50, $\beta\approx\pi^2/(2Pe)$. For Lagrangian coordinates, x must be replaced by vt, and v disappears from the solution:

$$\gamma_0 \approx \frac{T_m}{H}\left[1+2\exp\left(-\frac{\pi^2\kappa t}{H^2}\right)\right] \tag{9.32}$$

9.6 SOLUTIONS FOR A HALF-SPACE WITH A SINGULAR POINT

Let us look for a solution of equation (9.3) of the form $T=f(u)$, with $u^2=z^2/\kappa t$:

$$\frac{\partial u}{\partial z} = \frac{1}{\sqrt{\kappa t}}$$

$$\frac{\partial u}{\partial t} = -\frac{1}{2t} \tag{9.33}$$

The equation becomes: $f''f'=-u/2$, which can be integrated easily:

$$f' = C\exp\left(-\frac{u^2}{4}\right) \tag{9.34}$$

A general solution of the diffusion equation is therefore (see Appendix A.3):

$$T = T_0\,erfc\left(\frac{z}{2\sqrt{\kappa t}}\right) \tag{9.35}$$

This corresponds to a half-space with an initial temperature of zero, which remains zero at infinity, whereas the surface $z=0$ jumps to the temperature T_0 at the time $t=0$. A time interval of the order of $z^2/4\kappa$ is necessary to modify the temperature in a layer of thickness z. In the upper mantle, $\kappa\approx 25\,\text{km}^2/\text{Ma}$. One million years are therefore necessary for a temperature variation at the surface to reach a depth of 10 km, 30 Ma to reach 60 km. These are 'geological' time-scales. In geodynamics, *thermal diffusion processes are the first considered to explain phenomena occurring over millions of years without any apparent cause.* But this hypothesis must be supported by the estimation of orders of magnitude (see Section 14.8).

Faster warming of deep rocks is possible, following the intrusion of magmas, and faster cooling, following a freeing of water which allows melting (see Section 7.3).

The differentiation of equation (9.35) relative to z provides the solution of the following problem: warming of a terrain after irruption, at $t = 0$, of some heat quantity Q per square metre, at the level $z = 0$.

$$T = \frac{Q}{2\rho C\sqrt{\pi\kappa t}} \exp\left(-\frac{z^2}{4\kappa t}\right) \tag{9.36}$$

It is possible to model in this way the temperatures around a dyke, a sill or magma underplating. The intruding heat is mainly the latent heat of the magma. Its solidification happens at a rate controlled by the diffusion of heat.

9.7 TEMPERATURES IN A POLAR ICE CAP

The equation (9.16) is used to calculate the temperature field in a '*cold*' ice cap, i.e. a polar ice cap below melting point of ice, with the possible exception of the bed. (Glaciers entirely at the melting point of ice, like most glaciers in temperate regions, are called *temperate*.)

The flow there does not differ much from a simple shear, i.e. the velocities are sub-parallel to the x-axis. If the boundary conditions are varying very slowly with x and y, the same must be true of the temperatures. Then $\nabla^2 T$ is nearly reduced to $\partial^2 T/\partial z^2$. The component of the gradient $\partial T/\partial x$ is small compared to $\partial T/\partial z$, but v_x is large compared to v_z. The two terms of $\mathbf{v}\cdot\mathbf{grad}\, T$ must therefore be kept:

$$\kappa\frac{\partial^2 T}{\partial z^2} + \frac{Q}{\rho C} = \frac{\partial T}{\partial t} + v_x\frac{\partial T}{\partial x} + v_z\frac{\partial T}{\partial z} \tag{9.37}$$

Robin (1954) has considered the following model:

- The regime is steady.
- The viscous dissipation of heat Q is negligible.
- The temperatures are the same as if the whole surface layer of snow and firn were replaced by the same weight of compact ice. The thickness of the modified ice sheet is noted H.
- This thickness is uniform; the bed is plane (it is the plane $z = 0$).
- The temperatures do not depend on x. For $z = H$, the temperature equals T_S, the temperature measured 10 m below the surface, where the yearly oscillation is damped out.

- The vertical velocity is $v_z = -az/H$. As we assume a steady state, a is the surface accumulation, in metres of ice per year.

It is worth noting that, as ice is incompressible;

$$\frac{\partial v_x}{\partial x} + \frac{\partial v_y}{\partial y} = -\frac{\partial v_z}{\partial z} = \frac{a}{H} \tag{9.38}$$

Against the bed, the velocity v_x must therefore increase slowly downstream. Two cases are possible:

1. *Ice cap with a temperate bottom.* The melting point is reached at the bottom, which enables sliding on the bed and/or deformation of loose, water-saturated, unfrozen sediments below the ice. As the melting temperature decreases with pressure, it is of around −2°C at the bottom of an ice cap. It will be noted T_f.
2. *Ice cap with a cold bottom.* The ice sticks to the bed, but there is a very strong shear in a bottom boundary layer, where the model is no longer valid. In this case, the condition at the bottom is the conservation of the terrestrial heat flux. As the conductivity of ice is close to the conductivity of the bed, we will write the condition: $\partial T/\partial z = -G_0, G_0$ being the geothermal gradient.

The heat equation reads:

$$\kappa\frac{d^2T}{dz^2} = -\frac{a}{H}z\frac{dT}{dz} \tag{9.39}$$

Hence:

$$\frac{dT}{dz} = -G\exp\left(-\frac{az^2}{2\kappa H}\right)$$

$$T = T_S + G\sqrt{\frac{\pi\kappa H}{2a}}\left[erf\left(\sqrt{\frac{aH}{2\kappa}}\right) - erf\left(\sqrt{\frac{a}{2\kappa H}}z\right)\right] \tag{9.40}$$

If the bottom is cold, $G = G_0$ (Figure 9.3a). If it is temperate, G is determined by $T = T_m$ for $z = 0$.

$$T_m = T_S + G\sqrt{\frac{\pi\kappa H}{2a}}erf\left(\sqrt{\frac{aH}{2\kappa}}\right) \tag{9.41}$$

The heat excess, $\kappa(G_0 - G)$, is used to melt the ice (Figure 9.3b).

This model allows a rough estimate of the thickness of an old ice cap, which is interesting for climatologists modelling glaciations. Indeed, in central regions, the bottom is either cold, but close to T_m, or temperate, but with G close to G_0. It is possible to approximate erf by 1, as $aH/2\kappa > 1$. And, in degrees Celsius, T_m is small compared to T_S. We can then write:

$$H \approx \frac{2a}{\pi\kappa}\left(\frac{-T_S}{G_0}\right)^2 \tag{9.42}$$

This equation shows the factors controlling the thickness of an ice cap. The value of

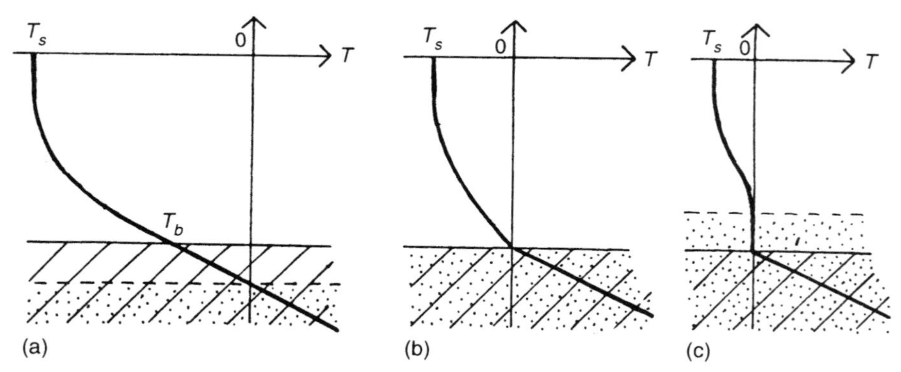

Figure 9.3. Temperature profiles in a steady polar ice cap. The bed corresponds to the hatched area. Dots indicate liquid water (water-saturated loose soil or temperate ice). (a) Cold bottom: the geothermal flux completely penetrates the ice. (b) Cold cap with temperate bottom: part of the geothermal flux enters the ice, the remainder is used to melt the ice. (c) Temperate bottom layer: no geothermal flux crosses the bottom layer and if a temperature gradient develops above, it is only due to heat dissipation by deformation of the ice.

κ for ice is $36.6\,\mathrm{m^2\,yr^{-1}}$. For a Precambrian basement, such as in East Antarctica or Greenland, $G_0 = 0.022\,\mathrm{K/m}$. This yields: $H \approx 18\,a\,T_s^2$. Thus, if we have $T_S = -30°\mathrm{C}$ and $a = 0.2\,\mathrm{m/yr}$ (as in central Greenland), or $T_S = -60°\mathrm{C}$ and $a = 0.05\,\mathrm{m/yr}$ (as in central Antarctica), the thickness is the same: $H = 3,240\,\mathrm{m}$, despite the very different horizontal dimensions.

This rule of thumb is not valid away from the central regions of ice caps. Ice streams with temperate bottoms appear at the edges, where G can become very small and even null (there appears then a temperate bottom layer, see Figure 9.3c). The ice cap profile is then determined first and foremost by the flow law of ice.

9.8 SUBSIDENCE OF THE OCEANIC LITHOSPHERE

The lithosphere is a rigid layer submitted to isostasy. At the axis of mid-ocean ridges, it consists only of the crust of basalts and gabbros. As the oceanic crust moves away, along with the underlying asthenosphere, everything cools down. The lithosphere thickens and includes more and more of the upper mantle. Simultaneously, as cooling slightly increases density, isostasy comes into play and the seafloor gets progressively lower.

This subsidence explains the formation of coral reefs and calcareous algae around 'high islands' (Figure 9.4a), the formation of atolls (Figure 9.4b) and the formation of guyots (Figure 9.4c).

A *high island* is a volcanic cone, which built up rapidly enough to avoid erosion by the waves. Otherwise, it forms a truncated cone, going deeper and deeper below the sea surface because of subsidence. If the waters are too cold to allow the development of corals, this forms a *guyot*. They are abundant in the north-east

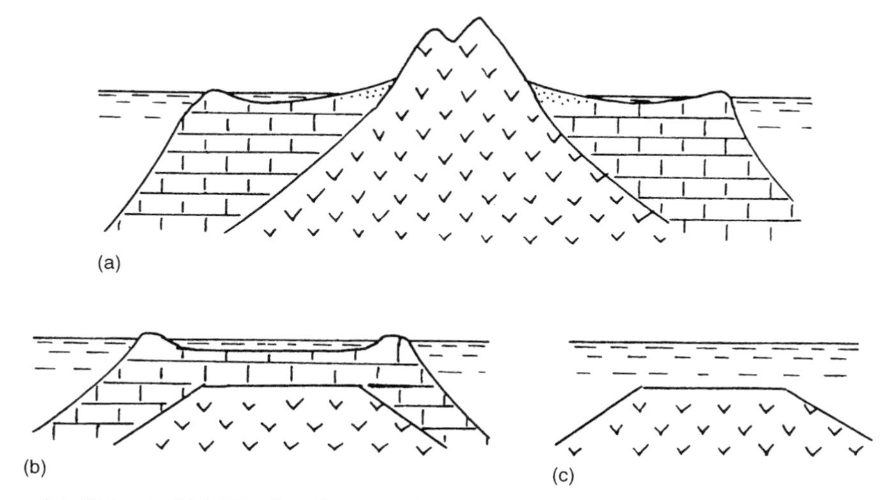

Figure 9.4. Volcanic 'high island', with a coral barrier; (b) atoll; (c) guyot.

Pacific. Their flat summit is currently 900 to 1,700 m below the sea surface; their slopes are of ca. 22°.

An *atoll* is a coral edifice on top of a guyot. Corals live only at less than 40 m from the surface, but their growth upward compensates for the subsidence. The growth is higher at the periphery, where the ocean renews the nutritive plankton. The atolls are therefore shaped like flat rings (cut with passes), surrounding a lagoon.

In fact, this is not as simple, because the sea level fluctuates with time at rates faster than the very slow subsidence of oceanic plates (these world-wide variations are called *eustatic*). The atoll is therefore subaerial at some periods, corals forming only on its flanks. During these emerged periods, the original aragonite and Mg-rich calcite are altered. They form a calcite cement with a low Mg content. The surface is corroded by rainwater, charged in carbonic gas, and a soil is created.

Past eustatic variations are most interesting in stratigraphy, because they brought about marine transgressions on continental platforms, conditioning sedimentation. The sea level has thus been estimated as higher by 220 ± 35 m during a large part of the Cretaceous, 100 to 80 Ma ago. The distinction between the relative contributions of eustatism and tectonic vertical movements is difficult, because of the lack of a fixed reference. (The highest amount of data is collected by oil companies drilling on passive continental margins, but these deepen progressively under the weight of sediments.) Atoll drillings reveal the periods of emergence, providing invaluable pieces of information. Such drillings were for example performed in 1970–1974 at Bikini and Enewetak, in the Marshall Island (the US nuclear tests were done at Bikini between 1946 and 1958, and at Enewetak[1] in 1969). The basaltic basement was reached at Enewetak in two

[1] Then spelled Eniwetok.

drillings, at 1,411 m and 1,287 m respectively. It dates back to the Palaeocene (54 to 65 Ma).

According to Lincoln and Schlanger (1991), the drillings show 3 sea-level drops, each larger than 30–50 m, during the Oligocene (35, 30 and 25 Ma ago), and 4 drops of similar magnitudes during the Miocene (15, 12, 10 and 5 Ma ago).

The only explanation of a 30–50 m sea-level drop seems to be the formation of a huge continental ice-sheet (12 to $20 \times 10^6\,\text{km}^3$ of ice, ignoring isostatic adjustments). To fix ideas, the actual Greenland ice sheet has a volume of $2.6 \times 10^6\,\text{km}^3$, and the actual Antarctic ice sheet of $30 \times 10^6\,\text{km}^3$. Glaciations in the northern hemisphere during the Oligocene and the Miocene can be ruled out, but the glaciation of Antarctica during these epochs is still very poorly known. (For more details, see Section 18.2.)

9.9 MODELLING THE COOLING OF OCEANIC PLATES

The cooling of the ocean floor and the resulting subsidence have been modelled very early. The ages t of the oceanic floors are well known, and the data collected all over the ocean can be used to look for a universal law $T(z, t)$ (where z is the depth below the seafloor). Its derivative relative to z will provide the total geothermal flux (through conductivity and hydrothermal circulation). And its integral will provide subsidence, as shown below.

Let D be the ocean's depth, ρ_w its density, H the compensation level (fixed), ρ_m the upper mantle density, α its volumic thermal dilatation coefficient (Figure 9.5). In this calculation, the crust is assimilated to the mantle, and the sedimentation is neglected. The indices 0 indicate values at the ridge axis.

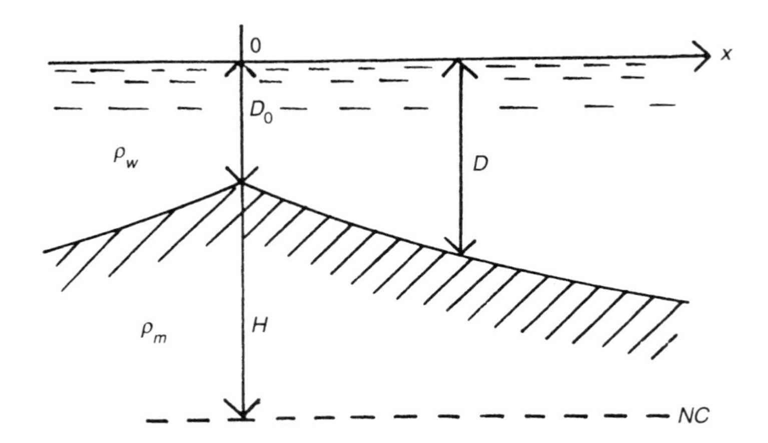

Figure 9.5. Notations used for the study of ocean floor subsidence.

As the temperature T_0 is independent from z, isostasy can be written as:

$$\rho_w D + \int_0^{H-D} \rho_m(1-\alpha T)\, dz = \rho_w D_0 + \rho_m(1-\alpha T_0)(H - D_0)$$

$$= \rho_w D_0 + \rho_m(1-\alpha T_0)[(H-D) + (D-D_0)] \quad (9.43)$$

Neglecting $\rho_m \alpha T_0$ compared to $(\rho_m - \rho_w)$, we get:

$$D = D_0 + \frac{\rho_m \alpha}{\rho_m - \rho_w} \int_0^{H-D} (T_0 - T)\, dz \quad (9.44)$$

Half-space model

The first physical model used was of a half-space, with $T(\infty, t) = 0$ and $T(z, 0) = T_0$, constant. It is justified only very close to the spreading centre. The cooling has affected only the shallowest layers, then, and the condition at the lower boundary is not important. But this model has been used much further.

Equation (9.35) is the solution. This model predicts a total geothermal flux decreasing as $t^{-1/2}$ and a subsidence increasing as $t^{1/2}$. Namely:

$$D - D_0 = \frac{\rho_m \alpha}{\rho_m - \rho_w} \int_0^{H-D} T_0\, erfc\left(\frac{z}{2\sqrt{\kappa t}}\right) dz$$

$$\approx \frac{\rho_m \alpha}{\rho_m - \rho_w} \int_0^{\infty} T_0\, erfc\left(\frac{z}{2\sqrt{\kappa t}}\right) dz$$

$$= 2\sqrt{\frac{\kappa t}{\pi}}\, T_0 \frac{\rho_m \alpha}{\rho_m - \rho_w} = b\sqrt{t} \quad (9.45)$$

With $\alpha = 3.1 \times 10^{-5}\,\mathrm{K}^{-1}$, $\rho_m/\rho_w = 3.33$, $\kappa = 25.35\,\mathrm{km}^2\,\mathrm{Ma}^{-1}$ and $T_0 = 1{,}335°\mathrm{C}$, we get a value of $b = 336\,\mathrm{m\,Ma}^{-1/2}$, i.e. a subsidence of 1,063 m for $t = 10$ Ma and of 3,274 m for $t = 80$ Ma. Let us note that the addition to this model of a constant geothermal flux, to make it more realistic, would not change the subsidence law.

'Thermal plate' model

This model was already described. There is still no geothermal gradient at $t = 0$, but there is one for $t = \infty$. We demonstrated that this was leading to:

$$T = T_m\left[\frac{z}{H} + \sum_{n=1}^{\infty} \frac{2}{n\pi} \exp\left(-\beta \frac{vt}{H}\right) \sin\left(\frac{n\pi z}{H}\right)\right] \quad (9.46)$$

The sines are small when z is close to H, and the integral of equation (9.44) is taken

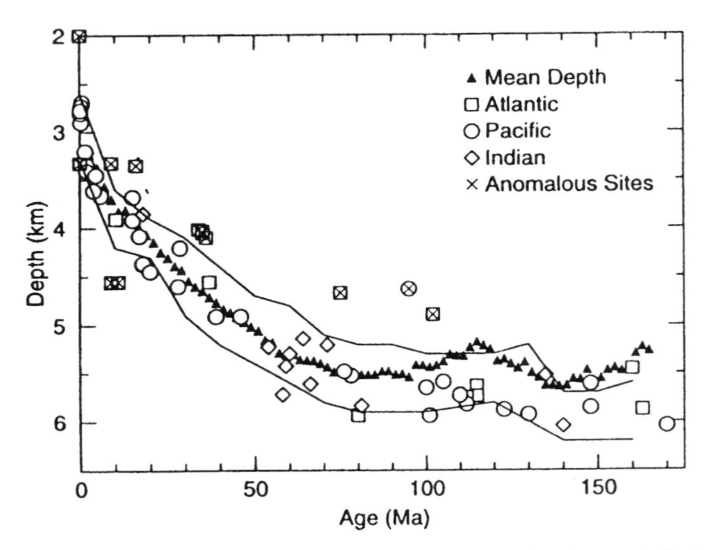

Figure 9.6. Mean oceanic depths as a function of age (black triangles). The other symbols show the depths of the basaltic basement below the sediments, obtained through deep-sea drilling. The continuous lines delimit a region regrouping 2/3 of the points in the North Pacific. From Carlson and Johnson (1994); © AGU.

to $z = H$ and not to $z = H - D$. Only the odd terms $(n = 2p + 1)$ remain. The resulting subsidence is:

$$D - D_0 = \frac{\rho_m \alpha T_m H}{2(\rho_m - \rho_w)} \left[1 - \frac{8}{\pi^2} \sum_{\rho=0}^{\infty} \frac{1}{(2\rho + 1)^2} \exp\left(-\beta \frac{vt}{H} \right) \right] \tag{9.47}$$

For $t = 0$, we rediscover a classical relation:

$$1 + \frac{1}{3^2} + \frac{1}{5^2} + \frac{1}{7^2} + \ldots = \frac{\pi^2}{8} \tag{9.48}$$

With $T_m = 1{,}450°\text{C}$, $H = 95\,\text{km}$, and the same values of ρ_m/ρ_w and α, for $t \to +\infty$, the subsidence tends toward a limit value of 3,050 m.

A careful study by Carlson and Johnson (1994) showed that neither model accounts properly for the subsidence at all ages, whatever the values given to the parameters. The parameters presented above are valid until 81 Ma, but not beyond. As shown in Figure 9.6, the mean depth of the oceans decreases instead of increasing between 83 Ma and 115 Ma, and again between 140 Ma and 165 Ma. This phenomenon is too general around the Earth to be attributed to hot spots. In my opinion, the physical models used are too simplistic. We need to better examine the conditions at the lower boundary of the lithospheric plate.

9.10 CROSSING OF A PARTIALLY MOLTEN LVZ BY THE GEOTHERMAL FLUX

The existence of a zone where the seismic velocities are reduced (the LVZ) was mentioned in Section 5.5. This zone extends below all the oceans, and part of the continents, deeper than 70–80 km. The consensus about its origin is that the upper mantle is very slightly molten (a few percents of magma inclusions, at most). But no modelist seems to have taken that into account.

The fluidity of a rock depends a lot more on the temperature than on the pressure. The lower boundary of the lithospheric plate is therefore assimilated to an isotherm. The LVZ belongs to the asthenosphere, and, as a first approximation, forms its upper layer in contact with the lithosphere.

The thermal evolution of oceanic lithospheric plates is therefore a problem with a 'free' boundary, moving with time, like the Stefan problem but more complex. The depth of the limit lithosphere-LVZ is one of the unknowns. But the problem is still mathematically determined because two conditions are obeyed at this limit, and not only one:

- The temperature must be the temperature of the rock solidus. The magma and different minerals must be locally in thermo–dynamical equilibrium.
- The geothermal flux is fixed. The temperature of the solidus depends on pressure, creating a fixed geothermal gradient in the LVZ. A latent heat flux is added to the corresponding conductibility heat flux, if magma forms continuously at the base of the LVZ, migrates upward and solidifies at the roof of the LVZ.

Unfortunately, these two fluxes cannot be expressed mathematically with the current knowledge. *If the chemical composition of the LVZ was completely fixed*, the Clapeyron relationship could be applied. Designating by L the fusion heat per unit mass, by ρ_L and ρ_S the respective densities of the liquid and the solid, by T the absolute temperature and by p the pressure, we should have:

$$\frac{L}{T} = \left(\frac{1}{\rho_L} - \frac{1}{\rho_S}\right)\frac{dp}{dT} \tag{9.49}$$

The pressure p varies with the depth z according to the Pascal Law ($dp/dz = \rho_S g$). Consequently, the geothermal gradient would have a fixed value:

$$\frac{dT}{dz} = \frac{dT}{dp}\frac{dp}{dz} = \frac{Tg}{L}\left(\frac{\rho_S}{\rho_L} - 1\right) \tag{9.50}$$

For a basaltic magma, $L = 4.7 \times 10^5\,\mathrm{m^2\,s^{-2}}$, $T = 1400\mathrm{K}$ to $1700\mathrm{K}$. With ρ_S/ρ_L between 1.12 and 1.15, the Clapeyron equation yields $dT/dz = 4.5 \pm 0.9°/\mathrm{km}$ (the adiabatic gradient, 10 times smaller, can be neglected). This fits the measured melting temperatures of a constant-composition peridotite at different pressures.

The corresponding geothermal flux, due to diffusion only, would then be approximately $14\,\mathrm{mW/m^2}$. The magma flux cannot be larger than 0.1 km/Ma, as otherwise seismologists would have detected a basalt layer in the upper mantle of old oceans, well below the crust. The corresponding maximal latent heat flux is $4.2\,mW/m^2$, which brings the total to $18.2\,mW/m^2$. However, at the surface of some old oceanic plates (144 to 163 Ma old), the measured fluxes reach $49.4 \pm 16.9\,mW/m^2$, 2 to 4 times larger. This heat can only come from below the LVZ. How did it cross it?

This paradox arises because *the Clapeyron relationship cannot be applied.* It is demonstrated by applying the Carnot theorem to a closed cycle described in a reversible way: melting at the pressure $p + dp$, solidification at the pressure p. But this is not a closed cycle anymore, if the magma migrates and the rock is modified. A correct calculation should start from the chemical potentials of the different components, which are functions of concentration and pressure. Laboratory measurements lead to errors, because the concentrations are not left to evolve until equilibrium at each pressure.

Because of this migration of magma inside the LVZ, its lower part becomes progressively sterile and its melting temperature increases. The geothermal gradient across the LVZ can therefore increase as much as necessary for the geothermal flux to cross. But this is limited in time. After a while, there is no more magma formation and migration. The cold from the surface solidifies all the magma inclusions still remaining, and the partially molten LVZ disappears.

The model of the 'thermal plate' assumes a constant temperature at a certain depth inside the LVZ. If the mean temperature actually increases with age, the subsidence could stop and even reverse.

9.11 ESTIMATION OF GEOTHERMAL FLUXES

9,684 measurements of the seafloor geothermal flux transmitted by conductibility were recently compiled and analysed (Pollack *et al.*, 1993; Stein and Stein, 1994). The measurements on continental shelves, or too close to be independent, were removed. The remaining 5,000 measurements are well spread over the oceans north of latitude 40°S. Mean values are calculated by age intervals. They fit reasonably well the empirical formula GDH1 of Stein and Stein (1994). Point values can sometimes be as far as 30% away from these mean values.

As shown in Figure 9.7, the theoretical model of the thermal plate with $T_m = 1{,}450°C$ and $H = 95\,km$ corresponds roughly to the data of subsidence along the ages, and to geothermal fluxes measured for $t > 55\,Ma$. This age of 55 Ma corresponds to the time needed for hydrothermal circulation to disappear, through 2 processes: (1) the connecting pores and fissures of the rock get obstructed by hydrothermal sediments; (2) the sediment layer deposited is thick and watertight enough. The GDH1 model predicts a geothermal flux (in mW/m^2):

$$\begin{cases} \varphi = 510\sqrt{1\,Ma/t} & (t \leq 55\,Ma) \\ \varphi = 48 + 96\exp(-t/36\,Ma) & (t \geq 55\,Ma) \end{cases} \tag{9.51}$$

For all oceans, the GDH1 model predicts a heat emission of 32.0 TW (1 terawatt equals 10^{12} W), whereas the heat transmitted through conductibility and measured is

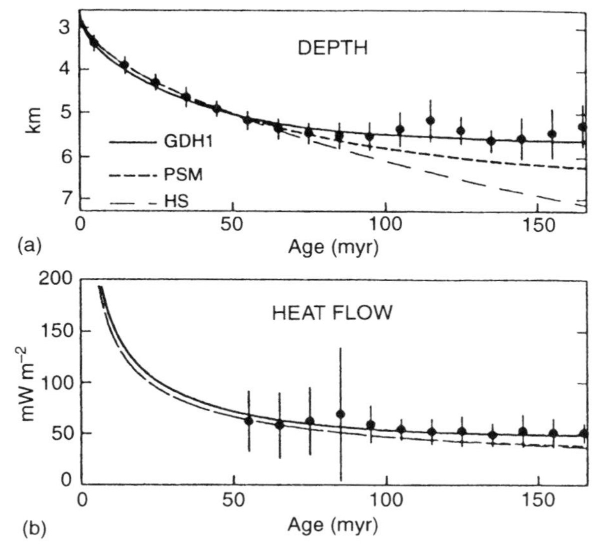

Figure 9.7. Depth and geothermal flux as a function of seafloor age. The averages and error bars are computed over 10 Ma. PSM: thermal plate model. HS: half-space model. From Stein and Stein (1994); © AGU.

only 21.0 ± 3.9 TW. The heat transmitted by hydrothermal circulation would therefore be around 11.0 TW.

Pollack *et al.* (1993) also analysed the geothermal fluxes measured on continents, as a function of their geology. Archean terrains yield the lowest values (51.5 mW/m^2 in average) and Cenozoic crystalline terrains yield the highest values (97.0 mW/m^2 in average). The weighted average for all the continents is 65 mW/m^2, and the total emission of heat by the continents is 12.2 TW.

The heat coming from inside the Earth would therefore be 44.2 TW.

These geothermal fluxes lower than 0.1 W/m^2 are completely negligible in the thermal balance at the surface of the Earth, which gives its temperature. The ground radiates in the infrared wavelengths like a black body, i.e. 315.6 W/m^2 at 0°C. Outside the atmosphere, the solar radiation brings 1,400 W/m^2.

But the heat from inside the Earth is huge, without common measure with the visible and spectacular heat provided by volcanic events away from mid-ocean ridges. The lava lake of the Niragongo volcano, 4 hectares in area, was radiating 6,000 MW, enough to cater for the energy needs of central Africa. But this is only 0.006 TW. The total energy dissipated by volcanism away from the ridges reaches at most 0.1 TW.

9.12 ORIGINS OF TERRESTRIAL HEAT

Most of the terrestrial heat comes from the natural **radioactivity** of rocks. Table 9.1 shows the main characteristics of radioactive isotopes (see also Figure 17.9).

Table 9.1. Characteristics of radioactive isotopes (1 pW = 1 picowatt = 10^{-12} W)

Radio-isotope	Content of the chemical element (in weight %)	Stable descendant	Half-life (in Ga)	Heat emitted (in pW/kg)
K 40	0.0119	A 40 + Ca 40	1.31	27.6
V 50	0.25	Cr 50 + Ti 50	> 1000	
Rb 87	27.85	Sr 87	47	1.15
La 138	0.089	Ce 138	200	
Ce 142	11.07	Ba 138		
Sm 147	15.07	Nd 143	1300	
Lu 176	2.6	Hf 176 + Yb 176	30	
Re 187	62.93	Os 187	43	
Th 232	100	Pb 208	13.9	26.4
U 235	0.7204	Pb 207	0.71	490
U 238	99.2739	Pb 206	4.51	93.8

Even if radioactive nuclei such as C14 or Be10 are currently forming in the atmosphere under the effect of cosmic radiation, most appeared well before the formation of the Sun and its planets, and have not stopped disintegrating since.

Let us designate with c_K% the mass content of a rock in potassium, $c_{Th} \times 10^{-6}$, $c_U \times 10^{-6}$ and $c_{Rb} \times 10^{-6}$ the mass contents in thorium, uranium and rubidium.

Accounting for the current isotopic abundances, the heat produced by radioactivity in a rock is: $(33.5c_K + 26.4c_{Th} + 98.0c_U + 0.32c_{Rb})$ pW/kg. (The other elements containing radio-isotopes do not produce enough heat, or are too rare in nature to need being accounted for.)

It is assumed that the silicate Earth (crust + terrestrial mantle) has globally the same composition than the most common type of rocky meteorites, the *chondrites*. It is, however, depleted in some moderately volatile elements: Na, K, Mn, Pb, Cr (see Section 17.12). These meteorites enclose small round masses, mainly of olivine or enstatite, called chondrules, in a fine-grained matrix of pyroxene, olivine and ferronickel. It is true that the different crustal rocks and the upper mantle (the only one that can be sampled) have compositions totally different from chondrites, but this results from segregations after the formation of the Earth and its general melting. Chondrites are thought to come from the fragmentation of much smaller proto-planetary bodies, which never reached this stage of melting and intense segregation.

The heat currently emitted by the ordinary chondrites varies from 5.4 to 6.3 pW/kg. The mass of the terrestrial crust and mantle is 4.06×10^{24} kg (at 1%). This hypothesis yields to an emission of heat by radioactivity inside the Earth of *21.9 to 25.6 TW*. It is possible that part of the potassium was captured by the core (contrary to Th and U, K dissolves in iron), but this does not change anything to our balance. As the Earth emits 44.2 TW, the origin of 18.6 to 22.3 TW must be explained.

Stored sensible heat

An important contribution to the geothermal flux comes from the stored sensible heat stored. This is not only the heat which appeared when the Earth formed by aggregation of planetary bodies, and their kinetic energies dissipated into heat. This particular heat would have disappeared in around one hundred million years, if there had been no radioactivity. But the heat produced by radioactivity could not be evacuated at the same pace.

During its formation, 4.55 Ma ago, the Earth contained 2 times more U238, 85 times more U235 (which formed then 23.4% of uranium), and 12 times more K40. According to Jackson and Pollack (1984), the heat Q dissipated by radioactivity was 5 times stronger then. But we shall see that thermal convection with internal heating ensures a surface geothermal flux proportional to $Q^{0.20}$ *or* $Q^{0.25}$. It would have been only 38% or 50% more than the current geothermal flux.

Therefore, the cooling of the Earth accelerates as it goes, because of the decreasing radioactivity. From the existence of olivine crystals in Precambrian rocks, McKenzie and Weiss (1975) infer that the upper mantle was hotter by 200°C, 3 Ga ago. If we assume it was the same all over the Earth, we can deduce that the cooling provided in average ca. 10 TW during the last 3 Ga. It could therefore provide ca. 20 TW now.

From a coarse cooling model and a few geological constraints, Davies (1980) estimated it to range between 14.4 and 22.6 TW. Schubert *et al.* (1980) only found 2.8 to 9.8 TW, but they were estimating the flux coming out of the mantle to reach 27.9 TW, and we know now that this was too small by ca. 9 TW.

9.13 CONCENTRATION OF RADIOACTIVITY IN GRANITIC TERRAINS

The amount of essential radioactive elements in the rocks (potassium, thorium and uranium) is highly variable.

Potassium is not well assimilated in olivines and pyroxenes, but can be present in small quantities in plagioclases. In mafic rocks, it is in noticeable quantity only if there has been fractionation of a basaltic magma into its ultimate terms, trachytes and phonolites. Conversely, potassium is abundant in alkali feldspars, amphiboles and micas from felsic rocks. In the latter minerals, its ionic radius (1.46 Å) enables it to replace OH (see Section 3.1).

The ions Th^{4+} and U^{4+} have a radius of 1.08 Å, close to the radii of Na^+ and Ca^{++}. But their charges and chemical affinities make them incompatible with most minerals. They are found diffusely at the limits of grains, in the micro-crystalline paste solidified last, in the microscopic inclusions of phenocrystals. Or in secondary minerals such as zircon (SiO_2–ZrO_2), sphene (SiO_2–TiO_2–CaO), apatite ($(PO_4)_3(F,OH,Cl)Ca_5$), monazite ($PO_4(Ce,Y,La,Th)$). However, U^{4+} gets oxidized easily into U^{6+} or $(UO_2)^{++}$, soluble forms which enable it to migrate. Close to the surface, it can be washed out at nearly 80%. This explains the formation of

Table 9.2. Average content in radioactive elements of some rocks. From Létolle and Cheminée (in Coulomb and Jobert, 1976)

	K(%)	Th(10^{-6})	U(10^{-6})
Oceanic peridotites	0.0054	0.042	0.019
Tholeiitic basalts	0.14	0.18	0.10
Alkali basalts	1.40	5.88	1.55
Andesites, dacites	0.84	2.32	0.91
Granites	3.5	26–56	5.7–15.0

Table 9.3. Average content in radioactive elements of the continental crust (Rudwick and Fountain, 1995).

Crust	K(%)	Th(10^{-6})	U(10^{-6})	Rb(10^{-6})	Heat (pW/kg)
Lower	0.25	1.2	0.2	11	63
Middle	0.83	6.1	1.6	62	365
Upper	1.41	10.7	2.8	112	640

pitchblende deposits, mainly made of UO_2. This case excepted, uranium is generally 2 to 4 times rarer than thorium.

Table 9.2 was borrowed from Létolle and Cheminée (in Coulomb and Jobert, 1976), and provides the average contents of a few rocks. Table 9.3 provides the average contents in radioactive elements of continental crust, compiled by Rudwick and Fountain (1995), and the resulting heat.

Radioactive elements visibly concentrated in the upper continental crust. Therefore, they are abundant in sediments and sedimentary rocks, and are caught by granitic magmas. Whereas the radioactivity of the oceanic crust is negligible (unless it is overlain by a thick layer of continental sediments), and does not contribute to the geothermal flux, the radioactivity of the continental crust provides more than half of the heat flux observed at the surface. According to Rudwick and Fountain (1995), it provides in average $37\,\mathrm{mW/m^2}$, i.e. *7.4 TW* for all continents ($2.01 \times 10^8\,\mathrm{km^2}$). If we accept the chondritic hypothesis, the continental crust makes only 1/135th of the Earth's volume and 1/267th of its mass, but encloses 1/4th of its radioactivity.

To model convection currents in the mantle, it is important to know the geothermal fluxes at the base of the lithosphere. This is called the *reduced flux*. From recent estimates, it would average $65 - 37 = 28\,\mathrm{mW/m^2}$ below the continents.

Birch and Roy (1968) remarked that, locally, there generally was a linear relationship between the geothermal flux φ and the surface production of radio-

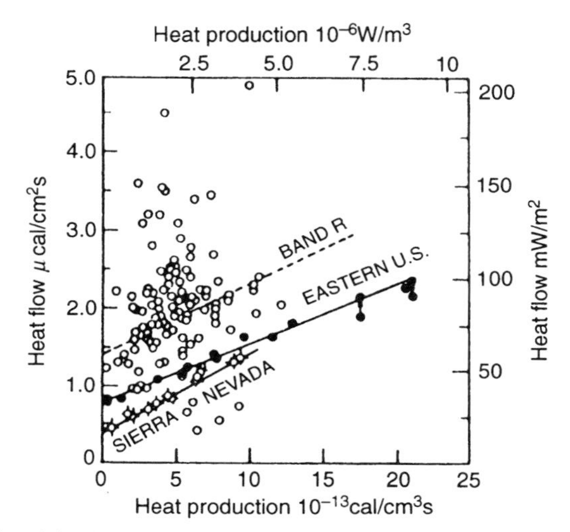

Figure 9.8. Heat produced by the radioactivity, per unit volume, of shallow rocks (in x), and surface geothermal flux (in y). There is a good linear correlation for the Sierra Nevada (crossed circles) and the Eastern US (black circles). But there is no correlation for the Basin and Range province, because of magmatic processes. From Sclater *et al.* (1980); © AGU.

active heat Q_0. This is shown in Figure 9.8 (which uses the old flux unit, the HFU (for: Heat Flux Unit), equal to $1\,\mu\text{cal}\,\text{s}^{-1}\,\text{cm}^{-2} = 41.86\,\text{mW/m}^2$).

$$\varphi \approx \varphi_r + bQ_0 \tag{9.52}$$

The coefficient b must depend on the distribution of radioactivity inside the column of crust, whereas φ_r must provide the reduced flux (if all the heat produced by the plate indeed comes from its radioactivity). According to Figure 9.8, the reduced flux equals $3.16\,\text{mW/m}^2$ in the Eastern US and $16.9\,\text{mW/m}^2$ below the Sierra Nevada (California). This law is not valid anymore for the Basin and Range province (Nevada), in extension, where the crust is warmed by magma intrusions.

The geothermal gradient is particularly high in the upper crust, and does not depend only on radioactivity, but also on thermal conductivity, much smaller in sedimentary rocks than in crystalline rocks.

The model below is for an 'average' crust and a lithosphere 120-km thick (Table 9.4). It shows which temperatures should be expected at depth. (Assuming each layer is uniform in composition, the heat flux increases linearly with altitude. Thus, the temperature difference between the base and the top of the layer is calculated from the mean heat flux in the layer.)

9.14 LITHOSPHERIC THICKNESS AND BASIN SUBSIDENCE

The geothermal flux in the oldest oceanic lithospheres (144–163 Ma old) is very close to the $48\,\text{mW/m}^2$ limit predicted by the GDH1 model of Stein and Stein (1994). The

Table 9.4. Geothermal flux for an average crust and a 120-km thick lithosphere

	Thickness (km)	Density (Mg/m^3)	Radioact. (mW/m^3)	φ ($\mu W/m^2$)	K ($Wm^{-1} K^{-1}$)	$<dT/dz>$ (K/km)	T (°C)
				65.2			10
Upper crust	12	2.65	1.70		2.1–2.4	26.2–39.3	
				44.8			324–481
Middle crust	14	2.8	1.02		2.3	16.4	
				30.5			553–710
Lower crust	14	2.8	0.18		2.5	11.7	
				28.0			717–874
Mantle lithosphere	80	3.2	0.01		3.1	8.9	
				27.2			1429–1456

reduced flux is practically the same. The reduced flux below the continents is much smaller, less than half for old platforms or cratons. This should not come as a surprise, because GDH1 is not a physical model. It is only a mathematical interpolation formula, valid (approximately) down to 163 Ma. The continents are much older. As we shall see in Chapter 17, they exist since at least 2,500 Ma. As long as orogenic events do not warm it, a continental lithosphere will continue cooling and thickening. If a partially molten LVZ at a depth of 100 km has disappeared after 150 Ma, the cooling will reach a depth of 200 km after an extra time of around $(200^2–100^2)/(4\kappa) \approx 300$ Ma.

Two facts attest this cooling: the rapidity of seismic velocities and the slow subsidence of continental basins.

Residuals of propagation times

Seismologists call *residual* the difference between the propagation time of a seismic wave from the source to the station, and the time predicted by the tables of Jeffreys-Bullen. This comes from the nature and the temperature of the terrains close to the source and close to the station, since, as a first approximation, the velocities very deep in the mantle only depend on the radius. Averaging the residuals in one station, for many large earthquakes, with foci scattered among several seismic regions in the world, the effects of terrains close to the source can be more or less removed. This average residual for a particular station provides an indication of seismic velocities below the station.

In 1979, Poupinet calculated these mean residuals for many seismic stations. The delays reach 1 second for stations in geologically young regions. The residuals are 0.5 to 1.2 second ahead for stations over older and older shields (up to 1.7 s for the Tiksi station, in Yakutia). Differences in the composition of the crust cannot explain that, and Poupinet attributed it to upper mantle temperature differences, i.e. to the thickness of the lithosphere. But the variation of seismic velocities with temperature is small. According to Bottinga and Allègre (1978), the variations for the P waves is:

$$dv_P/dt = -0.48 \times 10^{-3}\,\mathrm{km\,s^{-1}\,K^{-1}} \tag{9.53}$$

To explain residuals of a second, one must assume a cooling of several hundreds of degrees over a few hundreds of kilometres of the wave path, and therefore a lithosphere thicker than 200 km.

Since this pioneering work, the study of residuals has been greatly developed. This is *seismic tomography*, treated in Section 17.4 and in Section 19.11.

Subsidence of continental basins

Continents show large basins more or less circular where sediments have been deposited over hundreds of millions of years. They attest a continuous subsidence, extremely slow, of the order of 20 m/Ma (compare with the erosion of a high relief, 10 times faster). We saw in Section 8.9 that the weight of sediments amplifies the subsidence, by a factor 2 approximately, but that it does not cause it.

The best known example for French people is the *Parisian Basin*, 300 km in diameter. It formed during the Mesozoic and Palaeocene (225 Ma to 43 Ma ago), over a peneplanated Hercynian relief. The maximum sediment thickness (3.2 km) is reached 70 km east from Paris. At the beginning, during the Triassic, Belgium, the Armorican Massif and the Massif Central were emerged, and the water flowed eastward. The Triassic sediments therefore thicken from west to east, reaching 1.3 km in Lorraine. In the Cretaceous, SW Germany was also emerged, and the 'chalk sea' coincided with the current Parisian Basin. In detail, the aftermaths of peripheral events provoked local fluctuations of the sedimentation rate and of subsidence.

In North America, 6.5 km of sediments were deposited in the centre of the *Michigan Basin* (600 km in diameter), and 1.5 km were removed by erosion. This subsidence occurred 530 Ma to 150 Ma ago, with two episodic warmings in the Triassic and Cretaceous. In the *Williston Basin*, covering North Dakota, Eastern Montana and Southern Saskatchewan (1,200 km along its largest dimension), the subsidence was continuous from the Cambrian to the Jurassic, during 400 Ma. Between these two basins, the subsidence of the Iowa craton was of only 1.5 km.

The circular or elliptic hollow shape of these basins should be explained by: (1) the amplification of subsidence by sedimentation; (2) the regionalization of isostatic compensation, the greater as the lithosphere is thicker (see Chapter 13).

9.15 LVZ BELOW THE CONTINENTS

To account for propagation times, seismologists use a model of the upper mantle with several homogeneous layers superimposed, and they adjust the different seismic velocities by trial and error. But the solution found is not always the only one. Roughly, only the ratio of the thickness of a layer to its seismic velocity is fixed.

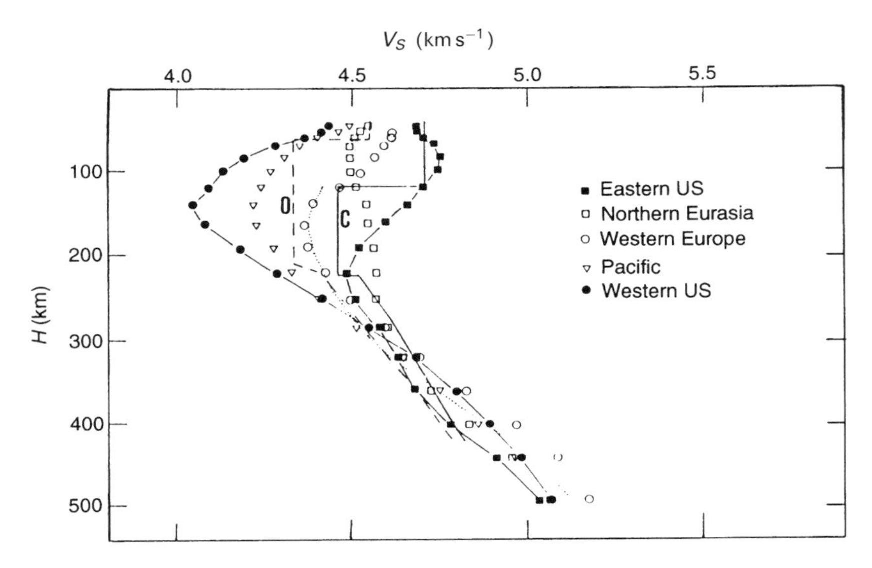

Figure 9.9. Models of seismic velocities obtained from the modes 0, 1 and 2 of the Rayleigh waves. O and C designate the earlier models of Dziewonski (1975) for the oceans and for the continents. From Cara, 1978.

After a time where every model had layers with lower velocities, one or several (up to 4) LVZ, the fashion ebbed away.

The best information comes from Rayleigh waves. In his thesis, Cara (1978) was the first to use the Rayleigh waves of periods between 20 and 100 s, which propagate along the upper mantle (see Figure 2.9). It was necessary to develop a method of spatial analysis of the records, to extract the harmonics, and an efficient method of data inversion. His results are reproduced in Figure 9.9. They show that the velocity variation is progressive, so much so that the limits of the LVZ and especially the lower limit, are very subjective.

An LVZ exists below the great plains of the US and Canada. It progressively deepens and becomes less pronounced as one goes from west to east. It is 70 km to 220 km deep below Arizona and New Mexico, between 115–130 km and 315–380 km below the Canadian shield.

The LVZ below Iberia is 80 to 180 km deep (Corchete *et al.*, 1995). Below the rest of Europe, it is rather between 90 km and 270 km. *But no LVZ is found below Northern Eurasia.*

Could it mean that it is not possible to talk of asthenosphere and isostasy for Northern Europe? Certainly not, the glacio-isostatic uplift of Fennoscandia shows that the concept is still valid. Simply, one must exclude any presence of magma in this asthenosphere. But it may be because the rock is sterile, depleted of the secondary minerals or water which would lower the temperature of its solidus. Nothing is opposed, then, to a temperature high enough to make the asthenosphere relatively fluid.

10

Elastic and isoviscous media

10.1 DEFINITION OF STRESS IN A CONTINUOUS MEDIUM

A *continuous medium* models real materials without considering their fine structure: grains of the rocks, soils or powders, micro-crystals in the metals, molecules or ions of the fluids, atoms of the solids, etc. The parameters defining the medium properties are supposed to vary continuously, if they vary, and to have spatial derivatives. It will therefore be possible to consider infinitely small volumes of matter, and to use differential calculus.

Any portion of a continuous medium, which we isolate in thought, is submitted to distant forces (gravity, electromagnetic forces), proportional to its volume, and to *contact forces*, acting on its boundaries and proportional to their surface. Let us consider an infinitely small portion of a plane, of area dS. Its orientation is defined by the normal unit vector $\boldsymbol{n}$. The two sides exert on each other directly opposed contact forces $d\boldsymbol{F}$. $d\boldsymbol{F}/dS$ is called the *stress* on the plane $\boldsymbol{n}$ at the point considered. Its orientation and, generally, its intensity, depend on $\boldsymbol{n}$. Except for a few privileged directions, its direction is not the direction of the normal vector $\boldsymbol{n}$. The component along $\boldsymbol{n}$ (normal stress, opposed to pressure) is supplemented by a component in the plane dS, the *shear stress*.

The stresses in a point are the vector components of a more complex mathematical structure, the *stress tensor*. A few notes about these definitions:

- There are 9 *components of stress*, corresponding to the components of the stress vectors on the 3 planes of reference.
- It is wrong to define stresses as cohesive forces (and not as contact forces): they exist within a powder or a gas.
- In fluids or solids, if a small volume of matter is removed, and forces are applied on the walls of the cavity to faithfully reproduce the *in situ* deformation, these forces are not exactly equal to the *in situ* contact forces, because of the apparition of *surface tension* forces. They are completely negligible in the usual measurements of material behaviour.

- In general, stresses are only accessible through the accompanying deformations. For instance, on metallic structures engineers glue carbon strips and measure their lengthening by the variation of their electric resistance. These are strain gauges, Stresses are next inferred from the measured elastic strains.
- If we adopt the model of a *rigid body*, with no deformation, the stresses inside are undetermined. It is not surprising if the mechanics of rigid bodies (improperly called 'solid mechanics') does not allow to determine, for example, the forces on the floor exerted by the 4 feet of a chair. These forces depend on the stresses inside the feet, and in the floor. In fact, feet and floor are deforming elastically.

10.2 ELASTIC STRAINS AND VISCOUS STRESSES

The mechanic study of continuous media developed during most of the 19th century by considering only two kinds of behaviour: the elastic solid and the (Newtonian) viscous fluid. In these two models, the stresses on one part, and the strain (for elasticity) or the strain rate (for viscosity) on the other, are linked by a *constitutive law* (or *rheological law*) which is linear and independent from time. The medium has no memory of past deformations or stresses.

An *elastic body* is a medium that deforms under local stresses, instantaneously and reversibly. If the stresses disappear, the body immediately returns to its original shape. This model is an improvement of the rigid body model.

A *viscous fluid* is a fluid where any deformation along time produces local shear stresses. This model is an improvement of the perfect fluid model, with no viscosity (*inviscid*). It allows solution of some problems encountered with the model of the perfect fluid. One such problem is the d'Alembert paradox: a perfect fluid would not oppose any resistance to a moving body, once it has started.

It would seem that stresses within an elastic body *cause* elastic strains, whereas viscous stresses *are caused by* the strains of the viscous fluid. In these ideal models, stresses and strains appear or disappear simultaneously. No criterion can be used to decide which is the cause and which the consequence. The only objective reality is the existence of 2 tensors, and of relations between their components. Saying that some are causes and others are consequences is only a way of putting it. It will only be when considering visco-elastic bodies, both viscous and elastic, that an explicit choice will be necessary, in order to add the consequences of an identical cause. If the stresses cause an elastic strain and a viscous strain rate, the deformations will be added (*Maxwell body*). If the strain causes both elastic stresses and viscous stresses, the stresses will be added (*Voigt body*, also called *Kelvin body*).

Lagrangian coordinates are used in elasticity problems, with a reference frame fixed to the material medium. The intervening stresses are Lagrangian stresses, or Cauchy stresses. They are null on a free surface (apart from the atmospheric pressure), whatever the deformation of this free surface. As we shall only consider small deformations, the distortion of the axes will only introduce second-order corrections compared to the deformations, and will therefore be negligible. Conversely, Eulerian coordinates are used in the problems of viscous flows.

10.3 STRESS TENSOR AND EQUILIBRIUM EQUATIONS

Let us consider an infinitely small parallelepiped located at a point (x, y, z) anywhere in the continuous medium, with edges parallel to the coordinate axes and of respective lengths dx, dy, dz (Figure 10.1). The 3 sides visible in the figure are submitted to stresses, whose components along the 3 coordinate axes are noted on the figure. In this book, we will adopt the notation of von Karman: σ_i replace τ_{ii}. Note that the arrows representing stresses exerted from outside on the volume are always drawn *outside* the volume, regardless of their direction. The stresses, not represented, that the inside medium exerts on the outside, should be drawn inside the volume.

In the case of Figure 10.1, all the stress components are positive. In particular, the normal stresses (which tend to stretch the volume) (traction) are positive. This is a convention present in all books and articles about mechanics or geodynamics. But the convention is opposite in soil mechanics: the normal stresses (which compress the volume) (compressions) are positive. *Beware*: to use the soil mechanics convention, the signs of all stresses should be changed in the following equations, *including shear stresses*.

In addition to the forces on the 3 visible sides, equal to the drawn stresses multiplied by $dxdy$, $dydz$, or $dzdx$, similar forces are applied to the 3 hidden faces. All these forces are infinitely small of order 2. And there is the gravity force, the inertia force and the Coriolis force, all three proportional to the mass of the volume $\rho dxdydz$. If the medium has a magnetization J, and there is a magnetic field $\boldsymbol{B}$, it is also subject to magnetic forces that are proportional to the volume. The resultant and the resulting moment must equal 0. We will write it for infinitesimals of the 2nd

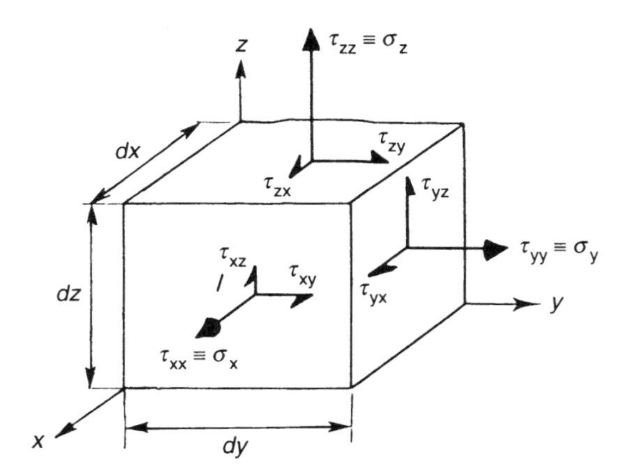

Figure 10.1. Stresses exerted by the medium on the elementary parallelepiped (notations of von Karman).

order, and then of the 3rd order. This is also true for infinitesimals of larger orders, but, as we have only 3 lengths *dx*, *dy* and *dz*, it is impossible to find out new independent relations.

The main term of the resultant, an infinitesimal of order 2, equals 0. This requires that the components of stresses on 2 opposite sides be equal and of opposite directions. They will therefore be designated with the same notations.

Let $\boldsymbol{J}$ with components (J_x, J_y, J_z) and $\boldsymbol{B}$ with components (B_x, B_y, B_z) be the magnetization vector and the magnetic field respectively. The moment of the orientation torque is $dx\, dy\, dz\, \boldsymbol{J} \wedge \boldsymbol{B}$. Because the resulting moment equals 0, we have:

$$\tau_{yz} - \tau_{zy} = J_y B_z - J_z B_y \tag{10.1}$$

Two similar equations are available, by cyclic permutation of the indices. Even in case of magnetite, in the Earth's magnetic field the moment of the magnetic torque would be 10 Pa at most. This is negligible in our problems. Therefore:

$$\tau_{yz} = \tau_{zy}; \qquad \tau_{zx} = \tau_{xz}; \qquad \tau_{xy} = \tau_{yx} \tag{10.2}$$

To write that the second term of the resultant of forces, an infinitesimal of order 3, equals 0, we need to introduce the stress variations over the lengths *dx*, *dy* and *dz*. There may be also a force per unit volume **grad**$(J \cdot \mathrm{B})$, but it is always negligible. In all the problems we will examine, the inertia and Coriolis forces will be completely negligible. Only the gravity (g_x, g_y, g_z) needs to be considered. The null condition along the 3 axes leads to:

$$\begin{cases} \dfrac{\partial \sigma_x}{\partial x} + \dfrac{\partial \tau_{xy}}{\partial y} + \dfrac{\partial \tau_{xz}}{\partial z} + \rho g_x = 0 \\[2ex] \dfrac{\partial \tau_{yx}}{\partial x} + \dfrac{\partial \sigma_y}{\partial y} + \dfrac{\partial \tau_{yz}}{\partial z} + \rho g_y = 0 \\[2ex] \dfrac{\partial \tau_{zx}}{\partial x} + \dfrac{\partial \tau_{zy}}{\partial y} + \dfrac{\partial \sigma_z}{\partial z} + \rho g_z = 0 \end{cases} \tag{10.3}$$

In mechanics, these relations are called the momentum equations, because the inertia force is the time derivative of the momentum, with its sign changed. But in geodynamics, we dropped this term and these relations are called the *equilibrium equations*.

The following conventions are often adopted:

- The indices *i*, *j* and *k* represent *x*, *y* or *z* indifferently.
- When the same index is present twice in the same term, it must be given all possible values (*x*, *y* and *z* in the present case), and the corresponding values of the term added (Einstein convention).
- A partial derivative like $\partial \tau_{ij}/\partial x$ is written $\tau_{ij,x}$.

The 3 equations of (10.3) can therefore be written in the condensed form:

$$\tau_{ij,j} + \rho g_i = 0 \tag{10.4}$$

Let us consider an orientation with unit vector $\boldsymbol{n}$, of components α, β, γ. By writing

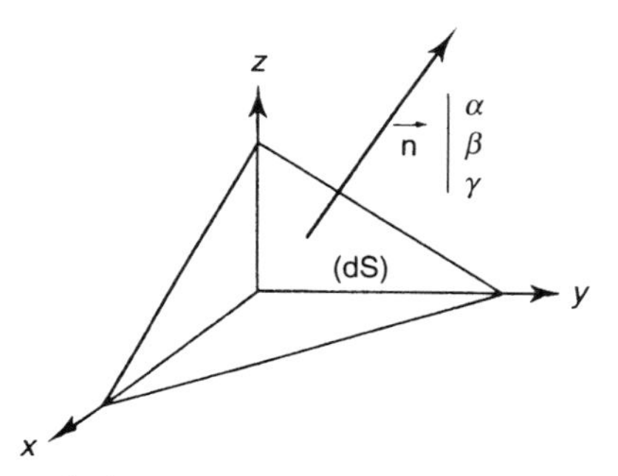

Figure 10.2. Infinitely small tetrahedron submitted to a stress $\boldsymbol{t}$.

the equilibrium equations of the infinitely small tetrahedron of Figure 10.2, at the 3rd order, we find that the 3 components along the coordinate axes of the stress $\boldsymbol{t}$ relative to $\boldsymbol{n}$ are:

$$\begin{cases} t_x = \sigma_x\alpha + \tau_{xy}\beta + \tau_{xz}\gamma \\ t_y = \tau_{yz}\alpha + \sigma_y\beta + \tau_{yz}\gamma \\ t_z = \tau_{zx}\alpha + \tau_{zy}\beta + \sigma_z\gamma \end{cases} \tag{10.5}$$

Using matrix notation, this can be written as:

$$\begin{bmatrix} t_x \\ t_y \\ t_z \end{bmatrix} = \begin{bmatrix} \sigma_x & \tau_{xy} & \tau_{xz} \\ \tau_{yx} & \sigma_y & \tau_{yz} \\ \tau_{zx} & \tau_{zy} & \sigma_z \end{bmatrix} \begin{bmatrix} \alpha \\ \beta \\ \gamma \end{bmatrix} \tag{10.6}$$

Any 3×3 matrix of coefficients which transform linearly every column vector as $\boldsymbol{n}$ into another column vector as $\boldsymbol{t}$ defines a second-order tensor, in the present case the *stress tensor*[1]. It is symmetrical, rigorously when the medium is not magnetic.

The normal stress (the pressure with the sign changed) on the plane with the normal $\boldsymbol{n}$ is:

$$\begin{aligned} \boldsymbol{t}\cdot\boldsymbol{n} &= t_x\alpha + t_y\beta + t_z\gamma \\ &= \sigma_x\alpha^2 + \sigma_y\beta^2 + \sigma_z\gamma^2 + 2\tau_{yz}\beta\gamma + 2\tau_{zx}\gamma\alpha + 2\tau_{xy}\alpha\beta \end{aligned} \tag{10.6}$$

Let us call the last expression $F(\alpha, \beta, \gamma)$. Then:

$$\begin{cases} t_x = \dfrac{1}{2}\dfrac{\partial F}{\partial \alpha} \\ t_y = \dfrac{1}{2}\dfrac{\partial F}{\partial \beta} \\ t_z = \dfrac{1}{2}\dfrac{\partial F}{\partial \gamma} \end{cases} \tag{10.7}$$

[1] *Tensor* comes from *tension*, the old term for stress. But for purists, stress is not a true tensor; it is a *tensor density*.

If we consider α, β and γ as Cartesian coordinates, not related by $\alpha^2+\beta^2+\gamma^2=1$ any more, then $F=constant$ represents an ellipsoid centred on the origin. t_x, t_y, t_z are the components of a normal to the surface at the point (α, β, γ). The same orientation relation exists between $\boldsymbol{t}$ and $\boldsymbol{n}$, than between the normal and the radius of the ellipsoid. The known symmetries of the ellipsoid show how intuitive is the following result (particular case of the analysis of symmetrical 3×3 matrices).

There are always at least 3 directions $\boldsymbol{n}$ perpendicular to each other (*principal directions*), for which $\boldsymbol{t}$ and $\boldsymbol{n}$ have the same direction. These stresses not including shear, only normal stress, are the *principal stresses*: $\sigma_1, \sigma_2, \sigma_3$. If the principal directions are taken as coordinate axes, all τ_{ij} are null. The stress matrix becomes diagonal, and the equation (10.5) becomes:

$$t_x=\sigma_1\alpha; \qquad t_y=\sigma_2\beta; \qquad t_z=\sigma_3\gamma \tag{10.8}$$

10.4 STRESS INVARIANTS AND DEVIATORIC STRESS

The general formulas to write the components of the stress tensor in any new system of reference are out of the scope of this work, but can be found in my book 'Very slow flow of solids', in Chapter 3. As there are six independent components and a rotation of the axes only provides three degrees of freedom, there are $6-3=3$ functions of the stresses, independent from each other, which are invariant through rotation of the frame of reference (*stress invariants*). Of course, any function of only these three invariants is also an invariant. The general theory of square matrices that define second-order tensors says that the trace (the sum of the diagonal elements) is an invariant. The first invariant will therefore be the mean normal stress (or octahedral stress, as it is found again for $\alpha=\beta=\gamma$):

$$\tfrac{1}{3}(\sigma_x+\sigma_y+\sigma_z)=\tfrac{1}{3}(\sigma_1+\sigma_2+\sigma_3)=\sigma_0=-p \tag{10.9}$$

Any stress state can be considered as the sum of a hydrostatic state, with the pressure p, and a *deviatoric stress*:

$$\begin{bmatrix}\sigma_x & \tau_{xy} & \tau_{xz}\\ \tau_{yx} & \sigma_y & \tau_{yz}\\ \tau_{zx} & \tau_{zy} & \sigma_z\end{bmatrix}=\begin{bmatrix}-p & 0 & 0\\ 0 & -p & 0\\ 0 & 0 & -p\end{bmatrix}+\begin{bmatrix}s_x & \tau_{xy} & \tau_{xz}\\ \tau_{yx} & s_y & \tau_{yz}\\ \tau_{zx} & \tau_{zy} & s_z\end{bmatrix} \tag{10.10}$$

The second matrix on the right-hand side has a trace of 0, and is the stress *deviator*. Its components s_{ij} (denoted τ_{ij} if $i\neq j$ and s_i if $i=j$) are the *deviatoric stresses*. The deviator is more often used to define a second stress invariant, the *effective shear stress*. This is a non-negative value, measuring the intensity of the deviator:

$$\tau=\left(\tfrac{1}{2}s_{ij}s_{ij}\right)^2=\left[\tfrac{1}{2}(s_x^2+s_y^2+s_z^2)+\tau_{yz}^2+\tau_{zx}^2+\tau_{xy}^2\right]^{1/2} \tag{10.11}$$

A third stress invariant can be the determinant of the stresses, or of the deviator. But it is useless in traditional elasticity or viscosity studies.

The mean pressure in the Earth is much larger than the absolute values of all deviatoric stresses. Therefore, when integrating the third equilibrium equation (10.3) along the descending vertical ($g_z = g$) from the surface down to the depth z, one finds:

$$-\sigma_2 \approx \int_0^z \rho g \, dz \tag{10.12}$$

This approximate value of $-\sigma_z$ is called the *lithostatic pressure*. If a hydrostatic pressure with the same value is subtracted from the stress tensor, the result does not exactly correspond to the deviatoric stresses, and the error can become very large close to the surface. These are called *tectonic stresses*.

10.5 SMALL STRAINS AND STRAIN RATES

Let $\boldsymbol{D}$ with components (U, V, W) be the displacement of a material point in a continuous medium. $\boldsymbol{D}$ is a function of the coordinates x, y and z. In a neighbouring point, at the distance $\boldsymbol{dM}$ (components: dx, dy, dz), the displacement along the x-axis is:

$$U(x+dx, y+dy, z+dz) = U(x, y, z) + \frac{\partial U}{\partial x} dx + \frac{\partial U}{\partial y} dy + \frac{\partial U}{\partial z} dz \tag{10.13}$$

This expansion may be written:

$$\begin{aligned} U(x+dx, y+dy, z+dz) = U &+ \frac{1}{2}\left(\frac{\partial U}{\partial y} - \frac{\partial V}{\partial x}\right) dy + \frac{1}{2}\left(\frac{\partial U}{\partial z} - \frac{\partial W}{\partial x}\right) dz \\ &+ \frac{\partial U}{\partial x} dx + \frac{1}{2}\left(\frac{\partial U}{\partial y} + \frac{\partial V}{\partial x}\right) dy \\ &+ \frac{1}{2}\left(\frac{\partial U}{\partial z} + \frac{\partial W}{\partial x}\right) dz \end{aligned} \tag{10.14}$$

The variations of V and W at the neighbouring point can be written in the same way, by cyclic permutation of U, V, W, and x, y, z. The 3 relations can be written with matrices (and with $U(x+dx,\ y+dy,\ z+dz) = U + dU$, etc.):

$$\begin{aligned} \begin{bmatrix} U + dU \\ V + dV \\ W + dW \end{bmatrix} = \begin{bmatrix} U \\ V \\ W \end{bmatrix} &+ \begin{bmatrix} 0 & -R_z & R_y \\ R_z & 0 & -R_x \\ -R_y & R_x & 0 \end{bmatrix} \begin{bmatrix} dx \\ dy \\ dz \end{bmatrix} \\ &+ \begin{bmatrix} \varepsilon_{xx} & \varepsilon_{xy} & \varepsilon_{xz} \\ \varepsilon_{yz} & \varepsilon_{yy} & \varepsilon_{yz} \\ \varepsilon_{zx} & \varepsilon_{zy} & \varepsilon_{zz} \end{bmatrix} \begin{bmatrix} dx \\ dy \\ dz \end{bmatrix} \end{aligned} \tag{10.15}$$

where:

$$\begin{cases} R_x = \dfrac{1}{2}\left(\dfrac{\partial W}{\partial y} - \dfrac{\partial V}{\partial z}\right) \\ \varepsilon_{xx} = \dfrac{\partial U}{\partial x} \\ \varepsilon_{yz} = \varepsilon_{xy} = \dfrac{1}{2}\left(\dfrac{\partial W}{\partial y} + \dfrac{\partial V}{\partial z}\right) \end{cases} \tag{10.16}$$

Six similar definitions can be obtained by cyclic permutation of (x, y, z) and (U, V, W).

The neighbourhood of the point (x, y, z) has therefore undergone:

- A *translation* D.
- A *rotation* $\boldsymbol{dM} \wedge \mathrm{R}$, where $\boldsymbol{R}$ is the vector of components R_x, R_y, R_z. Note that $2\boldsymbol{R} = \mathbf{curl}\,\boldsymbol{D}$.
- A *strain*, defined by the square symmetrical matrix of the ε_{ij}. The infinitesimal quantities of higher order are neglected, and this expression of deformation in a continuous medium is only valid for small deformations. In this linear case, it is possible to talk of a *strain tensor*.

As in any square matrix, representing a second-order tensor the trace is invariant when the system of reference is changed. It is the *volume dilatation*:

$$\varepsilon_{xx} + \varepsilon_{yy} + \varepsilon_{zz} = \frac{\Delta V}{V} = 3\varepsilon_0 \tag{10.17}$$

The strain can be decomposed into an isotropic dilatation, where distances increase by ε_0, and a *strain deviator*, of null trace, with the components:

$$e_{ij} = e_{ji} = \varepsilon_{ij} - \varepsilon_0 \delta_{ij} \tag{10.18}$$

Remember that δ_{ij} if $i = j$ and 0 if $i \neq j$.

The second deformation invariant is the *effective shear strain* γ, defined like the effective shear stress from the strain deviator, first multiplied by 2. By writing $2e_{ij} = \gamma_{ij}$:

$$\gamma = \tfrac{1}{2}(\gamma_{ij}\gamma_{ij})^2 = (2e_{ij}e_{ij})^{1/2} \tag{10.19}$$

The symmetries of the strain tensor and deviator are the same as for any symmetrical 3×3 matrix. There still are 3 principal directions, perpendicular to each other, and such that if they are chosen as Cartesian axes, the shear strains become 0. The corresponding strains are three relative stretchings $\varepsilon_1, \varepsilon_2, \varepsilon_3$ the *principal extensions* (or compressions).

If the displacements are progressive with time, as in the flow of a fluid, or the creep of a solid, the previous analysis is still valid if only the infinitesimal time interval dt is considered, during which the displacement is infinitesimal, and if everything is divided by dt. The velocity $\boldsymbol{u}$ of a point has the components:

$$\begin{pmatrix} u = \dfrac{dU}{dt} \\ v = \dfrac{dV}{dt} \\ w = \dfrac{dW}{dt} \end{pmatrix} \tag{10.20}$$

Its curl $\mathbf{curl}\,\boldsymbol{u} = \boldsymbol{\omega}$ is the *vorticity*. One gets:

$$\boldsymbol{u} + \boldsymbol{du} = \boldsymbol{u} + \tfrac{1}{2}\boldsymbol{\omega} \wedge \boldsymbol{dM} + \begin{bmatrix} \dot{\varepsilon}_{xx} & \dot{\varepsilon}_{xy} & \dot{\varepsilon}_{xz} \\ \dot{\varepsilon}_{yx} & \dot{\varepsilon}_{yy} & \dot{\varepsilon}_{yz} \\ \dot{\varepsilon}_{zx} & \dot{\varepsilon}_{zy} & \dot{\varepsilon}_{zz} \end{bmatrix} \begin{bmatrix} dx \\ dy \\ dz \end{bmatrix} \tag{10.21}$$

The ε_{ij} are the *strain rates*:

$$\begin{cases} \dot{\varepsilon}_{xx} = \dfrac{\partial u}{\partial x}; \quad \dot{\varepsilon}_{yy} = \dfrac{\partial v}{\partial y}; \quad \dot{\varepsilon}_{zz} = \dfrac{\partial w}{\partial z} \\ \dot{\varepsilon}_{yz} = \dot{\varepsilon}_{zy} = \dfrac{1}{2}\left(\dfrac{\partial w}{\partial y} + \dfrac{\partial v}{\partial z}\right), \quad \text{etc.} \end{cases} \tag{10.22}$$

To go from strain rates to the resulting large-amplitude finite strains, the rotation of the principal directions must be considered. In the case where a material segment, of length L_0 at the time t_0 and L_1 at the time t_1, remained always oriented along one principal direction, and its strain rate remained constant:

$$\dot{\varepsilon}(t_1 - t_0) = \ln\left(\frac{L_1}{L_0}\right) \tag{10.23}$$

10.6 ISOTROPIC ELASTICITY

We only consider the small elastic deformations of a continuous medium, isotropic both in its mechanical behaviour and for its thermal dilatation. The latter plays a role because modifying the stress produces or absorbs a bit of heat. If heat diffusion is very easy and the temperature of the medium does not vary, the deformation is *isothermal*. In the opposite case, more frequent, where there is no heat transfer, the deformation is *adiabatic*, or *isentropic*. Of course, all intermediate cases are also possible.

For a rigorous study, we should consider that the internal energy does not depend on two state variables (temperature and pressure) as in thermodynamics, but also on a third variable, the effective shear stress. Its effects are minimal and negligible, and we will not introduce this complication.

V designating the inverse of the volume mass, the *volume* thermal dilatation coefficient is:

$$\alpha = \frac{1}{V}\left(\frac{\partial V}{\partial T}\right)_p = -\frac{1}{\rho}\left(\frac{\partial \rho}{\partial T}\right)_p \tag{10.24}$$

The index p reminds that these are derivations at the mean pressure $p = -\sigma_0$ constant. In thermodynamics, it can be demonstrated that the temperature variation during an adiabatic deformation (with constant entropy S) is given by the coefficient:

$$\left(\frac{\partial T}{\partial p}\right)_S = \frac{\alpha T}{\rho C_p} \tag{10.25}$$

C_p is the heat capacity at constant pressure and T is the absolute temperature. Applying σ_0 therefore induces a temperature variation:

$$\Delta T = -\frac{\alpha T}{\rho C_p}\sigma_0 \tag{10.26}$$

The first law of elasticity (or, if one prefers, the first condition for the model of an elastic body) is:

$$\begin{cases} \text{for an isothermal deformation:} \\ 3\varepsilon_0 = \dfrac{\sigma_0}{k_T} \\ \text{for an adiabatic deformation:} \\ 3\varepsilon_0 = \sigma_0\left(\dfrac{1}{k_T} - \dfrac{\alpha^2 T}{\rho C_p}\right) = \dfrac{\sigma_0}{k_S} \end{cases} \tag{10.27}$$

For the materials of the Earth's crust, $\alpha = 2.5 \times 10^{-5}\,\mathrm{K}^{-1}$, and with the values given in Section 9.1, $\rho C_p = (2.5 \text{ to } 4) \times 10^6$ Pa. Consequently, $1/k_T$ and $1/k_S$ only differ by $(1 \text{ to } 4) \times 10^{-13}\,\mathrm{Pa}^{-1}$, while they are of the order of 10^{-11}. In the following, we will not make any difference and will omit the indices.

The second law of isotropic elasticity establishes the proportionality between the stress deviator and the strain deviator (which only provides 5 independent relations, because their traces equal 0). This similarity induces the identity of the principal directions:

$$\frac{s_{ij}}{2e_{ij}} = \frac{\tau}{\gamma} = \mu \tag{10.28}$$

The elastic parameter k is the *bulk modulus*. It measures the difficulty to modify the size of the crystalline lattice, without changing its shape. The parameter μ is the *shear modulus*, also called *Coulomb modulus*. It measures the difficulty of modifying the crystalline lattice by changing the valence angles. It is easy to foresee that these two parameters are independent. In a mixing, their inverses are obtained by weighted averages of the inverses for each constituent.

Relation between stresses and strains

Let us first express the stresses as a function of the strains:

$$\sigma_x = 3k\varepsilon_0 + 2\mu e_{xx} = k(\varepsilon_{xx} + \varepsilon_{yy} + \varepsilon_{zz}) + \frac{2\mu}{3}(2\varepsilon_{xx} - \varepsilon_{yy} - \varepsilon_{zz}) \tag{10.29}$$

Putting: $k - 2\mu/3 = \lambda$ (1st Lamé coefficient) it reads:

$$\begin{cases} \sigma_x = (\lambda + 2\mu)\varepsilon_{xx} + \lambda\varepsilon_{yy} + \lambda\varepsilon_{zz} \\ \tau_{yz} = 2\mu\varepsilon_{yz} \end{cases} \tag{10.30}$$

and 4 similar relations obtained by cyclic permutation of the indices.

Conversely, it is possible to express the strains as a function of the stresses:

$$\varepsilon_{xx} = \frac{\sigma_0}{3k} + \frac{s_x}{2\mu} = \frac{\sigma_x + \sigma_y + \sigma_z}{9k} + \frac{2\sigma_x - \sigma_y - \sigma_z}{6\mu} \tag{10.31}$$

By noting:

$$\frac{1}{3\mu} + \frac{1}{9k} = \frac{1}{E}; \qquad \frac{1}{6\mu} - \frac{1}{9k} = \frac{\nu}{E} \tag{10.32}$$

We find:

$$\begin{cases} \varepsilon_{xx} = \dfrac{\sigma_x}{E} - \dfrac{\nu}{E}\sigma_y - \dfrac{\nu}{E}\sigma_z \\ \varepsilon_{yz} = \dfrac{1+\nu}{E}\tau_{yz} \end{cases} \tag{10.33}$$

and 4 similar relations by cyclic permutation of the indices. E is the *Young modulus* and ν is *Poisson's ratio*. Knowing these 2 parameters, k and μ can be deduced from:

$$\begin{cases} k = \dfrac{E}{3(1 - 2\nu)} \\ \mu = \dfrac{E}{2(1 + \nu)} \end{cases} \tag{10.34}$$

Always, $n < 1/2$ and thus $E < 3\mu$.

Elastic energy

Let us calculate the work exerted by stresses acting on the faces of a unit volume, when stresses and strains progressively increase from 0 to their current value. Any volume can be divided into a series of infinitely small cubes, and doing the calculations for a cube will therefore prove sufficient. We can assume that its sides are parallel to the 3 principal directions: if the result is expressed with invariants, it is general. This work is worth:

$$\begin{aligned} W_{elast} &= \int_0 \sigma_1 d\varepsilon_1 + \sigma_2 d\varepsilon_2 + \sigma_3 d\varepsilon_3 \\ &= \int_0 (\sigma_0 + s_1)(d\varepsilon_0 + de_1) + (\sigma_0 + s_2)(d\varepsilon_0 + de_2) + (\sigma_0 + s_3)(d\varepsilon_0 + de_3) \end{aligned} \tag{10.35}$$

Considering that:

$$\begin{cases} s_1 + s_2 + s_3 = 0; \qquad de_1 + de_2 + de_3 = 0 \\ \dfrac{d(3\varepsilon_0)}{d\sigma_0} = \dfrac{1}{k} \\ \dfrac{de_1}{ds_1} = \dfrac{de_2}{ds_2} = \dfrac{de_3}{ds_3} = \dfrac{1}{2\mu} \\ s_1^2 + s_2^2 + s_3^2 = 2\tau^2; \qquad \tau = \mu\gamma \end{cases} \tag{10.36}$$

It is found:

$$W_{elast} = \frac{1}{2}\left(\frac{\sigma_0^2}{k} + \frac{\tau^2}{\mu}\right) = \frac{(3\varepsilon_0)\sigma_0}{2} + \frac{\gamma\tau}{2} = \frac{1}{2}[k(3\varepsilon_0)^2 + \mu\gamma^2] \tag{10.37}$$

The elastic energy is split into *compression energy* and *distortion energy*.

10.7 THE NAVIER EQUATION AND ITS CONSEQUENCES

By incorporating the equations (10.30) into the equilibrium relations (10.3), we obtain the 3 Navier equations, to which the 3 scalar functions U_x, U_y and U_z of x, y and z must obey. They can be written as one single equation to which the displacement vector $\boldsymbol{U}$ must obey:

$$\mu\left(\frac{\partial^2 U_x}{\partial x^2} + \frac{\partial^2 U_x}{\partial y^2} + \frac{\partial^2 U_x}{\partial z^2}\right) + (\lambda + \mu)\frac{\partial}{\partial x}\left(\frac{\partial U_x}{\partial x} + \frac{\partial U_y}{\partial y} + \frac{\partial U_z}{\partial z}\right) + \rho g_x = 0$$

$$\Leftrightarrow \qquad \mu\nabla^2\boldsymbol{U} + (\lambda + \mu)\nabla(\text{div}\,\boldsymbol{U}) + \rho\boldsymbol{g} = 0 \tag{10.38}$$

The following parameter appears:

$$\lambda + \mu = k + \frac{\mu}{3} = \frac{\mu}{1 - 2\nu} \tag{10.39}$$

Considering the expression of the vector Laplacian provided in Annex A.1, the Navier equation can be written as:

$$(\lambda + 2\mu)\nabla(\text{div}\,\boldsymbol{U}) - \mu\,\mathbf{curl}(\mathbf{curl}\,\boldsymbol{U}) + \rho\boldsymbol{g} = 0 \tag{10.40}$$

The curl is removed by calculating the divergence of this expression. When the contribution of the layer itself to the gravity $\boldsymbol{g}$ is negligible, $\text{div}\,\boldsymbol{g} = 0$. The resulting relation must be obeyed by the dilatation $\text{div}\,\boldsymbol{U} = 3k\varepsilon_0$:

$$\nabla^2(\text{div}\,\boldsymbol{U}) = 0 \tag{10.41}$$

The dilatation, and the mean normal stress $\sigma_0 = k\varepsilon_0$, are *harmonic* functions, with a Laplacian of 0. More interesting relations are obtained by calculating the vector

Laplacian of equation (10.38):

$$\begin{cases} \nabla^4 U_i = 0 \\ \nabla^4 \varepsilon_{ij} = 0 \qquad (i,j = x, y, z) \\ \nabla^4 \sigma_{ij} = 0 \end{cases} \tag{10.42}$$

Displacements, deformations and stresses are *bi-harmonic* functions. Let us consider the specific case of a plane deformation, for which $U_y = 0$ and the derivatives in y equal 0. The functions $f(x, z)$ to compute are s_x, s_z, τ_{xz}, U_x, U_z, which all verify:

$$\nabla^4 f \equiv \frac{\sigma^4 f}{\sigma x^4} + 2\frac{\sigma^4 f}{\sigma x^2 \sigma z^2} + \frac{\sigma^4 f}{\sigma z^4} = 0 \tag{10.43}$$

Let us investigate the case of a thick horizontal elastic layer, with a vertical z-axis. The partial differential equations of (10.43) can be transformed into equations in z only by using the Fourier transforms (see Appendix A.4). They become:

$$\frac{d^4\bar{f}}{dz^4} - 2\omega^2\frac{d^2\bar{f}}{dz^2} + \omega^4\bar{f} = 0 \tag{10.44}$$

The general solution is of the form:

$$\bar{f} = (A + B\omega z)e^{\omega z} + (C + D\omega z)e^{-\omega z} \tag{10.45}$$

A, B, C and D are functions of ω. *To satisfy the elasticity equations, the coefficients of the 5 Fourier transforms must be related. If* A, B, C and D are conserved for $f = U_z$, the others are linear expressions of these 4 coefficients. For instance:

$$\frac{\bar{\sigma}_x}{2\mu\omega} = [-A - (3 - 2\nu)B]e^{\omega z} - B\omega z e^{\omega z} + [C - (3 - 2\nu)D]e^{-\omega z} + D\omega z e^{-\omega z} \tag{10.46}$$

In layered media (which will be used as models in Chapter 18), apart from σ_x, the other 4 functions must be continuous when going from one layer to the next. Their Fourier transforms should behave similarly. They are considered as the 4 components of a column vector called the *Thomson-Haskell vector* $\theta(\omega, z)$. It is deduced by multiplying the column matrix of coefficients $c(\omega)$ by a square matrix $\boldsymbol{M}$:

$$\theta = \mathbf{Mc}; \qquad \theta = \begin{bmatrix} \bar{\sigma}_z \\ \bar{\tau}_{xz}/i \\ \bar{U}_x/i \\ \bar{U}_z \end{bmatrix}; \qquad \mathbf{c} = \begin{bmatrix} A \\ B \\ C \\ D \end{bmatrix}$$

$$\mathbf{M} = \begin{bmatrix} 2\mu\omega e^{\omega z} & 2\mu\omega(1 - 2\nu + \omega z)e^{\omega z} & -2\mu\omega e^{\omega z} & 2\mu\omega(1 - 2\nu - \omega z)e^{-\omega z} \\ 2\mu\omega e^{\omega z} & 2\mu\omega(2 - 2\nu + \omega z)e^{\omega z} & 2\mu\omega e^{\omega z} & -2\mu\omega(2 - 2\nu + \omega z)e^{-\omega z} \\ e^{\omega z} & (3 - 4\nu + \omega z)e^{\omega z} & -e^{-\omega z} & (3 - 4\nu - \omega z)e^{-\omega z} \\ e^{\omega z} & \omega z e^{\omega z} & e^{-\omega z} & \omega z e^{-\omega z} \end{bmatrix} \tag{10.47}$$

$\bar{\tau}_{xz}$ and $\bar{U}_x$ were divided by $\mathrm{i} = \sqrt{-1}$, because in the most frequently studied case, symmetrical in x, these are purely imaginary numbers, whereas σ_x and U_x are real numbers. Thus θ and $\boldsymbol{c}$ are real.

10.8 SOLUTION TO SOME ELASTIC PROBLEMS

Elastic waves

During the propagation of a P wave along the x direction, any transverse extension or compression is impossible: $\varepsilon_{yy} = \varepsilon_{zz} = 0$. From equation (10.30):

$$\frac{\sigma_0}{\varepsilon_{xx}} = \lambda - 2\mu = k + \tfrac{4}{3}\mu = \frac{(1-\nu)E}{(1+\nu)(1-2\nu)} = E_1 \tag{10.48}$$

It is better to use (10.33) to establish that:

$$\sigma_y = \sigma_z = \frac{\nu}{1-\nu}\sigma_x \tag{10.49}$$

During the propagation of an S wave along the x direction, the oscillations occur along the y direction, only the component V of the displacement does not equal 0, and is independent from y and z. The only non-zero strain is therefore $\partial V/\partial x = 2\varepsilon_{xy}$, and:

$$\frac{\tau_{xy}}{2\varepsilon_{xy}} = \mu = \frac{E}{2(1+\nu)} \tag{10.50}$$

Using the equilibrium equations, where gravity is neglected but with the inertia terms, we obtain a partial differential equation obeyed by waves propagating with the celerities:

$$v_P = \sqrt{\frac{E_1}{\rho}}; \qquad v_S = \sqrt{\frac{\mu}{\rho}} \tag{10.51}$$

To determine the elastic parameters of a material, one sends S or P waves through a thick slice, and looks for the resonance frequency. There is a stationary wave, then, and the thickness of the slide corresponds to its half-wavelength (or a multiple). The periods used are of the order of 10 µs, in the ultrasound domain. From these measurements, it is possible to infer the possible nature of the terrains with known seismic velocities, for infrasonic waves of period 1–10 s.

Thus, at the depth of the Moho below the continents, under a lithostatic pressure close to 1 GPa, the elastic parameters of dunite are $k = 140$ GPa, $\mu = 76$ GPa, $E = 193$ GPa and $\nu = 0.27$.

The elastic parameters found with ultrasonic or infrasonic (seismic) waves are used to model the lithospheric plates, during quasi-static deformations. Section 13.4 will explain that this is most probably only valid for the coldest layers.

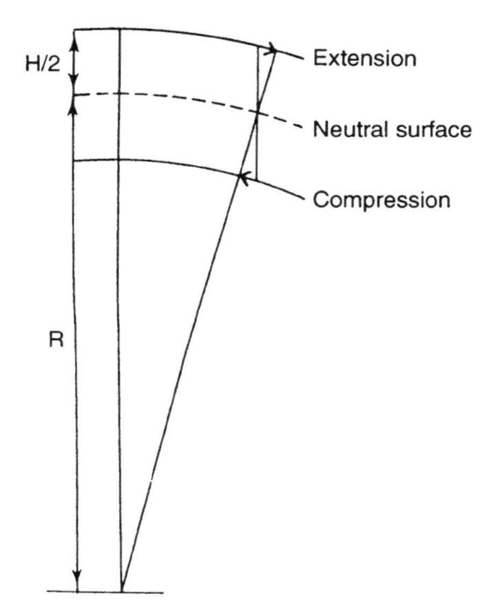

Figure 10.3. Plane bending of an elastic plate.

Plane bending of a thin plate

For the moment, we shall only examine the case where all the displacements are parallel to a plane (*plane strain*), either because we consider only the central portion of a very large plate, or because the geometry makes any transverse displacement impossible.

Let (x, z) be the deformation plane. Intuitively, ε_{xx} is positive on the convex side, negative on the concave side, 0 on a neutral surface nearly equidistant from the two faces of the plate, if it is homogeneous (Figure 10.3). We will adopt a Lagrangian perspective. Locally, the (x, y) plane is the neutral surface and the x-axis curves with it. As long as its curvature radius R is very large compared to the plate thickness H, the Cartesian coordinate formulas for deformation and equilibrium are still valid. The longitudinal extension is:

$$\varepsilon_{xx} = z/R \tag{10.52}$$

The stresses do not depend on x or y. If the pressure on the two free sides is 0, and gravity is negligible compared to the stresses, the 3rd equilibrium equation shows that $\sigma_z = 0$ for any z. It follows that:

$$\begin{cases} \sigma_y = \nu\sigma_x; \qquad \varepsilon_{xx} = \dfrac{1-\nu^2}{E}\sigma_x \\ \varepsilon_{zz} = -\dfrac{\nu(\sigma_x + \sigma_z)}{E} = -\dfrac{\nu}{1-\nu}\varepsilon_{xx} \\ W = -\dfrac{\nu z^2}{[1-\nu]2R} \end{cases} \tag{10.53}$$

The convex face moved closer to the neutral surface, and the concave face moved away, with distances negligible compared to $H/2$.

The *bending torque,* causing the flexion, is:

$$M = \int_{-H/2}^{+H/2} \sigma_x z\, dz = \frac{E}{1-\nu^2}\int_{-H/2}^{+H/2} \frac{z^2}{R}\, dz = \frac{\Gamma}{R}; \qquad \Gamma = \frac{EH^3}{12(1-\nu^2)} \tag{10.54}$$

$MR = \Gamma$ is the *flexural rigidity*, or, better, the *stiffness* of the plate, whether it is homogeneous (and equation (10.54) is valid) or not. We shall see in Chapter 13 how the stiffness of an oceanic lithosphere plate can be determined directly from the curvature it shows under the effect of isostasy.

Dislocations

An elastic body whose surface is not subjected to any force may however undergo strong internal stresses. They can result, for example, from an abrupt cooling of the surface. It is also possible to cut a ring along a radial plane, slide the two sides on each other and weld them back together in this 'unnatural' position. This produced a *dislocation*[2]. The vector defining the imposed slide is the *Burgers vector*. Any dislocation can be considered as the superimposition of a *screw dislocation*, with a Burgers vector parallel to the axis, and an *edge dislocation*, with a Burgers vector perpendicular to the axis (Figure 10.4).

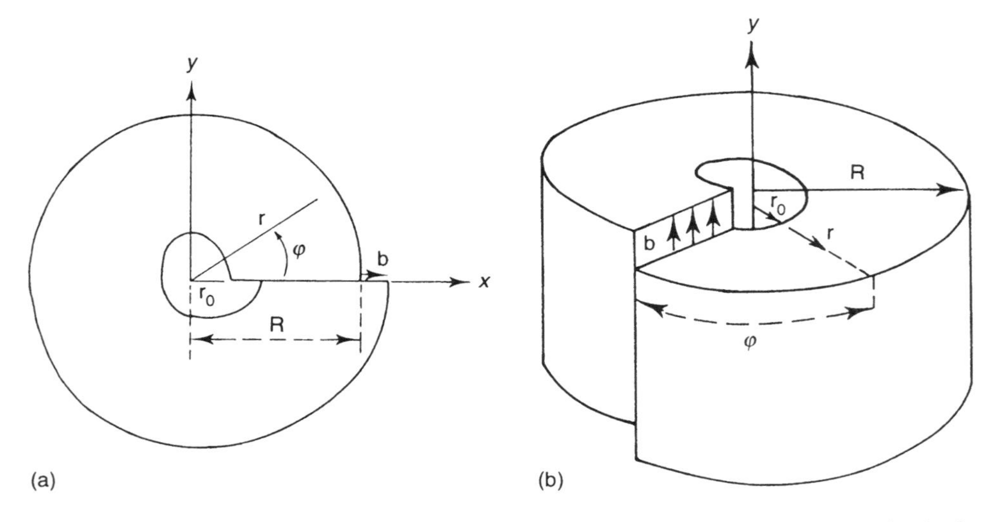

Figure 10.4(a). Edge dislocation; (b) screw dislocation. From Lliboutry (1987); © Kluwer Academic.

[2] *Dislocation* should be understood here in its etymological meaning of dis-location, the fact that an object is not in its place. The Italian mathematician Volterra studied elastic dislocations in 1907, calling them *distorsione*. The English scientist Love, in its elasticity treatise of 1927, translated *distorsione* by *dislocations*, and not by the more precise term of *distortions*.

A pipe without matter, of radius r_0, is necessary along the axis of the dislocation. This could also be a cylinder of highly deformed matter for which the elasticity law of continuous media is no longer valid. It is the *core* of the dislocation. The extremities of this core must be on the surface of the body, or close on themselves inside the body (this is a dislocation *loop*).

Analytical calculation of displacements and stresses around a dislocation is difficult. It is performed by successive approximations, adding solutions until correct boundary conditions are found. For the two types of dislocation, one can use the following approximate value of elastic energy per unit length (which is termed the *self-energy* of a dislocation):

$$\frac{dW_{elast}}{dz} \approx \frac{\mu b^2}{4\pi} \ln \frac{R}{r_0} \tag{10.55}$$

10.9 ISOTROPIC NEWTONIAN VISCOSITY

The isotropic Newtonian viscosity is the usual viscosity of gases and liquids. (Not the one of long macromolecules: the flow turns them to a single direction and the medium becomes anisotropic.) The viscous force dragging a layer in contact with another faster one has a different origin, depending on whether it is gas or condensed matter. In gases, molecules of one layer continuously cross into the other, the deeper their mean free path the higher, and therefore the higher the temperature. Viscous shear stress equals the flux of momentum exchanged between the two layers. It *increases* with temperature. In liquids, and even more in solids with a viscous behaviour, the molecules are not free. Two molecule planes sliding on each other are exerting a friction force, as if they were two rough planes with colliding bumps. Thermal agitation makes the sliding easier and therefore viscous shear *decreases* with temperature.

We shall only look at the condensed state. High-temperature liquids and solids with a viscous behaviour are Maxwell bodies (with $\mu = 0$ for liquids). The elastic component will be neglected. The deformation then occurs at constant volume:

$$3\dot{\varepsilon}_0 = \dot{\varepsilon}_{xx} + \dot{\varepsilon}_{yy} + \dot{\varepsilon}_{zz} = \operatorname{div} \boldsymbol{u} = 0 \tag{10.56}$$

The strain rate tensor is then a deviator. It is proportional to the stress deviator:

$$s_{ij} = 2\eta\dot{\varepsilon}_{ij} \tag{10.57}$$

η is the *viscosity* and its inverse $1/\eta$ is the *fluidity* of the medium. The SI unit of viscosity is the Pa s. For example, water has a viscosity of ca. 10^{-3} Pa s, castor oil a viscosity of ca. 1 Pa s. The official name of *Poiseuille* is not used much because of the possible confusion with the previous cgs unit, the *poise*, 10 times smaller. The asthenospheric viscosities are very high, and it is better to use the Ma MPa:

$$1 \text{ Ma MPa} = 3.1556 \times 10^{19} \text{ Pa s} = 3.1556 \times 10^{20} \text{ poise.}$$

The *effective shear strain rate* $\dot{\gamma}$ is an invariant measuring the intensity of the strain rate:

$$\dot{\gamma} = (2\dot{\varepsilon}_{ij}\dot{\varepsilon}_{ij})^{1/2} = [2(\dot{\varepsilon}_{xx}^2 + \dot{\varepsilon}_{yy}^2 + \dot{\varepsilon}_{zz}^2) + 4(\dot{\varepsilon}_{yz}^2 + \dot{\varepsilon}_{zx}^2 + \dot{\varepsilon}_{xy}^2)]^{1/2} \tag{10.58}$$

The similarity of deviators implies:

$$\tau = \eta\dot{\gamma} \tag{10.59}$$

The *energy dissipated per unit volume* (as heat) corresponds to the work of stresses which was giving the elastic energy stored by unit volume. But the factor 1/2 disappears from the equation:

$$\begin{aligned} \dot{W} &= (s_1 + \sigma_0)\dot{\varepsilon}_1 + (s_2 + \sigma_0)\dot{\varepsilon}_2 + (s_3 + \sigma_0)\dot{\varepsilon}_3 \\ &= s_1\dot{\varepsilon}_1 + s_2\dot{\varepsilon}_2 + s_3\dot{\varepsilon}_3 \\ &= \frac{\tau^2}{\eta} = \dot{\gamma}\tau = \eta\dot{\gamma}^2 \end{aligned} \tag{10.60}$$

When viscosity only depends on temperature and pressure, the viscosity is linear, or *Newtonian*. If the medium is isothermal, and the influence of pressure negligible, the material is *isoviscous*. This can only happen for very thin films or conduits, as isothermia supposes the dissipated heat is evacuated, which requires temperature gradients.

For an isoviscous, incompressible medium, the constitutive law is deduced from the law for an elastic body with k infinite (and therefore $\nu = 1/2$, and $E = 3\mu$), and replacing μ by the operator $\eta\partial/\partial t$. The Navier equation becomes the Navier–Stokes equation. The solution of an elasticity problem therefore provides the solution of a viscosity problem (*correspondence principle*). As the elastic solution is generally known with Lagrangian variables, the solution must first be converted to Eulerian variables. This will be shown in the example below.

10.10 LOAD DISTRIBUTED OVER AN ELASTIC OR ISOVISCOUS LAYER

Consider a horizontal elastic layer, with thickness H, and density ρ. What is the surface depression $\zeta(x)$ under the effect of a weight distribution $P(x)$? The problem is assumed to be independent of y, and symmetrical in x. It is then possible to use the expressions with the Fourier transforms of Appendix A.4.

Contrary to usage in elasticity studies, we will use Eulerian coordinates, fixed. *Without load*, the surface is $z = 0$, the base is $z = H$. The simple weight of the layer causes stresses and elastic displacements, but they are not taken into account. We will only calculate the modifications from this reference state.

The boundary conditions are reported to the levels $z = 0$ and $z = H$. On the free surface, $z = \zeta$, $\tau_{xz} = 0$ and $\sigma_z = -P$. At $z = 0$, they become $\tau_{xz} = 0$ and $\sigma_z = -P + \rho g\zeta$. At the bottom, 3 simple cases may be considered:

- The layer lies on a perfectly rigid basement, and 'sticks' to it. Then, for $z = H$, we have $U_x = U_z = 0$.
- The layer may slide freely over a perfectly rigid horizontal basement. Then, for $z = H, U_z = \tau_{xz} = 0$.
- The layer lies over a highly fluid medium, of density $\rho + \Delta\rho$. Then, for $z = H$, we have $\tau_{xz} = 0$ and $\sigma_z = -\Delta\rho\, g\, U_z$.

The Fourier transforms satisfy the same boundary conditions.

To perform the calculations by hand, it is convenient to write $e^{+\omega H} = \cosh(\omega H) + \sinh(\omega H) = \mathrm{Ch} + \mathrm{Sh}$, and to replace the unknown values A, B, C, D by $(A + C), (A - C), (B + D), (B - D)$. Indeed, we are looking for ζ, and in the three cases:

$$\bar{\zeta} = \bar{U}_z|_H = A + C = -2(1 - \nu)(B - D) \tag{10.61}$$

In the first case (fixed bottom), we find:

$$\bar{\zeta} = \frac{(\bar{P} - \rho g \bar{\zeta})(1 - \nu)}{\mu\omega} \left[\frac{(3 - 4\nu) ShCh - \omega H}{(3 - 4\nu) Ch^2 + (\omega H)^2 + (1 - 2\nu)^2} \right] \tag{10.62}$$

In the elastic problem, $\rho g \zeta$ is generally negligible compared to P.

The correspondence principle allows us to proceed to the case of an isoviscous layer on a rigid bottom (this is the first model proposed for glacio-isostatic rebound; we shall see in Section 18.6 that the presence of the lithosphere completely modifies the results). We write that $\nu = 1/2$ and μ is replaced by $\eta d/dt$:

$$\eta \frac{d\bar{\zeta}}{dt} = \frac{\bar{P} - \rho g \bar{\zeta}}{2\omega} \left[\frac{ShCh - \omega H}{Ch^2 + (\omega H)^2} \right] \tag{10.63}$$

In particular, if the load P disappears, but there is a depression ζ_0 at the time $t = 0$;

$$\frac{d\bar{\zeta}}{dt} = -\frac{\bar{\zeta}}{T}; \qquad \bar{\zeta} = \bar{\zeta}_0\, e^{-t/T}; \qquad T = \frac{2\eta\omega}{\rho g} \left[\frac{Ch^2 + (\omega H)^2}{ShCh - \omega H} \right] \tag{10.64}$$

For $\omega H \gg 1, T \approx 2\eta\omega/(\rho g)$. For $\omega H \ll 1, T \approx 3\eta/(\rho g H^3 \omega^2)$. For H fixed, T goes through a minimum T_m for $\omega = \omega_m$:

$$\omega_m = \frac{2.12}{H}; \qquad T_m = \frac{6.222}{\rho g H} \tag{10.65}$$

Let us insist on the fact that it is the Fourier transform of the profile $\zeta(x)$ which decreases exponentially with time. It is not the case of the profile itself, which equals:

$$\zeta = 2 \int_0^\infty \bar{\zeta} \cos \omega x \, d\omega = 2 \int_0^\infty \bar{\zeta}_0(\omega) \exp\left[\frac{-t}{T(\omega)} \right] \cos \omega x \, d\omega \tag{10.66}$$

The frequencies close to ω_m will disappear faster.

10.11 FLOW OF AN ISOVISCOUS FLUID IN A CYLINDER

The simplest flow is the flow of an isoviscous fluid in a cylindrical pipe, when the fluid 'wets' the walls (i.e. when the velocities at their contact equal zero), and the

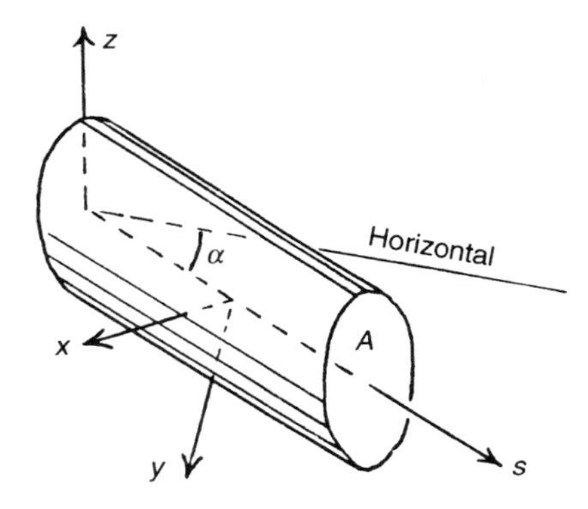

Figure 10.5. Viscous flow in a cylinder.

velocities are parallel. This is only a particular case of *laminar*, non-turbulent flows. In a turbulent flow, eddies are produced at the contact with the walls (where the shear strain rates are the highest) and carried away by the current.

Let us choose Cartesian axes (s, x, y), keeping the notation z for the vertical axis. The axis of the pipe will be the s-axis, oriented along the flow (Figure 10.5). The velocity $u(x, y)$ must be independent of s, as we assumed the transverse components of the velocity to be 0. Therefore:

$$\begin{cases} \sigma_x = \sigma_y = \sigma_s = \sigma_0 = -p \\ \tau_{xs} = \eta \dfrac{\partial u}{\partial x}; \qquad \tau_{ys} = \eta \dfrac{\partial u}{\partial y}; \qquad \tau_{xy} = 0 \end{cases} \tag{10.67}$$

The shear stresses are independent from s. The equilibrium equations become:

$$\begin{cases} \dfrac{\partial p}{\partial x} = \rho g_x \\ \dfrac{\partial p}{\partial y} = \rho g y \\ \eta \left(\dfrac{\partial^2 u}{\partial x^2} + \dfrac{\partial^2 u}{\partial y^2} \right) = \dfrac{\partial p}{\partial s} - \rho g_s \end{cases} \tag{10.68}$$

The transverse dimensions of the pipe, along x and y, are small compared to the length, and the variations of p in a cross-section are neglected. Let us now use the terminology of hydraulics (where this approximation is always made), although the flow in hydraulics is always turbulent. We designate as z the altitude relative to a reference datum. The height Z is the *hydraulic load*:

$$Z = \frac{p}{\rho g} + z \tag{10.69}$$

The loss of load per unit length is:

$$\varpi = -\frac{1}{\rho g}\frac{dp}{ds} - \frac{dz}{ds} = -\frac{1}{\rho g}\frac{dp}{ds} + \cos\alpha \tag{10.70}$$

This is the *hydraulic gradient*. The velocity in a cylindrical pipe must obey the following relation:

$$\frac{\partial^2 u}{\partial x^2} + \frac{\partial^2 u}{\partial y^2} = -\frac{\rho g \varpi}{\eta} \tag{10.71}$$

In the absence of flow, the hydraulic gradient equals 0 and we obtain the Pascal Principle.

We must have $u(x, y) = 0$ on the whole perimeter of a cross-section. The solution will be analytical if the curve $u(x, y) = 0$ forms an ellipse, 2 parallel lines (this would be for a fluid film very large compared to its thickness) or an equilateral triangle (the latter is the closest to capillary conducts in rocks at the junction of 3 crystals, where fluids can circulate).

For a *film of thickness H*, perpendicular to the y-axis,

$$u = u_0(1 - 4y^2/H^2) \tag{10.72}$$

Equation (10.71) provides the value of u_0:

$$u_0 = \frac{H^2}{8}\frac{\rho g \varpi}{\eta} \tag{10.73}$$

The discharge per unit width is:

$$q = \int_{-H/2}^{+H/2} u\,dy = \frac{\rho g H^3}{12\eta}\varpi \tag{10.74}$$

For an *elliptical pipe*,

$$\begin{cases} u = u_0\left(1 - \dfrac{x^2}{a^2} - \dfrac{y^2}{b^2}\right) \\ u_0 = \dfrac{a^2 b^2}{2(a^2 + b^2)}\dfrac{\rho g \varpi}{\eta} \end{cases} \tag{10.75}$$

By integrating u on the entire ellipse, of area $A = \pi ab$, we find that the discharge is:

$$Q = \frac{A^2}{8\pi}\left(\frac{2ab}{a^2 + b^2}\right)\frac{\rho g \varpi}{\eta} \tag{10.76}$$

In the case of a pipe of circular section, the factor between brackets equals 1, and the equation becomes the *Poiseuille formula*.

If the section is an equilateral triangle, of side a and area $A = \sqrt{3}a^2/4$, then:

$$\begin{cases} u = \dfrac{u_0}{a^3}(2\sqrt{3}x + a)(\sqrt{3}x + 3y - a)(\sqrt{3}x - 3y - a) \\ u_0 = \dfrac{A}{9\sqrt{3}}\dfrac{\rho g \varpi}{\eta}; \qquad Q = \dfrac{A^2}{20\sqrt{3}}\dfrac{\rho g \varpi}{\eta} \end{cases} \tag{10.77}$$

The average velocity $\bar{u} = Q/A$ can be expressed as a function of the *hydraulic radius* $R = A/P$, where P is the wetted perimeter of the section (here, the whole perimeter).

For a film of thickness H, we have $R = H/2$, and therefore:

$$\bar{u} = \frac{H^2}{12}\frac{\rho g \varpi}{\eta} = \frac{R^2}{3}\frac{\rho g \varpi}{\eta} \tag{10.78}$$

For a section which is a circle of radius a, $R = a/2$, and therefore:

$$\bar{u} = \frac{a^2}{8}\frac{\rho g \varpi}{\eta} = \frac{R^2}{2}\frac{\rho g \varpi}{\eta} \tag{10.79}$$

For a section which is an equilateral triangle, of side a, $R = a/(4\sqrt{3})$, and therefore:

$$\bar{u} = \frac{a^2}{80}\frac{\rho g \varpi}{\eta} = \frac{3R^2}{5}\frac{\rho g \varpi}{\eta} \tag{10.80}$$

10.12 PERMEABILITY AND POROSITY OF SOILS

When a soil contains a network of communicating pores or fissures filled with water, the water is at any point under a specific pressure p_w, called the *interstitial pressure*. The surface at which p_w equals the atmospheric pressure is the *piezometric surface*. It is determined by the water level in drill holes or wells. When the aquifer layer is in contact with the atmosphere, its surface is the saturation level. But above, there is a capillary layer of non-water saturated soil. The hydraulic pressure in one point of the microscopic pipe network equals the vertical distance to the piezometric surface.

At the macroscopic scale, the water discharge across a unit section of soil defines a vector $\boldsymbol{q}$, oriented along the flow direction. It has the dimensions of a velocity, but is different from the water mean velocity, much larger. This discharge follows the *Darcy Law*:

$$\mathbf{q} = -k_D \frac{\rho g}{\eta} \operatorname{grad} Z \tag{10.81}$$

The coefficient k_D (*permeability*) only depends on the geometry of the network of cracks, pores and fissures. It can be measured *in situ* by observing the water level fluctuations in a drill hole. If the permeability is high enough, the water level variations with the injection or pumping of a constant water flow can also be measured. Permeability used to be measured in Darcys, but is now rather given in μm^2 (1 Darcy $= 0.985\,\mu m^2$). In silts, $k_D = 1$ to $10^{-2}\,\mu m^2$. In clay, $k_D = 10^{-2}$ to $10^{-6}\,\mu m^2$.

The Darcy Law can be deduced from the laws of laminar flows (equations (10.78) to (10.80)):

$$\bar{u} = \frac{R^2}{m}\frac{\rho g \varpi}{\eta} \qquad (1.5 < m < 3) \tag{10.82}$$

The series of microscopic pipes through which a water molecule travels is not rectilinear and in the direction of $\boldsymbol{q}$: it is longer by a *tortuosity factor t*. With the same drop in p_w, water covers a distance t times longer. Furthermore, the component $\boldsymbol{U}$ of $\bar{u}$ along $\boldsymbol{q}$ is $\boldsymbol{t}$ times smaller. Therefore:

$$\boldsymbol{U} = -\frac{R^2}{mt^2}\frac{\rho g}{\eta}\,\mathbf{grad}\,Z \tag{10.83}$$

Let n be the *porosity* and Σ the *volumic surface*. n is the volume of pores per unit volume of soil. Σ is the surface of pores per unit volume of soil.

When looking for an average value of $\boldsymbol{U}$, one only needs to account for the pipes with large hydraulic radii. We will not talk of fissured and grainy soils, where they can be abnormally large. And we will admit that equation (10.83) can be used with the ratio n/Σ as the quadratic mean of R. We must multiply by n to go from $\boldsymbol{U}$ to $\boldsymbol{q}$, and thus:

$$k_D = \frac{n^3}{(mt^2)\Sigma^2} \tag{10.84}$$

With $mt^2 = 5$, this becomes the *Korzeny-Carman Law*, which is valid for sands and silts (but not for clays).

A few precisions about the parameters n and Σ. They are often replaced by the *void ratio e* and the *surface ratio S*:

$$\begin{cases} e = \dfrac{n}{(1-n)} = \dfrac{\text{volume of the pores}}{\text{volume of the solid particles}} \\ S = \dfrac{\Sigma}{(1-n)} = \dfrac{\text{surface of the particles}}{\text{volume of the particles}} \end{cases} \tag{10.85}$$

Because of stresses acting on the soil, its porosity may decrease (packing, compaction) or increase (*dilatance*) (see Section 12.7). But the porosity may also vary because the finer parts are carried by water from one area to another.

The surface ratio can be deduced from the particle size and shape distribution. For the sake of the argument, let us assume they are all spheres of diameter a. $V(a)$ is the volume of all grains with diameters smaller than a in the unit solid volume. We have:

$$S = \int_0^1 \frac{6}{a}\,dV \tag{10.86}$$

In sedimentology, the sizes are on a logarithmic scale. If $V = 1/2$ for $a = a_{1/2}$, we can use the variable $\log(a/a_{1/2}) = x$.

$$S = \frac{6}{a_{1/2}}\int_{-\infty}^{+\infty}\frac{dV}{dx}10^{-x}\,dx \tag{10.87}$$

Generally, the distribution function dV/dx is approximately an odd function of x. For the same size interval dx, because of the factor 10^{-x}, the smaller grains contribute most to S, and therefore, for a given porosity, they are the largest control on permeability.

11

Rock creep

11.1 PLASTICITY AND CREEP

Most mechanical tests to measure the behaviour of rocks are made by compressing a cylindrical sample along its axis at a constant rate, and recording the corresponding pressure $\sigma(t)$ along the axis. Proceeding in the opposite way, it would be difficult to control the rate at which the load is applied. Accounting for the elastic deformation of the machine, determined once and for all as a function of the load, the deformation $\varepsilon(t)$ of the sample is very well known, but the measurement of pressure is often less accurate.

The first concepts came from the study of metal behaviour, and we shall start with them, as the behaviour of rocks is more complex.

At temperatures far from the melting point (let us say absolute temperatures smaller by a half, to get an idea), the behaviour is first elastic, and the load σ increases proportionally to the deformation. Two cases are then possible:

Fracture

The sample breaks up. The material is said to be *brittle*. If the failure limit is low, the material is said to be *weak*. Fracture mechanisms will be examined in the next chapter. For the moment, let us only say that it results from the irreversible and very rapid growth of micro-fissures, pre-existing or caused under load by some microscopic defect.

Plastic deformation

When the material is *ductile*, and not brittle, above some load (the *yield strength*), there appears a permanent deformation. This strain is called *plastic* strain, and for a constant strain rate, the load increases only slowly (let us say 10 times slower, to get an idea) (Figure 11.1a). Plasticity may only appear first in parallel bands crossing the

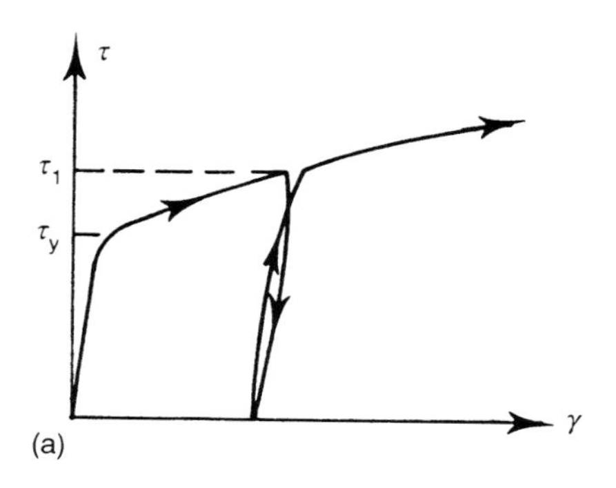

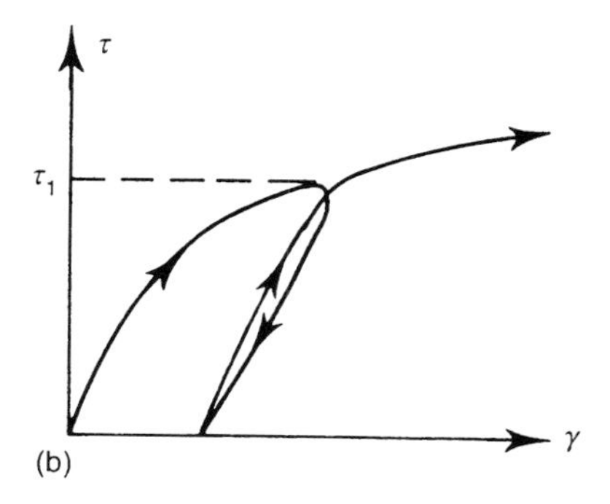

Figure 11.1. Stress–deformation curve: work-hardening. From Lliboutry (1987); © Kluwer Academic.

sample, the *Lüders stripes*. In pure annealed metals, the yield strength is negligible (Figure 11.1b).

Plastic deformation does not disappear at unloading (it is irreversible), but temporary unloadings do not affect the continuity of the curve $\sigma(\varepsilon)$, represented in Figure 11.1. Thus, previous plastic deformation increases the yield strength. This is the *work-hardening* process. The yield strength is a 'hidden variable', depending on the past plastic deformation of the sample.

Work-hardening can be suppressed by prolonged but moderate heating (the *recovery* process), or by a stronger heating that induces a complete recrystallization of the metal (the *annealing* process).

Creep

At a temperature closer to the melting point, another process becomes perceptible before a defined yield strength is reached: *creep*. Even with a constant load σ, the deformation goes on indefinitely. Just after loading (or after having increased the load), the strain rate decreases progressively (*transient creep*) until it stabilizes at a constant value, function of σ (*permanent creep*) (Figure 11.2). For an imposed strain rate, σ increases during transient creep up to a plateau during permanent creep.

At the permanent creep stage, the material behaves like a viscous fluid, but a non-Newtonian one. Creep can be said to result from the recovery destroying the work-hardening as it appears. To account for transient creep, one can either admit that work-hardening is much smaller at the start than in permanent regime, or that the recovery rate is much faster. But this is no proper explanation.

11.2 ROCK TEXTURES AND DEFORMATION

Rock deformation laws can be set forth only if the space and time scales are explicitly specified. In geodynamics, the interesting behaviours occur on the scale of the cubic kilometre or more. Conversely, in tectonophysics, the processes of interest occur in a

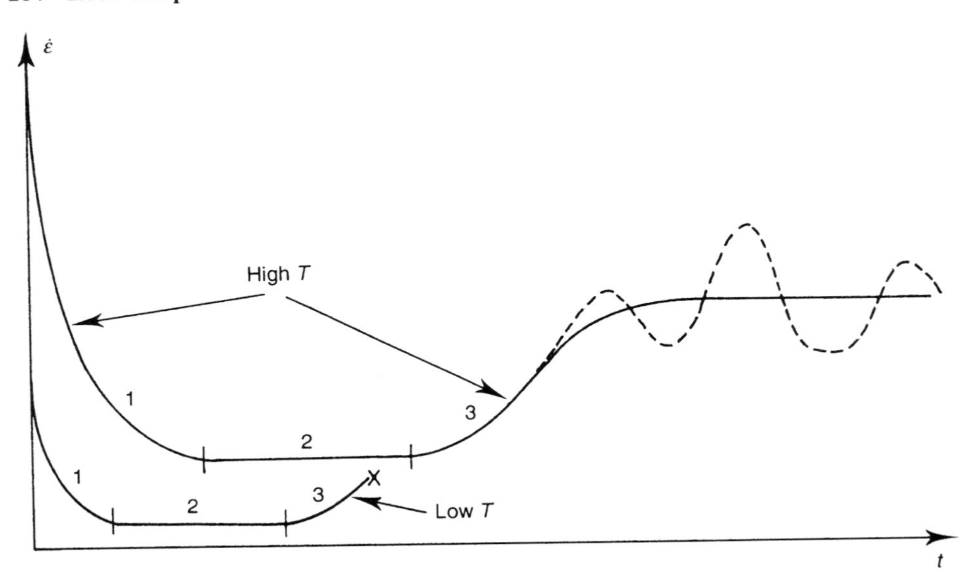

Figure 11.2. Evolution of the stress rate with time: creep. From Lliboutry (1987); © Kluwer Academic.

cubic metre or less. At this scale, there are no more faults and joints, nor bedding, but there are *textures*, resulting from the past deformations and which generally make the rock anisotropic. The main textures are represented in Figure 11.3.

a. With low metamorphism, lamellar minerals like mica may have taken a uniform orientation. It is proved that they are perpendicular to the previous main compression direction. The exact process creating this texture is, however, still discussed. This texture facilitates the quarrying of the rock into plates, and is called *slaty cleavage*.
b. Strong metamorphism and shear may have produced a flattening and a uniform orientation of the grains. This fluid texture forms the *flow cleavage*.
c. Metamorphism may also produce a segregation in parallel layers with differing compositions (*banding*). The processes (b) and (c) are generally associated. The texture is then called *schistosity* or *foliation*.
d. Fracture and slip may have occurred along parallel planes, followed by *healing* of the fractures. This is called *fracture cleavage*.
e. Fracture cleavage may have been accompanied by diffuse shear close to the slip planes. This is called *crenulation cleavage*.
f. The rock may enclose layers of grains finely ground and then re-welded together, forming a *mylonite*.
g. Finally, fissures may have appeared in the rock, often *en échelon* (showing that shear parallel to the fissures accompanied the extension). Another mineral fills these fissures, carried as a fluid solution.

These textures are not all reproducible in the laboratory, because they require very

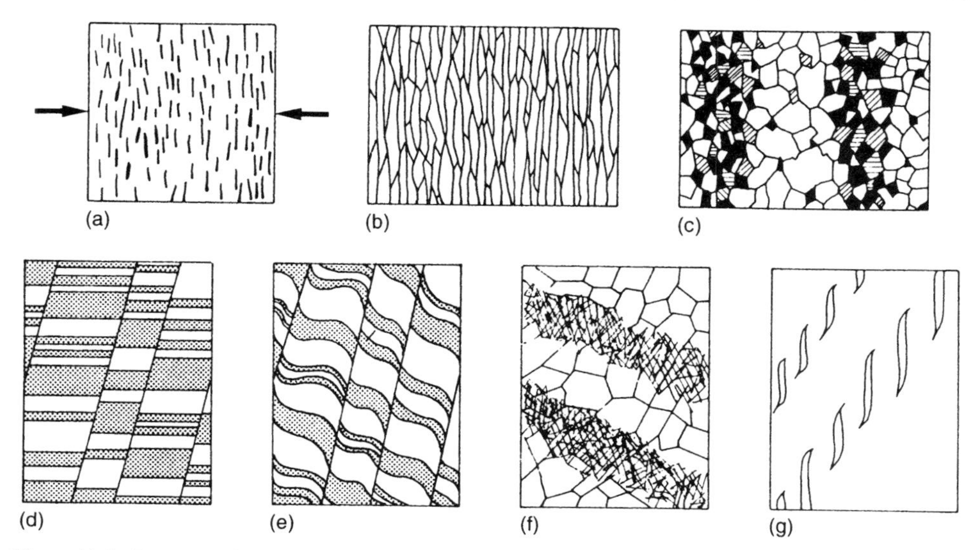

Figure 11.3. Textures of metamorphic rocks (see text for details). From Lliboutry, 1987.

large strains. Creep rates may be highly accelerated by increasing the temperature (without reaching the solidus temperature of the first material to melt). Then, at the best, strains become measurable, but they are still quite small.

Brittle behaviour

All rocks are *brittle* at the pressures and temperatures found in the first kilometres of the terrestrial crust, except rock salt (made of halite), limestone and marble (made of calcite). Therefore at shallow depth, in general, there are faults at any scale. Their movements, slow, or rapid during earthquakes, will account for the strains.

Kinking and twinning

Plastic deformation of rock minerals, i.e. large strains that appear almost instantaneously on loading, are the kinking of micas, the twinning of calcite, plagioclases and hornblende. Kinking divides lamellas into strips tilted by 25° to 120° from each other. Twinning transforms the crystalline lattice into its symmetrical relative to a particular plane; the corresponding shear is $\gamma = 0.275$ in plagioclases, 0.694 in calcite.

These finite deformations start at some nuclei, which increase very fast. They are observed at rather low temperatures. At high temperatures they may coexist with usual plastic creep (excepting hornblende, which remains brittle up to its melting).

Pressure solution creep

On the geological time-scale, both calcite and quartz can creep through *pressure solution*. This creep does not require water to circulate through the rock, but simply to impregnate it, forming a submicroscopically thin film between the grains. As the solubility of the mineral in water is proportional to the pressure, the mineral dissolves over the more compressed faces and precipitates on the less compressed ones. The grains change shape without deforming in their mass.

The law of pressure-solution creep can be easily found, without assuming a precise geometrical model of the grains or performing any calculation. It is only necessary to assume that the thickness of the liquid layer is independent from the mean size a of the grains. Between two faces of differing orientations, the concentration gradient (causing the diffusion) is proportional to the pressure gradient, and therefore proportional to a^{-1}. With this assumption, the flow from one face to the other is then independent of a. This flow ensures matter exchange between two faces of area a^2. The solution and precipitation velocities are therefore proportional to a^{-2}. The apparent deformation rate is obtained by division of this velocity by a. This yields a *Newtonian viscosity law with a fluidity inversely proportional to the size of the grains to the power 3.*

It is now thought that folded sedimentary layers like the calcareous Pre-Alps were deformed mainly through pressure-solution rather than plasticity.

Brittle or cataclastic creep

Brittle creep is another deformation process. The grains are crushed into fragments slipping onto each other, so that a global strain is possible. But these fragments later weld back together, and microfractures do not occur simultaneously, so that the rock remains coherent. Depending on the fragments' size, one obtains a *cataclasite* or a *mylonite*.

Brittle creep of feldspars and quartz can be observed in the laboratory from 300°C upwards, with pressures larger than 500 MPa (i.e. the lithostatic pressure around 20 km deep).

High-temperature, high-pressure creep

Non-brittle creep, distinct from pressure-solution creep, only exists at very high temperatures and very high pressures, as can be found in the lower crust or the mantle. The transition between brittle and non-brittle creep is not clear-cut, as the two processes may coexist: brittle creep for some minerals in the rock, non-brittle creep for other minerals. For example, in the Mont Blanc granite, there was cataclasis of feldspars and quartz, plastic creep (and chloritization) of the micas. Alpine geologists call such granite a protogine.

The many processes of non-brittle creep can be classified into two groups: the ones with strain inside the grains themselves (*plastic creep*, or *dislocation creep*), and the ones with superficial deformation of the grains (we have already seen one example, the pressure-solution creep).

Similar to plastic strain, plastic creep results from the displacement of linear defects, which are in fact *dislocations* of the crystalline lattice. It may begin by transient creep, of variable duration. When the creep becomes permanent, it generally follows a power-law of viscosity, defined later. When the cumulative deformation reaches 1%, creep is accompanied by the recrystallization of the material. The strain rate becomes constant again, but is several times higher. Creep followed by recrystallization was only observed and well documented in the laboratory for ice. In all cases, if the rock contains a quasi-single mineral (quartzite, dunite, etc.), the crystalline axes of the different grains progressively cluster along a few directions. This is called a *fabric*. Recrystallization (called *kinematic* or *dynamic* recrystallisation by metallurgists and *syntectonic* recrystallization by geologists) accelerates the formation of this fabric.

In a few rare cases, Newtonian viscosity could be proved, although it was a plastic creep proceeding by dislocations. This is *Harper–Dorn creep*, which will be detailed in Section 18.10 (as it may occur in the mantle).

There are at least four non-plastic, high-temperature creep processes, which should all lead to Newtonian-type viscosities:

a. Creep by diffusion of vacancies or interstitials.
b. Creep by slip at the grain joints (well known in some metals and improperly called *super-plasticity*).
c. Creep during grain growth, as observed in the shallow layers of polar ice caps.
d. Creep during metamorphism, demonstrated by the resulting fluidal textures, but of which nearly nothing is known.

11.3 IDEAL PLASTICITY AND N-POWER LAW ISOTROPIC VISCOSITY

In most physical models used in geodynamics, the Earth is modelled as a superposition of homogeneous layers with simple constitutive laws. These laws: (1) contain only 1 unknown parameter, whose adjustment accounts for the observations; (2) lead to mathematically solvable problems. This is the case of *isotropic elastic* or *isotropic isoviscous* layers. We can also add the model of the *perfect plastic body*.

The perfect plastic body (implicitly isotropic) is a continuous plastic medium (this means it cannot deform until the effective shear stress τ reaches a threshold k), with no work-hardening, so that in all deforming areas, $\tau = k$. The stress and strain rate deviators must be proportional, but their ratio may be anything (written for the different components, these are the *Lévy–von Mises relations*). This model is used to solve practical problems of stamping or punching in metallurgy or in soil mechanics. We shall come back to this model in Sections 12.6 and 13.9.

Laboratory studies of rock creep and the theory of creep by dislocations show that none of the preceding constitutive laws can account for permanent plastic creep. Creep obeys an *n-power viscosity law*, with n between 2 and 5.

An incompressible medium is viscous to the power n if the strain rates $\dot{\varepsilon}_{ij}$ are all multiplied by λ^n when the components s_{ij} of the stress deviator are multiplied by λ. If the behaviour is isotropic, the equations (10.57) and (10.59) also remain valid and:

$$\frac{2\dot{\varepsilon}_{ij}}{s_{ij}} = \frac{\dot{\gamma}}{\tau} = \frac{1}{\eta} = \varphi \tag{11.1}$$

The viscosity η and the fluidity φ must be invariant when the reference frame is rotated, as $\dot{\gamma}$ and τ. We assume that they are functions only of the effective shear stress τ, and, very weakly, of the mean pressure p, but not of the third stress invariant. This was proved experimentally only for ice, by Paul Duval and his team (Duval, 1981). But it can be justified in the general case with the following reasoning. Let τ_b be the shear on the active intracrystalline glide planes. As explained later, this shear on the microscopic scale should not differ very much from the shear on the macroscopic scale, considered in mechanics of continuous media. The deformation rates must be proportional to a weighted average of τ_b^n. For planes uniformly distributed in all the directions of space, the average of τ_b^n is τ^n, to a numeric factor depending on n.

When τ is multiplied by λ, $\dot{\gamma}$ is multiplied by λ^n, and therefore φ is multiplied by λ^{n-1}. Therefore, φ must equal τ^{n-1}, multiplied by a coefficient function of temperature and pressure whose form is given by thermodynamics. Finally, the isotropic n-power viscosity law reads:

$$2\dot{\varepsilon}_{ij} = A^* \exp\left(-\frac{H^* + pV^*}{RT} \right) \tau^{n-1} s_{ij} \tag{11.2}$$

R is the gas constant of thermodynamics ($R = 8.314\,\mathrm{J/mol}$), T the absolute temperature, $H^* = U^* - TS^*$ is the activation enthalpy (U^* = internal energy, S^* = entropy), V^* the activation volume and $G^* = H^* + pV^*$ the activation free enthalpy.

In a mono-axial compression test, with a strong confining pressure p added to avoid any brittle creep, the three main pressures are p, p and $(-\sigma_z + p)$ (compression being along the z-axis). The effective shear stress is $\tau = |\sigma_z|/\sqrt{3}$. As $\varepsilon_{xx} = \varepsilon_{yy} = -\varepsilon_{zz}/2$, the effective shear stress rate is $\dot{\gamma} = \sqrt{3}|\dot{\varepsilon}_{zz}|$. Investigators publish their results under the form:

$$|\dot{\varepsilon}_{zz}| = A|\sigma_z|^n \exp\left(-\frac{H^*}{RT} \right) \tag{11.3}$$

This neglects the activation volume. This formula must be identified with:

$$2\dot{\varepsilon}_{zz} = A^* \exp\left(-\frac{H^*}{RT} \right) \left(\frac{|\sigma_z|}{\sqrt{3}} \right)^{n-1} \left(\frac{2}{3}\sigma_z \right) \tag{11.4}$$

Therefore:

$$A^* = 3^{(n+1)/2} A \tag{11.5}$$

Numerical values for a granite and a peridotite are given below. The fluidities of quartz and olivine are highly increased by diffusion of water in the crystalline lattice, and therefore it will be specified whether the sample has been wetted for a long time or, conversely, dried by a prolonged heating.

Westerly granite (Rhode Island, USA), wet (Hansen and Carter, 1982)
$n = 1.9$; $H^* = 137\,\mathrm{kJ/mole}$; $A^* = 3.1 \times 10^{10}\,\mathrm{MPa}^{-n}\,\mathrm{Ma}^{-1}$

Mt. Burnet dunite, dry (Carter, 1976)
$n = 3.3$; $H^* = 465\,\mathrm{kJ/mole}$; $A^* = 4.3 \times 10^{17}\,\mathrm{MPa}^{-n}\,\mathrm{Ma}^{-1}$

These results can be extrapolated to lower temperature and effective shear stresses, such as can be found in the lower crust or the lithospheric mantle, respectively, and the following viscosities are obtained (in Ma MPa):

	Wet granite		Dry dunite		
	200°C	400°C	700°C	900°C	1,100°C
$\tau = 1\,\text{MPa}$	43×10^3	1.38	212×10^5	1.18×10^3	1.14
$\tau = 10\,\text{MPa}$	5.4×10^3	0.17	1.06×10^5	5.9	0.0057

If the extrapolation to geological intervals of time is valid, the isoviscosity ($n = 1$, $G^* = 0$) is manifestly an extremely rough model of the actual behaviour. This question will be discussed in Section 18.10.

Let us also provide the creep law with recrystallization of polycrystalline glacier ice, well determined by glaciologists from Grenoble (France). After recrystallization, this metamorphic rock presents a particular fabric (optical axes clustered in several directions, most often 4), which makes it isotropic macroscopically. (Conversely, the fabric before recrystallization often makes its macroscopic behaviour highly anisotropic.) At the melting point, it contains liquid water inclusions, mainly as lenses at the joints of grains. If $w\%$ is the content ratio of the liquid phase (around 1%);

$$\begin{cases} n = 3.0; \qquad H^* = 78 \pm 4\,\text{kJ/mol}; \qquad A^* = B\exp\left(\dfrac{H^*}{273.2R}\right) \\ B = 200 + 370w \pm 40\,\text{MPa}^{-3}\,\text{a}^{-1} \end{cases} \tag{11.6}$$

In glaciers, $\tau = 0.02$ to $0.15\,\text{MPa}$, and only the first 10 to 20 metres show crevasses, faults and brittle creep. Strain rates are approximately 10^6 times faster than in the terrestrial mantle. It is the most striking example of plastic creep in a crystalline solid.

The rheological law expressed in (11.2) can be written as:

$$2\dot{\varepsilon}_{ij} = \left(\frac{\tau}{k}\right)^{n-1} s_{ij}; \qquad k = (A^*)^{-1/(n-1)} \exp\frac{H^*}{(n-1)RT} \tag{11.7}$$

If n and A^* tend toward ∞, k remaining finite, τ/k must tend toward 1 and $(\tau/k)^{n-1}$ becomes undetermined. We come back to the model of the perfect plastic body. Theoretical mechanists are well embarrassed, as they call *fluid* any body where the stress deviator vanishes with the strain rate (such as the n-power viscous body), and *solid* any body with a plasticity threshold. It is their problem. In this book, the terms of *solid* and *fluid* will always correspond to their physical definition. Any crystallized material is a solid.

11.4 DISLOCATIONS IN CRYSTALS

No crystal is perfect, with 10^{12} to 10^{15} atoms exactly at their right place. The crystal contains point defects (missing atom, or *vacancy*, extra atom, or *interstitial*) and linear defects (*dislocations*, whose Burgers vector equal 1 inter-atomic distance).

A dislocation in a crystallographic plane may sweep this plane from one side of the crystal to the other, driven by the shear stress exerted on this *slip plane*. As Figure 11.4 shows, this provides a displacement of a whole side of the slip plane, by a distance equal to the Burgers vector. The distance between parallel potential slip planes is of the same order as the Burgers vector, and 1 over 1000 to 1,000,000 contains an active dislocation. The displacement of all dislocations that are active at a particular time therefore yields a global shear strain of 10^{-3} to 10^{-6}. But there are

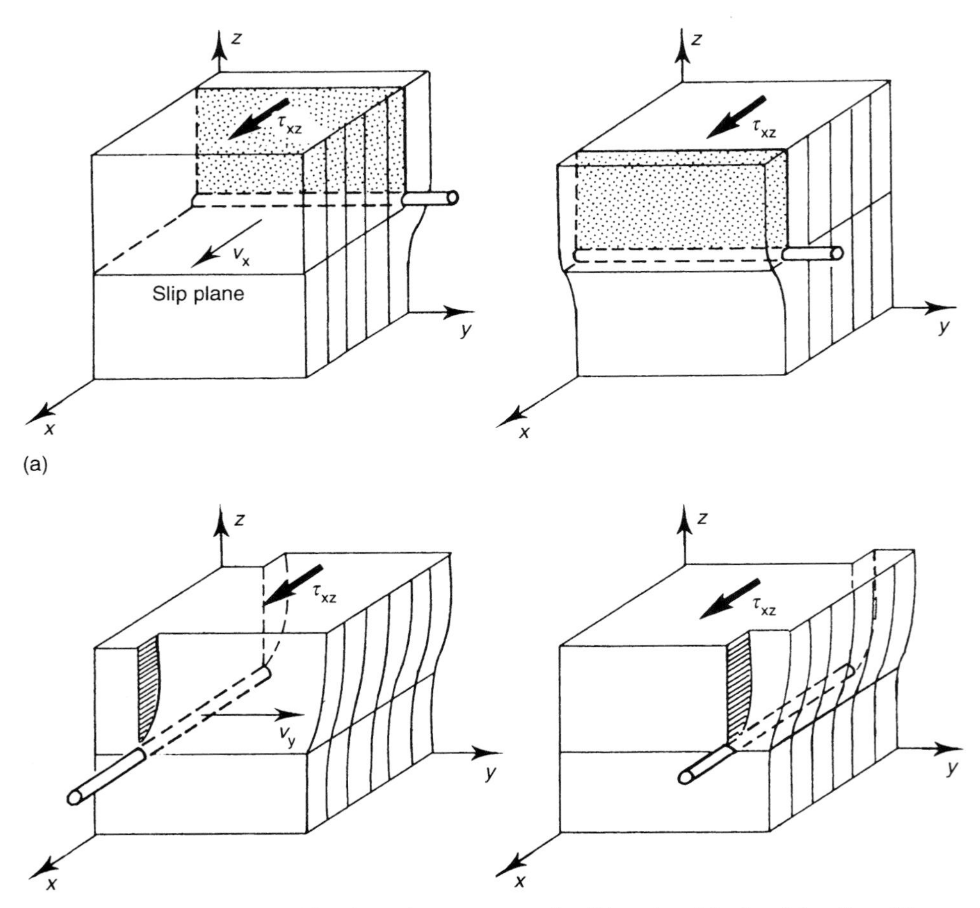

Figure 11.4. Simple shear resulting from the movement of a dislocation; (a) edge dislocation; (b) screw dislocation. From Lliboutry (1987); © Kluwer Academic.

dislocation sources, mainly at the joints of grains, which continuously produce new dislocations, so that shear strain can go on indefinitely.

Work-hardening comes from a blocking of dislocations by obstacles, mainly other dislocations crossing their slip plane. They pile up against them and stop being active. If recovery makes possible a permanent creep, this is because dislocations can escape from these 'traffic jams' (*pile-ups*). They may migrate on a parallel slip plane (*climbing* of the dislocation), or turn off to a slip plane with a different orientation (*cross-slip*).

Dislocations must form closed loops inside the crystal, or end at its limits, the grain boundaries. A slight attack on the crystal with acid makes *etch pits*, which are used to estimate the dislocation density (number of dislocations crossing a unit area). Even in an annealed, stress-free metal, there are 10^4 to 10^8 dislocations per cm^2. The stresses increase this number, and dislocations remain in the work-hardened material.

The dislocations in a transparent material can be made visible under a microscope, by *decorating* them. A particular impurity is diffused in the crystal at high temperature (e.g. manganese in olivine). At cooling, the impurity precipitates as extremely fine microcrystals along the dislocations. Dislocations can also be observed by X-ray diffraction, with a particular setting up of the rays and the sensitive plate (*Lang camera*).

This latter method allowed observing the displacement of dislocations under stress. Their speed is proportional to the shear stress on the slip plane. It is very fast in metals; at ambient temperature, $5.0\,\mathrm{m\,s^{-1}\,MPa^{-1}}$ for aluminium, $31\,\mathrm{m\,s^{-1}\,MPa^{-1}}$ for copper. It can be one million times slower in rock minerals. In ice at $-18°C$, it reaches $2\,\mathrm{\mu m\,s^{-1}\,MPa^{-1}}$. In metals, the shear strain rate is controlled by the rate at which the dislocations escape from the pile-ups (through climbing or cross-slip). On the other hand, in rock minerals, it is mostly controlled by the speed v at which the dislocations move along a slip plane. If ρ is the density of the active unblocked dislocations, and b is the Burgers vector, the shear strain rate is given by the *Orowan relation*:

$$\dot{\gamma} = b\rho v \tag{11.8}$$

The shear stress τ_b on a slip plane activates its dislocation sources. In the immediate neighbourhood of a dislocation, the elastic deviatoric stresses are of the order of $b\mu/2$. As the average distance between dislocations is $\rho^{1/2}$, there are some very local stresses roughly equal to $b\mu/\rho^{1/2}$. When they are much larger than the general shear stress τ_b, the latter does not have any more influence; the multiplication of dislocations has reached saturation. At saturation, the density of dislocations must therefore be proportional to $(\tau_b/b\mu)^2$. Assuming that permanent creep happens when dislocation saturation has been reached, and that the number of active dislocations is then proportional to their total number, the Orowan relation leads to a power law viscosity with exponent 3.

It is demonstrated that two dislocations on the same slip plane and with the same Burgers vector repel each other, with a force inversely proportional to their distance. This repulsion prevents the dislocations piled against an obstacle from being too

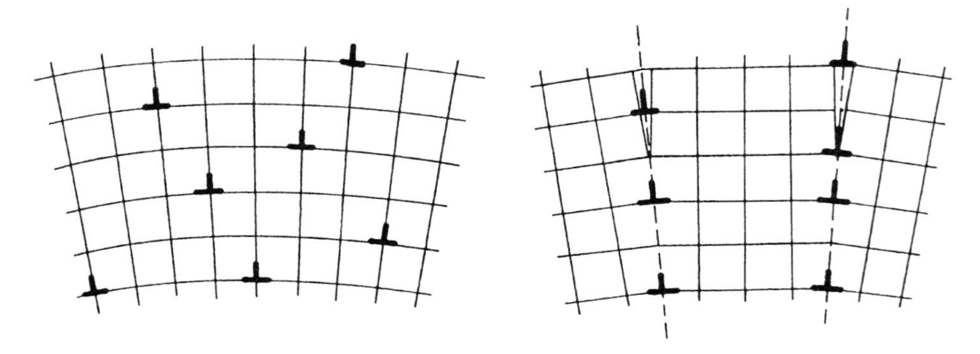

Figure 11.5. Polygonization of a crystal elastically curved in the plane of the figure. The lines are the traces of a few atom planes. Edge dislocations perpendicular to the figure's plane are represented by upside-down *T*, the vertical bar of the *T* points toward the half-plane of extra atoms. From Lliboutry (1987); © Kluwer Academic.

close, and limits their number. If they can be distributed before the obstacle over a length *L*, calculations show that their number and the total thrust against the obstacle (i.e. the thrust which allows the dislocations to cross it, one after the other) are proportional to $L^{1/2}$. If the piling-up occurs against grain joints, *a* being the mean grain size, the resulting crossing rate is proportional to $a^{1/2}$, and the shear rate is proportional to $a^{-1/2}$ (*Hall–Petch law*). But this law is only confirmed by observation for transient creep or for extremely fine grains. For medium-size or coarse grains, permanent creep does not depend on grain size. The reason is that most obstacles met by the dislocations are not grain joints, but obstacles inside the grains, in particular *grain sub-joints*, which are defined below.

Edge dislocations with the same Burgers vector and in two parallel slip planes may attract or repel each other. They are in stable equilibrium only if one is just below the other. An edge dislocation is the edge of a half-plane of extra atoms in the crystalline lattice (half-plane symbolized by the vertical bar of the *T* representing an edge dislocation in diagrams). In their stable disposition, all half-planes are side by side and form a highly acute dihedron (Figure 11.5). This pattern forms a *wall of dislocations*, or a *grain sub-joint*. The crystal is *polygonized*, and presents a mosaic structure. The blocks of the mosaic have sizes of ca. 10 μm, and the deviation between adjacent blocks (the dihedral angle) is of a few minutes only. Polygonization is very well observed by X-ray diffraction.

11.5 ROLE OF POINT DEFECTS IN CREEP

Interstitial water molecules in the crystalline lattice of silicates make them much more ductile. The strain rates under a given stress are more or less multiplied by 50,000 in quartz, 10,000 in olivine, but do not seem to change in diopside. These differences are probably explained by the varying possibility for water diffusion in the crystal. In the most sensitive mineral, quartz, at 900°C and 300 MPa, the saturating concentration of water is only 10^{-5} (in weight).

The concentrations in vacancies, or interstitials of the same nature as one of the elements forming the crystal, are of the same order close to the melting point (10^{-5} to 10^{-6}). They decrease when temperature decreases. Vacancies continuously form on the less compressed sides of the crystal, and they continuously disappear on the more compressed sides. The driving mechanism is the vacancies density gradient. For interstitials, the opposite happens. In both cases, the resulting creep is very slow and non-plastic, similar to pressure-solution creep.

Vacancy diffusion may occur through the bulk of the crystal. This is *Nabarro–Herring creep*, predicted since 1948 and observed in 1973 in extremely fine quartzite grains. If a is the grain size, the creep rate must be proportional to a^{-2}. Diffusion may also occur along the grain boundaries (*Coble creep*). The creep rate is then proportional to a^{-3} (the reasoning is similar to the reasoning for pressure solution creep). It is very unlikely that creeps by vacancy or interstitial diffusion occur in mantle creep: the high temperatures and very low deformation rates must favour the recrystallization into very large crystals.

But the point defects always intervene in plastic deformation, because they are essential for the displacement of dislocations. Creep is very sensitive to temperature, because the creation of point defects and their displacement are also sensitive to temperature.

Dislocations occupy specific positions in the crystalline lattice, which may be called ‘energy ditches’. Part of a dislocation may be in a ditch, another part in a neighbouring parallel ditch. The deviation from one to the other is called a *kink* if the two half-dislocations are in the same slip plane, and a *jog* if they are in two distinct parallel slip planes. In a slip plane, a dislocation does not jump as a block from one ditch to the next. This displacement is much easier if the kink occurs at one end and travels along the entire dislocation to the other end. Similarly, the climbing of a dislocation does not occur as a block, but by jogs migrating from one end to the other[1].

Point defects intervene in the propagation of a kink or a jog. Theories appeared in the 1970s using the mobility of an atom, or a small group of atoms, made possible by the presence of a vacancy. Silicate creep was explained by the mobility of oxygen atoms, enabling the movement of jogs and the climbing of blocked dislocations. It was inferred that the activation volume V^* should be the molar volume of oxygen, 11 cm^3. (Jaoul (1980) was however arguing for the mobility of SiO_2 groups, leading to $V^* = 30\,\text{cm}^3$). More detailed theories have appeared since, considering not only the crystalline structure but also the electronic densities in the lattice, as predicted by quantum physics. Point defects appear to be mainly electrical in nature. According to Hobbs (1983), olivine creep is controlled by the migration along dislocations of positively charged kinks. And if water facilitates silicate creep, this would be because

[1] This explanation is similar to the previous explanation of plane-over-plane slip. This latter theoretical prediction was confirmed by observation 30 years later. Kinks and jogs remain a theoretical explanation only, since they are not observable with the current techniques.

it brings protons, i.e. positive charges (see the review of the diverse theories on the role of H_2O in the *Journal of Geophysical Research*, vol. 80, no. B6, pp. 3991–4358, 1984).

11.6 FROM MICROSCOPIC TO MACROSCOPIC CREEP

The plastic deformation of a crystal consists in shear parallel to one or several slip planes, because of the shear stress acting on these planes. Let us call *macroscopic* the global stresses and strains in the polycrystalline rock and *microscopic* the local stresses and strains in each crystal (the rock grains). To go from the latter to the former, by calculus, is called *homogenization*. Without performing these calculations here, let us underline the problems they pose and the factors to be introduced. They are not all accounted for in homogenization and often make its results open to criticism.

The possible slip planes are specific crystallographic planes. If one is possible, others can be deduced by applying the crystal symmetries. For most metals, crystallized in a cubic system, the possible slip planes are thus tripled. But for ice, quartz (and cobalt), with a hexagonal system, the easy-slip planes are perpendicular to the hexagonal axis (also the optical axis). These planes are called *basal planes* and have only one orientation per crystal. Olivine is orthorhombic and its easy-slip planes have only two orientations, different whether the temperature is above or below 1,000°C. Since there are two or three possible slip directions in each slip plane, it must account for two degrees of freedom in the deformation. Keeping to easy deformations, to those not requiring microscopic shear stresses much higher than the macroscopic ones, there are 6 degrees of freedom in metals, 2 or 4 at most in rocks. But a constant-volume deformation requires 5 degrees of freedom (the values of the five independent components of the strain deviator). They are found in a metallic crystal, but not in rock grains.

If this geometric constraint were the only one, microscopic strain could be identical to macroscopic strain in metals, but not in rocks. Nevertheless, the microscopic strain rates must be proportional to the microscopic deviatoric stresses. And microscopic stresses in the polycrystal must obey the equilibrium conditions, which exclude large variations between two adjacent crystals. Therefore, in metals as in rocks, the microscopic strain rates and stresses fluctuate around their macroscopic values. The lack of degrees of freedom in rocks simply makes these fluctuations larger. Some grains, badly oriented at a particular time, will turn inside the polycrystal without deforming. This only lasts some time, because rotation soon gives them an orientation for which they can deform.

The compatibility of deformations of different crystals is also made possible by small slips on grain boundaries, and by *grain boundary migrations* (Means and Jessel, 1986). A plane case shows the mechanism (Figure 11.6). *A* and *B* are two adjacent cubic crystals and *C* another crystal close to the two. The slip planes, perpendicular to the figure plane and shown with hatching, enable the deformation of *A* and *C* while *B* can only rotate. This requires some slip at the grain boundaries *B–A* and

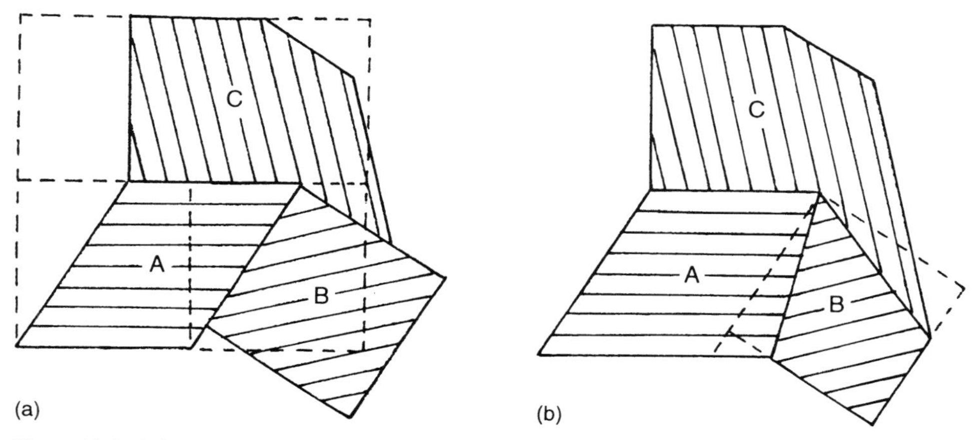

Figure 11.6. Anisotropic deformation of three crystals, made possible by the migrations of their interfaces. From Lliboutry (1987); © Kluwer Academic.

B–C (Figure 11.6a). Furthermore, because of the presence of other crystals around, kinks should not appear on the periphery. This is possible if these same grain joints migrate (Figure 11.6b). In fact, slip and migration occur simultaneously.

The migration of grain boundaries has another result. It destroys a large part of the dislocations' piling-ups. It must therefore be seen as an essential factor controlling the creep. An isolated crystal, submitted to a non-zero shear stress on its easy-slip plane, will deform tens of times faster than the corresponding polycrystalline rock, because there is no problem of compatibility of the deformation with its neighbours.

One must not confuse the simple migration of grain boundaries with *recrystallization*, the appearance of new grains growing at the expense of the old ones. The nuclei of the new crystals should be blocks from the mosaic structure at the edges of the grains, strongly deflected. The dislocation density in the new crystals is lower than in the old work-hardened crystals (this explains why they develop preferentially). But these dislocations are not piled-up yet, so that the polycrystal creep rate is higher when recrystallization appears. It appears in ice after a deformation of 1%, whatever the stress. There may be periodic phases of work-hardening and recrystallization, making the creep rate oscillate as in Figure 11.2 (dashed line). The same processes were observed with metals and some salts, but there is still no experimental study of syntectonic recrystallization in usual rocks (Drury and Urai, 1990).

11.7 FABRICS AND SUPERFICIAL ENERGY OF GRAINS

Fabrics appear after prolonged creep, i.e. large strains which cannot be described with the strain rate tensor at a given time. Ice is the simplest case. Observations with the Lang camera were made by Higashi and his team at the University of Hokkaido (Sapporo, Japan). They showed that ice is deforming nearly exclusively through the

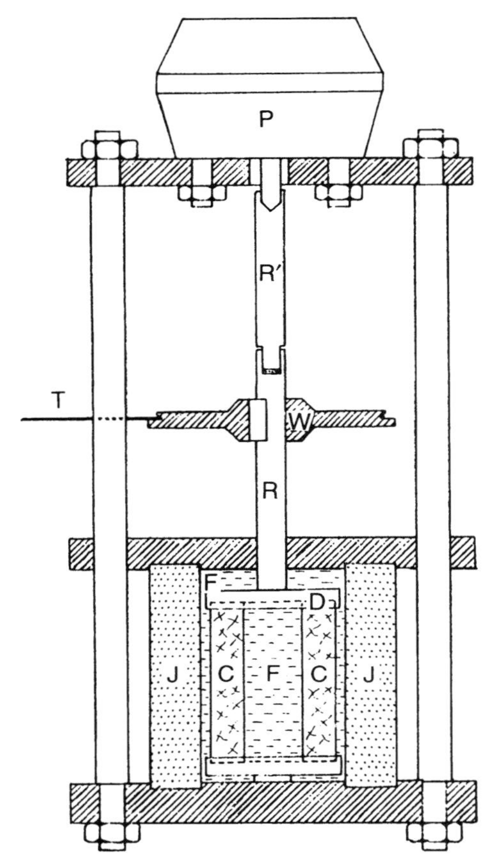

Figure 11.7. Duval's device, enabling torsion–compression tests on an ice core hollowed along its axis. From Lliboutry (1987); © Kluwer Academic.

C = hollowed ice core; F = confining fluid under pressure (pipe transmitting pressure not drawn); J = transparent jacket; D = disk transmitting mono-axial compression and torque; P = hydraulic jack transmitting the pressure through the rods R and R'; W and T = wheel and tangent cable (pulled by a weight) producing the torque. O-rings providing water-tightness, strain gauges and thermistors are not drawn.

displacement of screw dislocations in the basal planes. With the torsion–compression Duval's device (Figure 11.7), it is possible to submit a cylindrical ice core hollowed out along its axis to:

- Either a *simple shear*, by exerting a torque about the axis. The across-axis plane, materialized in the sample, always remains a plane of maximum shear. After a while, all basal planes are close to it, and the optical axes are therefore clustered close to the axis direction. This fabric 'with one maximum' only strengthens as the simple shear goes on, and facilitates it.
- Or a *mono-axial compression*, parallel to the axis of the core. The maximal shears occur on planes at 45° from the axis (see Chapter 12). A fabric appears,

with basal planes close to this optimal orientation, and therefore optical axes grouped in a circular ring around the compression axis, at 45° approximately from it. But, as mono-axial compression goes on, it squashes these basal planes sideways. With time, the optical axes of the grains get more or less perpendicular to the axis of the core.

- Or simultaneous torsion and compression. A third type of fabric appears then: the optical axes cluster around four directions. All the directions in space (i.e. all the points of a hemisphere) are usually represented by points in a plane inside a circle, in polar equal-area projection (*Schmidt projection*). The optical axes cluster at the summits of a diamond shape centred on the compression axis.

This curious fabric 'with four maxima' is always found in temperate glaciers and, often, in the bottom layer of ice sheets when it is recrystallized with very large grains.

The fabric with four maxima does not allow the fastest creep, for given stresses. But it is very stable. Simple shear or mono-axial compression do not destroy it to impose their associated fabrics. According to Higashi *et al.* (1978), this fabric would come from the formation of grain joints for which the two crystalline networks in contact have common sites (*coincidence site lattices*). The surface energy comes from the fact that the distances between atoms on each side of the grain boundary are not the optimal ones, as they are in the crystalline lattice. This energy is much lowered when there are 'coincident sites'. Such grain boundaries can migrate very easily while site coincidence remains.

Thus, some *interface energy* is stored at the grain boundaries, and it is smaller for coincident site lattices. It is usually higher than the energy between a crystal and a fluid at equilibrium with it, causing capillarity. The interface energy between two ice crystals is $0.065\,J/m^2$ in average, and is only $0.034\,J/m^2$ between ice and water.

Let us calculate a few orders of magnitude, by considering $1\,m^3$ of ice submitted to an effective shear stress of 0.1 MPa. For ice, $\mu = 3.45\,GPa$, and the macroscopic distortion elastic energy is $1.5\,J/m^3$. With 10^7 dislocations per cm^2, the cubic metre of ice contains 3×10^{11} of them, with an elastic self-energy of $10^{-9}\,J/m$. The self-energy of dislocations is therefore $300\,J/m^3$. With $10^3\,m^2$ of grain boundaries per m^3, the interface energy equals $65\,J/m^3$. Compared with these two energies, the macroscopic elastic energy is negligible and cannot play any role in the formation of the fabrics. The fabrics are *strain-induced*, and not *stress-induced.*

This is also the conclusion of tectonophysics investigations on silicate rocks (Nicolas and Poirier, 1976). Although we do not know up to which point the results obtained with polycrystalline ice, a particularly ductile rock because of its hydrogen links, are valid for silicate rocks.

The interface energy creates forces tending to reduce the total area of grain boundaries. This explains:

- The shape, usually equidimensional, of the grains. Without the 'surface tension' forces intervening during the grain boundaries' migrations required by the strain, the sliding along one single system of slip planes would infinitely flatten the grains, and flow cleavage would be the rule.

- The slow growth of the grain at high temperature, in the absence of deviatoric stresses. Theory says that the average grain size (a) must increase with the time t since the formation of a compact rock according to the law (K being an increasing function of temperature):

$$a^2 = a_0^2 + Kt \tag{11.9}$$

- The fast and very important grain growth by recrystallization after a strain, after the deviatoric stresses have disappeared (*post-kinematic recrystallization*). The interface energy is not the only one to decrease; so does the dislocation self-energy, as many dislocations piled up against the grain boundaries disappear during this growth. Nevertheless, should we not include the self-energy of dislocations at the interface (a microscopic concept) into the interface energy (a macroscopic concept)?

11.8 TRANSIENT CREEP

Transient creep has never been much studied, and only during a first loading, except for ice.

By applying a load σ of 10 to 100 MPa to a crystalline rock, at a temperature of 700–1,000°C, and maintaining it, we observe at the beginning a creep following the law:

$$\varepsilon = A(T)\sigma^m t^\alpha \qquad (1 \le m \le 2; \tfrac{1}{3} \le \alpha \le \tfrac{1}{2}) \tag{11.10}$$

Da Andrade established this law in 1910 for metals, with $\alpha = 1/3$. This transient creep is therefore called *Andrade creep*. In the case of ice, the creep after a first load obeys exactly the following law:

$$\dot{\gamma} = B\tau^3\left(1 + \frac{\tau^2}{\beta^2\gamma^2}\right) \tag{11.11}$$

The constant β, with the dimensions of a stress, does not depend on temperature (but on the grain size). At the beginning, $\beta\gamma \ll \tau$ and Andrade's law is found:

$$\gamma = \left(\frac{3B}{\beta^2}\right)^{1/3} \tau^{5/3} t^{1/3} \tag{11.12}$$

This creep varies with temperature like $B^{1/3}$. The apparent activation energy is the third of the energy inferred from permanent creep rates.

The exact integral of (11.11) is:

$$\gamma = \frac{\tau}{\beta}\arctan\frac{\beta\gamma}{\tau} + B\tau^3 t \tag{11.13}$$

After some time, relatively long, the strain becomes the one due to permanent creep strain plus a finite strain, $(\pi/2)$ (τ/β), *irreversible* and much larger than the elastic strain (80 times, in the experiments by Le Gac and Duval). In my opinion, it does not

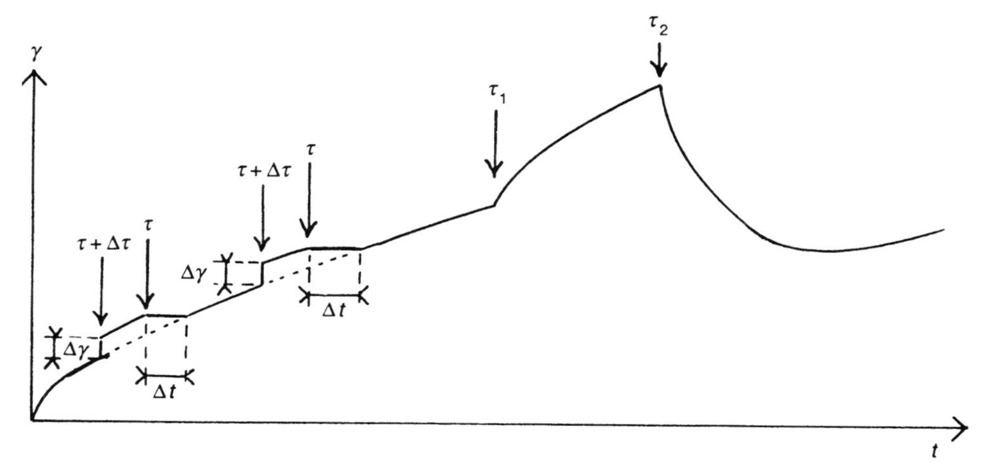

Figure 11.8. Strain versus time under varied successive loads, applied at the times shown by arrows and kept constant until the next arrow.

come from a process distinct and independent from permanent creep. Simply, the number of active dislocations is not proportional to τ^2, but to $[\tau^2 + \tau^4/(\beta^2\gamma^2)]$.

Unfortunately, because of an insufficient number of experiments, no general law could be established to give transient creep after any change of the stresses. Only experimental facts can be mentioned, which any general law should encompass (Figure 11.8).

Reversibility of the strain due to a small load increment

A small increment $\Delta\tau$ of the load applied rapidly provides an increment $\Delta\gamma$ of the strain. The ratio $\Delta\tau/\Delta\gamma = h$ *(work-hardening coefficient)* is the same whether in permanent or transient creep.

A small decrement $-\Delta\tau$ does not induce a temporary decrease of γ (other than elastic). The deformation stops for a certain time Δt (the ratio $\Delta\tau/\Delta t$ is the *recovery rate*), followed by permanent creep at the rate corresponding to the new load. In particular, if we go back to the initial τ after an increment $\Delta\tau$, the new curve $\gamma(t)$ is a prolongation of the initial curve. For small stress increments, the strain is *recoverable*. Microscopic elasticity, linked to the dislocation network, should provide the return forces.

Transient Andrade creep after a large change in load

Conversely, a large increase in load induces another transient Andrade creep, which is irreversible.

A large decrease in the load (more than 20%) induces transient creep in the opposite direction. The deformation rate decreases in absolute value, reaches zero, becomes positive and increases until the permanent creep rate corresponds to the new load.

Unloading and logarithmic creep

When the load is entirely removed, the creep is in the opposite direction and follows a logarithmic law:

$$\gamma = \frac{\tau_0}{\mu_1} \ln\left(1 + \frac{t}{t_1}\right) \tag{11.14}$$

The parameter t_1 is of the order of one second, and the pseudo-elastic modulus μ_1 is noticeably smaller than the work-hardening coefficient h ($h = 885\,\text{MPa}$; $\mu_1 = 166\,\text{MPa}$). The stress τ_0 is the one that disappeared. The creep in the opposite direction should be caused by the energy stored in the dislocations. No theory could be established leading to this logarithmic law, with γ proportional to the past value τ_0.

During the time where the measures were still collected, the logarithmic creep during the unloading only suppressed a small part of the deformation undergone during loading. This reverse creep should stop with post-kinematic recrystallization, which erases any memory of former loading in the material.

Non-simultaneous torsion and compression

Let us consider a macroscopically isotropic sample of ice, submitted to a fixed torque in the Duval device. Once permanent creep has been reached, and no pronounced fabric has had time to form, if we add some compression, there appears irreversible Andrade creep, in compression *and in torsion*. The same occurs if compression is applied first, and next torsion.

It is unfortunate that these experiments are not continued and that no transient creep law can be provided to scientists modelling the mantle's behaviour. It might indeed be possible that transient creep is more relevant to the uplifts and subsidences linked to isostasy, taking place over thousands of years, and that permanent creep is more relevant to the convection currents moving the plates over hundreds of millions of years (Weertman, 1978).

One single fact is, however, certain. The transient behaviour of rocks at high temperature after a stress variation has nothing to do with the models built by mathematicians by coupling Maxwell and Kelvin bodies, in series or in parallel. This is the reason why these models are not described here, contrary to other books on this matter. These models lead to differential equations of time with relatively high orders, but always linear. Therefore, they predict transient creep rates which are sums of decreasing exponentials of time, never the simple laws in $t^{-2/3}$ of Andrade creep or in $1/t$ of logarithmic creep. And, except in the models where perfect plasticity is added, they only yield reversible transient deformations, while they generally are irreversible.

11.9 ANISOTROPIC MATERIALS

This chapter cannot be ended without talking about anisotropic materials, because most rocks have a texture and their constitutive minerals have a fabric. (The case of ice with a fabric with four maxima is a fortunate exception; the macroscopic behaviour remains isotropic).

Anisotropic elasticity is well known. (It exists in the upper mantle, as shown by the slight differences in seismic velocities, depending on the azimuth). The stress deviator and the small strain deviator are not proportional then. The only thing that can be asserted is that, if the material has no memory, there exists function of stress, the *elastic potential* $X(\tau_{ij})$, such that:

$$\varepsilon_{ij} = \frac{\partial X(\tau_{ij})}{\partial \tau_{ij}} \tag{11.15}$$

This function must be invariant for the different symmetries of the anisotropic medium. If we assume that X is a second-degree homogeneous function of the τ_{ij}, the six ε_{ij} are linear functions of the six τ_{ij} with a symmetrical 6×6 matrix of coefficients (*compliance matrix*).

The τ_{ij} and ε_{ij} are not represented by 2nd order tensors (T and E) in the 3-D space any more, but by a vector (1st order tensor) in a 6-D space. There is another approach: keeping the tensors T and E in the 3-D space and introducing an *anisotropy tensor* A. We must then write $E = MT$, M being a 4th order tensor function of A. This approach allows us to study separately anisotropy and its evolution with time. But it is much more complicated and until now, does not seem to have been extended to anisotropic viscosity.

In an anisotropic medium with a viscous behaviour, there is a *dissipation function* $\Phi(\tau_{ij})$ and the strain rates are functions of its partial derivatives. But, if the medium is uncompressible, they differ from (11.15) because there are only five independent deformation rates and not 6. Let us note them γ_i ($i = 1$ to 5). The small role played by the average pressure is included in the coefficients of the constitutive law. The five rates are therefore functions of the five deviatoric stresses independent of each other, s_i. If the γ_i are built from the $\gamma_{ij} = 2\varepsilon_{ij}$, in the same way as the s_i are built from the s_{ij}, then we must have:

$$\dot{\gamma}_i = \frac{\partial \Phi(s_i)}{\partial s_i} \tag{11.16}$$

The definition of n-power viscosity, as given in Section 11.3, holds for an anisotropic material. The constitutive law is such that when all deviatoric stresses are multiplied by λ, the strain rates are multiplied by λ^n. Thus, $\Phi(s_i)$ is a homogeneous function of degree $(n + 1)$, with the symmetries of the medium. The s_i can only figure through invariants for these symmetries (J_k). Equation (11.16) can then be written:

$$\dot{\gamma}_i = \sum_k \frac{\partial \Phi(J_k)}{\partial J_k} \frac{\partial J_k}{\partial s_i} \tag{11.17}$$

We assume that the J_k are integer algebraic functions of the s_i, and that $\Phi(J_k)$ is an integer algebraic function of the J_k. (The creep rates measured and analyzed with a non-integer n could as well have been analyzed as the sum of two or three creeps with different integers n, considering the inaccuracy of the measurements). For example, one could take τ^2 as an invariant, but not τ. It is then possible to write, for a given kind of anisotropy, the more general expression of $\Phi(J_k)$.

Let us remark that the dissipation function is not an integer algebraic function of strain rates. The hypothesis made suppresses the mathematical symmetry between deviatoric stresses and strain rates which existed in linear elasticity and viscosity. But it does not conflict any logical imperative or physical law. This symmetry is already suppressed when adding viscous and elastic strains for a same stress, when choosing a Maxwell body and not a Kelvin body.

Let us consider in particular an anisotropic incompressible medium, with mechanical properties symmetrical about the plane xOy (*orthotropy*), and about the z-axis (*transversal isotropy*). The only first-degree invariant is s_z. Let us take rather:

$$s_{ax} = \left(\frac{\sqrt{3}}{2}\right) s_z \tag{11.18}$$

There are only two independent invariants of degree 2:

$$\begin{cases} \tau_{\|}^2 = \tau_{zx}^2 + \tau_{zy}^2 \\ \tau_{\perp}^2 = \left(\dfrac{s_x - s_y}{2}\right)^2 + \tau_{xy}^2 \end{cases} \tag{11.19}$$

For the third-degree invariant, rather than the determinant of the deviator, we introduce K_3, which has a physical significance (Lliboutry, 1987, p. 449):

$$\begin{aligned} K_3 &= \left(\frac{s_x - s_y}{2}\right)(\tau_{xz}^2 - \tau_{zy}^2) + 2\tau_{yz}\tau_{zx}\tau_{xy} \\ &= \det[s_{ij}] + \frac{2}{\sqrt{3}} s_{ax} \left[\frac{s_{ax}^2}{3} + \frac{\tau_{\|}^2}{2} - \tau_{\perp}^2\right] \end{aligned} \tag{11.20}$$

As we start with five deviatoric stresses, independent from each other, and as a rotation about Oz affords only 1 degree of freedom, there are $5 - 1 = 4$ invariants independent from each other. And it is impossible to find one that cannot be built with these four. For instance:

$$\tau^2 = s_{ax}^2 + \tau_{\|}^2 + \tau_{\perp}^2 \tag{11.21}$$

If $n = 1$ (anisotropic Newtonian viscosity), Φ is of degree 2 and cannot include K_3. If $n = 3$, Φ is of degree 4 and can include $s_{ax}K_3$. Its general form must be:

$$\begin{aligned} \Phi = {} & B_{ax}s_{ax}^4 + B_{\|}\tau_{\|}^4 + B^4\tau_{\|}^4 + 2A_{\|\perp}\tau_{\|}^2\tau_{\perp}^2 \\ & + 2A_{\perp ax}\tau_{\perp}^2 s_{ax}^2 + 2A_{ax\|}s_{ax\|}^2\tau_{\|}^2 - Cs_{ax}K_3 \end{aligned} \tag{11.22}$$

(I chose the coefficient of s_{ax} and the fourth invariant K_3 so that in the isotropic case, all the B_i and A_{ij} are equal to B and $C = 0$, so that $\Phi = B\tau^4$). The calculation of the seven coefficients in (11.22) was made by homogenization (Lliboutry, 1993), with the following model:

- The microscopic stresses are identical to the macroscopic stress.
- Only sliding in the basal planes intervenes.
- The shear rate parallel to a basal plane is proportional to the cube of the effective shear stress on this plane, and, for a given temperature and effective shear stress, *depends on the fabric.*

This way, we get the ratios between the seven coefficients as a function of the statistical distribution of optical axes around Oz. For the ice cores extracted from the polar ice caps, only this fabric and one of the coefficients are easily measured. But the evolution of the fabric during the flow cannot be predicted, and the main problem of glaciology cannot be solved correctly: how do the flow and dimensions of an ice cap vary with the climate?

The development of any fabric and anisotropy can be calculated with the 'self-consistent' method developed by Molinari *et al.* (1987). The microscopic stresses in the crystal are still supposed uniform. The other surrounding crystals are replaced by a medium microscopically homogeneous, where stress and strain rates are no longer uniform. The problem in mechanics can then be solved rigorously for successive time steps.

This method requires very heavy numerical calculations; it was used by Castelnau *et al.* (1996) for cold polar ice, taking into account the slip on planes other than the basal planes. Even if this slip would become important only for strong strain rates, its modification of the fabrics is noticeable.

12

Rock fracture and earthquake prediction

12.1 NORMAL STRESS AND SHEAR STRESS ON ANY PLANE

Let σ_1, σ_2 and σ_3 be the 3 principal stresses at a point M in a continuous medium, numbered in increasing order. The corresponding principal directions are chosen as x-, y-, z-axes. Consider a plane going through M, with the unit normal vector $\boldsymbol{n}$. The spherical coordinates of $\boldsymbol{n}$ are:

$$\begin{cases} \alpha = \sin\theta\cos\varphi \\ \beta = \sin\theta\sin\varphi \\ \gamma = \cos\theta \end{cases} \tag{12.1}$$

Only a quarter of the hemisphere is shown on Figure 12.1a. The normal stress N and the shear stress T on this plane are given by:

$$\begin{cases} N = \mathbf{t}\cdot\mathbf{n} = \alpha^2\sigma_1 + \beta^2\sigma_2 + \gamma^2\sigma_3 \\ T^2 = \mathbf{t}^2 - N^2 = (\alpha\sigma_1)^2 + (\beta\sigma_2)^2 + (\gamma\sigma_3)^2 - N^2 \end{cases} \tag{12.2}$$

Let us replace α, β, γ by their values and eliminate either φ or θ. A simple but laborious calculation leads to:

$$\begin{cases} \left(N - \dfrac{\sigma_1+\sigma_2}{2}\right)^2 + T^2 = \left(\sigma_3 - \dfrac{\sigma_1+\sigma_2}{2}\right)^2\cos^2\theta + \left(\dfrac{\sigma_2-\sigma_1}{2}\right)^2\sin^2\theta \\ T^2 = [\sigma_3 - N]\left[N - \dfrac{\sigma_1(\sigma_3-\sigma_1)\cos^2\varphi + \sigma_2(\sigma_3-\sigma_2)\sin^2\varphi}{\sigma_3 - \sigma_1\cos^2\varphi - \sigma_2\sin^2\varphi}\right] \end{cases} \tag{12.3}$$

In the plane (N, T), the curves $\theta =$ constant are arcs of circles centred on the N-axis at $N = (\sigma_1 + \sigma_2)/2$. The curves $\varphi =$ constant are arcs of circles centred on the N-axis and passing through $N = \sigma_3$ (Figure 12.1b). The angles θ and φ are found on this *Mohr diagram*, as shown on the figure. Because of the symmetries of the stress

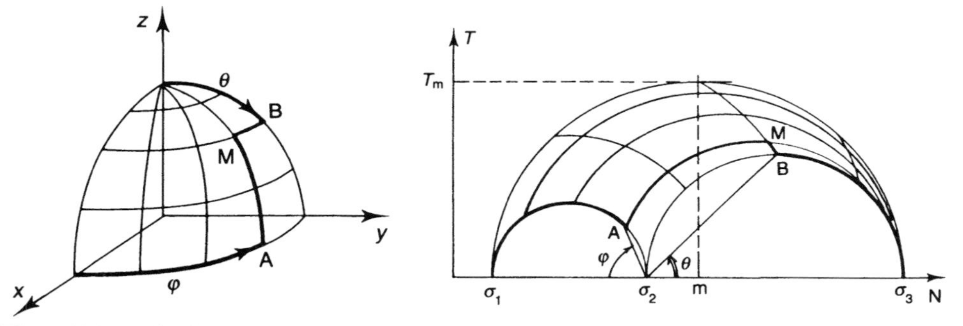

Figure 12.1. Mohr diagram: (a) angles defining a direction OM in space; (b) normal stress N and shear stress T on the planes perpendicular to OM. From Lliboutry (1987); © Kluwer Academic.

tensors, all couples (N, T) fill the curvilinear triangle whose sides are half-circles. The maximal value of T is:

$$T_M = \frac{\sigma_3 - \sigma_1}{2} \tag{12.4}$$

It is reached for $\theta = 45°$, $\varphi = 0°$ or $\varphi = 180°$, i.e. *on the bisecting planes of the principal directions* σ_1 *and* σ_3. These planes intersect along the direction of the intermediate principal stress σ_2. However, if $\sigma_2 = \sigma_1$, φ is undetermined and one has a shear stress T_M on all planes at 45° from σ_3.

12.2 DIFFERENT STATES OF STRESS

The 3 parameters, independent of each other and determining a state of stress, may be:

- either the principal stresses σ_1, σ_2, σ_3
- or $N_m = (\sigma_1 + \sigma_3)/2$, T_M and the *Lode parameter* Λ:

$$\Lambda = \frac{\sigma_2 - N_m}{T_M} = \frac{2\sigma_2 - \sigma_1 - \sigma_3}{\sigma_3 - \sigma_1} \tag{12.5}$$

- or (and this is only interesting in an isotropic medium), the mean normal stress $\sigma_0 = (\sigma_1 + \sigma_2 + \sigma_3)/3$, $r > 0$, and the angle ψ, such that:

$$\begin{cases} \sigma_1 - \sigma_0 = s_1 = r \sin(\psi - 120°) \\ \sigma_2 - \sigma_0 = s_2 = r \sin\psi \\ \sigma_3 - \sigma_0 = s_3 = r \sin(\psi + 120°) \end{cases} \tag{12.6}$$

Let S be a point in the plane, with polar coordinates (r, ψ), and O be the origin. The 3 principal deviatoric stresses s_1, s_2, s_3 are the components of OS along the 3 axes at 120° shown on Figure 12.2. If $\sigma_1 < \sigma_2 < \sigma_3$ is imposed, then $-30° < \psi < +30°$. The 5 other 60°-sectors delimited by the 3 axes correspond to different orders. The same

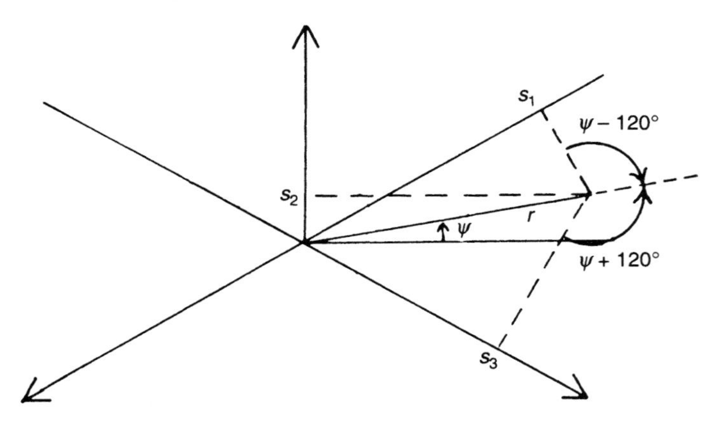

Figure 12.2. Deviatoric stresses and parameters r, ψ.

diagram is obtained by projecting the point with Cartesian coordinates $(\sigma_1, \sigma_2, \sigma_3)$ on the plane of equation $\sigma_1 + \sigma_2 + \sigma_3 = 0$.

The relations between these different parameters, the effective shear stress τ or the octahedral shear stress τ_{oct} (either of which can be chosen as a second stress invariant) and $s_1 s_2 s_3$ (determinant of the deviator = third stress invariant), are as follows:

$$\left\{\begin{aligned}
\tau^2 &= \tfrac{3}{2}\tau_{oct}^2 = \tfrac{1}{2}[(\sigma_1 - \sigma_0)^2 + (\sigma_2 - \sigma_0)^2 + (\sigma_3 - \sigma_0)^2] \\
&= \tfrac{1}{3}[\sigma_1^2 + \sigma_2^2 + \sigma_3^2 - \sigma_2\sigma_3 - \sigma_3\sigma_1 - \sigma_1\sigma_2] \\
&= \tfrac{1}{6}[(\sigma_1 - \sigma_2)^2 + (\sigma_2 - \sigma_3)^2 + (\sigma_3 - \sigma_1)^2] \\
&= T_M^2[1 + \Lambda^2/3] \\
r &= \frac{2}{\sqrt{3}}\tau = \sqrt{2}\tau_{oct} \\
T_M &= \frac{\sqrt{3}}{2} r\cos\psi = \tau\cos\psi \\
\Lambda &= \sqrt{3}\tan\psi \\
\det[s_{ij}] &= s_1 s_2 s_3 = \left(-\frac{r^3}{4}\right)\sin 3\psi
\end{aligned}\right. \tag{12.7}$$

Octahedral shear stress is the shear stress on a plane with identical slopes along the 3 principal directions. It is only a curiosity which does not justify using it instead of the effective shear stress. We always have:

$$\tau_{oct} \leq T_M \leq \tau \tag{12.8}$$

Material tests are nearly always *axial compressions* or extensions, performed on a section of a core or on a cylindrical rod. The sides of the sample are often submitted to confining pressure from a fluid, or, for very high stresses at very high temperature, from a highly ductile salt. In soil mechanics, this is called a 'tri-axial test'. This is not a proper term, as it wrongly leads us to think that the stresses were fixed on 3 axes independently, while in fact 2 of them remain equal. In all these tests, the Mohr curvilinear triangle reduces to a half-circle, and:

$$\begin{cases} \Lambda = \pm 1 \\ \psi = \pm 30^\circ \\ T_M = \pm 0.866\tau = \pm 1.061\tau_{oct} \end{cases} \tag{12.9}$$

A homogeneous stress state, with any Λ, can only be obtained with a complex and costly press. It is reached approximately (in the sense that the stresses are not exactly homogeneous) with the compression–torsion device described in Chapter 11. In cylindrical coordinates, the stress deviator is:

$$[s_{ij}] = \begin{bmatrix} -\dfrac{\sigma_z}{3} & 0 & 0 \\ 0 & -\dfrac{\sigma_z}{3} & 0 \\ 0 & \tau_{z\varphi} & \dfrac{2\sigma_z}{3} \end{bmatrix} \tag{12.10}$$

To compute the principal stresses, it is only necessary to calculate the invariants τ^2 and $\det[s_{ij}]$, and then use the 5th and 7th Equations (12.7):

$$\begin{cases} \tau = \sqrt{\dfrac{\sigma_z^2}{3} + \tau_{z\varphi}^2} \\ \sin 3\psi = -\dfrac{\sqrt{3}}{2}\dfrac{2\sigma_z^3/9 + \tau_{z\varphi}^2\sigma_z}{(\sigma_z^2/3 + \tau_{z\varphi}^2)^{3/2}} \end{cases} \tag{12.11}$$

Λ is a function of $\sigma_z/\tau_{z\varphi}$. It varies monotonously from 0 for $\sigma_z = 0$ to ± 1 for $\tau_{z\varphi} = 0$.

But selecting the Lode parameter is not sufficient, when studying important deformations which may lead to a fabric or texture. Thus, in the cases where $\Lambda = 0$ (*pure shear*), it is not indifferent whether the plane of maximal shear stress, materialized in the medium, remains a maximal shear stress plane (*simple shear*) or not.

12.3 TENSILE STRENGTH IN MECHANICAL TESTS

May the deformation of a ductile substance lead to fracture? We shall only consider here the case of 'ductile fracture', observed during the extension of a cylindrical rod. Such breaking does not correspond to an intrinsic property of the material; it is an artefact due to the mechanical testing procedure.

Indeed, a uniform normal stress s is not applied all along the rod. It is the force σS which is uniform, S being the area of the cross-section. If at some point this area is slightly different $(S + dS)$, the local stress becomes $\sigma + d\sigma$, with:

$$\frac{d\sigma}{\sigma} + \frac{dS}{S} = 0 \tag{12.12}$$

Because of work-hardening, when s increases, the relative (logarithmic) lengthening ε is:

$$\varepsilon = C\sigma^m; \qquad \frac{d\varepsilon}{\varepsilon} = m\frac{d\sigma}{\sigma} \tag{12.13}$$

with m between 3 and 10. An additional extension $d\varepsilon$ will occur only if $d\sigma > 0$ and therefore if $dS < 0$. Assuming a small negative variation dS_0 in a place, it will amplify and become dS. The plastic deformation occurs with volume conservation:

$$\begin{cases} \dfrac{dS - dS_0}{S} = -d\varepsilon = -m\varepsilon\dfrac{d\sigma}{\sigma} = m\varepsilon\dfrac{dS}{S} \\ dS = \dfrac{dS_0}{1 - m\varepsilon} \end{cases} \tag{12.14}$$

When ε reaches the value $1/m$, the stretching is unstable. There appears a neck where S tends toward 0, σ and ε tend toward ∞. Ductile fracture follows. In fact, it occurs before the apparent S gets null, because growing voids appear in the rod at the level of the neck. There is internal stretching between these small voids.

The reasoning remains valid in case of creep, $\dot{\varepsilon}$ replacing ε and n replacing m.

Such ductile fractures are devoid of interest for tectonophysics, except for fractures during the extension of a thin layer less ductile than the surrounding ones. This corresponds to a *boudinage* of this *competent* layer. It is segmented and the outcrop looks like a string of sausages (*boudin*, in French). In the following, we shall only look at *brittle fracture*, without proper plastic deformation (even though it may occur, highly localized at the extremities of microcracks). This event limits the elastic domain, and replaces the yield strength.

12.4 THEORY OF THE MOHR ENVELOPE

This theory is based on the following approximate law:

1. There is no fracture, and one remains inside the elastic domain, as far as the curvilinear Mohr triangle remains inside a particular domain of the plane (N, T), limited by a particular curve, specific to the medium considered, called the *Mohr envelope* (Figure 12.3).
2. When the Mohr triangle touches the envelope, there is fracture along a plane of orientation defined by the contact point. This point is on the largest Mohr circle, corresponding to $\varphi = 0$, and therefore the fracture plane always contains the intermediate principal direction.

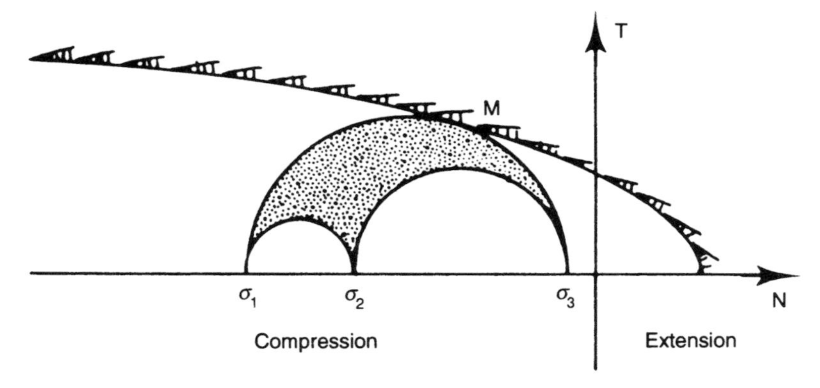

Figure 12.3. Mohr envelope. From Lliboutry (1987); © Kluwer Academic.

3. The elastic domain is open on the side of the very high normal pressures, a hydrostatic pressure (as large as possible) not being able to fracture a compact medium. (This excludes for the moment the case of porous media, considered in soil mechanics.) If the domain extends on the side of $N > 0$ (extensions), the envelope must cross the N-axis perpendicularly, because *cleavage* fractures with negligible shear stress on the fracture plane may occur.

The large Mohr circle corresponds to $\varphi = 0$, $\alpha = \sin\theta$, $\beta = 0$, $\gamma = \cos\theta$ in the system of equations (12.2). One finds:

$$\begin{cases} N = \dfrac{\sigma_1 + \sigma_3}{2} + \dfrac{\sigma_3 - \sigma_1}{2}\cos 2\theta = N_m + T_M \cos 2\theta \\ \\ T = \dfrac{\sigma_3 - \sigma_1}{2}\sin 2\theta = T_M \sin 2\theta \end{cases} \tag{12.15}$$

The largest shear stress, T_M, is obtained for $\sin 2\theta = 1$ and thus $\theta = 45°$ (see Section 12.1). If the largest Mohr circle and its envelope show at their contact point a slope $-\tan\phi$ (and $\phi > 0$, as extensions are counted positively), we need to have: $(N - N_m)/T = \tan\phi$, i.e. $\text{cotan}\, 2\theta = \tan\phi$, which can be expressed as:

$$\theta = 45° - \phi/2 \tag{12.16}$$

The fracture plane differs from the maximal shear stress plane by an angle $\phi/2$. It becomes closer to the plane submitted to the minimal principal pressure.

As far as this theory is valid, the fracture limit and the orientation of the fracture plane do not depend on the intermediate principal stress, contrary to the yield strength of ductile materials (where τ intervenes). This should not come as a surprise: plastic deformation is an average of deformation through sliding between the crystallographic planes with all orientations, whereas fracture results from the development of only one or a few cracks, with very specific orientations.

In fact, it could be experimentally demonstrated that the yield strength depended on the value of τ, and not on the value of T_M. But similar experiments for fracture were not conclusive, because they could not be reproduced well enough. The results are broadly dispersed, whereas the ratio $T_M/\tau = \cos\psi$ can only vary between 0.866 and 1.

12.5 MICROSCOPIC FRACTURE MECHANISMS

Engineers have performed many fracture tests, for a long time, but the study of its mechanisms only started in 1947. It first used an idea presented by Griffith in 1920. Every material would present microcracks. A macroscopic stress imposed to the material will increase some of these. If the decrease in elastic energy, resulting from the widening of a crack, is larger than the cohesion energy (i.e. larger than the surface energy created), the widening will be irreversible and fast. The microscopic cracks spread everywhere will meet, and the whole sample will break.

This theory led to some nice elasticity calculations. The widening of a crack can be split up into three fields of displacement and microscopic stresses, mutually independent of each other (Figure 12.4):

- **Mode I** (*cleavage*, or *opening mode*). The two sides of the crack move away from each other. Displacements and microscopic stresses can be determined by adding those resulting from edge dislocations suitably distributed along the crack, with their Burgers vectors perpendicular to the crack.
- **Mode II** (*sliding mode*). The two sides slide along each other, perpendicularly to the edge of the crack. This is equivalent to edge dislocations suitably distributed, with Burgers vectors aligned with the sliding.
- **Mode III** (*tearing mode*). The two sides slide on each other, parallel to the edge of the crack. This corresponds to screw-dislocations parallel to this edge.

In all 3 cases, if r is the distance to the edge of the crack, the microscopic stresses are in $r^{-1/2}$. The singularity for $r = 0$ is suppressed by assuming that, very locally, there

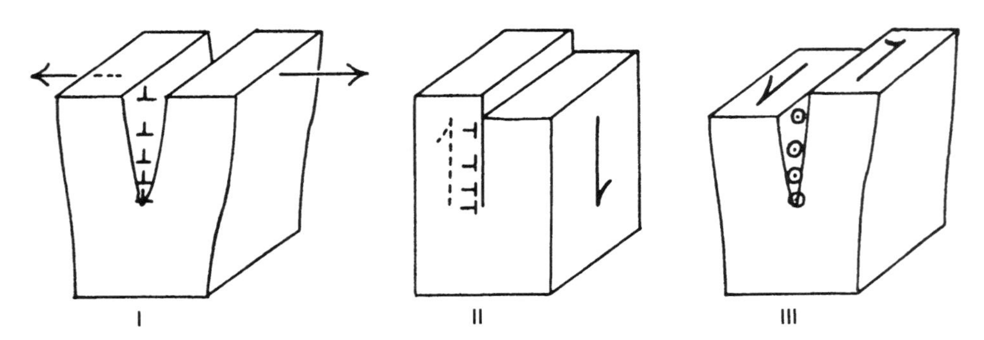

Figure 12.4. Fracture modes I, II and III, and equivalent edge/screw dislocations for the calculation of elastic stresses (their distribution along the fracture planes being suitably chosen).

is plastic deformation and blunting of the tip of the crack. But the resulting error in the calculation of the total elastic energy, which is the only important one, is negligible.

These microscopic stresses and strains around the crack tip are determined by the calculation mentioned earlier, to a multiplicative constant. To compute it, one needs to consider the larger scale and determine the stresses around the crack as a function of stresses 'infinitely far' (noted with the exponent ∞). The analytical calculation is possible for an ellipsoid infinitely flattened, with a radius a, or for an ellipsoidal cylinder of width $2a$. Finally, the elastic energy per unit area disappearing when the crack develops is (for a crack in the xOz plane, with its edge along the x-axis):

$$\begin{cases} \text{In mode I:} & W = \beta \dfrac{\pi a}{E} (\sigma_y^\infty)^2 \\ \text{In mode II:} & W = \beta \dfrac{\pi a}{E} (\tau_{yz}^\infty)^2 \\ \text{In mode III:} & W = \dfrac{\pi a}{2\mu} (\tau_{xy}^\infty)^2 \end{cases} \tag{12.17}$$

The factor β is close to 1. If Γ is the surface energy per unit area, W must be larger than 2Γ for the crack to develop. For a given stress, as W is proportional to the size of the crack, the cracks of size larger than a critical value will grow at an increasing rate.

This theory by Griffith-Irwin has greatly evolved since the 1980s. It has been acknowledged that 3 distinct processes had to be recognized, depending on different parameters:

- The appearance of microcracks under non-hydrostatic stresses.
- The possible instability of these microcracks.
- The localization of these microcracks on a fracture surface. They can be unstable without localization; this leads to a *brittle creep*.

Appearance of microcracks

Microcracks are not forced to pre-exist in the material before the load is applied. They often exist only on the surface of the sample, and therefore the fracture limit found during mechanical tests depends on the size and surface state of the sample. Microcracks are generally produced by the non-hydrostatic stress. They appear at grain boundaries or at inclusions with an elasticity very different from that of the matrix. They are more or less perpendicular to the weak principal pressure. This microcracking is accompanied by a slight volume increase (*dilatancy*). Non-communicating porosity appears, not sensitive to the pressure of fluids inside the rock, at least at the beginning.

It seems that, at least at the start, they are not microcracks, but spherical voids, of diameters smaller than 10 μm. They take some time to appear, maybe because they result from a grouping together of vacancies in the crystalline lattice.

Growth of microcracks

A microcrack does not necessarily grow in its initial plane, because the presence of a crack modifies the stresses locally, in a larger and larger region. This explains the curved, conchoidal fractures of glasses. A crack may also stop growing, and start branching in a different direction. In particular, a sliding fracture (mode II) may give rise to a cleavage fracture (mode I) starting at 70° from the original. For high loads, this *wing-crack* tends to be parallel to the largest macroscopic principal pressure.

In a grained rock, the crack generally stops at the first grain boundary. It may go on propagating along this boundary. To singularities of the microscopic stress field at the edge of microcracks, one must add 'super-singularities', much more intense, at the junction of 3 crystals with anisotropic elasticity (or of different nature and elasticity). They must therefore influence the propagation.

Localization of microcracks

The localization and the macroscopic fracture must depend on the following factors:

- The friction coefficient between the sides of a crack, when it grows in mode II or III;
- The dilatancy, which makes fracture in mode I easier;
- The work-hardening coefficient. It was defined in Section 11.8, when studying transient creep, but its definition does not always imply creep. It is always worth $h = \Delta\tau/\Delta\gamma$. The localization of microcracks on a macroscopic fracture surface, during an axial compression, requires a negative work-hardening coefficient (*strain-softening*). In rocks, this softening may be linked to the formation of a schistose texture.

The current state of research precludes being more definitive. In any case, it is clear that the study of microscopic processes does not justify the theory of the Mohr envelope. It is only an empirical law, relatively simplistic, but still very useful in geodynamics.

12.6 PLASTIC BODIES IN MECHANICS

In solid physics, the term *plasticity* is reserved to ductility resulting from the movement of dislocations. It excludes fracture. For creep, plastic creep is contrasted with brittle creep (and other types of creep). But, in soil mechanics, when the ground is not deforming elastically, there is no distinction between plasticity (macroscopic aspect), and fracture (microscopic aspect; there is indeed fracture of the links between grains). Rock mechanics came last, and uses concepts from both previous specialities. This may increase confusion for the beginner.

We shall provide here definitions relative to plasticity, from the macroscopic and abstract point of view of mechanics. Mechanics builds up models which are logically consistent, and complete to allow the solution of mechanical problems. But these models are only a more or less faithful image of reality. To clearly show this, we shall talk of a plastic *body*, and not of a plastic 'material' or 'medium'.

Perfect plastic body

A body is plastic when stresses imposed and then kept fixed may cause a fixed and irreversible strain. It is a perfect plastic body if there exists a well-defined function of the stresses, noted $f(\tau_{ij})$ and called the *loading function*, such that the body does not deform when $f < 0$, and that the case $f > 0$ is excluded. For $f = 0$, the deformation is plastic, not entirely arbitrary, but with an undetermined amplitude.

The loading function must be a function of only the stress invariants corresponding to the symmetries of the body. Thus, for an isotropic body, we only need to consider the space $(\sigma_1, \sigma_2, \sigma_3)$. To enable us to solve problems where these principal stresses vary continuously from one point to another, they must be numbered independently of their order of size. The planes (r, ψ) are equally inclined on the 3 axes, and must cross the loading surface along a closed curve with the projections of the 3 axes as symmetry axes (Figure 12.5). If $f = T_M - k$ (perfect plastic body obeying the *Tresca criterion*), this section is a regular hexagon and the loading surface is a prism. If $f = \tau^2 - k^2$ (*von Mises criterion*), this is a circle and the loading surface is a cylinder.

In fact, the perfect plastic body must be a Maxwell body; otherwise, stresses would not be defined. The same stresses τ_{ij} cause an elastic strain ε_{ij}^e and a plastic strain ε_{ij}^p. The total strain is the sum of both.

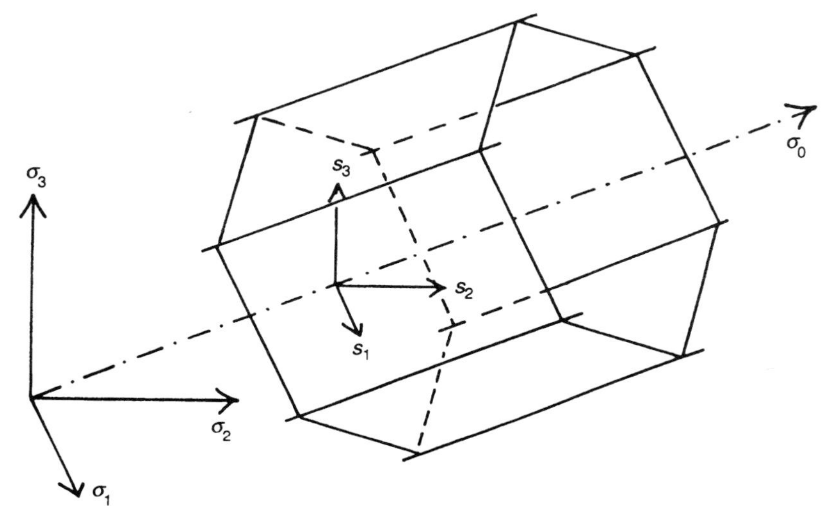

Figure 12.5. Loading surface of a perfect plastic body with a yield strength given by the Tresca criterion.

Standard perfect plastic body

The perfect plastic body is a standard one when the plastic strains obey the *normality condition*:

$$d\varepsilon_{ij}^{p} = d\lambda \frac{\partial f(\tau_{ij})}{\partial \tau_{ij}} \tag{12.18}$$

In particular, with the criterion of von Mises:

$$f(\tau_{ij}) = \tau^2 - k^2 = \tfrac{1}{2}\tau_{ij}\tau_{ij} - k^2 \tag{12.19}$$

The partial derivative relative to τ_{ij} is τ_{ij}, and one finds the equation of Lévy–von Mises again:

$$d\varepsilon_{ij}^{p} = d\lambda\, \tau_{ij} \tag{12.20}$$

A standard perfect plastic body whose plastic strain does not entail any volume variation ($d\varepsilon_{ii}^{p} = 0$) is a *Prandtl body*. If the elastic deformation is negligible, it is a *Saint-Venant body*. The stresses in a Saint-Venant body can be determined and computed only by considering it as a Prandtl body, and then moving the elastic moduli toward ∞ (a method called *limit analysis*).

Non-perfect plastic body

In a non-perfect plastic body, the loading function depends on previous plastic deformations. For instance, in a metal, the loading surface grows with each new plastic strain, encompassing the previous loading surfaces. This is proper *work-hardening*. This term is sometimes used abusively in soil mechanics, although a previous plastic deformation may have broken the links between grains, loosened the soil, and therefore decreased the elastic domain.

The assumption is often made that the plastic history of a material, its 'work-hardening', can be defined with only 1 variable (e.g. e, measuring its porosity). We shall see later that this is only true for one particular class of plastic histories. This assumption has two consequences:

- The loading surfaces are a family, i.e. they form only one simple infinity.
- The future plastic deformation does not depend on the way in which the current τ_{ij} and ε_{ij}^{p} have been reached. A *plastic potential* $g(\tau_{ij})$ can therefore be defined in the whole region where loading surfaces exist:

$$d\varepsilon_{ij}^{p} = d\lambda \frac{\partial g(\tau_{ij})}{\partial \tau_{ij}} \tag{12.21}$$

But the exact expression of g is partly arbitrary, as $d\lambda$ is undetermined. The plastic potential does not have any theoretical basis.

For a standard perfect plastic body, the loading function plays the role of plastic potential. This is not the case here any more. The vector of component $d\varepsilon_{ij}^{p}$ is normal to an equipotential surface $g =$ constant, and no longer to a loading surface $f = 0$.

12.7 SOIL PLASTICITY, CONSOLIDATION AND DILATANCY

Loose soils where particles are 2 to 1/16 mm in diameter are called *sand*. When the particle sizes range between 63 and 4 µm, the soil is called *silt*. A soil with even finer particles is improperly called *clay*, whereas in mineralogy, this name only applies to some hydrated phyllosilicates (the proper term would be *pelite*). The soil mechanics specialist often has an outrageously simplistic view of it, ignoring the water or ions adsorption processes, pH, flocculation, etc. The limited concepts he uses are only sufficient for the study of sands.

In this section, the normal stresses σ will be counted positively if they correspond to pressures, to follow the convention of soil mechanics, and contrarily to the convention of the rest of this book. But we will keep the order $\sigma_1 < \sigma_2 < \sigma_3$.

The mechanical tests are mostly axial compressions. The results can then be plotted on a plane (x, y). In general, the choice is:

$$x = \sigma_0 = \frac{\sigma_3 + 2\sigma_1}{3}; \qquad y = \sigma_3 - \sigma_1 \tag{12.22}$$

It follows that:

$$\sigma_1 = x - \frac{y}{3}; \qquad \sigma_3 = x + \frac{2y}{3} \tag{12.23}$$

The principal stresses can therefore be found on the diagrams by referring to these oblique axes.

In the case of a soil soaked with water at the pressure p_w, when water is allowed to flow out a consolidating soil or to enter a dilatant soil, the behaviour has been proved to depend only on the difference $\sigma - p_w$, called the *effective pressure* (this notion does not apply to clays, that are too impermeable). To simplify the notation, s_1, s_2 and s_3 will correspond in the following to the effective principal pressures.

The elastic domain is limited by a line (denoted (1) in Figure 12.6), of slope M, going through the origin. In the case where the grains are welded together, it does not:

$$y = C_0 + Mx \tag{12.24}$$

This *Mohr–Coulomb law* can be written as:

$$T = C + N \tan\phi \tag{12.25}$$

Comparing with equations (12.23) and (12.16), and taking into account the change of sign of N, T, and the σ_i, equation (12.15) reads:

$$T = \frac{y}{2}\cos\phi; \qquad N = \left(x + \frac{y}{6}\right) - \frac{y}{2}\sin\phi \tag{12.26}$$

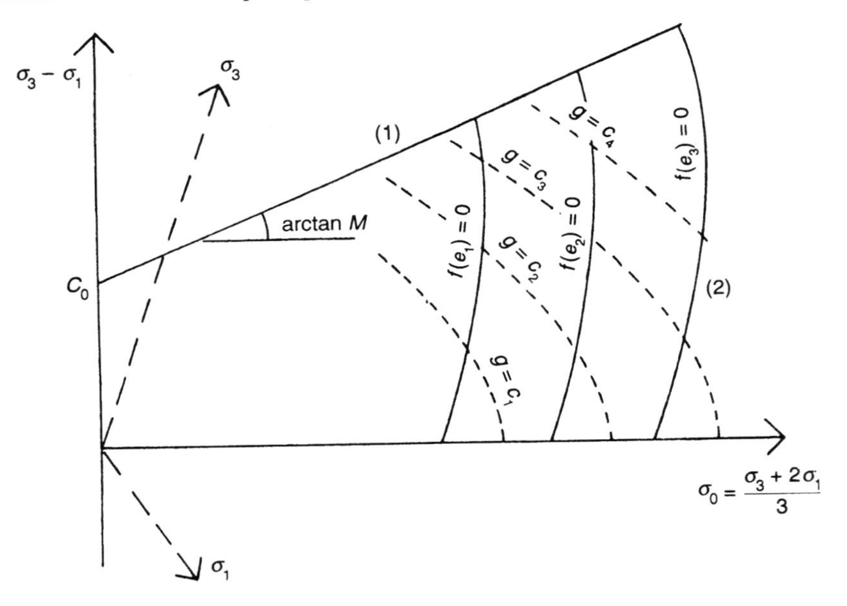

Figure 12.6. Plasticity thresholds for decreasing void ratios e_1, e_2, e_3, and plastic equipotentials in the case of a clay. Simplified from Wong and Mitchell (1975). The shape of the curves differs greatly with the authors, and this cannot be always explained by the diversity of the soils studied. A review can be found in (Monnet, 1983).

Equations (12.25) and (12.26) yield a linear relationship between x and y, which must be identical to equation (12.24). It follows that:

$$\begin{cases} \sin\phi = \dfrac{3M}{6+M} \\ C = C_0\left(\dfrac{1}{2\cos\phi} - \dfrac{\tan\phi}{6}\right) \end{cases} \tag{12.27}$$

The yield strength (equation 12.24) may therefore be likened to a Mohr envelope (equation 12.25), which is a fracture threshold. In fact, micro-fractures (through cleavage) only account for C. The remainder consists in micro-friction between the sand grains. $\tan\phi$ is therefore called the *inner friction coefficient*.

When $C = 0$, the angle ϕ equals the maximum inclination of the surface α_M (*angle of repose*). Indeed, α denoting the slope of the surface, if the x-axis is chosen along the steepest slope and the z-axis points downward (Figure 12.7), at a depth small enough, the stresses only depend on z. The equilibrium equations yield:

$$\tau_{xz} = \rho g z \sin\alpha; \qquad \sigma_z = \rho g z \cos\alpha \tag{12.28}$$

These are the values of T and N on planes parallel to the surface. The ratio $T/N = \tan\alpha$ cannot increase to more than $\tan\alpha_M$.

But the elastic domain of a soil is also limited on the side of large pressures, because of irreversible volume variations. This plasticity threshold is given by the

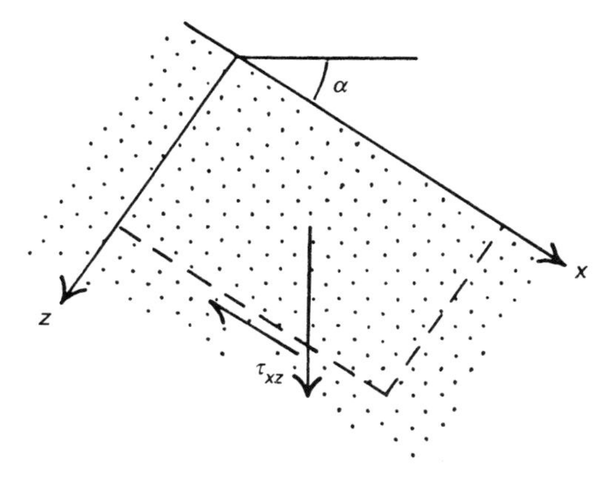

Figure 12.7. Calculation of stresses due to the surface slope, when they are independent of x.

curve denoted (2) in Figure 12.6. Its position depends mainly on the porosity, and *has nothing to do with a Mohr envelope.*

These curves (2) move away from the origin as porosity decreases. They seem to meet the x-axis (hydrostatic stresses) parallel to the σ_3-axis and not to the y-axis (which is the s_3-axis). In this part of the diagram, the plastic equipotentials lay as indicated on the figure (dashed lines). Although plasticity is caused by volume changes, the first plastic deformation is more shear than volume diminution. Some authors even extend plastic equipotentials obliquely on to the x-axis, assuming that the intersection is a singular point where the direction of $d\varepsilon_{ij}^p$ is not determined. As with most authors, they are drawn here meeting the x-axis perpendicularly, a simple hydraulic compression not producing any shear.

Porosity is not a single function of current stresses. The first compression of a loose soil is irreversible. Calling p the average pressure ($p = \sigma_0$ in this section), the void ratio e decreases according to the law:

$$e = e_0 - 0.5 \log_{10} p \tag{12.29}$$

The soil is then *consolidated.* If p then decreases, and increases again without reaching the previous consolidation pressure, the volume variations become reversible (Figure 12.8). They follow the law:

$$e = e_1 - 0.3 \log_{10} p \tag{12.30}$$

If the stress state is not hydrostatic, the law (12.29) is no longer followed. Beyond some threshold, further loading increases the volume, a behaviour termed *dilatancy.* With very high deviatoric stresses, dilatancy is no longer reversible, the soil is loosened and its consolidation suppressed.

Let us come back to the limit (1) of the elastic domain, due to micro-fractures with inner friction. If it is not sensitive to p for high values of p, and therefore for small porosities, this is no longer the case for large porosities. The limit curve goes down as porosity increases. This is the case when micro-fractures cause dilatancy. There is

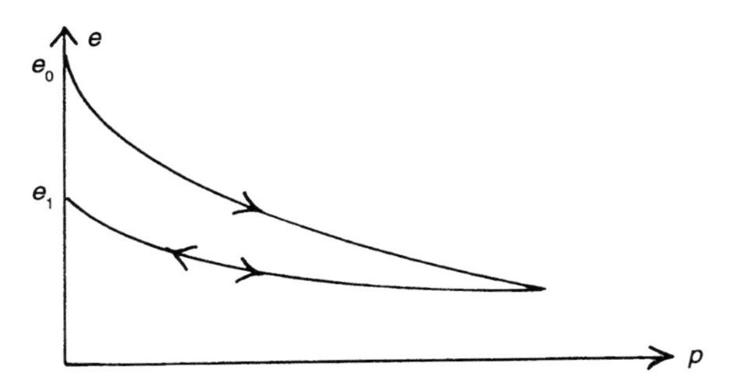

Figure 12.8. Consolidation, followed by a reversible elastic compression, under the effect of an isotropic pressure.

fracture *localization*, and one or several macroscopic decollement surfaces appear. The soil divides into purely elastic blocks sliding one on another. From a mathematical point of view, the continuous solution of the plasticity problem, without localized fracture, is *unstable* and cannot be calculated numerically.

12.8 ROCK FRACTURE, JOINTS AND FAULTS

Not considering the foci of intermediate or deep earthquakes observed in subducted plates, all seismic foci appear in the first 20 km on continents (sometimes less, e.g. above 15 km in California), and in the first kilometres on the ocean floor. On continents, they come most often from the *reactivation of an old fault*. It is a problem of abrupt sliding with friction, and not of fracture of a sound rock. New faults, which may appear in surface during an earthquake result from the strike-slip movement of a pre-existing deep, fault and are only an epiphenomenon.

This statement only sends back the problem into the past. Laboratory experiments show that the differential stress $(\sigma_3 - \sigma_1)$ inducing the fracture is of 200 ± 40 MPa for granite or gneiss, and 250 ± 100 MPa for basalt. Now, regional tectonic stresses related to the movement of plates always seem to be less than 100 MPa (see Section 14.9). The faults are not the only ones that need explanation. They are not very numerous compared to all the fractures without throw (in mode I), the *joints* omnipresent in rocky terrains. It is likely that any fault was originally a joint, where sliding occurred because the orientation was favourable. Later, the movement of the fault often causes new joints, which are branching away from it. Despite their ubiquity, joints have not been studied much and only get a few lines in geology manuals.

On large regions, joints have 2, often 3 orientations more or less perpendicular to each other. A rocky outcrop is therefore often an assembly of rocks with roughly parallelepiped shapes, with sizes of a few metres to a few decimetres. One system or the other may be more pronounced. All the possibilities are present, between the large extension fracture, empty ('dry') or filled with *gouge* (clay or other deposits), and the simple, oriented microcracking, the beginning of a joint, invisible to the naked eye but looked for in quarries as it makes the rock-cutting easier.

In sedimentary rocks, one of the joints systems is parallel to the bedding (*bedding joints*). When metamorphism produced a cleavage texture, another system corresponds to this schistosity. But joints also exist in thick, homogeneous calcareous layers or in non-metamorphosed crystalline rocks.

Four mechanisms are known to lead to high local deviatoric stresses, and to a mode I fracture:

1. *Contraction during the cooling* of a lava flow. Columnar basalts are perpendicular to the cooling surface (even when it is highly inclined on the horizontal). The same process seems to work when a pluton is cooling.
2. *The folding of a competent layer*. A strong extension appears at the convex side of the fold.
3. *A very pronounced relief*. A large vertical lithostatic pressure exists within the foot of a cliff, whereas the horizontal pressure perpendicular to the wall equals 0. This results in joints parallel to the wall.
4. *Frost-cracking*, i.e. the action of frost on a slightly permeable rock, whose micro-pores are filled with water. Contrary to what is written everywhere, many freezing-thawing cycles are not necessary; a deep and persistent freezing is sufficient. Water migrates toward the freezing front and feeds any ice vein forming in a microcrack. It is the same process as the formation of ice lenses in soils easily cracked by frost, albeit much more slow (Hallet *et al.*, 1991). Note that the water in microcracks is adsorbed water tied to the rock by molecular forces, but which can migrate along the surface. It freezes only around $-1.5°C$, absorbing less cold than free water.

To these processes, one may add the very likely hypothesis that on geological time-scales, orogenic stresses alone, of 10 MPa at most, are sufficient to induce a very slow growth of microcracks in the rock, and its fracture (Atkinson, 1984).

However, *hydraulic fracturing*, as artificially induced in the oil industry, cannot be but an enlargement process of cracks already there. It is true that, in a volume bounded by impermeable barriers such as shale layers or impermeable veins cross-cutting the layers, the interstitial pressure p_w of the captive aquifer ranges between the extreme principal pressures of the rock σ_1 and σ_3, and therefore $(p_w - \sigma_1)$ may be very high. But this effective pressure remains less than $(\sigma_3 - \sigma_1)$, a differential stress already too small to crack sound rock.

12.9 FOCAL MECHANISM AND REAL FRACTURE MECHANISM

Only at the end of the 19th century, earthquakes were no longer attributed to mysterious underground explosions, but rather linked to faults and orogenesis. In 1910, Reid explained that they were related to the elastic rebound after the break of the rock.

As explained in Section 2.11, the observation of P waves received in many stations shows that that there are two opposite quarters of the Earth where the first movement, the impetus, is directed toward the source, and two quarters where it points away from the source. They are delimited by two perpendicular planes, the nodal planes. The first idea, intuitive but wrong, was that the fault which moved was in one of these two nodal planes (Figure 12.9a, b). But when it became possible, with the WWNSS network, to determine exactly the impetus of S waves, it was discovered that their orientations did not fit with this idea. Two faults would work at the same time, one on each nodal plane, which is difficult to imagine (Figure 12.9c).

In fact, the elastic waves recorded at large distances, with wavelengths of a few kilometres, do not 'see' in detail the displacements over the fault and its exact orientation. They depend on the stress drop (or, if one prefers, on the corresponding elastic rebound), at a certain distance from the source, on a sphere where the medium remains continuous. Before and after the earthquake, these stresses are balanced and do not create any moment. The maximal shear stress T_M is exerted on two perpendicular planes, bisecting the two extreme principal directions (see Section 12.1). These are the nodal planes. Seismologists wrongly continue calling one the 'fault plane', the other the 'virtual fault plane', and their orientation the 'focal mechanism', when in fact it is the 'stress drop around the focus' and its symmetries that are obtained.

When the fault is observable in the field, it makes an angle of ca. 15° with one of the nodal planes, resulting in a lower pressure perpendicular to the fault plane (Stauder, 1962). This is in accordance with the theory of the Mohr envelope, with $\phi/2 = 15°$, if $\tan\phi$ is the slope of this envelope.

The freeing of sticking spots and the sliding between the two sides of the pre-existing fault start in one point and propagate along the fault at a speed nearly half the speed of the Rayleigh waves, i.e. a few km/s. The fracture stops after a few kilometres because of some obstacle on its way. It would be as realistic (but as imprecise) to say that it stops where the sides of the fault are better welded together.

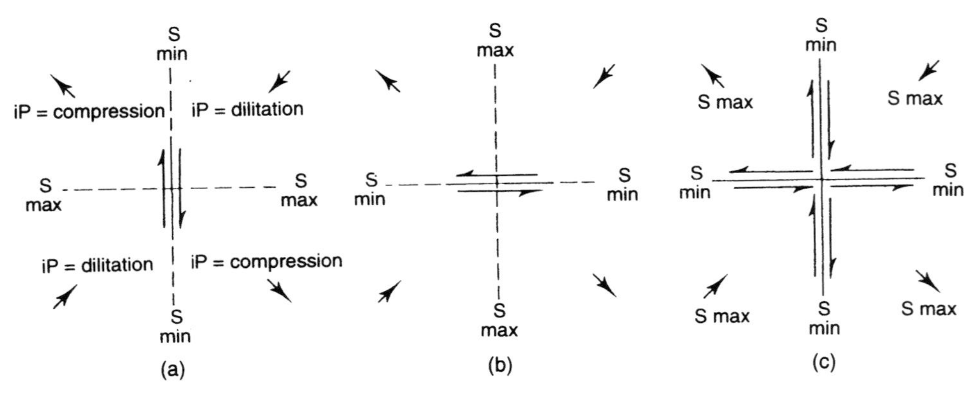

Figure 12.9. Simplistic model of elastic rebound for a focal mechanism: (a) dextral fault; (b) sinistral fault; (c) double fault.

The duration of an earthquake on the fault is thus of the order of one second. A readjustment of the surrounding ground follows it. At depth, this can be a ductile strain, expressed at the surface by an additional strike-slip, the *post-seismic strain*. In the brittle crust, the stress modifications produce after-shocks, generally with the same focal mechanism. However, when a fracture stops propagating, it may start again further away, on the same fault or a neighbouring parallel fault. This produces a strong earthquake, of magnitude 7 or higher.

The idea of a single fault between two continental blocks is indeed simplistic. Mostly, there are several neighbouring parallel faults. For example, the famous San Andreas fault, in California, is doubled around San Francisco by three other faults (San Gregorio, Hayward and Calaveras), also active, as shown by the displacement field (Figure 12.10).

The new networks of digital seismometers made possible the study of seismic radiation at high frequency (1–10 Hz). As the resolution is increased, the movements appear more complex and random in detail. Rupture propagation is far from being as regular as in the models. This comes from the fact that the large faults are not discrete surfaces, but a series of intertwined small sliding surfaces of hectometric sizes, in the middle of a highly fractured zone often altered by hydrothermalism (e.g. Chester *et al.*, 1983).

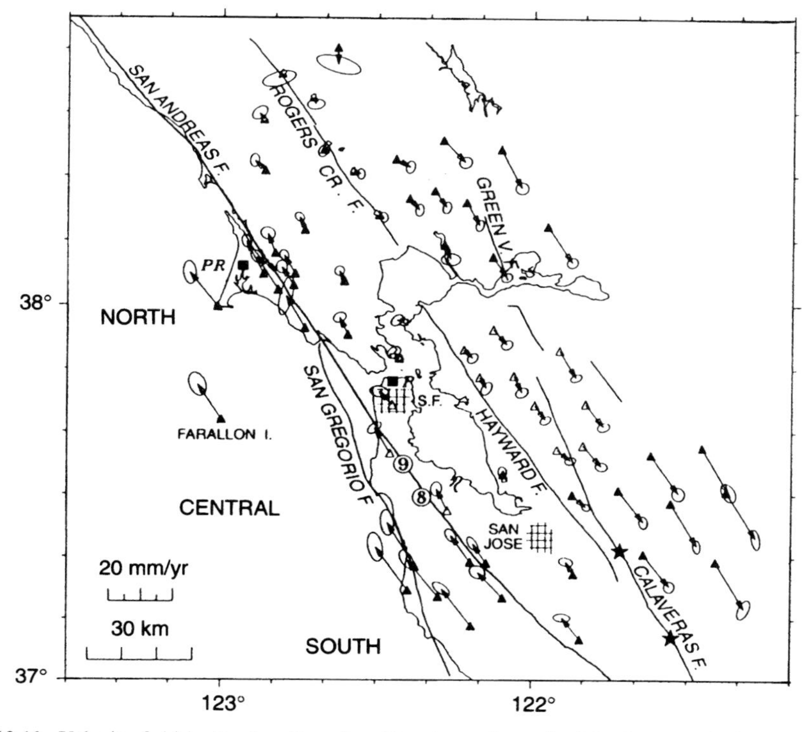

Figure 12.10. Velocity field in the San Francisco Bay Area, determined by laser geodetic studies. From Lisowski *et al.*, 1990.

12.10 STICK-SLIP MOTION AND EARTHQUAKE RECURRENCE

Laboratory experiments on the sliding between two solids show that it follows, as a first approximation, the Coulomb law of sliding (also called the Admonton law by US scientists). This should not be confused with the Coulomb fracture criterion, although their mathematical expressions are the same:

$$T = fN \tag{12.31}$$

T is the shear stress, and N is the effective pressure, exerted by the two surfaces on each other. This is a macroscopic law; at the microscopic scale, the actual contacts only concern small fractions of the surface, with a total area proportional to N. The microscopic processes will be different, depending on whether the bumps are overcome through elastic or plastic deformation, or through fracture. It is not necessary to study them for the moment, especially as other processes also act in seismic faults.

The coefficient f has a value at standstill (*static friction coefficient*) larger than during sliding (*kinetic coefficient*). There are 2 possibilities for a given value of the effective pressure N; either the shear stress equals T_S, and there is no relative movement, or it equals $T_C(< T_S)$, and there is sliding with an undetermined speed (different from 0).

The two often alternate when sliding of one solid against the other (or simply of a wet finger on a windowpane) is attempted. The resulting jerky movement is called *stick-slip*. Jerks are larger as N increases (and therefore as T_C and T_S increase). The important factor is in fact the corresponding elastic deformation of the system (solids + driving mechanism). During a jerk, the elastic energy accumulated between the jerks is partly dissipated through friction on the fault, partly through the emission of elastic waves (cf. the theory of Section 2.6, where T_C and T_S are respectively called T_C and T_0).

In the 1970s, earthquake recurrence on the same fault was explained by this model, the relative velocity of lithospheric plates very far from the fault being imposed. With stress drops of the order of the MPa on a 50-km width around the fault, and a shear modulus of the order of 50 GPa the predicted throw for each earthquake is of the order of 1 m. This is very reasonable, if sliding only occurs during the earthquakes. But this was observed to not always be the case. As we shall see later, a slow sliding may occur on the fault, outside earthquakes. In this case, there may be episodes of highly accelerated sliding, called 'slow earthquakes'.

For this reason, and to apply the standard mathematical theory of stability, the model of a friction coefficient with only 2 distinct values was abandoned. It was assumed that the coefficient was a continuous function of sliding velocity. It must then decrease rapidly enough when velocity increases, for an instability to appear. This instability appears if the velocity far from the fault is perturbed, or if the far shear stress is perturbed (the latter seems unrealistic in the case of earthquakes). Such a decrease of friction with sliding velocity was indeed found by Dieterich (1979) during laboratory experiments.

To represent analytically Dieterich's results, Ruina (1983) proposed a complicated mathematical model (i.e. not based on some specific physical process). Friction would be the sum of an explicit function of the sliding velocity (with 3 parameters) and of several hidden variables, continuous and decreasing with time, following laws with 3 parameters for each hidden variable. But, although the multiplication of parameters makes adjustments easier, experimental results or realistic stress drops cannot be reproduced at all (Weeks, 1993). Furthermore, even if an exponential acceleration of the velocity simulating the triggering of an earthquake can be obtained, the earthquake cannot be stopped without introducing yet another complication. Past some critical sliding velocity, friction should become an increasing function of velocity.

But Ruina's mathematical model was not abandoned, as a clear conception of scientific research ought to have done! Some scientists tried to complete it, by taking into account the filling of faults (*gouge*). Friction experiments with an additional gouge were many in the 1980s. Their results are complex and contradictory.

This approach therefore seems fruitless. As in many domains of geodynamics, sophisticated mathematical models can be completely on the wrong path, when they are not based on correct field observations, on the right time and space scales.

In the 1990s, the importance of fluids was realized (e.g. water for the upper crust). They are useful not only to decrease the effective pressure between the sides of a fault, but also to weld them together. This welding happens mainly through pressure solution, in a time smaller than the recurrence period of large earthquakes on a particular fault (Blanpied *et al.*, 1991). Studies were then focused on hydrothermal processes in seismic zones (Hickmann *et al.*, 1995, and the next articles). Many models were suggested to explain earthquake recurrence, even foreshocks and aftershocks, but these models are too conjectural enough to be detailed here.

12.11 ASEISMIC SLIP AND SILENT EARTHQUAKES

Continuous sliding may coexist on a fault with the very fast sliding (a few seconds) which produces earthquakes. More precisely, it seems that diffuse shear occurs on a strip of a few tens to hundreds of metres, often recognizable in the field (Mattauer, 1973, calls it a 'plastic fault'). Seismic displacement on a fault must then be considered as a *localization* of this shear.

In California, the North American and Pacific plates show a strike-slip movement of 47 ± 2 mm/year. It follows the very long San Andreas Fault (Figure 6.8) and several other secondary, parallel faults (Figure 12.10). Geodetic measurements show that aseismic slip is very important in some portions of the San Andreas Fault, and is sufficient to ensure the whole strike-slip (Figure 12.11). These aseismic zones cut the fault into segments with independent seismic histories.

The proportion of aseismic slip seems to differ in the subduction zones of Chile and Kuriles-Japan.

In Chile, series of very large earthquakes in the same month (which may be considered as foreshocks and aftershocks of the most important one), were

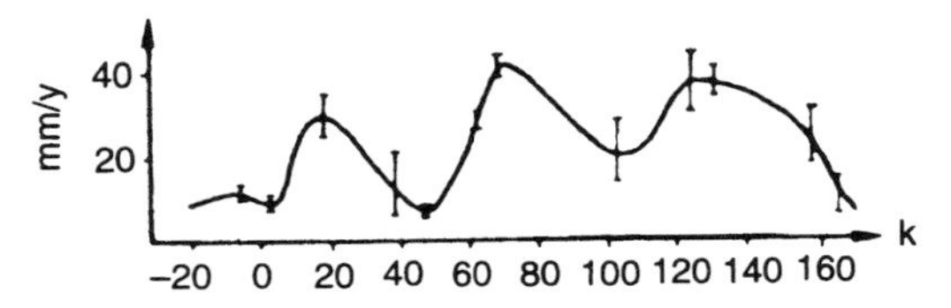

Figure 12.11. Aseismic strike-slip movement, measured along the San Andreas Fault between 1967 and 1971. From Wesson *et al.*, 1973.

experienced in 1960, 1837, 1737 and 1575, i.e. every 100 to 150 years. During the last series, geodetic measurements near Lebu, at the edge of the area affected, show a horizontal throw of 10–12 m. From the magnitude, it must have been twice as large in the central portion. The subducted Nazca plate moves at 11 m per century, relative to the South American plate. It is therefore reasonable to think that all the relative displacement between the two plates is seismic.

Conversely, the Kuril Islands and Japan undergo series of large earthquakes over 2 to 32 years. Each earthquake affects a specific portion of the island arc (Figure 12.12). Series are separated by intervals of 90–150 years. The relative movement of the two plates (Pacific and North America) is 9 m per century, whereas the horizontal displacement during these large earthquakes is 2–3 m. Three-quarters of subduction therefore occurs through aseismic slip (some authors even argue for a higher proportion). In this region, sudden movements are also recorded; they are too slow to generate seismic waves, but fast enough to generate tsunamis. They are called *silent earthquakes* (or *Sanriku-type earthquakes*).

According to Uyeda and Kanamori (1979), these contrasting behaviours would come from the differences in pressure between the subducted plate and the overlaying plate, depending on the presence or absence of a back-arc basin. They distinguish the following cases (Figure 12.13):

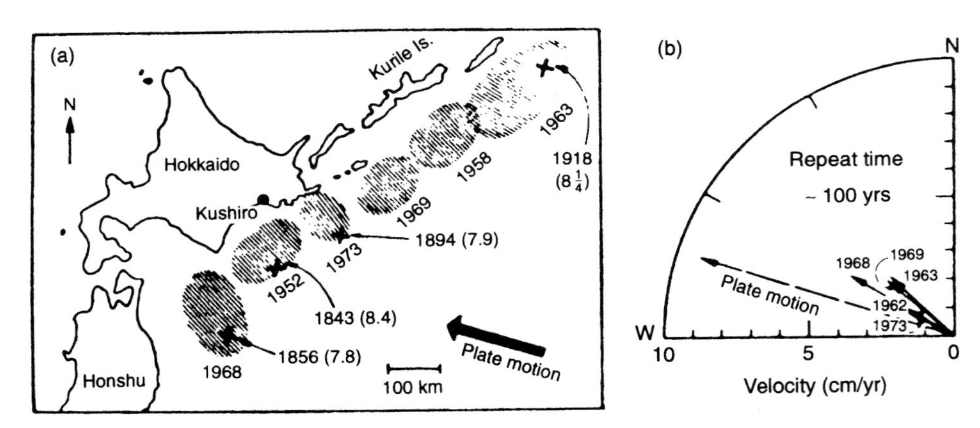

Figure 12.12. Seismicity related to the Kurile Trench: (a) areas affected by the different earthquakes; (b) average velocity related to the earthquakes (assuming their recurrence period is 100 years, compared to the subduction velocity deduced from magnetic anomaly stripes). From Kanamori, in Talwani and Pitman (1977).

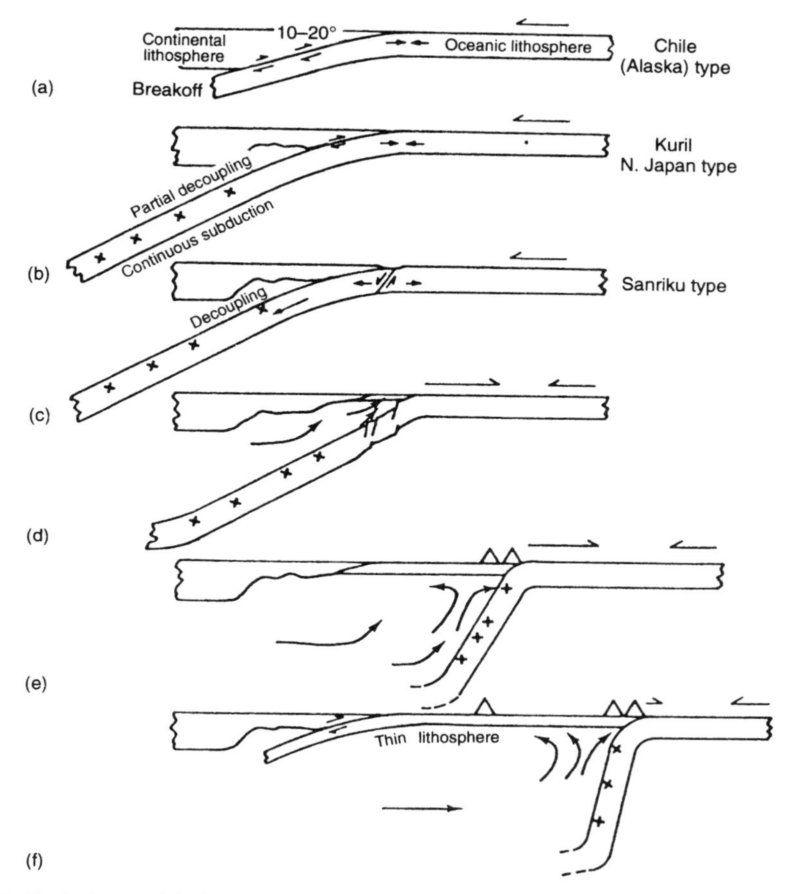

Figure 12.13. Evolution model of a subduction zone. From Uyeda and Kanamori (1979).

- Case (a): In Chile or Alaska, the subducted oceanic plate is in friction against a thick continental lithosphere. This causes large earthquakes, and, according to the authors, a periodic rupture of the subducted plate at shallow depths.
- Case (b): A thinned continental lithosphere close to the subduction zone decreases friction and partially decouples the two plates. The earthquakes are less important, the subducting plate remains continuous down to great depths. This would be the case of the Kuril Islands and Japan.
- Case (c): Silent earthquakes are observed as the decoupling between the plates increases. The subducted plate is in extension, at least in its upper part.
- Case (d, e, f): When the weight of the plate is even higher, the diving point (i.e. the oceanic trench) moves away from the continent. Oceanic crust appears in a back-arc basin (at this time, Uyeda and Kanamori did not know of the appearance of a mid-ocean ridge in this basin, and of its relatively short life; see Section 7.4).

12.12 INTERMEDIATE AND DEEP EARTHQUAKES

The sources of intermediate and deep earthquakes in subducted plates were enigmatic for a long time. The Coulomb friction law seems to preclude any movement of a fault, the normal stress N being very high. But there is no sudden volume diminution, an implosion resulting from the transformation of some metastable mineral in a denser allotropic variety. Focal mechanisms always show the movement of a fault.

In fact, the Coulomb law is not valid any more. First because it was established with moderate values of N, for which the effective contact is limited to small portions of the surfaces. Elasticity means that the area of effective contact is proportional to N. But this law is limited when contact is everywhere effective. Second, because friction is highly reduced in the presence of interstitial fluids, or when the faults are filled with a very ductile material similar to a fluid.

During allotropic transformation, a polycrystalline mineral becomes very ductile for a short time (Vaughan and Coe, 1981). Would that be the explanation? Olivine in peridotites and in mantle lithosphere undergoes many allotropic transformations around 410 and 520 km deep (see Section 17.2). It could subsist in a metastable state, but why would that be only in fault gouge? Some scientists suggested that serpentine in the oceanic crust could get dehydrated and become glassy, but the moving fault is, for sure, located in mantle lithosphere.

On June 9, 1994, an earthquake of magnitude $M_W \approx 8.3$ occurred, with a source 650 km below Bolivia. (Despite its high magnitude, this earthquake did not cause any damage, but was felt as far as Montreal.) It was recorded by 4 seismic networks and very well analysed (Wallace, 1995, and the next articles). The subducted Nazca plate was broken, along a sub-horizontal fault of 30 km × 50 km. The rupture took 45 seconds to travel along this fault. A foreshock on January 10 had a source 600 km deep, and the strongest aftershock, on August 8, was 590 km deep. There is therefore a strong global shear at depths of 630–590 km below Bolivia. On this occasion, it was realized that the '660-km' discontinuity below the Bolivian Altiplano is in fact at 700 km, definitely deeper than the deep earthquake zone.

The conclusion is that deep earthquakes should happen within layers at the bottom of the upper mantle with large vertical gradients of the horizontal velocity. These layers should be the bottom boundary layers of convection cells in the upper mantle, as advocated in Section 19.11.

12.13 SEISMIC HAZARD ESTIMATION FROM HISTORICAL EARTHQUAKES

The assessment of seismic risk is important for communities, which must use it in building regulations and make emergency plans for natural hazards. This is less important for insurance companies; when earthquakes are not excluded from policies, they still take no risk. In California, they ask for premium fees 100 times higher than the compensations they expect to make. But the estimation of seismic hazard is difficult.

Statistical earthquake studies did not wait for the studies of the mechanisms mentioned above. They led to the *Gutenberg–Richter law*, suitably verified for a particular regional seismic network. The average number N of earthquakes with magnitudes $\geq M$, recorded per year by the network, is a decreasing exponential function of M:

$$\log N = a - bM \tag{12.32}$$

For example, in the Dead Sea area, between 1919 and 1963, $a = 3.25$ and $b = 0.8$.

Of course, it is not possible to use highly negative values of M and apply this rule to micro-seismic activity. For large values of M, the Gutenberg–Richter law may not be obeyed, as it is the extrapolation of an empirical law. Looking at the whole Earth and at earthquakes of magnitudes larger than 6 (which were all taken into account), Madariaga and Perrier (1991) find that the law is still quite valid, by changing b beyond $M = 7.5$, and using the magnitude M_W inferred from the seismic moment (Section 2.7), as shown in Figure 12.14.

The need to choose different values of a and b depending on the region was proved by the old estimations of *seismicity* of the different countries. To compare them, N was computed for the same surface of 100,000 km^2. Using all the earthquakes felt (i.e. with very low M), Japan appears as the most seismically active country. But if only the destructive earthquakes are considered (i.e. with high M), Greece holds this sad record, with 57 destructive earthquakes per century and 100,000 km^2, followed by Italy (24), and Japan is only in 10th position. This destructive seismicity is less important for France ($10/5.5 = 1.8$) than for the whole Earth, oceans included ($13{,}900/5{,}102 = 2.7$).

The risk can be better assessed by considering large structural regions. The Alps and the Pyrenees are very little active, the Parisian Basin or Normandy not at all. But there were destructive earthquakes in the 12th and 13th centuries in Rouen and in 1775 in Caen (both in Normandy). France is not seismic enough to give any credit to maps assessing seismic hazard for one district or the other, on a detailed scale.

The situation is completely different at plate boundaries, where the numerous earthquakes come from the continuous activity of subduction or transform faults. On a particular segment, the large earthquakes seem to occur more or less periodically, which should enable medium-term predictions (i.e. for the next few years). Earthquakes will be predicted for a zone with no large earthquake for a long time, a *seismic gap* along the plate boundary. But the existence of a seismic gap does not imply necessarily a risk, as this could be a zone of aseismic slip. There is risk only if earthquakes have already occurred on this segment.

Unfortunately for prediction, the periodicity is far from strict. For example, along the San Andreas Fault, there were earthquakes of magnitude 6 in 1857, 1881, 1901, 1922, 1934 and 1966 near Parkfield (half-way between Los Angeles and San Francisco). In 1985, two seismologists estimated the recurrence interval to be 22 ± 3 years, and announced there were 95% chances for a new earthquake before

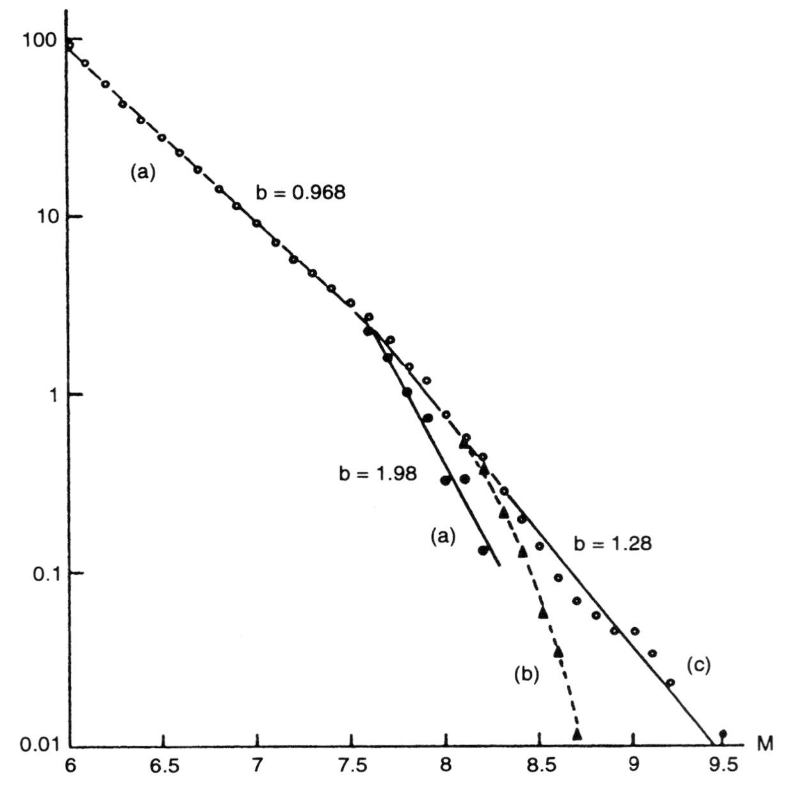

Figure 12.14. Annual number of earthquakes (N) as a function of the magnitude (M), and values of the parameter b of the Gutenberg–Richter law. (a) Earthquakes with magnitudes $M_S \geq 6$ during 1975–1989; (b) Earthquakes with magnitudes $M_S \geq 8$ during 1904–1989; (c) Earthquakes with $M_S \geq 7.6$, replaced with M_W. From Madariaga and Perrier, 1991.

1993. (A more rigorous calculation even provides me with 95.4% chances that the 7th earthquake occurs before 1991.5). A very important monitoring and study programme was immediately started, but the earthquake had still not occurred in 1996 (Roeloffs and Langbein, 1994).

Furthermore, the notion of 60-km long fault segments with independent seismic histories is not valid for the larger earthquakes, which may affect several segments at the same time.

An extreme case is the Chile earthquake of May 22, 1960, with a magnitude of 9.5 and a source offshore from the coastal town of Valdivia. We will limit to the foreshocks and aftershocks of magnitudes larger than 6.5, which are already large earthquakes. On May 21 and May 22, there were foreshocks of magnitudes 7.6, 6.5, 7.0, 7.4, 7.5. The source of the first one was 250 km further north, at Lebu, on the coast (100 km south of Concepcion), and the ensuing sources moved SW. On May 25, an aftershock of magnitude 6.9 originated from 200 km offshore Puerto Aisen, i.e. nearly 900 km south of Lebu. On June 6, an earthquake of magnitude 6.9

occurred at Puerto Aisen itself. On June 20, there was another earthquake of magnitude 6.9 with a source close to Lebu, and another of magnitude 6.9 at half-way. The Nazca plate subduction induced repeated earthquakes over 1 month and along 900 km.

A *similarity law* exists when only one fault segment is concerned. The seismic moment is roughly proportional to the cube of the length of the fault that moved:

$$M_0 = CL^3; \qquad C = 2 \text{ to } 20 \times 10^6 \text{ Pa} \tag{12.33}$$

Applied to the previous earthquake, this law would lead to $M_W = 10.1$ to 10.8, and not 9.5. This is easy to understand: the domain concerned could not extend deeper than the lithosphere.

12.14 PALAEOSEISMICITY

An old earthquake can only be known about if it occurred in a populated region, with learned people who recorded the facts in archives or chronicles. And even then, it is generally impossible to find in the field the fault that worked. Our knowledge of past earthquakes is therefore non-existent for most of the Earth, and rarely goes back very far into the past (only 200 years in California).

When a major fault is known, scientists now want to determine the series of earthquakes it produced, by geomorphological observations. Geologists were long to start on this ground, because of the lack of dating techniques, but also because they long took the uniformitarian principle (Section 1.2) too strictly. Now, along the Alaska coastline, 6 elevated beaches are witnesses of past earthquakes, which elevated the ground over the last 4,500 years (the last one occurred in 1964). One has to realize that, in geology, a 'current' process may only happen every 1,000 years!

Times have changed. In 1994, a paleoseismology workshop was attended by 89 scientists from 15 different countries (Yeats and Prentice, 1996, and the next 26 articles of the *Journal of Geophysical Research*).

Apart from the elevated beaches, observations are made in trenches dug perpendicularly to the fault. The indices of a paleo-earthquake are:

- Stratigraphic discordance. Multiple faults activated during the earthquake are not extending into the upper layers, more recent. Or inclined and folded beds are covered with beds much less perturbed.
- Colluvial wedges at the foot of inverse faults.
- Markers of the liquefaction of sands during the earthquake (the compaction it induces increases the interstitial pressure): dykes filled with sand, or, at the bottom of old lakes, thin layers where the lacustrine sediments are mixed with sand.

The dates are obtained through C^{14} or thermoluminescence. Let us note a few striking results from these studies:

- The Eastern Mediterranean had a 'tectonic paroxysm' between 450 and 650 AD, approximately. Beaches were elevated from 0.5 to several metres, from the

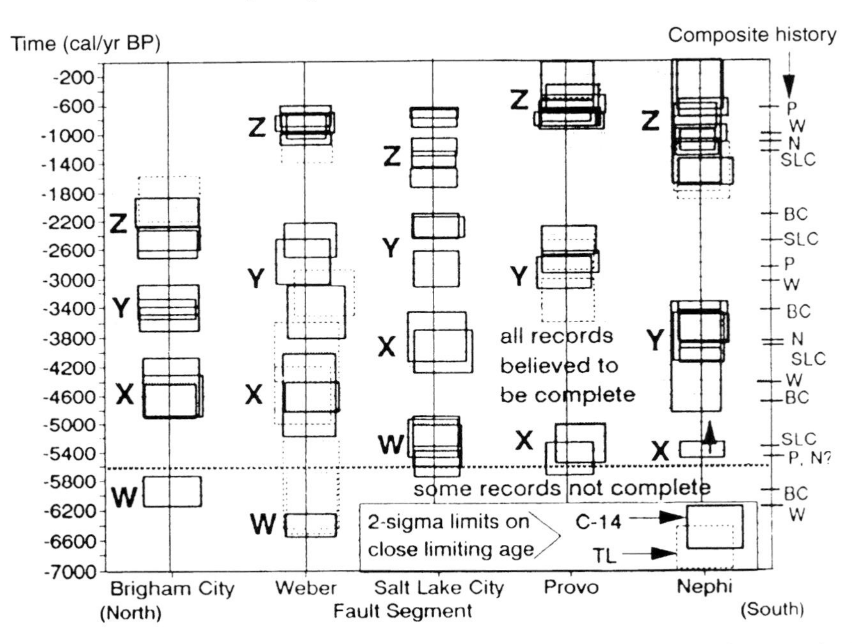

Figure 12.15. Space–time diagram of the large earthquakes ($M > 7$) along the Wasatch Mountains fault. The list is thought to be complete for the last 5,600 years. From McCalpin and Nishenko (1996); © AGU.

Ionian Islands to Lebanon. The maximum was a rise of 9 m at the SW extremity of Crete.

- The normal fault along the Wasatch Chain (Utah), 300-km long, can be divided into 5 segments with independent seismic histories. Stratigraphy shows there were 16 earthquakes of magnitude larger than 7 in the last 5,600 years (Figure 12.15). The possibility of such an earthquake threatens Salt Lake City, which never knew anything like that since it was founded.
- The Red Sea spreading ridge separates the Arabic and African plates, and extends into a sinistral transform fault in the Jordan valley. Israeli scientists were able to reconstitute the seismic history for the last 50,000 years, using the mixed clay and sand layers deposited in the Dead Sea (which was much larger before). This time, dating was performed with Th^{230}–U^{234}. There were alternatively periods with large earthquakes ($M > 6.5$) and quieter periods, each lasting around 10,000 years.

12.15 LONG- AND MEDIUM-TERM SEISMIC PREDICTION

The prediction of an earthquake in the medium-term (next few years) or on a longer term, is only based on the existence of a seismic gap along the fault, and on the periodicity of large earthquakes. We saw that these criteria are not sure. Furthermore, palaeoseismicity studies showed that, on some faults, the large earthquakes were 2 to 10 centuries apart, and even more. A prediction should only be made for the next few human generations.

Another hazard comes from hidden active faults. Large normal faults generally show up at the surface, but inverse faults are only visible from the folding of superficial terrains. And deep seismic surveys revealed the possible existence of active sub-horizontal faults at depth.

This happens in the Los Angeles Basin (Shaw and Suppe, 1996). This small basin extends over 70 km along the Pacific Ocean, 90 to 50 km from the San Andreas Fault. Below Pleistocene sediments, up to 3 km thick at the centre, there are 4 km of Miocene sedimentary rocks containing oil (oil exploration was essential to discover the deep structure of the basin), and then Mesozoic schists. In the Miocene, the basin was submitted to extension perpendicular to the San Andreas Fault, and therefore subsidence. Since 4 Ma, compression replaced the stretching. But it is expressed as overthrusting, not as a general deformation. The sedimentary basin and the first 3 km of Mesozoic schists took off the basement and progress toward the Pacific Ocean as a wedge, limited by the subhorizontal decollement fault and an inverse fault. The former is therefore completely hidden. It has been speculated that it could produce an earthquake of magnitude 6.7 every 200 years, or, as well, an earthquake of magnitude larger than 7 every 13,000 years or more.

12.16 SHORT-TERM SEISMIC PREDICTION

Short-term seismic prediction presented huge successes and huge failures as well. In China, the February 4, 1975, earthquake of Haicheng was predicted in time, and the population evacuated. But the earthquake of July 28, 1976, in Tangshan, in the same province east of Beijing, was not predicted and was the most murderous of the century.

These predictions are based on the observations of phenomena that might be earthquake precursors.

1. *Foreshocks*. Repeated earthquakes, more and more frequent (with sometimes a strange complete stop during the last half-day) are the most accurate pointer to imminent danger. Thus, before the Japan earthquake of November 15, 1930, there were:
   - the 3rd week before the earthquake, a few recorded tremors;
   - the 2nd week, 100 tremors recorded, 10 were felt by the population;
   - the week before, 700 tremors were recorded, 70 were felt.

   Less and less frequent earthquakes, on the other hand, are aftershocks, and show that the danger is receding. There may, however, be a cluster of small earthquakes without a main earthquake, or, unfortunately quite often, no foreshocks before the large earthquake. (An isolated earthquake does not mean anything, and it is only seen as a foreshock with hindsight.)

2. *Variation of the seismic velocity ratio* v_S/v_P, in the weeks before the earthquake. This is attributed to dilatancy when the sliding along the fault accelerates. This effect occurs on a few faults only.
3. *Acceleration of local deformations.* This is difficult to observe, because it is very local and may not appear at the surface. Tilting and vertical movements of the coastline are more accessible.
4. *Hydrological phenomena* (water level variations in wells, changes in springs).
5. *Radon emission.* It is continuously produced in granitic terrains, but a change in ground porosity enhances its release.
6. *Variations of telluric currents.* This is the base of the VAN method, which is much questioned.
7. *Abnormal behaviour of animals.* The Chinese include these observations, always very subjective, amongst the pointers to an imminent danger.

Not all these phenomena always occur before an earthquake. When they do, except for foreshocks, they may have other causes than coming earthquakes. Only the accumulation of such hints can be used for a prediction, and there will always be unpredictable earthquakes.

13

Mechanics of lithospheric plates

13.1 HOMOGENEOUS ELASTIC PLATE MODEL

The simplest model of a lithosphere shell is a homogeneous elastic plate, with a thickness H very small compared to the involved horizontal distances. In Section 10.8, its bending was computed. We shall now redo the computation in the general case, with three dimensions.

The vertical displacement relative to the reference state, positive upwards, is denoted $\zeta(x, y)$. It is assumed negligible relative to the Earth's radius. The slopes $\partial\zeta/\partial x$ and $\partial\zeta/\partial y$ are assumed to be very small. In these conditions, the neutral surface is very close to the middle of the plate (this was demonstrated for the 2-D case), and this distance can be neglected. At a distance z (positive upwards) from the neutral surface, the displacement of a material point (x, y, z) has the components $(-z\,\partial\zeta/\partial x, -z\,\partial\zeta/\partial y, \zeta)$. For a neighbouring point at the same distance z, the displacement differs by (U, V, W):

$$\begin{cases} U = -z\left(\dfrac{\partial^2\zeta}{\partial x^2}dx + \dfrac{\partial^2\zeta}{\partial x\,\partial y}dy\right) \\ V = -z\left(\dfrac{\partial^2\zeta}{\partial x\,\partial y}dx + \dfrac{\partial^2\zeta}{\partial y^2}dy\right) \\ W = 0 \end{cases} \tag{13.1}$$

These differences are the cause of the local elastic strain:

$$\begin{cases} \varepsilon_{xx} = \dfrac{\partial U}{\partial x} = -z\dfrac{\partial^2\zeta}{\partial x^2} \\ \varepsilon_{yy} = \dfrac{\partial V}{\partial y} = -z\dfrac{\partial^2\zeta}{\partial y^2} \\ \varepsilon_{xy} = \dfrac{1}{2}\left(\dfrac{\partial U}{\partial y} + \dfrac{\partial V}{\partial x}\right) = -z\dfrac{\partial^2\zeta}{\partial x\,\partial y} \end{cases} \tag{13.2}$$

The elastic constitutive relations (from Chapter 10) lead to the stresses:

$$\begin{cases} \sigma_x = \dfrac{E}{1-\nu^2}(\epsilon_{xx} + \nu\varepsilon_{yy}) + \dfrac{\nu}{1-\nu}\sigma_z \\[2ex] \sigma_y = \dfrac{E}{1-\nu^2}(\nu\varepsilon_{xx} + \varepsilon_{yy}) + \dfrac{\nu}{1+\nu}\sigma_z \\[2ex] \tau_{xy} = \dfrac{E}{1+\nu}\varepsilon_{xy} \end{cases} \tag{13.3}$$

Using equations (13.2) and (13.3), the equilibrium equations read:

$$\begin{cases} -\dfrac{zE}{1-\nu^2}\left(\dfrac{\partial^3\zeta}{\partial x^3} + \dfrac{\partial^3\zeta}{\partial x\,\partial y^2}\right) + \dfrac{\nu}{1-\nu}\dfrac{\partial\sigma_z}{\partial x} + \dfrac{\partial\tau_{xz}}{\partial z} = 0 \\[2ex] -\dfrac{zE}{1-\nu^2}\left(\dfrac{\partial^3\zeta}{\partial x^2\,\partial y} + \dfrac{\partial^3\zeta}{\partial y^3}\right) + \dfrac{\nu}{1-\nu}\dfrac{\partial\sigma_z}{\partial y} + \dfrac{\partial\tau_{yz}}{\partial z} = 0 \\[2ex] \dfrac{\partial\tau_{xz}}{\partial x} + \dfrac{\partial\tau_{yz}}{\partial y} + \dfrac{\partial\sigma_3}{\partial z} = \rho g \end{cases} \tag{13.4}$$

The lateral variations of σ_z are negligible compared to the variations of σ_x and σ_y, and thus the terms in $\partial\sigma_z/\partial x$ and in $\partial\sigma_z/\partial y$ are neglected.

Two components of the *shear force* may be defined:

$$\begin{cases} S_x = \displaystyle\int_{-H/2}^{+H/2} \tau_{xz}\,dz \\[2ex] S_y = \displaystyle\int_{-H/2}^{+H/2} \tau_{yz}\,dz \end{cases} \tag{13.5}$$

The shear stresses τ_{xz} and τ_{yz} are zero for $z = \pm H/2$, and therefore:

$$\begin{cases} \displaystyle\int_{-H/2}^{+H/2} \frac{\partial\tau_{xz}}{\partial z} z\,dz = [z\tau_{xz}]_{-H/2}^{+H/2} + \int_{-H/2}^{+H/2} \tau_{xz}\,dz = -S_x \\[2ex] \displaystyle\int_{-H/2}^{+H/2} \frac{\partial\tau_{yz}}{\partial z} z\,dz = -S_y \end{cases} \tag{13.6}$$

Multiplying the first two equations of (13.4) by z and integrating from $-H/2$ to $+H/2$, one gets:

$$\begin{cases} \Gamma\dfrac{\partial}{\partial x}\left(\dfrac{\partial^2\zeta}{\partial x^2} + \dfrac{\partial^2\zeta}{\partial y^2}\right) + S_x = 0 \\[2ex] \Gamma\dfrac{\partial}{\partial y}\left(\dfrac{\partial^2\zeta}{\partial x^2} + \dfrac{\partial^2\zeta}{\partial y^2}\right) + S_y = 0 \qquad \left(\Gamma = \dfrac{EH^3}{12(1-\nu^2)}\right) \end{cases} \tag{13.7}$$

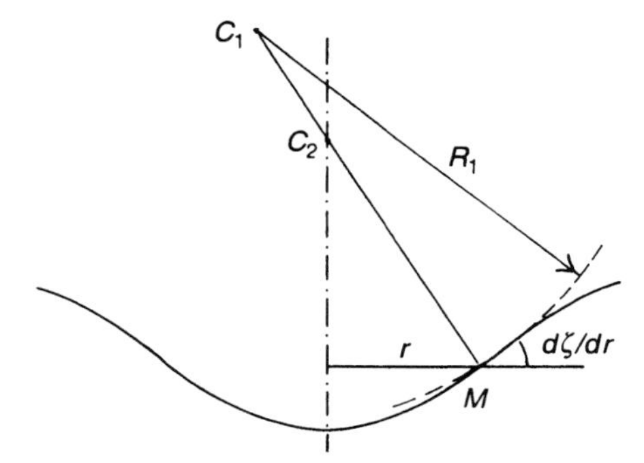

Figure 13.1. Principal curvature centres (C_1, C_2) at a point M of the axisymmetric surface $\zeta(r)$. If $|d\zeta/dr| \ll 1$, $1/R_1 = d^2\zeta/dr^2$ and $1/R_2 = (1/r)\, d\zeta/dr$.

The following quantity, termed the total curvature of the surface $\zeta(x, y)$, is invariant when the orientation of the x- and y-axes changes:

$$\frac{\partial^2\zeta}{\partial x^2} + \frac{\partial^2\zeta}{\partial y^2} = \frac{1}{R_1} + \frac{1}{R_2} \tag{13.8}$$

For an axisymmetric system, with cylindrical coordinates (r, φ), ζ is function of r only and:

$$\frac{\partial^2\zeta}{\partial x^2} + \frac{\partial^2\zeta}{\partial y^2} = \frac{d^2\zeta}{dr^2} + \frac{1}{r}\frac{d\zeta}{dr} \tag{13.9}$$

The two terms correspond to the curvature in a meridian plane and to the curvature in a plane normal to the surface and to the meridian plane, respectively (Figure 13.1).

S_x and S_y are removed from equations (13.7) by using the third equilibrium equation of (13.4), and integrating over the whole plate thickness:

$$-\left(\frac{\partial S_x}{\partial x} + \frac{\partial S_y}{\partial y}\right) = \Gamma\left(\frac{\partial^2}{\partial x^2} + \frac{\partial^2}{\partial y^2}\right)\zeta = [\sigma_z]_{-H/2}^{+H/2} - \int_{-H/2}^{+H/2} \rho g z\, dz \tag{13.10}$$

But two sorts of problems need to be distinguished here:

1. *Deformation is only due to forces and torques exerted at the periphery* (Figure 13.2). For example, the problem is plane, $\zeta = \zeta(x)$ with $x > 0$, the boundary forces and couples only exist for $x = 0$. Or the problem is axisymmetrical, with a load applied in $r = 0$. Let us assume this is an oceanic plate. ρ_a and ρ_w are the densities of the asthenosphere and the ocean, respectively. The deflection ζ changes $\sigma_z(H/2)$ by $\rho_w g\zeta$ and $\sigma_z(-H/2)$ by $\rho_a g\zeta$. Equation (13.10) then reads:

$$\Gamma\left(\frac{\partial^2}{\partial x^2} + \frac{\partial^2}{\partial y^2}\right)^2 \zeta + (\rho_a - \rho_w) g\zeta = -\int_{-H/2}^{+H/2} \rho g z\, dz \tag{13.11}$$

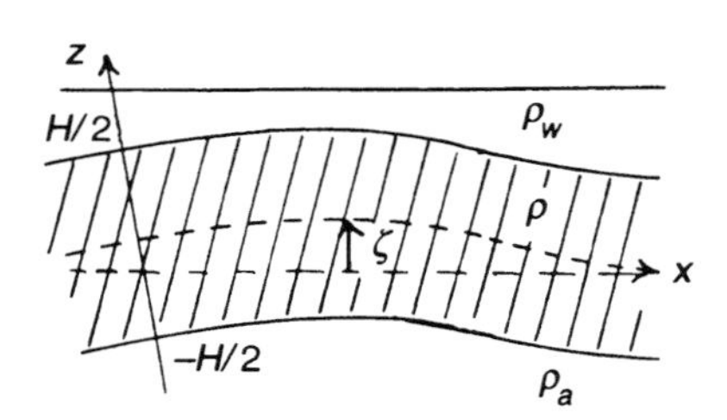

Figure 13.2. Plate deformed by the torques and forces acting at its extremities.

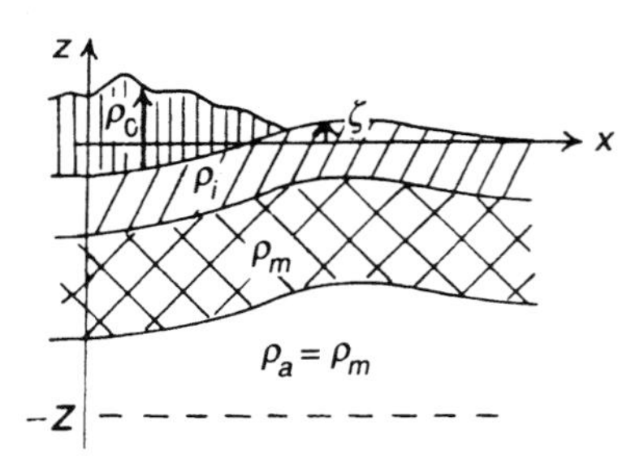

Figure 13.3. Plate deformed by a load distributed over its surface.

With a suitable origin of ζ, the term on the right-hand side can be removed. This means that the isostatic equilibrium in the absence of plate curvature is chosen as the reference state.

2. *Deformation is due to a load distributed all along the plate* (Figure 13.3). The vertical displacements coming from all local loads must be added, but this can only be achieved with a fixed ζ origin, and not with the Lagrangian coordinates used until now. The small slope of $\zeta(x, y)$ makes the two coordinate systems almost parallel, and only the σ_z values are modified. The third equation of (13.4) is then integrated along z, between a fixed compensation level $-Z$ in the asthenosphere and the level $z = 0$. Assume the existence of a continent, with an upper crust of density ρ_0, a lower crust of density ρ_l and a mantle lithosphere with the same density $\rho_m = \rho_a$ as the asthenosphere. The elastic parameters are assumed to be the same in the whole lithosphere, so that the equations written remain valid. $h(x, y)$ being the altitude of the surface, for $z = 0$, $\sigma_z = -\rho_0 g h$, and for $z = -Z$, $\sigma_z = -P$, a constant. The integral of ρg does not equal $-P$ anymore, as it did in the reference case. At the Moho, over a distance z, mantle replaced the lower crust. The displacement of the lower limit of the plate did not introduce any change. Writing $\rho_m - \rho_i = \Delta\rho$, the result is:

$$\Gamma\left(\frac{\partial^2}{\partial x^2} + \frac{\partial^2}{\partial y^2}\right)^2 \zeta + \Delta\rho g \zeta = -\rho_0 g h \tag{13.12}$$

If the flexural rigidity Γ were zero, the vertical displacement of the Moho in the Airy theory (no regional compensation) would be found.

3. *Deformation is due to variations $\delta(x, y)$ of the Moho level* (Figure 13.4). This is equivalent to a load $\Delta\rho g \delta$ at the base of the crust. This time, the vertical displacement ζ of the plate creates a relief, and therefore $\sigma_z(0) = -\rho_0 g \zeta$. In the ρg integral from $-Z$ to 0, it is necessary to account for a variation $\Delta\rho$ over a distance δ. It is found:

$$\Gamma\left(\frac{\partial^2}{\partial x^2} + \frac{\partial^2}{\partial y^2}\right)^2 \zeta + \rho_0 g \zeta = -\Delta\rho g \delta \tag{13.13}$$

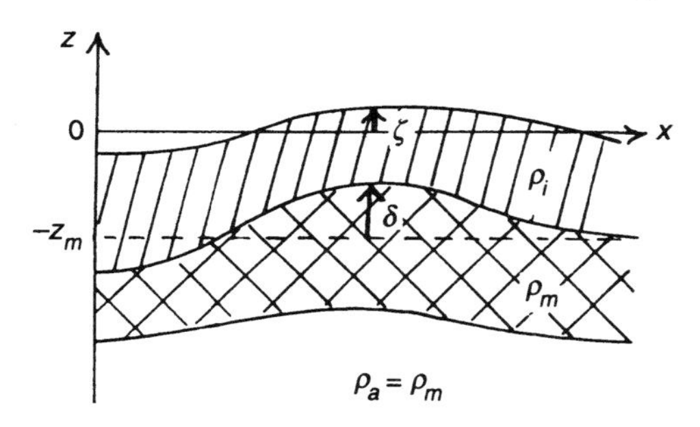

Figure 13.4. Plate deformed by Moho depth variations.

13.2 BENDING OF AN OCEANIC PLATE

We use equation (13.11) and write:

$$\left[\frac{\Gamma}{(\rho_a - \rho_w)g}\right]^{1/4} = a \tag{13.14}$$

Axisymmetrical problem

Let us consider an isolated seamount or an island, whose weight deforms the plate and causes a hollow. Assimilating it to a punctual load, applied in $r = 0$, the profile $\zeta(r)$ must obey the equation:

$$\Gamma\left(\frac{d^2}{dr^2} + \frac{1}{r}\frac{d}{dr}\right)^2 \zeta + (\rho_a - \rho_w)g\zeta = 0$$

$$\frac{d^4\zeta}{dr^4} + \frac{2}{r}\frac{d^3\zeta}{dr^3} - \frac{1}{r^2}\frac{d^2\zeta}{dr^2} + \frac{1}{r^3}\frac{d\zeta}{dr} + \frac{\zeta}{a^4} = 0 \tag{13.15}$$

The general solution of (13.15) is a linear combination of the Kelvin functions $ber(r/a)$, $bei(r/a)$, $ker(r/a)$ and $kei(r/a)$. But only the last one has limits for $r = 0$ and for $r = \infty$ and can be kept (Abramowitz and Stegun, 1965).

$$\zeta = C\, kei(r/a) \tag{13.16}$$

At the origin, $kei(0) = -0.7854$ and the derivative is null. The first root ($\zeta = 0$) occurs for $r = 3.9147a$. The function then oscillates around zero, with amplitudes decreasing very rapidly. The first maximum, for $r = 4.9a$, only reaches 1/70th of the central depression.

Plane problem

The equation (13.11) becomes:

$$\frac{d^4\zeta}{dx^4} + \frac{\zeta}{a^4} = 0 \tag{13.17}$$

The general solution is a linear combination of terms in $\exp(p\,x/a)$, where $p^4 + 1 = 0$, i.e. $p = (\pm 1 \pm i)/\sqrt{2}$. As ζ must tend toward 0 at ∞:

$$\zeta = C \exp\left(-\frac{x}{\sqrt{2}a}\right) \cdot \sin\left(\frac{x - x_0}{\sqrt{2}a}\right) \tag{13.18}$$

This yields the slope, the flexural moment M (defined in Section 10.8) and the shear force S_x:

$$\begin{cases} \dfrac{d\zeta}{dx} = \dfrac{C}{a}\exp\left(-\dfrac{x}{\sqrt{2}a}\right) \cdot \cos\left(\dfrac{x - x_0}{\sqrt{2}a} + \dfrac{\pi}{4}\right) \\ M = \Gamma\dfrac{d^2\zeta}{dx^2} = -\dfrac{\Gamma C}{a^2}\exp\left(-\dfrac{x}{\sqrt{2}a}\right) \cdot \cos\left(\dfrac{x - x_0}{\sqrt{2}a}\right) \\ S_x = -\Gamma\dfrac{d^3\zeta}{dx^3} = -\dfrac{\Gamma C}{a^3}\exp\left(-\dfrac{x}{\sqrt{2}a}\right) \cdot \sin\left(\dfrac{x - x_0}{\sqrt{2}a} + \dfrac{\pi}{4}\right) \end{cases} \tag{13.19}$$

First application: forebulges at oceanic trenches

100 to 500 km before the trench, oceanic plates show a bulge of 300 to 500 m above the abyssal plains, e.g. the *Hokkaido Rise* before the Japan Trench, or the rise supporting the Loyauté Islands coral edifices before the Tonga Trench. The first was modelled as early as 1971 (Hanks, 1971), neglecting the forces from convection currents in the asthenosphere. The plate undergoes a shear force at $x = 0$, due to the overthrusting plate, but there is no bending torque (Figure 13.5).

This assumption implies that $x_0/(\sqrt{2}a) = \pi/2$. Therefore, if $\zeta(0) = \zeta_0$, the first maximum of ζ occurs for $x = (3\pi/4)\sqrt{2}a = 3.322a$, and equals $-\zeta_0/14.92$.

The height of the forebulge is correct, but its distance to the trench leads to $a = 30$ to 45 km. With $\rho_a - \rho_w = 2.3\,\text{Mg/m}^3$, it leads to $\Gamma = (2 - 9) \times 10^{22}\,\text{Nm}$. As $E \approx 10^{11}\,\text{Pa}$ and $\nu \approx 0.25$, the resulting elastic lithosphere thickness is $H = 13.1$ to 21.6 km only. Let us remember that the LVZ, considered in plate tectonics to be just below the lithosphere, only appears at 70-km depths.

Since this first study in 1971, the bathymetric profiles before the trenches have improved in resolution. The model must account not only for bathymetry, but also for gravimetric anomalies. Lewis and Sandwell (1995) have studied 117 profiles. As

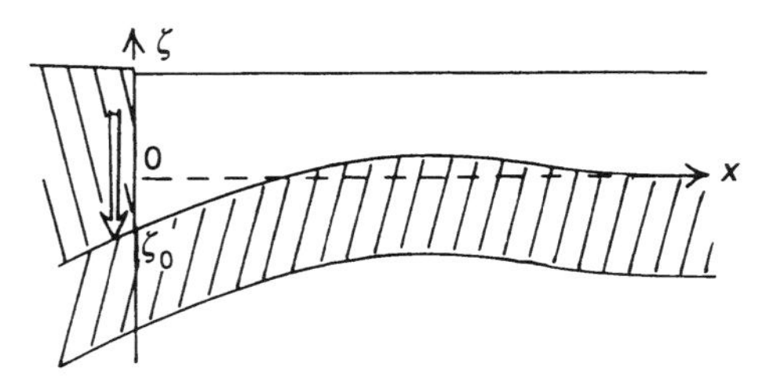

Figure 13.5. Forebulge close to an oceanic trench. The heights are greatly exaggerated and the diagram is not to scale.

the scattering of results is very large, as no new concept is introduced, and as the qualitative result of the initial coarse model remains valid, these later developments will not be detailed here.

The main reason of this very small elastic thickness is that the plate mainly bends through the action of faults. Chapple and Forsyth (1979) demonstrated this by studying the earthquakes around the Japan Trench:

1. Independently of earthquakes occurring at the interface between subducted and overthrusting plates, shallow earthquakes are observed before the trench. They are caused by the bending of the subducted plate, in a place mobile relative to it. Focal mechanisms show horizontal traction, perpendicular to the trench, 6 to 25 km below the seafloor. In two cases (out of the 15 earthquakes kept because their sources were very well localized), there was a horizontal compression perpendicular to the trench, 42 km deep. The neutral surface is therefore between 25 and 42 km below the seafloor, and not between 6.6 and 11 km, as predicted with the mere elastic model.
2. In the layer in extension, the cumulative throw of these faults is at least equal to the one necessary to ensure bending.
3. The largest seismic activity is not 70–120 km away from the trench, as required by the purely elastic model, but only 30 km away.

Chapple and Forsyth (1979) can account for these facts and for the profiles before the trench with an *elasto-plastic* model of the plate, with two layers (perfect plasticity simulating the deformation through fault movements). Their model consists in an upper layer, 20-km thick and with a yield strength of 100 MPa, and a lower layer, 30-km thick and with a yield strength of 600 MPa. As the plate moves toward the trench, the yield strength is reached at increasing depths. Close to the trench, another plastic layer develops at the lower limit (Figure 13.6).

Second application: subsidence under the weight of a chain of volcanoes

Let us assimilate the chain of volcanoes to a rectilinear load in $x = 0$. The slope of 0 in this point implies (because of equation (13.19)) that $x_0/(\sqrt{2}a) = 3\pi/4$. Therefore: $\zeta_0 = -C/\sqrt{2}$. The first maximum of ζ is reached for $x = 4.443a$ and equals $-\zeta_0/23.14$.

As shown in Figure 13.7, the Hawaiian Chain causes a bulge approximately 200 km away from its axis. With the purely elastic model, $a \approx 45$ km, $\Gamma \approx 10^{23}$ Nm and $H = 22$ km. This small value can be explained partly as above: far from the neutral surface, the plate deformation occurs through fault movements. But the main aspect is the *local warming* from magmatism. Results from deep seismics and gravimetry are given in Figure 13.8. Basaltic magmas were underplated, and their solidification produced heat at the base of the crust. Heated lithosphere became asthenosphere.

A hot *plume* in the asthenosphere and this warming produced a swelling of the seafloor all around the Hawaii Chain on widths larger than 1,000 km, the *Hawaiian*

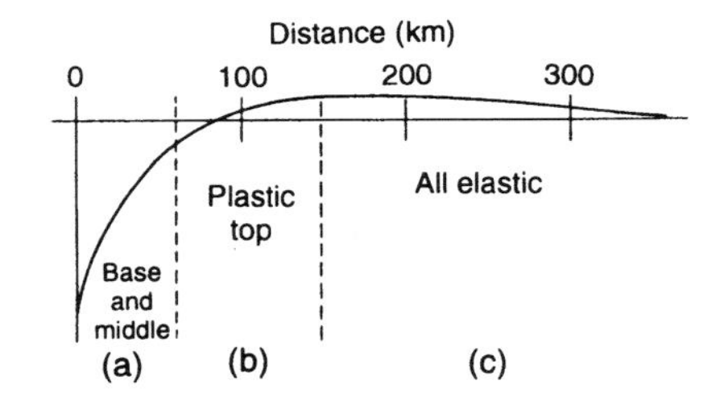

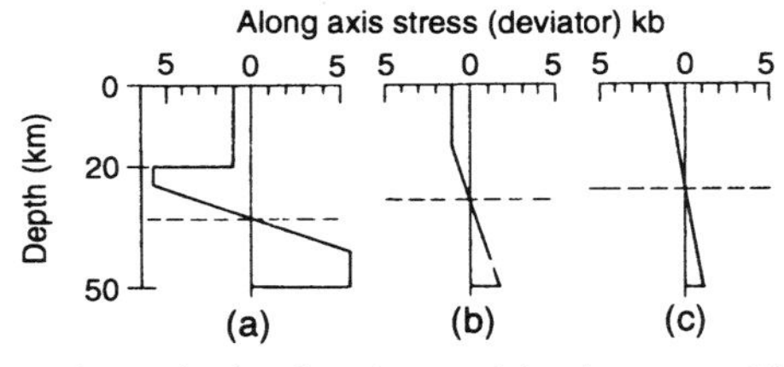

Figure 13.6. Bending of an elasto-plastic plate in a subduction zone. The elasticity parameters are $E = 65\,\text{GPa}$, $\nu = 0.25$. The yield strength is $1\,\text{kb} = 100\,\text{MPa}$ in the upper 20 km, 600 MPa in the lower 30 km. In area (*c*), the deformation is elastic only. In area (*b*), an upper layer deforms plastically. In area (*a*), a bottom layer also deforms plastically, but there remains an intermediate layer where the yield strength is not reached. Redrawn from Chapple and Forsyth, 1979.

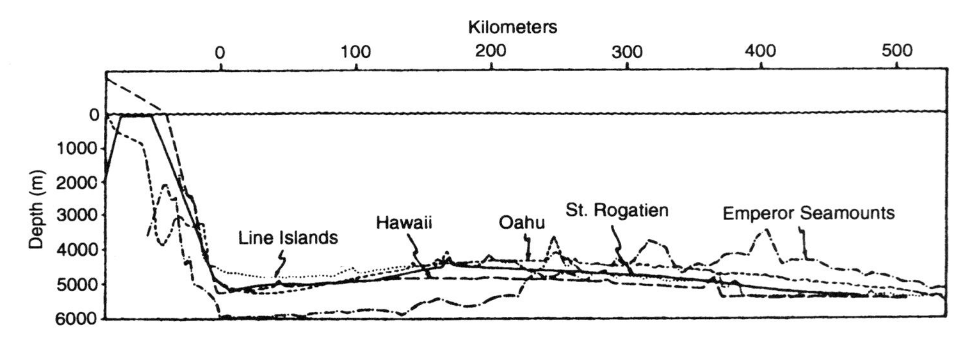

Figure 13.7. Bathymetric profiles perpendicular to the Hawaii Chain and the Emperor Chain. From Menard (1964); © McGraw-Hill, 1964.

Swell. Over this vast swell, the geothermal flux is 20% to 25% larger than its normal value in oceanic crust of the same age (Detrick *et al.*, 1981).

13.3 CORRELATION BETWEEN TOPOGRAPHY AND BOUGUER ANOMALY

The deflection ζ of a continental lithosphere is not directly accessible, but it can be inferred from the Bouguer anomaly, defined in Section 8.6. This anomaly comes

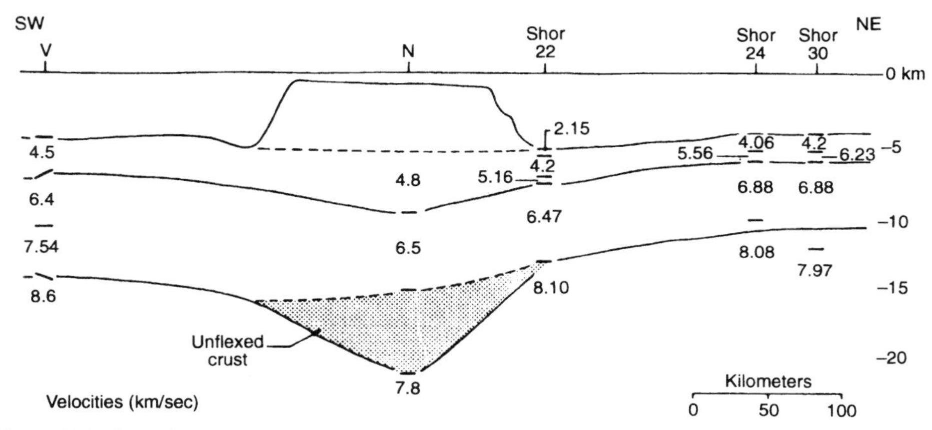

Figure 13.8. Oceanic crust structure around the Oahu guyot (Hawaii). The seismic wave velocities come from Furumoto *et al.* (1968). In grey: the crust formed after sinking, not in elastic extension. From Suyenaga (1979); © AGU 1979.

from the fact that we omitted from computation of 'normal' gravity the density change at the Moho, created by this deflection.

In the first computations, it was assumed that the deflection and the Bouguer anomaly were only coming from the surface load, from topography. In fact, they also come from local Moho depth variations apart from isostasy, i.e. from a load at the base of the crust. In the elastic lithosphere model, the corresponding equations are (13.12) and (13.13). Because they are linear, the two effects can be added.

The following notation will be adopted:

- If the surface load acted alone, once isostasy is reached, the altitude of the relief would be h_s and the Moho elevation, equal to the plate elevation, would be ζ_s.
- If the load at the bottom of the crust acted alone, the relief elevation, equal to the plate elevation, would be h_b and the Moho elevation (original + due to the plate elevation) would be ζ_b.
- When the two loads act together, the relief elevation is $h = h_S + h_b$ and the Moho elevation is $\zeta = \zeta_s + \zeta_b$.

The equations (13.12) and (13.13) then read:

$$\begin{cases} \left[\Gamma\left(\dfrac{\partial^2}{\partial x^2} + \dfrac{\partial^2}{\partial y^2}\right)^2 + \Delta\rho g\right]\zeta_s = -\rho_0 g h_s \\ \left[\Gamma\left(\dfrac{\partial^2}{\partial x^2} + \dfrac{\partial^2}{\partial y^2}\right)^2 + \rho_0 g\right]\zeta_b = -\Delta\rho g h_b \end{cases} \tag{13.20}$$

The gravity potential, at level 0, created by the excess mass per unit area $\Delta\rho\zeta$ in the Moho equals (noting z_M the depth of the Moho):

$$V(x, y) = G\Delta\rho \int_{-\infty}^{+\infty}\int_{-\infty}^{+\infty} \frac{\zeta(x', y')\, dx'\, dy'}{[z_M^2 + (x - x')^2 + (y - y')^2]^{1/2}} \tag{13.21}$$

The Bouguer anomaly (B) is the downward component of its gradient:

$$B(x,y) = -\frac{\partial V}{\partial z_M} = G\Delta\rho \int_{-\infty}^{+\infty}\int_{-\infty}^{+\infty} \frac{z_M \zeta(x',y')\, dx'\, dy'}{[z_M^2 + (x-x')^2 + (y-y')^2]^{3/2}} \tag{13.22}$$

To components ζ_s and ζ_b of ζ correspond Bouguer anomalies B_s and B_b respectively. It is worth computing the 2-D Fourier transforms (FT), defined in Annex A.4. By deriving the term in the sum, the FT of $\partial f/\partial x$ is $i\omega_x \bar{f}$, the FT of $\partial^2 f/\partial x^2$ is $-\omega_x^2 \bar{f}$, etc. Writing that $\omega_x^2 + \omega_y^2 = \omega^2$, the equations (13.20) become

$$\begin{cases} (\Gamma\omega^4 + \Delta\rho g)\bar{\zeta}_s = -\rho_0 g \bar{h}_s \\ (\Gamma\omega^4 + \rho_0 g)\bar{\zeta}_b = -\Delta\rho g \bar{h}_b \end{cases} \tag{13.23}$$

Equation (13.22) gives the Bouguer anomaly as a *convolution* (in two dimensions) of the functions $G\,\Delta\rho\,\zeta(x,y)$ and $F(x,y) = z_M[z_M^2 + x^2 + y^2]^{-3/2}$. The Fourier transform of the convolution of two functions is the product of the Fourier transforms of these two functions. To compute the Fourier transform of F, the axes (x,y) may be oriented so that $\omega_x = \omega$, $\omega_y = 0$, and thus:

$$\bar{F} = \int_{-\infty}^{+\infty} e^{-i\omega x}\, dx \frac{z_M\, dy}{[z_M^2 + x^2 + y^2]^{3/2}} \tag{13.24}$$

Integration relative to y can be performed by the substitution: $y = (z_M^2 + x^2)^{1/2} \tan\theta$. Next, integration relative to x is easily made by the theorem of residues.

$$\bar{F} = \int_{-\infty}^{+\infty} \frac{2z_M \exp(-i\omega x)}{z_M^2 + x^2}\, dx = 2\pi \exp(-\omega z_M) \tag{13.25}$$

Therefore:

$$\bar{B} = 2\pi G\,\Delta\rho\,\bar{\zeta}\exp(-\omega z_M \bar{\zeta}) \tag{13.26}$$

Topography yields h, hence $\bar{h} = \bar{h}_s + \bar{h}_b$. Gravimetry yields B, hence $\bar{B} = \bar{B}_s + \bar{B}_b$. If Γ were known, equations (13.23) and (13.26) would allow $\bar{h}_s$, $\bar{h}_b$, $\bar{B}_s$ and $\bar{B}_b$ to be obtained separately. But Γ is not known, it is exactly what we are looking for.

The problem is still solvable, as shown by Forsyth (1985), because the Fourier transforms are complex numbers giving the module and the phase:

$$\bar{h} = r\,e^{ia}; \qquad \bar{B} = b\,e^{i\theta} \tag{13.27}$$

The moduli r and b and the phases α and θ are real numbers, functions of ω_x and ω_y. If we designate with an asterisk the complex conjugate, $\bar{h}\bar{h}^* = r^2$ is the *spectral power of h*, $\overline{BB}^* = b^2$ is the *spectral power of B*, and $\frac{1}{2}(\bar{h}\bar{B}^* + \bar{h}^*\bar{B}) = rb\cos(\alpha - \theta)$ is the *spectral correlation of h and B*. The *coherence* of B, and h is the dimensionless quantity:

$$\gamma^2 = \frac{[\frac{1}{2}(\bar{h}\bar{B}^* + \bar{h}^*\bar{B})]^2}{(\bar{h}\bar{h}^*)(\bar{B}\bar{B}^*)} = \cos^2(\alpha - \theta) \tag{13.28}$$

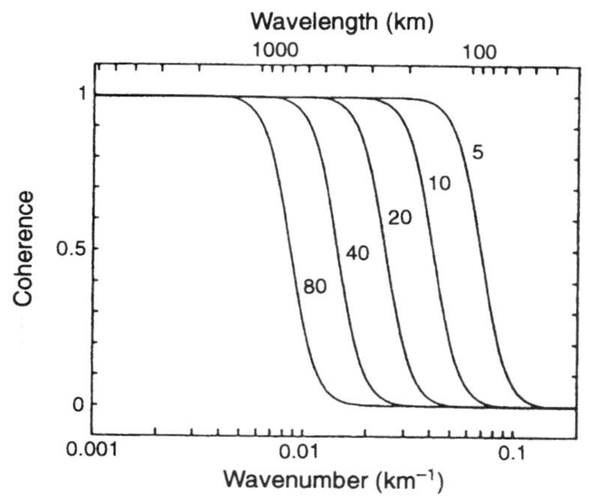

Figure 13.9. Theoretical coherence between Bouguer gravity anomaly and topography as a function of wavelength, when the load is equally distributed between the surface and the Moho (around 30 km deep). The numbers on the different curves represent the adopted elastic thickness, in kilometres. From Forsyth (1985); © AGU 1985.

The same definitions are possible with h_s and B_s, h_b and B_b. Let us write $\omega_x = \omega \cos \varphi$ and $\omega_y = \omega \sin \varphi$. And let us make the essential assumption that, in a ring of the plane (ω_x, ω_y), corresponding to the interval $(\omega, \omega + \Delta\omega)$, the crossed phase shifts $(\alpha_s - \theta_b)$ and $(\alpha_b - \theta_s)$ are completely random. In that case, averages over the ring (noted with angle brackets) make these random terms disappear:

$$\begin{cases} \langle \overline{h}\,\overline{h}^* \rangle = \langle \overline{h}_s \overline{h}_s^* \rangle + \langle \overline{h}_b \overline{h}_b^* \rangle \\ \langle \overline{B}\,\overline{B}^* \rangle = \langle \overline{B}_s \overline{B}_s^* \rangle + \langle \overline{B}_b \overline{B}_b^* \rangle \\ \langle \overline{h}\,\overline{B}^* + \overline{h}^* \overline{B} \rangle = \langle \overline{h}_s \overline{B}_s^* + \overline{h}_s^* \overline{B}_s \rangle + \langle \overline{h}_b \overline{B}_b^* + \overline{h}_b^* \overline{B}_b \rangle \end{cases} \tag{13.29}$$

The theoretical computation of γ^2 as a function of ω is now possible. The regionalization of isostatic compensation, brought by the plate, makes the coherence tend toward 0 for $\omega = 2\pi/\lambda$ high, and toward 1 for very small values of ω. As demonstrated in Figure 12.9, the value of ω for which $\gamma^2 = 0.5$ depends strongly on Γ, i.e. the elastic thickness H of the plate (computed with $E = 10^{11}$ Pa and $v = 0.25$). When H goes from 5 to 80 km, the wavelength λ for which $\gamma^2 = 0.5$ goes from 90 km to 714 km. It is therefore possible: (1) from an assumed value of Γ, to compute the values of γ^2 for successive intervals $\Delta\omega$, from the Fourier transforms of the field values; (2) to examine on which theoretical curve $\gamma^2(\omega)$ the points are found, and deduce a better value of Γ; (3) to start again until convergence.

13.4 ELASTIC LITHOSPHERE THICKNESS

Forsyth's method, as detailed above, was applied by Doucouré *et al.* (1996) to the determination of the elastic thickness H of the African cratons. They adopted values

Table 13.1. Ages of the last tectogenesis and elastic thickness of lithosphere.

Cratons	Age (Ga)	*H* (km)	Craton	Age (Ga)	*H* (km)
N. Australia	1.7–3.1	134	Grenville	1.0–1.2	49
W. Australia	2.7–3.5	132	Namagua-Natal	1.0–1.2	38–48
Rio de la Plata	1.0–2.0	115			
W. Brazil	1.0–2.0	100			
Kasai (Zaire)	3.0	100	**Other terrains**		
Upper Great Lake	2.7	39–143	E. Central Australia	<0.7	30–80
Central Australia	1.2–1.8	88	Central Altiplano	<0.1	50
Kaapval	2.6–3.7	72	Colorado Plains and Rocky Mountains	<0.1	16–32
Slave Lake	2.8–3.0	45–88			
Tanzania	3.0	64	W. Australia	<0.1	16–26
S. India	2.7–3.0 and 0.5	50	W. USA	<0.1	4–16

of $\rho_0 = 2.62$ to $2.72\,\mathrm{Mg/m^3}$, $\Delta\rho = 0.67$ (which I think is too high), $E = 10^{11}$ Pa, $\nu = 0.25$. Table 13.1 sums up their results, compared to the values found for other places.

The elastic lithosphere thickness reaches 50 to 143 km when no orogenesis has occurred for 1.2 Ga, 16 to 50 km when it was more recent. Between 3.7 and 1.2 Ga continental crust was exuding progressively from the mantle (see Section 17.13). Thus cratons of Kaapval, Slave Lake and Tanzania might have been warmed and thinned without orogenesis.

According to Doucouré *et al.* (1996), the geothermal flows show in all cases that the lower boundary of the elastic lithosphere corresponds more or less to the 600°C isotherm. The weakness of geothermal flow below cratons is not only the consequence of a very long cooling (in 1.2 Ga, cold can diffuse down to 350 km deep; see Section 9.6). In general, the highly radioactive granitic upper crust has been removed by erosion.

This 600°C isotherm is the exact level below which earthquakes disappear, below the oceans (Figure 5.8). But the elastic lithosphere of gravimetry should not be assimilated to the 'seismic lithosphere' of seismologists. To the contrary, the existence of earthquakes shows that the elastic limit has been crossed, and the model of a perfect plastic body is more appropriate. It is curious that Forsyth, who introduced this model for oceanic plates in 1979, did not account for the shallow fracturation of continental plates in 1985. It seems to me better to admit that the upper 10 to 15 km, highly faulted, have virtually no flexural rigidity and that the elastic plate begins below. This would not modify in any way the technique and its results, and would only increase the temperature at the base of the elastic lithosphere to about 730°C. And even more if using the results of Section 14.9.

At least, one is very far from lithospheric thickness values intervening in plate theory and in mantle convection currents, Jordan's *tectosphere*, which might be called the *geodynamics lithosphere*. As explained in Section 5.5, its base is around the 1,280°C isotherm, i.e. 70 km deep below the oceans, far from mid-ocean ridges, and between 200 and 300 km deep below cratons.

It has been said in Section 5.5 that the non-elastic lithosphere of geodynamics, between the 730°C and 1,208°C isotherms, has 'some fluidity'. Let us be more precise. Personally, I think that this is no permanent creep, as occurs in convection currents, but a *reversible transient creep*. Its effect is analogous to the effect of elasticity, with $\nu = \frac{1}{2}$ and μ much smaller than for seismic waves (around 80 times for ice; there are no data for peridotites or any other rock at these temperatures and pressures). The flexural rigidity due to such a pseudo-elasticity would be completely negligible.

13.5 MEASURING STRESSES IN THE CRUST

In mining or civil engineering, it is often interesting to determine the stresses in a rocky outcrop. In the immediate neighbourhood of a free surface, a principal stress is perpendicular to this surface and equals 0, and the other two are parallel to the surface and perpendicular to each other. When the direction of such a principal compression is roughly known, it can be measured by replacing this normal stress by the pressure of a fluid. Three strain gauges are glued to the rock's surface, to measure the extensions or compressions in three directions at 60° from each other. A narrow and deep groove is emptied, perpendicular to the free surface. It is then closed and some fluid is injected. Its pressure is adjusted so that the strain gauges come back to the values when they were glued.

Tectonic stresses in the crust can be measured this way, by choosing a horizontal surface at the bottom of a quarry or in a mine deep enough that the two stresses are compressions. (Elsewhere, lithostatic pressure must be added.) But the direction of the horizontal principal stresses is not known beforehand. Therefore, measures must be made along several azimuths, and the direction found is not very precise. Indeed, if the groove hollowed out is not along one of the two principal directions, a simple pressure cannot re-establish exactly the stresses and elastic strains before the hollowing. To this must be added the perturbation to general tectonic stress brought about by the presence of the quarry or the tunnel, of neighbouring joints and faults, as well as the terrain anisotropy.

Figure 13.10 shows results collected in France and Bavaria with this method.

Measurements in boreholes are more reliable. The best technique is *overcoring*. The diameter of the hole is measured precisely at different levels and along different directions. Concentric drilling from the surface isolates a hollow rock cylinder, and allows its elastic expansion before measuring again. For large depths, overcoring is very expensive. Engineers then perform a simple coring, and the elastic expansion when retrieving the core is measured. But this is less precise.

These measurements have showed large variations in tectonic stress intensity for the same drilling, depending on the depth and the nature of the layers crossed. *The maximal shear stress increases more or less linearly with depth down to 5 km at least.* The reason is the presence in rocks of faults and joints which decrease stresses. The cohesion between the sides is provided only by solid friction. According to

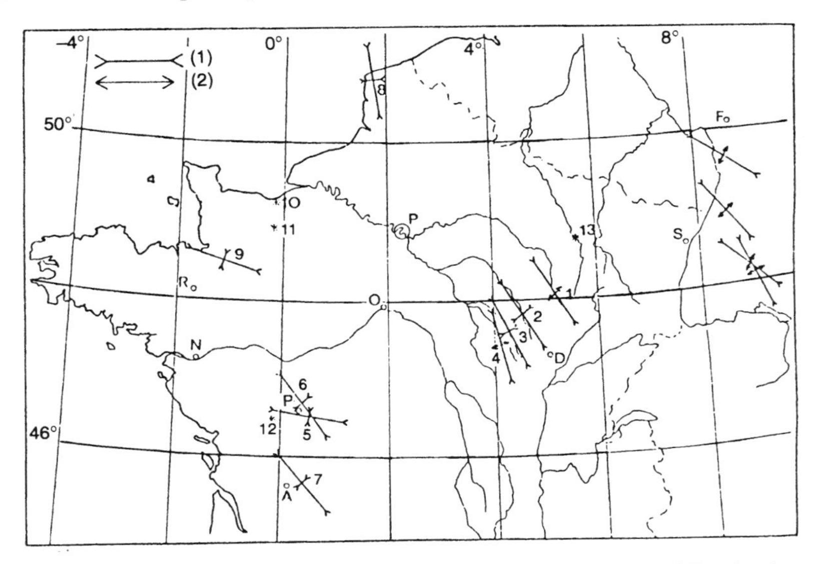

Figure 13.10. Principal stresses measured in quarries by Paquin, Froidevaux and Souriau (unpublished data). 1: compression; 2: extension (the intermediate principal stress is vertical). These stresses occur in thick Jurassic layers, except in site 9 (Armorican shield basalt). The maximal tectonic stress would be of only 1.1–2.45 MPa in Burgundy, but would reach 17.7 MPa at site 9. The azimuthal precision is variable, from $\pm 5°$ to $\pm 26°$.

Coulomb's law, it is proportional to normal pressure, and therefore more or less proportional to depth (Jamison and Cook, 1980).

This friction cannot be larger than shear in a sound rock, and the maximal shear stress must therefore stop increasing at large depths. Rock fracture also disappears, along with all permeability. The opposite had been advanced for granodioritic gneiss found 12-km deep during the Kola Peninsula scientific drilling, but microcracks had appeared in the cores extracted. Measured under *in situ* pressure, their permeability goes from 10^{-5} to 10^{-10} μm^2 (Vernik *et al.*, 1994).

Rougher estimates can be made from seismic studies, when the extent of the fault that moved is known. According to a review article of Soviet studies (Gzovsky, 1967), the maximal shear would be of ca. 10 MPa in continental platforms. It would decrease in shields (2–11 MPa in Scandinavia, 2–4 MPa in Ukraine) and would reach 100 MPa in seismic areas.

The simple hydraulic fracture of a drilling hole itself gives details about the direction of principal stresses. Similarly, the orientation of past stresses can be determined through the orientation of dykes.

13.6 DRIVING MECHANISM OF LITHOSPHERIC SHELLS

Soon after the movements of lithospheric shells were established, geophysicists investigated the possible driving mechanisms. One possible mechanism, long

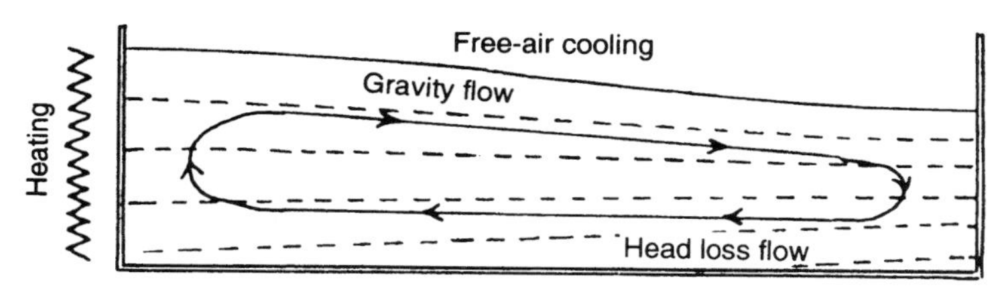

Figure 13.11. Example of a thermal convection roll, in case of side heating. The dashed lines correspond roughly to isobaric surfaces (in fact the horizontal and vertical pressures are different).

announced by geologists, is the existence of thermal convection currents in the mantle. An early model, broadly popularized but simplistic, assumes a *thermal convection roll* under each plate, driving it passively.

A simple example of a thermal roll is found in large tanks of molten glass, during its fabrication (Figure 13.11). As the liquid is highly viscous, there is no turbulence and the flow is *laminar* (which does not mean that velocities at neighbouring points are parallel). The tank is heated on one side, which dilates the liquid and raises its surface. But this rise is limited by the *gravity flow* toward the opposite side, and the liquid cools down during this flow. If there were no flow, and only a temperature difference between the two sides, the pressure from the fluid on the bottom of the tank would remain the same on both sides. Because of the flow, there is extra matter on the cool side and thus a higher pressure at the bottom. Although the increase in horizontal pressure is less than the increase in vertical pressure, there is a return current at the bottom of the tank, by *head loss*.

Mantle convection (the subject of the last chapter of this book) is very different, but the gravity flow at the surface and head loss at the bottom remain. First, mantle convection is different because heating is applied to the bottom (by the core) and inside the medium (by radioactivity and by taking some of the stored heat), and not on one side. It follows that convection only starts if heating reaches some threshold. The second difference is that lithosphere shells are not *simple thermal boundaries*, where the transverse temperature gradient is very high. Their rigidity prevents the surface horizontal velocity from increasing progressively, when moving away from an upwelling current, or from decreasing, when getting closer to a downwelling current. Surface or subducted, the plates are part of the convection and modify it considerably.

The first forces recognized as acting on the plates were (Isacks *et al.*, 1968): (1) the traction from subducted plates; (2) the drag from asthenosphere, slowing down or driving them according to the orientation of the plate/asthenosphere relative velocity. Let us look at their orders of magnitude.

Pull by a subducted plate

The computation from McKenzie (1969) will be reproduced. Let us consider a true sensible plate, of constant thickness H, and not a plate whose thickness decreases as it warms up. An adiabatic temperature gradient is assumed for the mantle. As the plate warms up through adiabatic elastic compression, there is no need to consider these adiabatic temperature variations.

The z-axis is perpendicular to the subducted plate and oriented downwards. The boundary conditions are $T(x,0) = T(x,H) = T_m$, a constant; and at the trench, $T(0,z) = T_m z/H$. The computation is similar to the one for the horizontal 'thermal plate' in Section 9.6. With the same notations, we get the equations:

$$T = T_m \left[1 - \sum_{n-1}^{+\infty} \frac{2}{n\pi} \exp\left(- \beta \frac{x}{H} \right) \sin \frac{n\pi z}{H} \right]$$
$$\approx T_m \left[1 - \frac{2}{\pi} \exp\left(- \frac{\pi^2 \kappa x}{H^2 v} \right) \sin \frac{\pi z}{H} \right] \tag{13.30}$$

This temperature is lower than the mantle temperature T_m, and makes the subducted plate denser than the asthenosphere (assimilated to a fluid in isostatic equilibrium). There follows a negative Archimedes thrust, of which only the horizontal component is interesting. By noting φ the subduction angle, this horizontal traction by the subducted plate is, per unit area and averaged over H:

$$\langle \sigma_x \rangle = \frac{1}{H} g \sin \varphi \int_0^{+\infty} dx \int_0^H dz \rho \alpha (T_m - T)$$
$$= \frac{4}{\pi^2} \rho g \alpha T_m \sin \varphi \int_0^{+\infty} \exp\left(- \frac{\pi^2 \kappa x}{H^2 v} \right) dx$$
$$= \left(\frac{4 \times 24}{\pi^4} \right) \frac{\rho g \alpha T_m H^2 v}{24 \kappa} \sin \varphi \tag{13.31}$$

Accounting for all terms, from $n = 1$ to $+\infty$, McKenzie finds that the factor in brackets, equal to 1.015, must be suppressed. With $\rho = 3.3\,\mathrm{Mg/m^3}$, $g = 10\,\mathrm{m/s^2}$, $\alpha = 3.1 \times 10^{-5}\,\mathrm{K^{-1}}$, $\kappa = 25\,\mathrm{m^2\,a^{-1}}$, $T_m = 800°\mathrm{C}$, it is found:

$$\langle \sigma_x \rangle = 340 \left(\frac{H}{50\,\mathrm{km}} \right)^2 \left(\frac{v}{100\,\mathrm{km/Ma}} \right) \sin \varphi \quad \mathrm{MPa} \tag{13.32}$$

The McKenzie model does not account for the possible drag of subducted plates by convection currents, or the possible delays in allotropic transformations at depth of olivine. The presence of an oceanic crust, with a density smaller than the mantle's by about $0.5\,\mathrm{Mg/m^3}$, is ignored. However, part of it melts and migrates upward, another must transform into granulite near 40 km deep, and in eclogite near 63 km (according to laboratory experiments by Green and Ringwood, 1967). Finally, crust melting and allotropic transformations absorb or emit heat. Clearly, equation (13.32) can only provide an order of magnitude.

Drag or driving by the asthenosphere

This force is still poorly known, as current mantle convection currents have not been realistically modelled. A shear strain rate of ca. $1\,\mathrm{Ma^{-1}}$ below the plate seems possible. Viscosity should be around the Ma MPa (see Chapter 18), so that shear stresses of the order of 1 MPa should be expected at the bottom of the plate. They cumulate over distances equal to tens of the plate thickness. It is therefore probable that this driving or drag increases or decreases the thrust from the subducted plate at the trench by 10 to 50 MPa.

Driving mechanism due to mid-ocean ridges

In 1969, I pointed out two other forces which could act on the plates. Later mid-ocean ridge exploration disproved the first one.

- The magmagenic zone (*mush zone*, see Section 5.7) and the magma pockets of mid-ocean ridges could push on mantle lithosphere. They contain fluids in hydrostatic equilibrium rising thousands of metres above abyssal plains. As altitude increases, pressure decreases less quickly than lithostatic pressure in neighbouring plates. The fluid should therefore be over-pressured compared to the plates. This reasoning is wrong, because the fluid is not in a rigid container. The plate has almost no strength at the ridge axis. Two facts prove that my idea was false: (1) we now know that ridges are covered over a large width by normal faults showing ridge-normal extension, and not compression; (2) mid-ocean ridges are locally in isostatic equilibrium; the free-air anomaly is rarely more than 10 mgal.
- There must be some gravitational flow of the whole plate over a broad swell that thermal convection requires. This same year of 1969, independently, Hales was also highlighting this mechanism. This swell (relative to the geoid) is unknown, as it is masked by the seafloor rise consecutive to thermal dilatation, much larger (see Section 9.9 and Figure 9.6). To deduce it from free-air gravity anomalies, thermal convection should be known with precision, as the variation of g results from two opposite effects. The swell creates a positive anomaly $2\pi G\rho = 0.1258$ mgal per metre, and the hot plume below creates a negative anomaly.

Let us remark that the thrust at the trench, due to gravitational flow, depends only on the swell height at the ridge (ΔZ), not on the distance from the ridge to the trench. It equals $(\rho - \rho_w)g\Delta Z$. If $\Delta Z = 500$ m, the thrust equals 10 MPa. This is most likely a maximum.

For a long time, these two effects were not well understood and confused by the scientific community. It has been written that 'plates glide over the ridge flanks' and ΔZ has been taken for the ridge height above the abyssal plains, which multiplies by 5 or 8 the above estimate. Furthermore, this thrust was improperly called the *ridge push*, as if it were the first mechanism, whereas it is the 'useful' component of the weight of the plate, increasing progressively from the ridge to the trench.

13.7 EQUILIBRIUM OF THE FORCES ACTING ON PLATES

The push from contiguous plates, or shear stress when the boundary is a transform fault, should be added to the horizontal forces acting on a lithospheric plate (any subducted part excluded). The forces acting on a plate must balance each other. As it is in fact a spherical shell and not a plane plate, it is the moment resulting from these forces about the centre of the Earth that must be zero. This gives three scalar equations, but remains insufficient to determine the different forces precisely, and they are all rough estimates.

A few speculations are allowed, however, inasmuch as the balance of forces is respected. It may be imagined that a strong convection roll in the asthenosphere causes a swell and a ridge, and later a subduction zone. The driving forces would then be the driving by a faster convection current and the useful component of weight (Figure 13.12a). Convection may decrease with time, and its driving be replaced by a braking, but the pull by the subducted plate may take its place (Figure 13.12b).

Two different developments followed these first models. In 1972, I assumed that convection was negligible in the lower mantle, so that it could be assimilated to a rigid mesosphere. Three superposed layers modelled the whole Earth: rigid shells of uniform thickness and two layers in the upper mantle, respectively ensuring the inward and outward convection currents. Four perturbations from the fixed state must be considered. They all are continuous functions over the sphere: the elevation ζ relative to the geoid, the shear τ at the bottom of the plates, and the densities ρ_1 and ρ_2 of the two asthenospheric layers (Lliboutry, 1974). At any point over the Earth, the plate velocities relative the mesosphere are linear functions of these four perturbations. From the continuity of perturbations, it is deduced that the resulting moment of all *velocities* relative to the mesosphere must be null. The lower mantle then constitutes the NNR absolute-velocity referential defined in Section 6.5.

Another more ambitious way was followed, mainly by Harper (1989). The forces acting on the different lithospheric plates were estimated separately. For example, the traction by a plate subducted into a trench is proportional to $H^3 v \sin \varphi$, according to McKenzie's formula (equation 13.31). According to the half-space

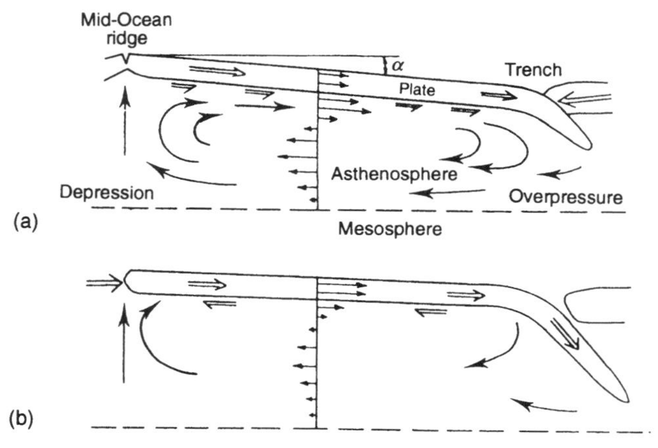

Figure 13.12. Driving forces (white arrows) acting on a lithospheric plate, in the case of a shallow and minimal convection: (a) driving by a convection roll; (b) driving by the subducted part of the plate.

model, the plate thickness H should be proportional to the square root of its age t_f. It was therefore assumed that the traction at trenches was proportional to $t_f^{3/2} v \sin \varphi$, a known quantity. The so-called 'ridge push' was assumed proportional to the height of this ridge, etc. The proportionality coefficients were adjusted so that the forces were balanced for all the plates. Despite the sophistication of the computations (using spherical harmonics), it is dubious that this approach has any predictive value, because one starts from very coarse models, and there are too many adjustable parameters. When the forces for the plate configuration of 49 Ma ago were computed using the same proportionality coefficients, the results were nonsensical.

13.8 FORCES ACTING ON THE SUBDUCTED PLATES

The main source of error in the above theories should come from the fact that the complex processes occurring in subduction zones are not well accounted for. The vertical shear force exerted by the overthrusting plate on the subducted plate, the pull due to the excess weight of the latter, and the drag from the overthrusting plate are not the only forces. Overpressure or underpressure created by asthenospheric currents on the subducted plate must also be taken into consideration.

These pressures ensure the *unbending* of a subducted plate, after its plastic folding at the trench (see end of Section 7.2 and Figure 7.8). There must be some overpressure on its lower side, which unbends it. It does not need to be large, as the lever arm is very long.

If the asthenospheric current flows next below the subducting plate, this overpressure increases the horizontal traction at the trench that was computed by McKenzie (Figure 13.13a).

If the subducted plate penetrates in a lower mantle which is poorly mobile, the trench will be 'anchored' and more or less fixed relative to the lower mantle. If the overthrusting plate, because of other forces, moves away from the trench, there results a *suction force* (Forsyth and Uyeda, 1975) (Figure 13.13b). This will be independent from intrusions of basaltic magmas or the appearance of a small mid-

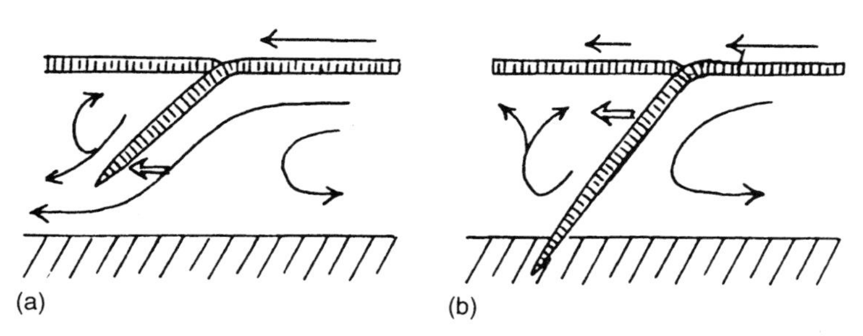

Figure 13.13. (a) Fixed overthrusting plate, the subducted plate is driven by a faster asthenospheric current below; (b) subducted plate penetrating into the lower mantle, whereas the overthrusting plate goes back. A suction force is exerted on the subducted plate. The white arrows represent forces.

oceanic ridge in the back-arc basin. These facts are the consequence, and not the cause, of the opening of the basin.

13.9 LATERAL EXTRUSIONS AND THE PERFECT PLASTIC MODEL

The forces acting at the periphery of a plate may move it laterally or rotate it. The concept of *lateral extrusion* was introduced by Tapponnier and Molnar (1976), following their satellite imagery studies of large faults in Central and Southeast Asia (Molnar and Tapponnier, 1975). To simplify, after the collision of the Indian continent with Asia 40 Ma ago, Indochina and the neighbouring regions (a block sometimes called *Sundaland*) moved southward relative to Asia, with a thousand-kilometre long sinistral strike-slip along the Red River faulted zone. Today, a sinistral strike-slip movement occurs along the Altyn Tagh fault, and there is a dextral strike-slip movement along the same Red River fault. Both are much smaller, and allow ESE extrusion of Southern China. This lateral extrusion produces EW extension and NS grabens in Tibet (Figure 13.14).

The driving force remains the SN push of India against Asia. It cannot be due to the traction by some subducted plate, which does not exist. It must imply some large and long-lived convection current. India acts on Central Asia as an engraving tool, an *indenter*.

Metal punching and stamping are computed by assimilating the ductile metal to a *perfect plastic body*. Recall that such a body can deform only when the maximal shear stress reaches a fixed value k. For a plane strain, it is possible to compute the *slip-lines*, curves tangent in any point to the plane of maximal shear stress. As there are two such directions, normal to each other, there are two perpendicular families of slip-lines. These lines obey the following theorems, which can be easily demonstrated:

- Where the body is deforming, shear stress on the slip-lines equals k. But the inverse is not true. Shear stress on the slip-lines may reach the yield strength k in domains with boundaries of fixed shape, which precludes any deformation.
- The boundaries between rigid zones (because k is not reached, or because they are confined) and plastically deforming zones, must be slip-lines. This comes from the fact there is no extension or contraction in the direction of the slip-lines.
- When some specific slip-lines are crossed, the velocity component in the direction of the slip-line may change abruptly. Physically, if this slip-line remains fixed during the deformation, relative to the medium, this is a fault with a friction k between its sides. If this discontinuity moves relative to the medium, this is the front of a *plastic wave*.

Let us apply these concepts to the **punching of a half-infinite plastic plate by a plane indenter**.

If the indenter is *perfectly rough* (no friction at its contact), the solution provided by Prandtl in 1920 is reproduced in Figure 13.15a. The indenter drags a rigid triangle

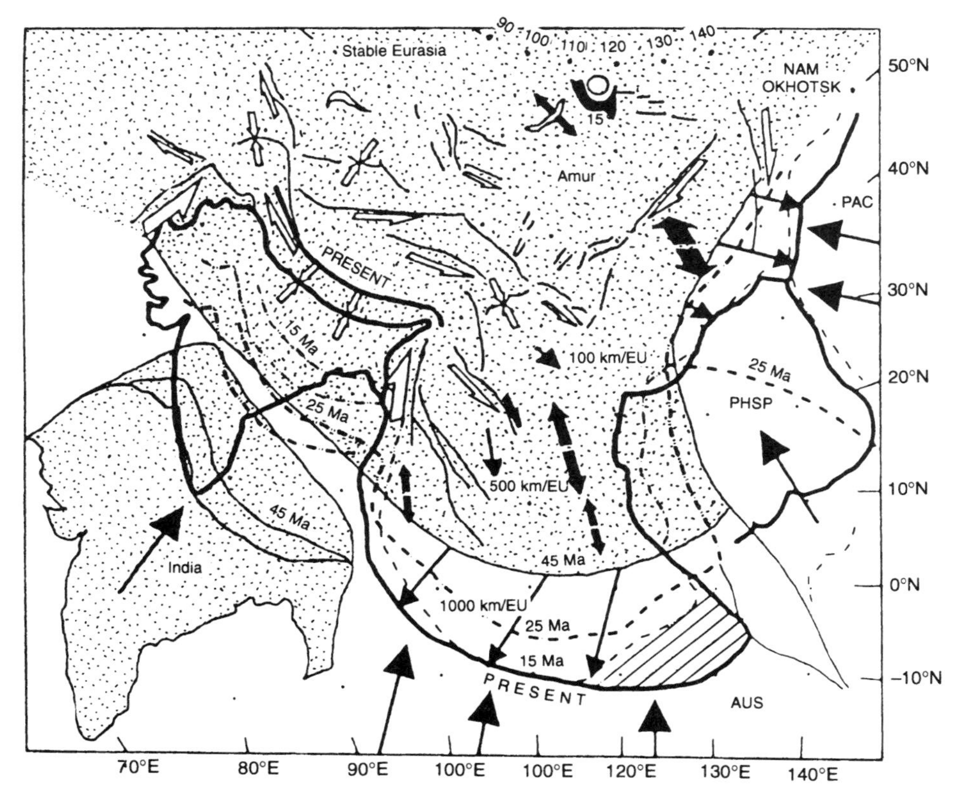

Figure 13.14. Reconstruction of the shape of Asia before India collided, and approximate displacements of its margins since 45 Ma. From Jolivet *et al.* (1994); © AGU 1994.

$AA'C$, where stresses are indeterminate. Everything takes place as if this triangle was the indenter. Two other rigid triangles, ADE and $A'D'E'$, are moved laterally. Two sectors deform plastically between these and the indenter-triangle. They are limited by two radii and one circumference which must be slip-lines.

If the indenter is *perfectly smooth* (no friction), the previous solution remains valid. But this time, the yield strength is reached everywhere in the triangle $AA'C$, and stresses can be computed. But *this solution is not the only one possible*. In particular, Hill (1980) gave another solution, shown in Figure 13.15b. The indenter drags two rigid triangles ABC and $A'BC'$ and a confined plastic zone, behaving like a rigid body, bounded by a curve (Σ) (which is not a slip-line, and of poorly known shape). Two rigid triangles ADE and $A'D'E'$ are extruded. They are twice smaller, but move twice faster than previously.

This theory proved to be disappointing when the punching of Central Asia by India was investigated, because theory assumes a medium continuous everywhere, and a gradual progression of the deforming zones. In fact, large permanent faults may appear in some places, and they modify the deformation of the whole system

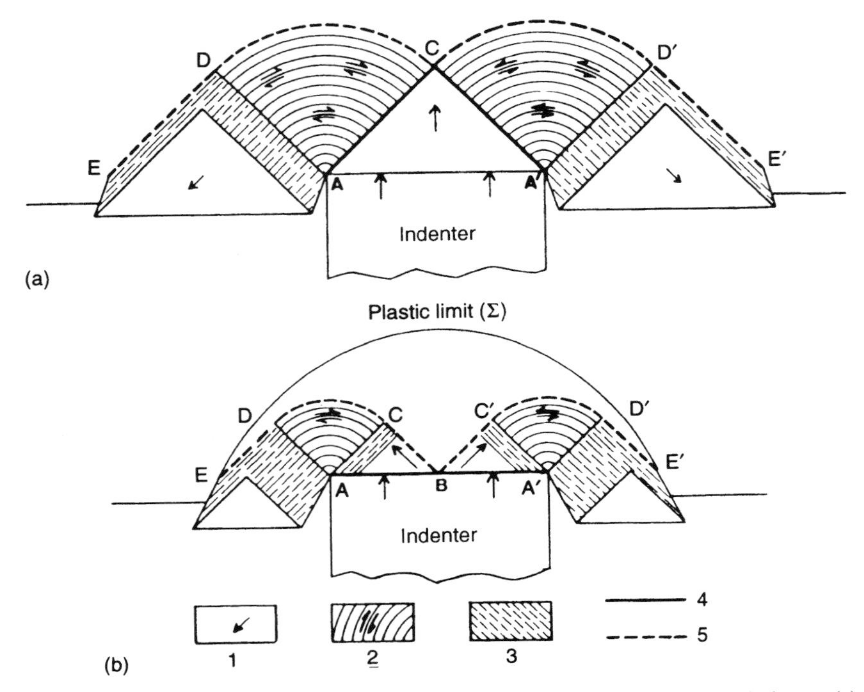

Figure 13.15. 2-D punching of a perfect plastic body, nearly infinite upward, by a rigid indenter: (a) rough indenter; (b) smooth indenter. The last symmetrical solution is found if accounting for elasticity. The numbers denote: (1) the rigid zones and their velocities; (2) the deforming zones and shear; (3) the zones currently rigid but having deformed plastically; (4) the faults; (5) the plastic wavefront, without rupture.

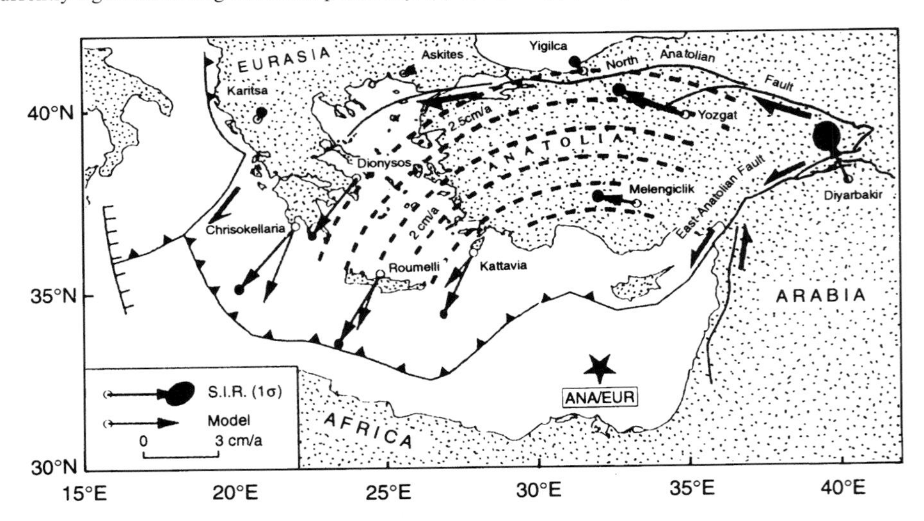

Figure 13.16. Extrusion and rotation of the Anatolian Plate, Eurasia being fixed. SLR: velocities obtained through spatial geodesy, and confidence ellipses (with a 1-s threshold). The best-fitting model of undeforming plate is indicated with a black star for its Eulerian pole. The black triangles are the front of the accretion prism. From LePichon *et al.* (1995); © AGU 1995.

(for example, the strike-slip movement changed direction on the Red River fault). The introduction of faults in numerical studies is very difficult. Lateral extrusion is better simulated with plasticine models, where permanent faults may appear, even if the stresses cannot be accessed.

Tapponnier suggested many lateral extrusions in the evolution of the Mediterranean Basin. But only one is without doubt, proved by recent spatial geodesy measurements, the extrusion of Asia Minor, or *Anatolia*. The convergence of the Arabian Plate and Europe moves it westward, and activates two diverging faults; a dextral fault in the North and a sinistral fault in the South (Figure 13.16). The first is more active, with many strong earthquakes, so that westward extrusion is accompanied by an anticlockwise rotation relative to Europe. The Aegean seafloor belongs to this continental microplate, so that rotation has formed a graben along the Macedonian coastline. (Before GPS measurements, the whole Aegean Sea was thought to stretch in the N–S direction).

14

Orogenic processes

14.1 DATING OF EVENTS

The aim of this chapter is to isolate, describe and explain the many tectonic processes which operate in the continental crust. From its formation at mid-ocean ridges to its subduction, oceanic crust knows only one single evolution process. But continental crust exists since the Archaean and has always undergone multiple tectonic processes, which generations of geologists endeavour to untangle.

Their main task is to establish the chronology of events. One must distinguish clearly:

1. *The age of sedimentary deposits*, obtained through the presence of specific fossils. The *stratigraphic scale* thus defined encompasses around one hundred *stages*, of variable duration, as a larger number of well-preserved fossils and a broader diversification of species enables a finer dating of the more recent terrains. These stages are grouped into *epochs*, which are grouped into *periods*, themselves grouped into *eras* (see the chronostratigraphic table at the end of the book). Only the names of the 15 periods of fossiliferous times (Phanerozoic) will be supposed known to the reader. The lengths of the 9 periods of the Palaeozoic and Mesozoic are comparable (34 to 80 Ma), and comparable to the whole length of the Cenozoic era: 65 Ma. Cenozoic is divided into Paleogene (three first periods) and Neogene (Miocene, Pliocene and Pleistocene, which lasted 18.6, 3.3 and 1.9 Ma respectively). Holocene or Recent, the epoch since the last Ice Age, began only 0.01 Ma ago. Geological history is known with sufficient precision to be used as a basis for geodynamics, only for Mesozoic and Cenozoic.
2. *The age of metamorphism and the age of crystalline rocks*, obtained from the ratios of radiogenic or slightly radioactive isotopes, and the laws of radioactivity (Sections 17.7–8). This is the age of the last metamorphism having affected the rock. For example, the outer crystalline massifs of the French Alps (Mont Blanc, Belledonne, Pelvoux, Argentera) come from schists which

should have deposited in the Lower Carboniferous (Kulm or Dinantian stage, between 345 and 320 Ma ago), but in part suffered metamorphism much later. Radiochronology provides ages around 350 Ma and 300 Ma for the intrusive granites of Mont Blanc. (This corresponds to the Briton and Asturian phases of the Hercynian orogeny.) The other ages provided are around 250–120 Ma, a period without orogenesis in Western Europe, and 41 Ma. The latter age shows a 'rejuvenation' of terrains, with formation of greenschist-facies minerals, simultaneous to or immediately before intense orogenesis in the Inner Alps, further east.

3. *The age of tectogenesis*, formation of folds, faults and large overthrusts. It is often dated (especially if it is very old) through the age of a metamorphism thought to be contemporaneous. The example above shows that this is not always sound. Dates based on geological reasoning are more reliable.
4. *The age of orogenesis strictly speaking*, i.e. the age of the *epeirogenic (global) uplift* which created the relief. Isostatic uplift always accompanies the crustal thickening produced by tectogenesis, and the term of 'orogenesis' is therefore commonly used to designate tectogenesis and orogenesis strictly speaking together. But it is often necessary to make a distinction between the two processes. For example, the uplift of the large Western Alps massifs only occurred in the Miocene. It started only 10 Ma after the paroxysmal tectogenesis phase, and it might therefore be argued that it results from a completely different and independent process.

The age of uplifts and subsidences is deduced from the nature of deposits, and from stratigraphy. Two facies are particularly significant:

Flysch – deep-water deposits (alternate layers of shales and claystones), enclosing *turbidite* layers, sediments enriched in coarse elements of continental or littoral origin. The latter spread out in a submarine delta during landslides, the *turbidity currents*. Flysch proves the existence of strong submarine slopes, and signals the beginning of an orogenesis.

Molasse – thick layers of sandstone, shales and marls, without regular alternation, enclosing large layers of conglomerates. Molasse shows the destruction of a relief by erosion, and signals the end of an orogenesis.

14.2 THE OLD GEOSYNCLINE CONCEPT

The *geosyncline theory* dominated geological thinking for one century. It was introduced by Hall (1859), Dana (1873) and Haug (1900). It became progressively more complicated as the diversity of mountain chains in the World was discovered. The theory found its culmination in the book '*Geosynclines*' by Auboin (1965), before being renounced by Auboin himself a few years later. Even if this theory is now completely abandoned, it must be known to understand the older geology articles. Patiently accumulated field observations, of persistent interest, are presented in these with the terms of this outdated theory.

This theory assumes a fixed chain of events during an *orogenic phase*:

1. First, a large submarine trough would form, deep (several kilometres), narrow (one or several hundreds of kilometres) and long (one or several thousands of kilometres). Sediments would accumulate there during a long period. This trough is called the *geosyncline*. The geosyncline would be divided in two along-axis zones: the *miogeosyncline*, with thick series of sediments without volcanics, deposited in deep water, and, separated by a ridge, the *eugeosyncline*, encompassing oceanic deposits, intrusive volcanics (including ophiolites, with were then thought to be such), and volcano-detrital sediments (*greywackes*).
2. In a later stage, the sides of the geosyncline would move closer, causing folding. Deformations could affect the deep old layers, the basement, and not only the recent sedimentary cover. General metamorphism and appearance of granitic plutons would also happen during this phase. This would be the tectogenesis stage.
3. Sometimes simultaneously, but more often immediately after tectogenesis, and always extending afterwards, there would be an epeirogenic uplift of the geosyncline (orogenesis strictly speaking).
4. Finally, in the post-geosyncline period, continuous deformations would stop and the whole would become brittle. Vertical faults would occur then, enabling the vertical movement of blocks and post-geosyncline volcanism.

Several orogenic phases followed each other in parallel and neighbouring geosynclines, separated by quiet periods during which erosion destroyed a large part of the recently created reliefs. The migration of tecto-orogenic activity would always happen in the direction eugeosyncline → miogeosyncline. This direction would also be the one of overthrust and large thrust sheets (*nappes*) movements. By definition, it is called the inner–outer direction in the mountain system (by analogy with the case of the Western Alps arc).

As we said in Section 1.2, these are simple regularities, often violated. There are very long basin subsidences without consecutive tectogenesis. There are folds not followed by any uplift (e.g. in the northern Pyrenees, below the Aquitaine Neogene). There are large uplifts without any tectogenesis (e.g. the Colorado Plateau). Similar orogenic phases do not always repeat themselves, and the *neotectonics* of seismologists and geodetists may differ completely from the previous tectonics.

Contrary to the exigencies of the geosyncline theory, many mountain systems formed in regions constantly emerged (e.g. Pyrenees, High Atlas). Geologists named them *intracontinental chains*, by opposition to *geosyncline chains* built on old continental margins and obeying the theory. But the Andes are a typical example of mountains built on a continent margin, and they do not show much oceanic sediment.

Plate theory, firmly established with the oceanic portions of plates, brought important clarifications:

- *There never were any eugeosynclines*. Their supposed deposits are either mixes of sediments brought toward the trench and compressed by subduction against the continent, or ophiolites resulting from obduction. The mountain chains supposed to come from these eugeosynclines are either chains built on an ocean margin, in a subduction zone with abundant volcanism (*marginal chains*), or chains built during the closing of an ocean by subduction (*collision chains*), but with the setting up of ophiolites by obduction.

- *The required transverse compressions are due to the relative movements of lithospheric shells carrying the continents.* (A century ago, geologists thought that these transverse compressions were due to the simple cooling inside the Earth.) These relative movements are due to vast convection currents in the mantle, where the energy liberated is much larger than necessary for orogenesis. Lithospheric plates are relatively decoupled from asthenosphere, and carry orogenic stresses over huge distances. There is, however, a remaining problem: if orogenic stress exists in the whole plate, why do thrusts, folds and other tectonic movements concentrate in a narrow orogenic belt? The easiest explanation is that, at the beginning, the lithospheric plate was thinner there. The first tectonic phase weakened it, and this belt long remains a privileged location for later phases.
- *Continent splitting with formation of an ocean, or union of two continents through the disappearance of the intermediate ocean, were frequent in geological history.* Therefore, one can find collision chains on the side of an ocean (Hercynian chains of Brittany and Iberia), or, conversely, marginal chains in the middle of a continent (Transhimalaya).

However, plate theory on its own did not succeed in explaining most vertical movements. They will therefore be studied in detail later.

Furthermore, plate theory, in its starting simplistic form, made believe that there always was an ocean with subduction at the margin of an orogenesis. It was therefore discussed whether, below the Alps, there had been a plate subducted northward or southward. In fact, there indeed was southward subduction of a Valaisian ocean, but only for 10–20 Ma, at the Jurassic–Cretaceous limit, much before the Western Alps orogenesis.

14.3 OROGENIC PHASES AND OROGENIES

The only important notion that remains from the geosyncline theory is the notion of tectonic phase, or orogenic phase in its largest sense. This phase is limited in time (10 Ma at most, sometimes no more than 1 Ma). This temporary activity, with a sudden start, can be understood if it is the consequence of a continental collision, but such an event is not always obvious. The lithosphere can indeed carry forces over large distances, but, for example, the main orogenic phases (*paroxysmal*) did not occur at the same time in the Hellenides–Dinarides, the Eastern Alps (east of the Upper Rhine), the Western Alps and the Pyrenees. Each time, the other regions only experienced a very attenuated repercussion. It is therefore very dubious that a same cause yield tectonic phases simultaneously in parts of the Earth very far from each other. Of course, two completely independent collisions may occur at nearly the same time, purely by chance, and dating is not accurate enough to reject a possible simultaneity.

It is surprising, however, that distinct phases can be recognized in the building of the Andes (about five, not simultaneous from one end of the Cordillera to the other), in the last 100 Ma, whereas the evident cause, the subduction of the neighbouring oceanic plate, persisted without interruption for much more than 100 Ma.

In Southern Europe, eight distinct orogenic phases can be recognized in the last 140 Ma, and each affected only part of the region. They were separated by quiet phases, or distension phases. A temporary ocean appeared in Liguria-Piedmont, and another one in Valais. There was also appearance of basins in the Western Mediterranean (see below, and the next chapter). Tectonic activity (in the broad sense: formation of troughs and basins as well as mountains) occurred almost continuously in the Mediterranean area. This is called *diastrophism*; a rather chaotic series of events, very far from the orderly succession described in the geosyncline theory.

This diastrophism period affecting Southern Europe, creating the Alps and the Mediterranean, forms the *Alpine orogeny*. The orogenic phases occurring in Europe, from Harz to Iberia, during the Carboniferous, form the *Variscan* or *Hercynian orogeny* (the adjective is derived from Harz). The phases which occurred in Northern Europe during the Upper Ordovician and the Silurian form the *Caledonian orogeny* (Caledonia is the former name of Scotland).

There is no need to extend these orogenies, defined in one region, to the whole Earth. Thus, the western North American cordillera formed during the *Nevadan orogeny* in the Jurassic and Lower Cretaceous, and the Rocky Mountains formed during the *Laramide orogeny* in the Cretaceous and Lower Cenozoic. These cycles are not phases of the Alpine orogeny, contrary to what some people wrote. It is absurd to say that the Hindu-Kush is Hercynian, when an ocean separated Europe and Asia at the time. These expressions come from a time where it was thought possible that the large system of convection cells postulated in the mantle would change globally, from time to time. These turmoils would have induced simultaneous orogenic cycles all around the Earth, separated by long quiet periods. This old idea still reappears sometimes in the articles from some geophysicists, but their theoretical speculations do not correspond to the well-established geological facts. For example, in a period believed to be quiet, between the Alpine and Hercynian orogenies, China experienced the most orogenic phases, following three successive collisions of continental blocks.

14.4 FOLDING

The layers affected by folds are said *competent*, in contrast with neighbouring layers deforming irregularly to conform to the shape of folds. When a competent layer is

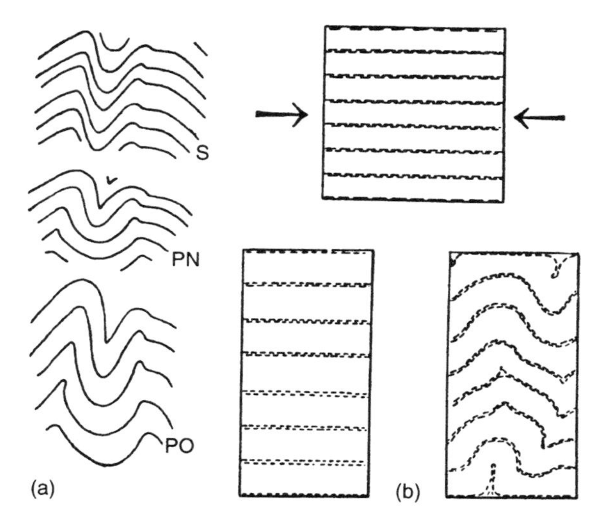

Figure 14.1. (a) Similar folds (S) and isopachous folds (PN). In a cross-section (PO), the isopachous folds do not appear isopachous; (b) The deformation of a stratified terrain by horizontal compression may produce a homogeneous thickening of the layers or isopachous folds through local sliding at the interfaces. From Goguel, 1952.

thin, it usually keeps a uniform thickness after folding. Such a fold is called *concentric* or *isopachous*. In a cross-section, an isopachous fold is limited by parallel, non-identical curves (Figure 14.1). If several competent layers are superposed, their isopachous folding can only occur through local sliding at the interfaces (*bedding-plane slip*).

In detail, the strain may concentrate at the hinges of folds and produce fractures, in extension on the external face and in compression on the internal face (Figure 14.2a). Deformation may conversely concentrate on the sides of the folds, and reduce their thickness. The fold becomes *similar*.

Contrary to earlier ideas, folds do not always form at great depths, where high lithostatic pressures makes breaking difficult, and high temperatures make creep possible. The frequent absence of any metamorphism is a confirmation. The process is more often pressure solution creep (see Section 11.2). Water soaks the rock and is able to carry calcite or silica over very short distances. The rock then behaves like a Newtonian viscous body. The same sedimentary rock may contain incompetent and competent layers, because of their diverse porosities and grain size.

Gravity impedes the folds to rise much higher than the surface; the higher-amplitude ones become recumbent. A high-amplitude fold can reach a level where faulting becomes possible. All intermediates are possible between the *recumbent fold*, the *fold fault* and the *overthrust* (Figure 14.3).

A second type of folds may appear during large overthrusts on a subhorizontal fault plane, if it includes a ramp: *kink folds*. In this case, the length of the ramp determines the fold geometry (Figure 14.4).

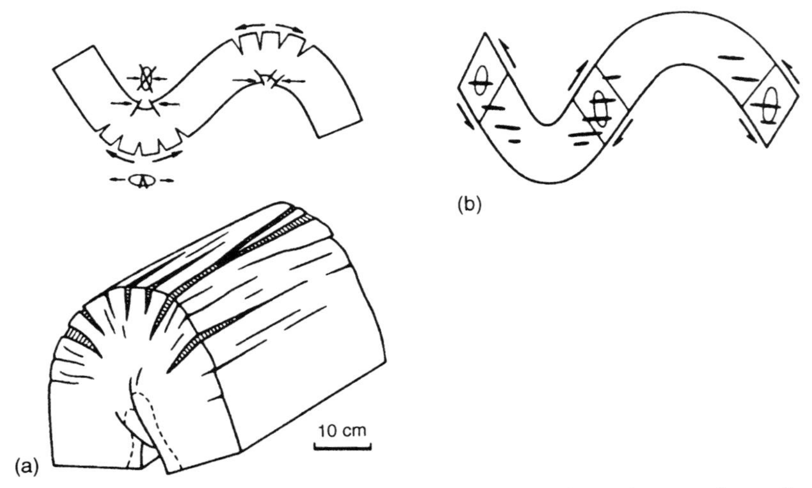

Figure 14.2. (a) Cracks appearing at the hinges of isopachous folds; (b) cracks appearing at the sides of similar folds. From Mattauer, 1973.

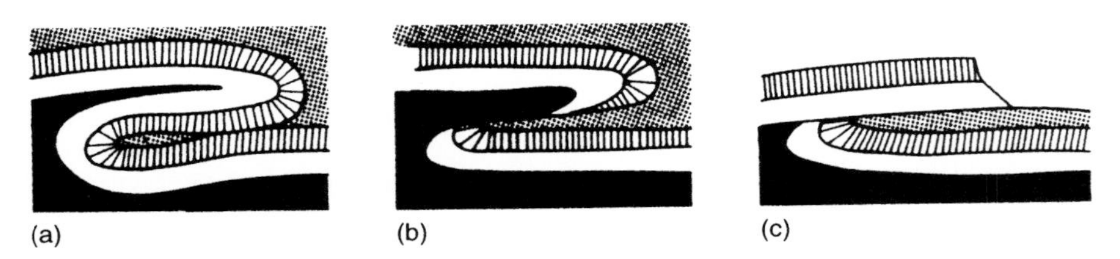

Figure 14.3. Cross-sections of a recumbent fold (a), a fold fault (b) and an overthrust (c).

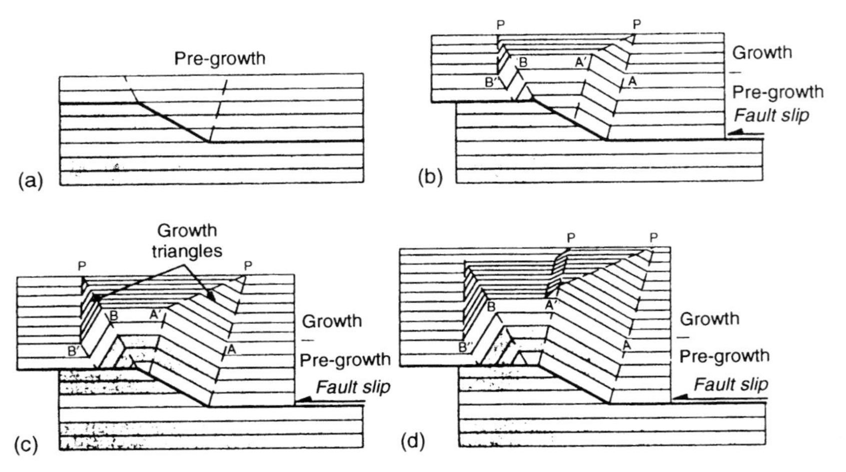

Figure 14.4. Successive stages in the formation of a kink fold, with continuing sedimentation. From Shaw and Suppe (1996); © AGU 1996.

14.5 THEORY OF FOLDING INITIATION

Biot (1961) suggested a basic theory of folding with the following model:

- The competent layer is unique and its thickness h is very small compared to the wavelength λ of the appearing folds.
- The layers on each side of the competent layer are isoviscous layers with the same density, the same viscosity η_1, and thickness very large compared to λ so that they can be considered as half-spaces. The folds modelled therefore appear at depth.
- The amplitude a of the folds is small compared to λ. *One studies only the appearance of the folds*, the instability of transversal compression.

Because the incompetent layer is very far from the surface, the surface does not deform. Thus, lithostatic pressure does not change, and can be removed from the equation; gravity is not accounted for.

Competent elastic bed

We established in Section 13.1 the equilibrium equation for the profile $z(x)$ of a thin elastic layer. It reads:

$$\begin{cases} \Gamma \dfrac{d^4\zeta}{dx^4} + \sigma_z|_{z=0} = 0 \\ \Gamma = \dfrac{Eh^3}{12(1-\nu^2)} = \dfrac{\mu h^3}{6(1-\nu)} \end{cases} \tag{14.1}$$

But this formula was established without longitudinal stress along the competent layer. Let us assume that a *compression* σ can be transmitted, i.e. a force σh per unit width. At the two extremities of a slice dx, the two forces σh are not directly opposite. The small angle between them is $(d^2\zeta/dx^2)dx$. The resultant per unit area is $\sigma h d^2\zeta/dx^2$; it adds to the pressure or traction from the viscous layers, to counteract the flexural rigidity. Equation (14.1) must then be replaced by:

$$\Gamma \frac{d^4\zeta}{dx^4} + \sigma_z|_{z=0} + \sigma h \frac{d^2\zeta}{dx^2} = 0 \tag{14.2}$$

The Fourier transform of this pressure or traction σ_z from a viscous layer on to the competent layer is computed with equation (10.47), where $\nu = 1/2$ and $\mu = \eta_1 d/dt$. As the medium is half-infinite, $C = D = 0$. Therefore:

$$\begin{cases} \bar{\zeta} = \bar{U}_z|_{z=0} = A \\ \bar{\sigma}_z|_{z=0} = 2\omega\eta_1 \dfrac{dA}{dt} \end{cases} \tag{14.3}$$

There are two layers, one on each side of the competent layer, and this last value must be doubled. The Fourier transform $\bar{\zeta}(\omega)$ finally obeys the equation:

$$\frac{\mu h^3}{6(1-\nu)}\omega^4\bar{\zeta} - \sigma h\omega^2\bar{\zeta} + 4\omega\eta_1\frac{d\bar{\zeta}}{dt} = 0$$

$$\frac{d\bar{\zeta}}{dt} = \left[\sigma\omega h - \frac{\mu(\omega h)^3}{6(1-\nu)}\right]\frac{\bar{\zeta}}{4\eta_1} = \frac{\bar{\zeta}}{T(\omega)} \tag{14.4}$$

As far as the expression between brackets is positive, i.e. for ω small enough, $\bar{\zeta}$ will increase exponentially with time. For all minute sine perturbations always present in an initial profile $\zeta(x)$, the perturbation developing the fastest and dominating in the end is the one that minimizes $T(\omega)$, and maximizes the expression in brackets. This occurs when:

$$\begin{cases} (\omega h)^2 = 2(1-\nu)\dfrac{\sigma}{\mu} \\ \lambda = \pi h\sqrt{\dfrac{2\mu}{(1-\nu)\sigma}} \\ T = \dfrac{3\eta_1}{\sigma^{3/2}}\sqrt{\dfrac{2\mu}{(1-\nu)}} \end{cases} \tag{14.5}$$

As we assumed the folds were not very pronounced ($a/\lambda \ll 1$), the horizontal strain $\varepsilon_{xx} \approx -a^2\omega^2/4$ is very small. So is the corresponding stress in the viscous layers. In this model, the elastic competent layer transmits most of the stress. The tectonic force σ could be deduced from the measurement of λ/h, but this elastic model seems very unrealistic.

Viscous competent bed

Instead of an elastic competent layer, let us assume it is isoviscous, with a viscosity η much larger than η_1. The layer is compressed horizontally with the same rate ε_{xx} than the layers above and below, with viscosity η_1. For this plane problem, the corresponding deviatoric stress is $(\sigma_x - \sigma_z)/2$, or, as gravity is not accounted for, $\sigma_x/2$. The horizontal tectonic stresses (σ in the competent layer and σ_1 elsewhere) verify:

$$\sigma = 4\eta|\dot{\varepsilon}_{xx}|; \qquad \sigma_1 = 4\eta_1|\dot{\varepsilon}_{xx}| \tag{14.6}$$

The global horizontal stress increases the amplitudes of any sinuosity of the layer, at a rate inversely proportional to horizontal contraction. It can therefore be said that this is the effect of σ_1 on the incompetent layers, and subtract this solution from our problem. To calculate only the folding by instability, we must account only for an additional compression $(\sigma - \sigma_1)$ transmitted by the competent layer.

Let us apply the correspondence principle to the solution for the elastic competent layer. Equation (14.4) becomes:

$$\begin{cases} \dfrac{\eta h^3}{3}\omega^4 \dfrac{d\bar{\zeta}}{dt} - (\sigma - \sigma_1)h\omega^2\bar{\zeta} + 4\omega\eta_1 \dfrac{d\bar{\zeta}}{dt} = 0 \\ \dfrac{d\bar{\zeta}}{dt} = \dfrac{4|\dot{\varepsilon}_{xx}|(\eta - \eta_1)}{\dfrac{\eta(\omega h)^2}{3} + \dfrac{4\eta_1}{\omega h}}\bar{\zeta} = \dfrac{\bar{\zeta}}{T(\omega)} \end{cases} \tag{14.7}$$

We see that folds develop exponentially for any value of ω (in the elastic layer case, this was only for the smallest values of ω). The wavelength minimizing $T(\omega)$ will always dominate:

$$\begin{cases} (\omega h)^3 = 6\eta_1/\eta \\ \lambda = 2\pi h\left(\dfrac{\eta}{6\eta_1}\right)^{1/3}; \qquad T = \dfrac{\eta^{1/3}(6\eta_1)^{2/3}}{4|\dot{\varepsilon}_{xx}|(\eta - \eta_1)} \end{cases} \tag{14.8}$$

With deformations due to the pressure solution process, viscosity is inversely proportional to permeability, and the ratio η/η_1 may be very high. For example, if $\eta/\eta_1 = 100$, $\lambda = 53h$ and $1/T = 26|\varepsilon_{xx}|$.

14.6 GRAVITATIONAL NAPPES

Stratigraphy usually distinguishes between the *cover* of sedimentary rocks, or sometimes volcanics, lying in discordance over a metamorphic terrain, affected by previous orogeneses but later eroded and transformed into a peneplain, the *basement*. In Western Europe, the basement corresponds in general to all terrains older than the Hercynian orogeneses, i.e. before the Carboniferous. For instance, in the middle of the Western Alps, this basement is masked by a very thick cover from the Carboniferous and Permian, almost devoid of metamorphism. But in the outer (west) region, there are decollements along beds of the Upper Triassic (Keuper). The basement then encompasses all terrains older than the Keuper.

An *overthrust* is any large horizontal displacement (10 to 200 km), which brought a set of *allochthonous* terrains over the *autochthonous* terrains. The overthrust may affect only the cover, or the basement as well. In the latter case, the fault where decollement occurs cannot respect the stratigraphy. It is more proper to talk of a *thrust sheet*, or a *slab*. In a genuine *nappe*, the decollement occurs at a stratigraphic level with a plastic layer of clay, marl, gypsum or rock salt, which acts as a lubricant.

Sometimes, close to the front, the nappe looks like a recumbent fold, and the normal stratigraphic series is duplicated below by the same series in the inverse order. This is the case of the *Pennine Nappes* in the Alps. But the old idea that nappes would be huge folds formed at depth and become recumbent is totally wrong: they would have undergone metamorphism, which is not the case.

The nappe thickness is variable, from a few kilometres to a few hundreds of metres only for the Helminthoïdes flysch nappe. The *ultra-Helvetic* nappes, at the bottom of the piling of Alpine sheets, are themselves a pile of sheets hundreds of metres thick each (this process is called *branching*). During their progression, the nappes were often highly stretched longitudinally, or even broke into segments. All these facts exclude the setting of nappes by some push from behind. It can only be some *gravitational gliding* (Gignoux, 1948).

Let us look at the magnitudes involved. Considering the length of a nappe, the slope $\tan\alpha$ over which it slid should not be larger than 1/20. If h is its thickness and ρ its density, when there is no thrust from the back or traction from the front, the bottom shear stress (the friction over the bed) is $\rho g \sin\alpha$ (see Section 12.7 and Figure 12.7). One finds its value should be 0.5 to 5 MPa. The sheet slid over a ductile layer, at a velocity of ca. 10 km/Ma. If this layer is 10 m thick, its shear strain rate must have been of around 1,000 Ma^{-1}, and therefore its viscosity should have been around 10^{-3} Ma MPa, which is very small for a rock.

The lubricating layer must have behaved like a ductile fault, that localization processes may have transformed into a real fault in places. Anyway, the sliding is conceivable only if there was *interstitial water under pressure*, and remaining under pressure despite the dilatancy accompanying deformation.

14.7 LARGE OVERTHRUSTS OF THE BASEMENT

In a continental collision, the shortening of the crushed zone reaches several hundred kilometres. Folds and inverse faults can only be a modest contribution. Mostly, there is overlapping and *imbrication* of large thrust sheets, limited by subhorizontal faults. Sometimes, there is at depth *wedging* of a portion of crust or upper mantle.

The first recognized case of such an overthrust (in 1863) was found in the mines of the French–Belgian coal field. It dates back to the Hercynian orogenesis. The Devonian red sandstones were carried over the Carboniferous coal terrains.

The second recognized case was in the Eastern Alps. Its central part is mainly made of a large slab of crystalline basement, the *Silvretta-Oetztal thrust sheet*, 3–5 km thick and carried northwards over sedimentary series from the Mesozoic. The latter are visible in the 'windows' of Lower Engadine and Hohe Tauern, opened by erosion. This overthrust occurred in the Middle and Upper Cretaceous, in three phases (110, 80 and 67 Ma), in response to getting closer to Africa and Europe (at that time, the African continent extended up to this region).

Himalaya is the most grandiose case of collision. Its Nepalese part has been studied for a long time, thanks to the deep valleys cutting the chain. Three large subhorizontal faults, emerging in the south, have moved successively. The Main Central Thrust first made the Precambrian and Palaeozoic basement overlap by 100–200 km a folded marine series from the Upper Jurassic and Lower Cretaceous. This upper slab forms the Upper Himalaya (some high summits, however, are more recent intrusive granites). Further south, and later, the Lower Himalaya overthrust the Indian continent (Main Boundary Thrust). Finally, in the Garwhal, a third overthrust further south made molasse from the Upper Cenozoic overlap that from the Lower Cenozoic, forming the Siwalik Range (Figure 13.5, left side).

The structural study of Tibet only started with the French–Chinese geophysical expedition of 1980–81. The very high altitude of this large plateau (ca. 5,000 m) had given rise to much speculation. For Argand, the first geologist to account for continental drift in his vision of Asia, the Indian continent crust would have slid below the crust of Tibet, doubling its thickness. For Dewey and Burke (1973), Tibet would have compressed homogeneously in the N–S direction (and there would be no real lithosphere in this region). The investigations of Chinese geologists and a large deep-seismics expedition (INDEPTH programme) showed that there were in fact a huge quantity of overthrusts, which thickened the crust in a very irregular way.

Figure 14.5 shows a N–S cross-section of the southern part of Tibet. It still belongs to the Indian continent, up to the suture, well marked by the *Indus-Yarlung* belt of ophiolites (or *Indus-Zangbo*: Yarlung and Zangbo respectively are the Chinese and Tibetan names of the Upper Brahmaputra). This suture was made between India and the *Lhassa block* (rightmost in the figure). This is continental passive margin of India, a continental platform over which sediments were deposited during the whole Mesozoic. Flysch at the end of the series indicates that the collision started around 60–50 Ma, and molasse indicates the beginning of the Himalayan chain destruction.

On the southern margin of the Lhassa block, the *Gangdise Chain* (formerly called *Transhimalaya* by European explorers and culminating at 7,096 m) shows uninterrupted magmatism between 110 and 40 Ma. This indicates that the subduction of the ocean between India and the Lhassa block (called Eastern Tethys[1] by French geologists) was happening northward, below the Lhassa block.

The top of Figure 14.5 shows a palinspastic reconstruction (from the Greek words *palin*, walk back, and *spân*, stretch). This is in fact a restoration of the horizontal layers before overthrust (or folding). One must start from a *balanced* cross-section in which the layers are given their original thickness. This reconstruction shows that the cross-contraction changed the width of this strip from 270 km to 131 km.

Of course, all intermediates exist, between the large Himalayan overthrusts and the limited movement of many reverse faults. For example, further west, the mere movement of reverse faults (indicated by a strong seismicity) in the Zagros chain, south of the Iran plateau, and in the Alborz chain in the north, allow the Arabian plate to move toward Eurasia at the considerable velocity of 35 mm/year. There is currently no subduction below any of these chains.

In overthrusting, the fault moves against gravity, and must stop when the overlying slab has become high enough. One can wonder why, when tectonic compression stops, gravity does not induce sliding in the opposite direction, the fault becoming normal. This is because, even after the end of tectogenesis, there remains a transversal compression, because the sides of the fault weld together, and erosion and a process studied in Section 14.11 decreases the height of the slab. The

[1] The term of Tethys is clarified in Section 15.4. To avoid any confusion and be clearer, it is better to call this palaeo-ocean the *Himalayan Ocean*.

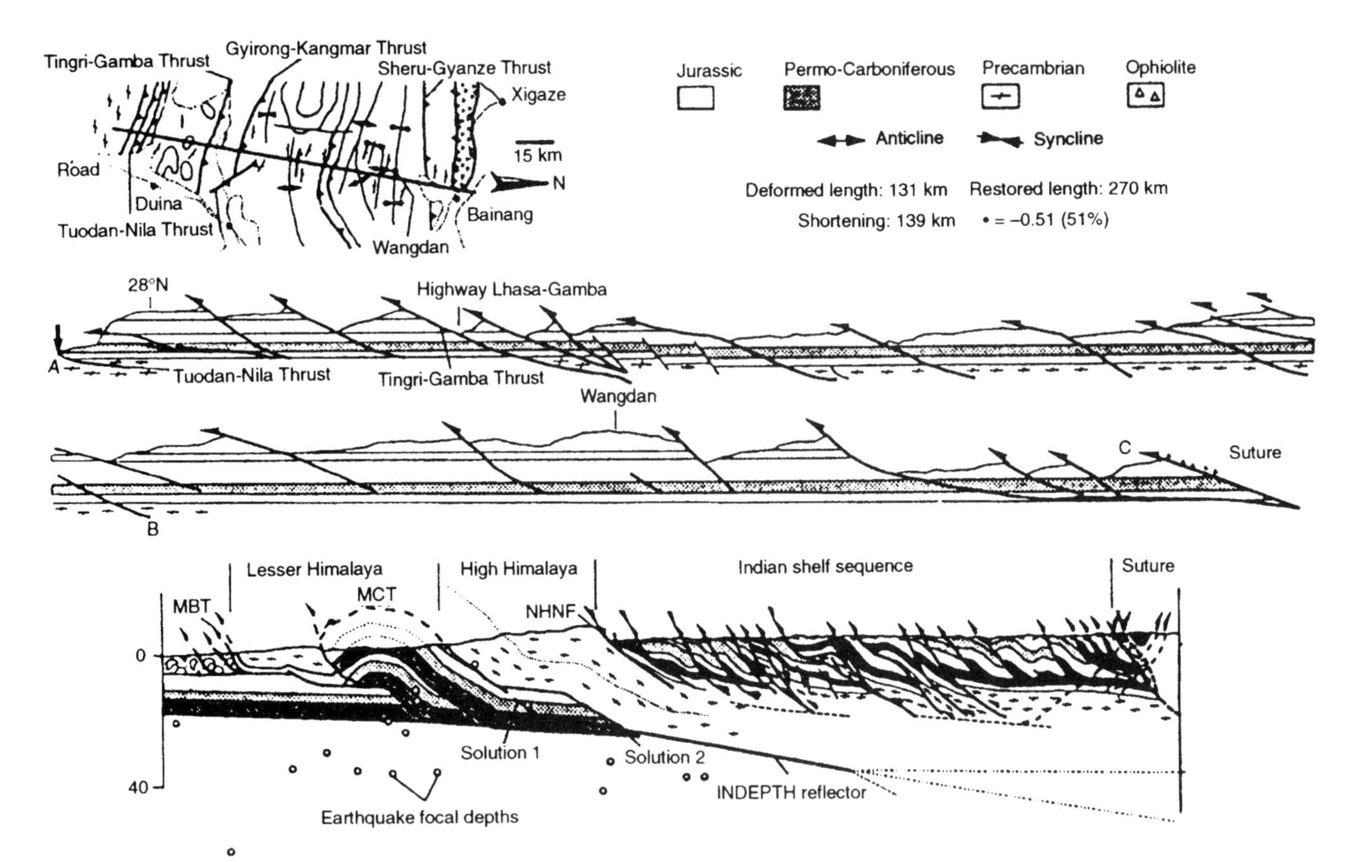

Figure 14.5. (bottom) Cross-section of Himalaya and South Tibet, south of the Zangbo suture between the Indian continent and the Lhassa block, made from the map by the Xizang Geological Survey and other Chinese studies. (top) Balanced reconstruction, before the continental collision, at the same scale. From Ratschbacher *et al.* (1994); © AGU 1994.

fault remains a weak point of the crust, and can move again during another orogenic phase.

14.8 WARMING OF A PLATE SUDDENLY SHORTENED HORIZONTALLY

Tectogenesis (by folding and overthrusting) contracts a plate horizontally; therefore it thickens it in the vertical direction. Its bottom is relatively rapidly immersed into a hotter asthenosphere. It warms very slowly, which makes it less dense and should make it emerge by isostasy. This effect was supposed to be the cause of the post-orogenic uplift predicted by the geosyncline theory. (In fact, we saw that the main uplift of the Western Alps, in the Miocene, took place 10 Ma later).

The following simple calculation seems never to have been published. It shows that this emergence by warming is insignificant in any case, and smaller than the erosion rate.

Let us note H the plate thickness *after* tectogenesis, at time $t = 0$. The z-axis pointing downward, with the origin at the surface, let us note $T(z, t)$ the (negative) temperature elevation relative to $T_0(z)$, the temperature back to normal and reached for $t = \infty$ (Figure 14.6). If ρ_m is the density of the mantle (lithosphere or asthenosphere), ρ_c the density of the upper crust, and α the coefficient of volume dilatation, the uplift Δh at time t is given by isostasy:

$$\rho_m \alpha \int_0^H [T(z,t) - T(z,0)]\, dz = \rho_c\, \Delta h \tag{14.9}$$

If γ is the geothermal gradient at the bottom of the lithosphere, and H_0 its thickness before tectogenesis, $T(H,0) = -\gamma(H - H_0)$, and it can be admitted that, even if vertical stretching is not totally uniform:

$$T(z,0) = \frac{z}{H} T(H,0) = -az \tag{14.10}$$

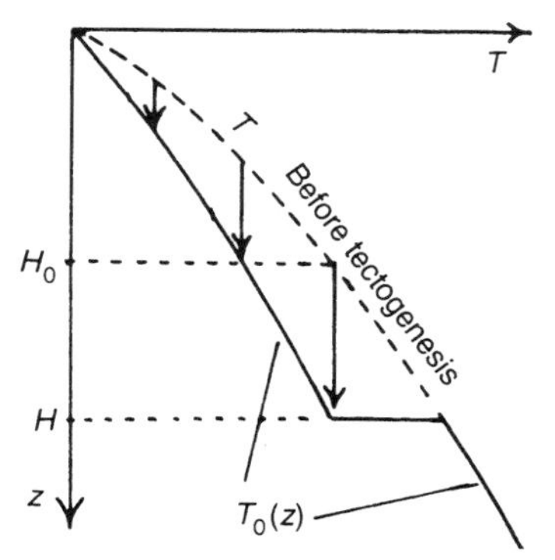

Figure 14.6. Temperature profile before and just after a tectogenesis, bringing the plate thickness from H_0 to H.

Heat conservation during tectogenesis requires:

$$\rho C \int_0^H \gamma z\, dz = \rho C \int_0^H (\gamma - a)\, z\, dz \implies a = \gamma\left(1 - \frac{H_0^2}{H^2}\right) \tag{14.11}$$

Relative to the normal state, a negative amount of heat $C\rho T(z', 0) = -C\rho a z'$ has been introduced at time $t = 0$ in each infinitesimal layer $(z', z' + dz')$ of the plate $(0 < z' < H)$. This cold diffuses downward as well as upward, following the law expressed in equation (9.36). The heat equation is linear, and the different inputs can be summed (i.e. $T(z, 0)$ and the function of equation (9.36) can be convoluted):

$$\begin{aligned}
T(z,t) &= \int_0^H \frac{-az'dz'}{2\sqrt{\pi\kappa t}} \exp\left[-\frac{(z'-z)^2}{4\kappa t}\right] \\
&= -\frac{a}{\sqrt{\pi}} \int_{-z/2\sqrt{\kappa t}}^{(H-z)/2\sqrt{\kappa t}} (z + 2\sqrt{\kappa t}\, u) e^{-u^2}\, du \\
&= -\frac{az}{2}\left[erf\left(\frac{H-z}{2\sqrt{\kappa t}}\right) + \text{erf}\,\frac{z}{2\sqrt{\kappa t}}\right] \\
&\quad + a\sqrt{\frac{\kappa t}{\pi}}\left[\exp\left(-\frac{(H-z)^2}{4\kappa t}\right) - \exp\left(\frac{-z^2}{4\kappa t}\right)\right]
\end{aligned}$$

$$\begin{aligned}
T(z,t) - T(z,0) &= T(z,t) + az \\
&= \frac{aH}{2} erfc\left(\frac{H-z}{2\sqrt{\kappa t}}\right) + \frac{a}{2}\left[z\, erfc\frac{z}{2\sqrt{\kappa t}} - (H-z)\, erfc\frac{H-z}{2\sqrt{\kappa t}}\right] \\
&\quad + a\sqrt{\frac{\kappa t}{\pi}}\left[\exp\left(-\frac{(H-z)^2}{4\kappa t}\right) - \exp\left(\frac{-z^2}{4\kappa t}\right)\right]
\end{aligned} \tag{14.12}$$

Integrating from 0 to H, the integral of the first term is the only one remaining:

$$\int_0^H [T(z,t) - T(z,0)]\, dz = aH\sqrt{\frac{\kappa t}{\pi}}\left[1 - \exp\left(\frac{-H^2}{4\kappa t}\right)\right] + \frac{aH^2}{2}\, erfc\left(\frac{H}{2\sqrt{\kappa t}}\right) \tag{14.13}$$

As H equals ca. 100 km, and $\kappa \approx 25\,\text{km}^2\,\text{Ma}^{-1}$, only the first term is important if $t < 50$ Ma. Bringing this value into equation (14.10), we get:

$$\frac{\Delta h}{H} \approx \frac{\rho_m}{\rho_c} \alpha\gamma\left(1 - \frac{H_0^2}{H^2}\right)\sqrt{\frac{\kappa t}{\pi}} \qquad (t \ll H^2/4\kappa) \tag{14.14}$$

With $\rho_m/\rho_c = 3.32/2.7$, $\alpha = 3 \times 10^{-5}\,\text{K}^{-1}$, $\gamma = 9\,\text{K km}^{-1}$ and $\kappa = 25\,\text{km}^2\,\text{Ma}^{-1}$, it is found (the value of t being expressed in Ma):

$$\frac{\Delta h}{H} \approx 0.94 \times 10^{-3}\left(1 - \frac{H_0^2}{H^2}\right) \tag{14.15}$$

Even if the bottom of the lithosphere has gone down by 50 km, its warming will only lift the ground up by 70 m in 1 Ma, and 314 m in 20 Ma. This is lower than the erosion rate by one order of magnitude at least.

14.9 MECHANICAL PROPERTIES OF CONTINENTAL LITHOSPHERE

The existence of earthquake foci down to 12 km shows that the upper continental crust is brittle. Numerous hydraulic fractures in oil drillings have been used to determine the fracture limit, the Mohr envelope (see Section 12.4). Since the slope of this envelope is small, the shear stress on the fracture plane can be taken as $T_M = (\sigma_3 - \sigma_1)/2$, and the pressure normal to this plane (N) can be confused with the lithostatic pressure $p = \rho g z$, with $\rho g \approx 27\,\mathrm{MPa/km}$.

The results vary a lot from one borehole to the next, mainly because of joints and faults in the rock. The fracture limits found in the laboratory with sound rocks (200 to 250 MPa; see Section 12.8) are rarely found. According to McGarr (1980), the average values for hard rocks (granite, quartzite) are:

$$T_M = 5.0 + 6.6\,z = 5.0 + 0.244\,p\ (\mathrm{MPa}, z \text{ in km}) \tag{14.16}$$

The two coefficients are smaller for soft rocks (schists, sandstones).

In geodynamics, on scales of a kilometre or more, fractures always are pre-existent in the upper 5 km, and the apparent fracture limit is in fact friction between the sides of faults or joints. This friction (τ) depends on the *effective pressure* (p_e), the pressure between the sides minus the water interstitial pressure. According to Byerlee:

$$\begin{cases} \tau = 0.85\,p_e & (p_e < 240\,\mathrm{MPa}) \\ \tau = 60 + 0.6\,p_e & (240 < p_e < 1700\,\mathrm{MPa}) \end{cases} \tag{14.17}$$

(This excludes some very ductile layers and pressure solution creep, mentioned earlier.)

Further down, at higher temperatures, brittle creep and permanent dislocation creep appear. Often, some minerals undergo creep, some remain rigid. Among the many laboratory measurements, only those made with marble, granite, granodiorite or diorite can give an idea of intermediate crust viscosity. For the lower crust, the only data come from pyroxenites and granulites.

For the upper mantle, peridotite data are quite abundant. Data from water-soaked rocks are the most pertinent for the upper crust, whereas dry rock measurements are more pertinent for the upper mantle. This mantle exuded basaltic magmas in the past, and it has been shown that magmas catch nearly all the water present.

It is doubtful, however, that laboratory data, obtained with strain rates of 10^{-5} to $10^{-4}\,\mathrm{s}^{-1}$ (i.e. 3×10^8 to $3 \times 10^9\,\mathrm{Ma}^{-1}$), can be extrapolated to strain rates of the order of $1\,\mathrm{Ma}^{-1}$, as may exist in continental lithosphere. Maps of controlling creep mechanisms according to temperature and maximal shear stress (or strain rate) were established to decide whether this extrapolation was valid. But these maps do not include essential factors, such as recrystallization and mainly metamorphism, which

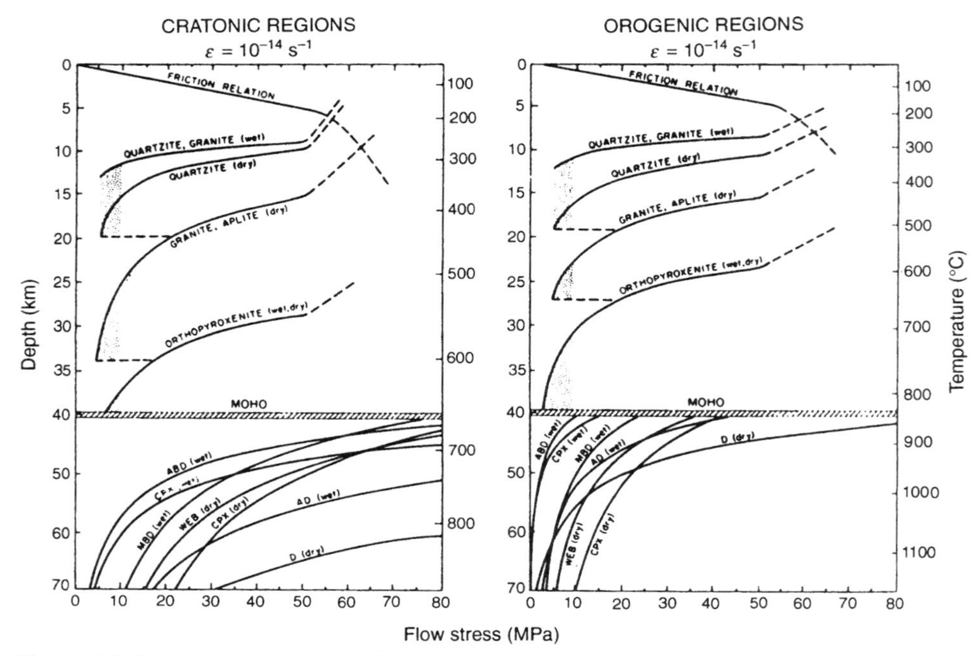

Figure 14.7. Stress $(\sigma_3 - \sigma_1)$ inducing a permanent strain rate ε_3 in crustal or upper mantle rocks (clinopyroxenite CPX, dunite D and other peridotites), dry or wet. The temperatures are the ones commonly found in cratons (left) or in orogens (right). The upper curve corresponds to deformation by sliding with friction on faults, conforming to Byerlee's law. From Carter and Tsenn, 1987; © Elsevier.

increase fluidity. The results of such a study, made by Carter and Tsenn in 1987, are however reproduced in Figure 14.7.

All these studies show that the lower continental crust must be able to creep at a significant rate, when subject to deviatoric stresses not high enough to make either the upper and intermediate crust or the mantle lithosphere creep significantly. The lower continental crust is not only *ductile*, which means that permanent deformations are easy. It does not 'harden' following such deformation, and shows *fluidity*. Continental lithosphere can therefore be assimilated to a sandwich of two rigid plates (elastic, in fact), separated by what I suggest calling a *crustal asthenosphere*. It was recently realized that this model should be adopted to account for many geodynamical processes. The inverse problem, adjusting thickness and viscosity for a quantitative fit, is currently a major domain of research.

14.10 CONSEQUENCES OF THE EXISTENCE OF A CRUSTAL ASTHENOSPHERE

Crustal asthenosphere exists only in continents (and it is not proved that it exists everywhere); this does not change what was said about the relative displacements of large lithosphere shells. Only limited and local displacements of the upper crust relative to the mantle lithosphere are possible. One cannot drift away from the other. It is, however, possible to envisage the rotation of a small massif, of diameter smaller than the lithosphere thickness; only the upper crust would turn. It was absurd to assume the rotation of some microplate, a cylinder whose height would be greater than its diameter. (Thus, palaeomagnetism showed that, before the Permian, the Pelvoux massif, 30 km in diameter, turned 30°–35° anticlockwise, relative to Europe.)

The calculation of the elastic thickness of a plate from its flexural rigidity should be reviewed, and temperature estimates at the base of the elastic part should be modified.

A crustal asthenosphere is probably too thin for its extrusion out of a loaded zone or its intrusion into an unloaded zone to be noticeable over a few thousand years. Furthermore, the density contrast between upper and lower crust is very small. The crustal asthenosphere should not play any significant role in isostatic readjustments. But, at a scale of a million years, matter fluxes in the crustal asthenosphere may become important.

The best model to date looks at the centripetal flow of crustal asthenosphere in a region thinned by stretching, the Halloran Hills, east of the Mojave Desert in California (USA) (Kaufman and Royden, 1994). Stretching ended around 9 ± 2 Ma. The progressive return to a crustal asthenosphere of normal thickness tilted the sedimentary deposits in the lakes posterior to stretching, which means that we know the thickness variation with time.

Kaufman and Royden (1994) started by giving the crustal asthenosphere a 3rd power law viscosity, depending on temperature, like the Westerly granite (Section 11.3). The numerical solution shows that the flow is significant only in a lower layer some 10-km thick. They adopted a second model, with an *isoviscous* crustal asthenosphere, with the thickness $H = 10$ km in the normal state. This makes it possible to obtain an analytical solution, accounting this time for the upper crust flexural rigidity. To be able to adjust their model, they have to assume that viscosity dropped by a factor 10 to 100 during the last 8 Ma, implying a warming of 75–100° at the Moho, and that its current value is $\eta = 10^{19}$ Pa s ≈ 0.3 Ma MPa.

The decoupling between upper crust and mantle lithosphere gives an answer to an embarrassing question.

When two continental crusts meet in a subduction zone, there is collision and imbrication of the upper crusts. But what befalls the mantle lithosphere that was subducted? It is difficult to think that the mantle convection currents, to which it belonged, come to a stop immediately. One idea was that subduction goes on, the crust being stripped off, a process called *off-scraping*, *stripping away*, or, sometimes abusively, *delamination* (the genuine delamination, as first defined, is the block detachment of the crust, once subducted; we pointed out the absurdity of this process in Section 8.8). With a ductile lower crust, this off-scraping does not present any difficulty.

However, in the Alps as in Tibet, deep seismic exploration shows that *overthrusts also occurred in the mantle lithosphere*. We reproduce in Figure 14.8 the inter-

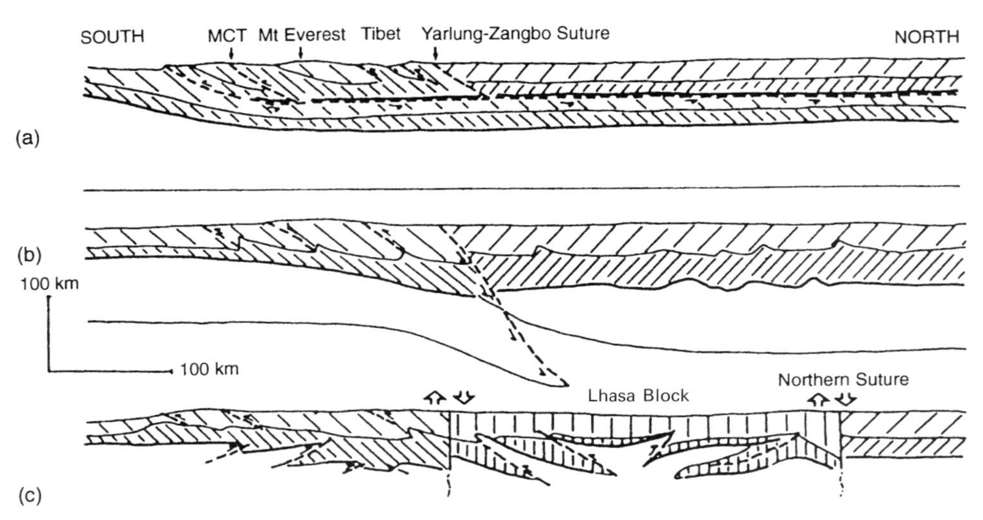

Figure 14.8. Schematic cross-section of Himalaya and South Tibet. (a) Older Argand hypothesis (superposition of the Indian and Asiatic crusts); (b) older hypothesis of Dewey and Burke (homogeneous thickening of the two crusts, without superposition); (c) the actual structure, revealed by the INDEPTH seismic expedition, and interpreted by Hirn (1988).

pretation by Hirn (1988) and others of seismograms collected in Tibet, not leaving place to any doubt. With such overthrusts, the mantle lithosphere subduction is reduced and could even disappear. The upper crust tectonics is doubled by an upper mantle *hypotectonics*, hidden and still to be discovered.

14.11 POST-OROGENIC SUBSIDENCE

When the relative ductility of the lower crust was discovered, some authors concluded that the upper crust of a high region should stretch and subside, because it was not held at the bottom. This would be accompanied by the formation of normal faults and grabens. These authors even talked of 'gravitational collapse' of mountains. But:

- either the subsidence is posterior to orogenesis because it is very slow, and the term 'collapse' is not appropriate (I shall call this 'post-orogenic subsidence');
- or the weakening is as fast as orogenesis. In this case, it counteracts and limits the ground uplift, but there is no subsidence.

The truth must lie between these two possibilities. At first, orogenesis lifts the ground up, more and more counteracted by the subsidence. Then, after the time $t = 0$, orogenesis stops and there is only subsidence. We shall model this second stage, by assuming that the traditional isostatic adjustment, due to mantle asthenosphere flow, is much faster than the one due to crustal asthenosphere, and can therefore be considered as instantaneous.

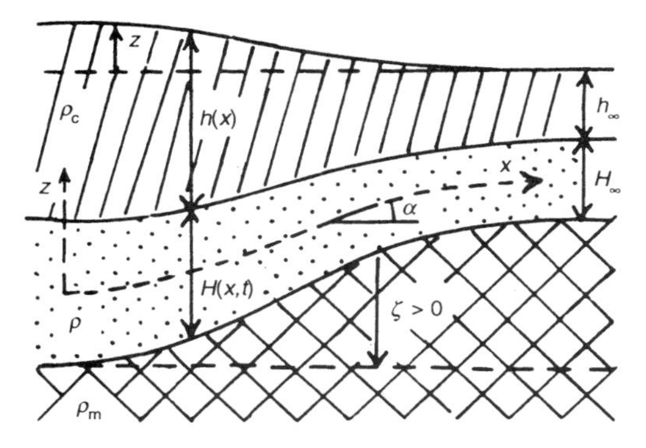

Figure 14.9. Notation for a plate model with a fluid lower crust (crustal asthenosphere).

Consider the plane problem, in the plane xOz, with a z-axis pointing upward (Figure 14.9). The elevation is $Z(x,t)$, tending toward 0 at $+\infty$. The crustal lithosphere has a density $\rho_c = 2.73\,\mathrm{Mg/m^3}$, and a thickness $h(x,t) = h_\infty + h_1(x,t)$.

The crustal asthenosphere has a density $\rho = 2.80\,\mathrm{Mg/m^3}$, a thickness $H(x,t) = H_\infty + H_1(x,t)$, and an isoviscous behaviour with a constant viscosity η. The mantle lithosphere has a density $\rho_m = 3.20\,\mathrm{Mg/m^3}$, and its thickness does not intervene in the computation. The Moho is at ζ above a fixed level, and tends toward ζ_∞ at infinity. The latter is given by:

$$\zeta + H + h - Z = \zeta_\infty + H_\infty + h_\infty$$

$$\zeta_\infty = \zeta - Z + h_1 + H_1 \tag{14.18}$$

The profiles are smoothed enough for ignoring the stiffnesses of the upper crust and of the whole traditional lithosphere which spread isostatic compensations. The 'instantaneous' isostatic equilibrium imposes:

$$\rho_c h + \rho H + \rho_m \zeta = \rho_c h_\infty + \rho H_\infty + \rho_m \zeta_\infty \tag{14.19}$$

Comparing with equation (14.18):

$$Z = \left(1 - \frac{\rho_c}{\rho_m}\right) h_1 + \left(1 - \frac{\rho}{\rho_m}\right) H_1 \tag{14.20}$$

Within the crustal asthenosphere, the 'hydraulic' head as defined in Section 10.11 is:

$$\frac{\rho_c}{\rho} h + H + \zeta = \frac{\rho_c}{\rho} h_1 + Z - h_1 + \text{a constant} \tag{14.21}$$

Thus the 'hydraulic' gradient is:

$$\varpi = \left(1 - \frac{\rho_c}{\rho}\right)\frac{\partial h_1}{\partial x} - \frac{\partial Z}{\partial x}$$

$$= -\left(\frac{\rho_c}{\rho} - \frac{\rho_c}{\rho_m}\right)\frac{\partial h_1}{\partial x} - \left(1 - \frac{\rho}{\rho_m}\right)\frac{\partial H_1}{\partial x} \tag{14.22}$$

Equation (10.47) gives then the volume discharge of crustal asthenosphere per unit width:

$$\begin{cases} q = -D\left[\dfrac{\rho_c}{\rho}\dfrac{\partial h_1}{\partial x} + \dfrac{\partial H_1}{\partial x}\right] \\ D = \left(1 - \dfrac{\rho}{\rho_m}\right)\dfrac{\rho g H^3}{12\eta} \end{cases} \tag{14.23}$$

Volume conservation in the crustal asthenosphere yields:

$$\frac{\partial H_1}{\partial t} = -\frac{\partial q}{\partial x} = D\left[\frac{\rho_c}{\rho}\frac{\partial^2 h_1}{\partial x^2} + \frac{\partial^2 H_1}{\partial x^2}\right] + \frac{dD}{dx}\left[\frac{\rho_c}{\rho}\frac{\partial h_1}{\partial x} + \frac{\partial H_1}{\partial x}\right] \tag{14.24}$$

Assuming that H is an even function of x, the second term on the right-hand side may be ignored. On the other hand, we assume that the erosion rate is proportional to the altitude:

$$\frac{\partial h_1}{\partial t} = -\lambda Z \tag{14.25}$$

To find an analytical solution, the further approximation $\rho = \rho_c$ will be made. Then equations (14.20) and (14.24) read:

$$\begin{cases} Z = \dfrac{\rho_m - \rho_c}{\rho_m}(h_1 + H_1) \\ \dfrac{\partial H_1}{\partial t} = D\dfrac{\partial^2 (h_1 + H_1)}{\partial x^2} \end{cases} \tag{14.26}$$

Comparing equations (14.25) and (14.26) it is found that:

$$\begin{cases} \dfrac{\partial Z}{\partial t} = D\dfrac{\partial^2 z}{\partial x^2} - \lambda' Z \\ \lambda' = \dfrac{\rho_m - \rho_c}{\rho_m}\lambda \end{cases} \tag{14.27}$$

Putting $Z = F(x,t)\,\mathrm{e}^{-\lambda t}$, the function F obeys the classical diffusion equation, which has many well-known solutions. In particular assume that at $t = 0$ the cross-sectional profile of the mountain range is the Gaussian curve:

$$Z(x,0) = Z_0 \exp\left(-\frac{2x^2}{L_0^2}\right) \tag{14.28}$$

Then the solution of (14.27) is:

$$Z(x,t) = \frac{Z_0 L_0}{\sqrt{L_0^2 + 8Dt}} \exp\left(-\frac{2x^2}{L_0^2 + 8Dt} - \lambda' t\right) \tag{14.29}$$

The coefficient λ is of the order of $0.1\,\mathrm{Ma}^{-1}$ (see Section 15.3), and thus λ' is of the order of $10^{-2}\,\mathrm{Ma}^{-1}$. With $H = 10\,\mathrm{km}$ and $\eta = 1\,\mathrm{Ma\,MPa}$ (or with $H = 20\,\mathrm{km}$ and $\eta = 8\,\mathrm{Ma\,MPa}$), $D \approx 280\,\mathrm{km}^2\,\mathrm{Ma}^{-1}$. Assume that the width of the mountain range at $t = 0$, before subsidence and spreading, is $2L_0 = 100\,\text{km}$. Then, time being expressed in Ma, the maximum height is:

$$Z(0, t) = \frac{Z_0}{\sqrt{1 + 0.9t}} \exp(-10^{-2}t) \tag{14.30}$$

With the assumed values, the outflow of lower crust is more efficient than erosion in erasing the relief by one order of magnitude. It takes several million years, however. This process is very slow, compared with the isostatic adjustment by outflow of the mantle asthenosphere, not because the viscosity, η, is much larger, but because the thickness H is much smaller.

14.12 POST-OROGENIC TECTONIC STRESS

Using the model above, let us calculate the horizontal tectonic stress in the elastic upper crust, where it should be the most modified by post-orogenic subsidence, at $x = 0$. For this purpose, the balance of horizontal forces between $x = 0$ and very large values of x will be drawn up.

Over the section $x = 0$, where $h = h_0$, there is lithostatic pressure plus a *tectonic traction* of average value σ. The total corresponding force is:

$$\frac{\rho_c g h_0^2}{2} - \sigma h_0 \tag{14.31}$$

At ∞, lithostatic pressure combines with a *tectonic pressure* of average value p_∞ (it decreased after tectogenesis, but did not vanish). The total force is:

$$-\frac{\rho_c g h_\infty^2}{2} - p_\infty h_\infty \tag{14.32}$$

At the base are acting:

(1) a hydrostatic pressure, nearly equal to $\rho_c gh$, tilted at $\partial(h - Z)/\partial x$ from the vertical. Using equation (14.20), the resultant of its horizontal component is:

$$\int_0^\infty \rho_c gh\left[\frac{\rho_c}{\rho_m}\frac{\partial h}{\partial x} - \left(1 - \frac{\rho}{\rho_m}\right)\frac{\partial H}{\partial x}\right] dx$$
$$= -\frac{\rho_c^2}{\rho_m} g \frac{(h_0^2 - h_\infty^2)}{2} - \left(1 - \frac{\rho}{\rho_m}\right)\rho_c g \int_0^\infty h \frac{\partial H}{\partial x}\, dx \tag{14.33}$$

(2) a shear stress, due to the crustal asthenosphere flow. A similar shear exists at the base of this asthenosphere, and both balance the force due to the head loss. The driving force over half the crustal lithosphere is therefore:

$$\int_0^\infty \varpi \rho g \frac{H}{2}\, dx = -\frac{\rho g}{2}\left(1 - \frac{\rho}{\rho_m}\right)\int_0^\infty \left[\frac{\rho_c}{\rho_m}\frac{\partial h_1}{\partial x} + \frac{\partial H_1}{\partial x}\right] H\, dx \tag{14.34}$$

The sum of the 4 forces expressed in (14.31) to (14.34) must equal 0. The calculation is simpler by keeping the approximation $\rho = \rho_c$. Integrating $\int h(\partial H/\partial x)\,dx$ by parts, it is found:

$$\sigma h_0 = -p_\infty h_\infty + \frac{\rho_c g}{2}\left(1 - \frac{\rho_c}{\rho_m}\right) \times \left[h_0^2 - h_\infty^2 + \frac{H_0^2 - H_\infty^2}{2} + 2(h_0 H_0 - h_\infty H_\infty) + \int_0^\infty H \frac{\partial h_1}{\partial x}\,dx\right] \tag{14.35}$$

Assume that the ratio $H/h = a$ is independent of x. (This should be true at the start, before the post-orogenic subsidence.) Then the initial thickening of the crust at $x = 0$ is $M = H_0/H_\infty = h_0/h_\infty$. It follows that:

$$\int_0^\infty H \frac{\partial h_1}{\partial x}\,dx = -\frac{a}{2}(h_0^2 - h_\infty^2)$$

$$\sigma = -\frac{p_\infty}{M} + \frac{\rho_c g h_0}{2}\left(1 - \frac{\rho_c}{\rho_m}\right)\left(1 - \frac{1}{M^2}\right)\left(1 + \frac{a^2}{2} + \frac{5a}{2}\right)$$

$$= \frac{1}{M}\left[-p_\infty + \frac{\rho_c g}{2}\left(1 - \frac{\rho_c}{\rho_m}\right)(h_\infty + H_\infty)(M^2 - 1)\frac{2 + 3a + a^2}{2(1 + a)}\right] \tag{14.36}$$

Assuming $h_\infty + H_\infty = 36$ km, and expressing σ and ρ_∞ in MPa:

$$\sigma = \frac{1}{M}\left[-p_\infty + 62(M^2 - 1)\frac{2 + 3a + a^2}{2(1 + a)}\right] \tag{14.37}$$

For a increasing from 0.1 to 0.4, the last ratio only increases from 1.05 to 1.20. Thus the main unknown is the transverse tectonic stress still acting, p_∞ (see Section 13.5). Let us remark that the almost unknown viscosity of the crustal asthenosphere does not enter the calculation: stresses in the upper crust depend only on the geometry.

For $M > M_c$, σ becomes positive, and normal faults can be activated at the axis of the range. With $p_\infty = 100$ MPa, $M_c \approx 1.56$, and with $p_\infty = 10$ MPa, $M_c \approx 1.07$. Since the maximum thickening of the crust in a mountain belt such as the Alps is about 50% ($M = 1.5$), the case $\sigma \geq 0$ should be quite general. The conclusion is that *the fluidity of lower crust should be the main factor stopping mountain building, although the transverse tectonic force p_∞ has not vanished.*

Some have said that a graben appears at the surface of the chain's axis, but this is a step that should not be crossed without irrefutable examples. This is not the case of the Limagnes area, examined later. The appearance of a graben requires a thin crustal lithosphere, when, to the contrary, it was thickened by orogenesis.

14.13 EXHUMATION BY RETRO-SUBDUCTION

The fluidity of the lower crust seems to bring an explanation to the exhumation of crustal rocks from depths of 100 km or more, which cannot be explained by traditional plate theory (Chemenda *et al.*, 1995). This enigma was noted in Section 3.8.

It has been observed that this exhumation is frequently associated with large *normal* faults, subparallel to the reverse faults of overthrusts. For example in Himalaya, north of the large reverse fault of the Main Central Thrust, the North Himalayan Normal Fault runs parallel to it over 1,500 km. A metamorphic rock belt nearby has been exhumed from 20–30 km deep in a very short time, probably less than 3 Ma, ca. 20 Ma ago. It seems therefore that the High Himalaya slab was subducted northward, before 'going backward' and southward. I propose to call this a *retro-subduction* (Figure 14.10).

The whole plate does not come back up, only a slab of upper crust. This is made possible by the lower crust's ductility. This is not enough to allow the mantle asthenosphere to come inside the plate and cause delamination (see Section 8.7). But it allows the light part of the plate to slide up on the heavy part.

To understand this, let us consider a portion of plate completely immersed in traditional mantle asthenosphere. With the notation of Figure 14.11, the equilibrium of the upper layer along ox is written:

$$\frac{\partial \tau_{xz}}{\partial z} = \rho_c g \sin \varphi - \frac{\partial \sigma_x}{\partial x} = -(\rho_m - \rho_c) g \sin \varphi \tag{14.38}$$

Integrating from $z = 0$ to $z = h$ (where $\tau_{xz} = 0$), it appears that shear at the base of the upper layer equals:

$$\tau = (\rho_m - \rho_c) g h \sin \varphi \tag{14.39}$$

If $h = 30$ km and $\sin \varphi = 1/2$, $\tau = 75$ MPa.

For an exhumation from 30 km deep, i.e. 60 km counted along the plate, to occur in 3 Ma at most, when the plate was subducting at a rate of 50 km/Ma, the sliding velocity of the upper crust relative to mantle lithosphere must have been larger than 70 km/Ma. The shear rate in the crustal asthenosphere was therefore larger than (70/8) Ma^{-1} (if it is 8-km thick, as Chemenda *et al.* (1995) say). The 75 MPa must be divided by this value to get viscosity: $\eta < 8.6$ Ma MPa.

Another mechanism (*unroofing*) was proposed for the massif of Dora Maira (Piedmont). It will be explained in Section 15.5.

14.14 OROGENIC PROCESSES PROPER TO MARGINAL CORDILLERAS

Now we have examined the tectonic processes in collision chains, let us study other processes typical of the cordilleras, at the margin of an oceanic plate subduction.

The vocabulary needs to be clarified. A range is a chain of summits more or less aligned and coalescent, i.e. where two slopes can be unambiguously recognized. It is called *sierra* in Spain (Sierra de Guadarrama, Sierra Nevada). To designate a vast set of more or less parallel ranges, with at least the dimension of the Alps (850 km long,

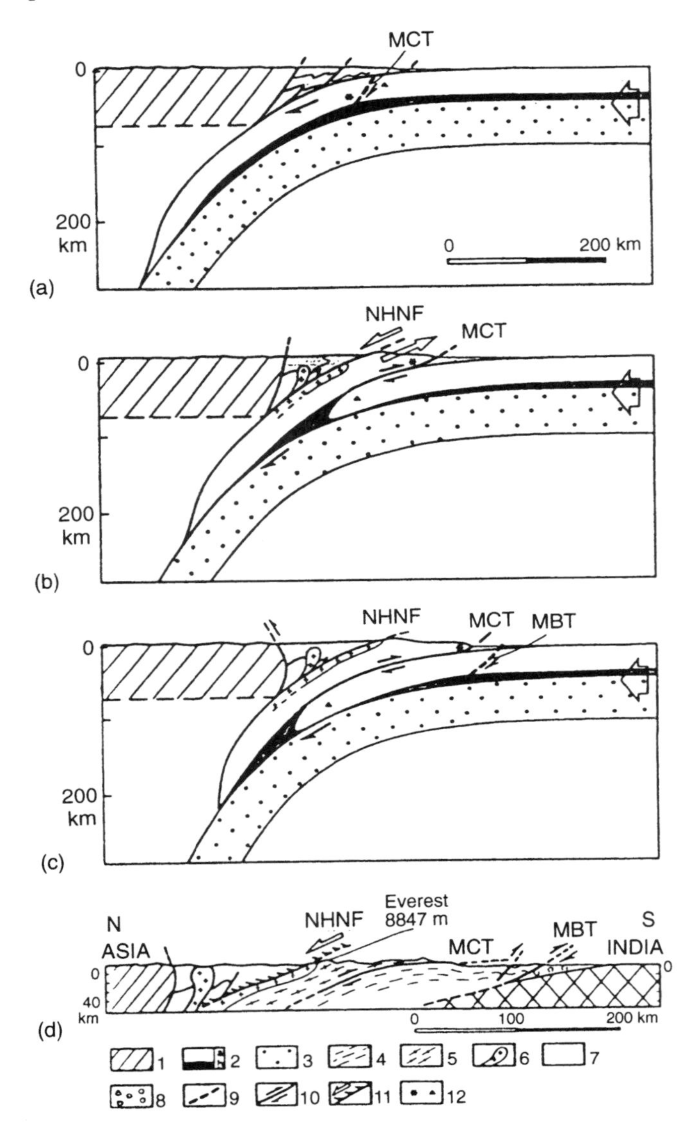

Figure 14.10. Evolution of the Himalayan Main Central Thrust (MCT), from Chemenda *et al*., (1995); © Elsevier; (a) subduction of the Indian crust, until the MCT appears; (b) the ductile lower crust allows this subducted crust to rise, and its later exhumation along the North-Himalayan Normal Fault (NHNF); (c) exhumation stops after a few Ma, and the subduction below stops when another overthrust appears (Main Boundary Thrust, MBT).

100–150 km wide), I shall use the word *cordillera* (even if Chileans also call simple ranges cordilleras).

Marginal cordilleras are a set of ranges, valleys and plateaus parallel to subduction. Let us be more specific with three E–W traverses across the American

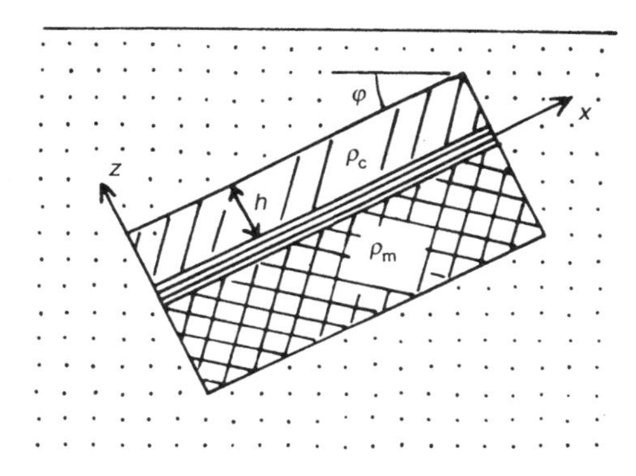

Figure 14.11. Rough estimate of the shear between crust and mantle lithosphere, during exhumation of the former.

Pacific coastline (satellite images and maps of the first two can be found in my book *Sciences Géométriques et Télédétection*, Masson, Paris, 1992).

At the latitude of Santiago (Chile), i.e. 33.5°S, we find from West to East:

- The Coast Range (*Cordillera de la Costa*), culminating at 2,281 m.
- A flat valley 50 km wide and at an altitude of 500 m (*Valle Central*).
- The Andes Cordillera, 100 km wide, with summits higher than 5,000 m, in general not volcanoes, and culminating with the Aconcagua (6,954 m). It is possible to recognize a peneplain around 3,000 m, deeply cut by valleys. These erosion valleys follow most often NNW–SSE tectonic directions, and the cordillera therefore comprises 3 to 6 parallel ranges.
- The Argentinian Pampa, of altitude at the foot of the Andes ranging between 800 and 1,400 m (the Indian name of *pampa* designates any flat area, whatever its altitude).

At the latitude of Antofagasta (23.5°S), we find:

- The Coast Range, with summits between 2,000 and 2,600 m. It contains plutons from the Permian to the Lower Cretaceous, and volcanics from the Triassic to the Cretaceous, of age decreasing with distance to the shore.
- A flat valley at altitude 1,000 m, the *Pampa del Tamarugal*.
- A very long range with summits ranging from 3,200 to 4,500 m, which are not volcanoes, the *Cordillera Domeyko*.
- Another flat valley around 2,200 m, with large saline playas (*salares*), in particular the *Salar de Atacama*.[2]
- The *Western Cordillera*, in fact a concentration of recent volcanoes, close to 6,000 m, over a plateau at 3,700–4,200 m.

[2] Atacama is the name of the vast province of colonial times, bordering the shore and containing the port of Antofagasta. It was conquered by the Chileans from the Bolivians, allied to Peruvians, during the Pacific Wars (1836–39 and 1879–84).

- The rest of this plateau, with other *salares*. This is the *Puna de Atacama*, prolongation in Chile and Argentine of the Bolivian *Altiplano*.
- The *Eastern Cordillera*. Whereas the previous 6 topographic units are around 60 km wide each, this cordillera is 250 km wide and is made of several ranges.

The last traverse goes at 40°N from Cape Mendocino to Denver. It contains: a coastal range, a coastal valley (the Sacramento Valley), a narrow western cordillera (the Californian Sierra Nevada), a plateau at around 2,000 m over 700 km (the *Basin and Range Province*), and a very wide eastern cordillera (600 km).

The eastern cordilleras look more like collision cordilleras: no recent volcanism, overthrusts and a few nappes in the East. Excluding these, 4 facts are characteristic of proper marginal mountain systems:

1. *The abundance of volcanics* and recent volcanism in the western cordillera.
2. *The absence of large overthrusts.*
3. *A high plateau between the two cordilleras*, when there are two of them.
4. The uninterrupted presence, in northern California as in Chile, of a *longitudinal valley* 50–100 km wide, separated from the oceanic trench by a *non-volcanic coastal chain*. (In Atacama, there are two of these pairs.)

The first two characteristics are of course results from the subduction below the edge of the continent, and of the absence of collision with another continent. The coastal chains, where reverse faults are many, are explained by upper crust compression. The plate subducted below has still not reached the depth at which it starts emitting magmas.

The high plateaus are the consequence of a very thick crust, even if a free-air gravity anomaly of up to 50 mgal in the centre of the Altiplano shows the role played by a contemporaneous epeirogenic uplift, not related to isostasy. 10 km of molasse is found at the Altiplano surface, resulting from the destruction of cordilleras. But this is not sufficient to explain the crust thickness. One must call upon *ablative subduction*, already discussed in Section 7.6. The edge of the continent was ablated and carried by subduction.

The longitudinal valleys are still enigmatic. Their generality and regularity excludes the possibility that they are old closed back-arc basins, as suggested by some. The subsidence relative to the cordillera in the east reaches 3–4 km, with subvertical faulting. They cannot be grabens due to transverse extension, because the close alternation, every 60 km, of compression in the coastal chain, extension in the valley and compression again in the cordillera, is mechanically impossible.

Neotectonics, however, confirms this rapid alternation of uplift and subsidence. During the very large earthquakes of Chile in 1960 and Alaska in 1964, measurements were made of strike-slip displacements of the oceanic plate below the continental plate of 10 to 20 m, an uplift of 11.3 m on the continental platform in Alaska, and a subsidence reaching 2.3 m in Alaska, 150 km away from the trench (Figure 14.12). But the emplacement of these opposite vertical movements does not coincide with the coastal chain and the central valley; they moved closer to the trench.

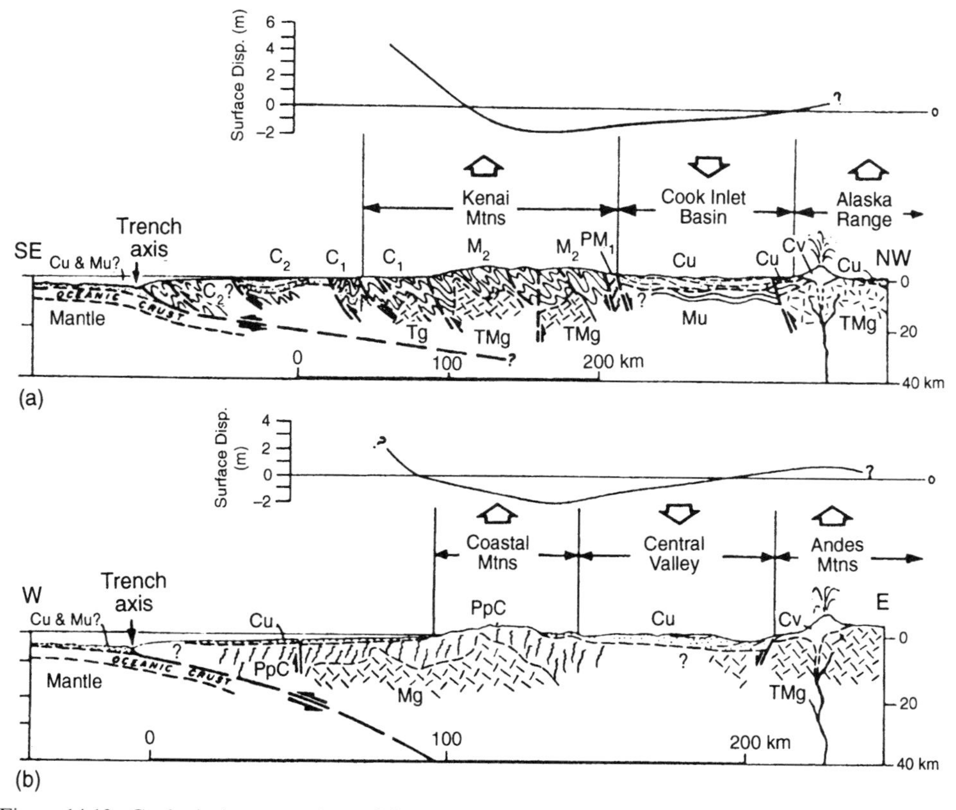

Figure 14.12. Geological cross-sections of the Pacific coastline from Alaska and Chile, with indications of the faults linked to subduction. Above each cross-section are plotted the vertical movements of the ground, during very large recent earthquakes. From Plafker (1972); © AGU 1972.

I see only one explanation, which cannot unfortunately be proved directly. Below the trough, there would be *ablative subduction of a corner of mantle lithosphere*. The longitudinal valleys of subduction zones would therefore be genuine collapse grabens, caused by the disappearance of matter below.

There remains another enigma of marginal cordilleras, already mentioned. Whereas subduction has been continuous for 200 Ma or much more, and ablative subduction must be continuous on the 20-Ma scale, there were in a particular region only a few orogenic phases widely spaced in time. One must also explain the very broad epeirogenic uplift of the Andes, from Peru to Central Chile, progressively decreasing southward and replaced by subsidence in Patagonia (where the Central Valley becomes the Chiloe Sea).

It seems that another major process must intervene: *without changes in the location of the trench, modifications in the subduction occur*, in the location of upwelling mantle currents and in the geometry of the subducted plate. But why these very slow changes would suddenly induce an orogenic phase remains an enigma.

14.15 PASSIVE GRABENS

It seems that three kinds of grabens must be distinguished: (1) the genuine collapse grabens, such as the central valleys of marginal cordilleras; (2) the ones improperly called collapse grabens, but in fact following transverse stretching of the crust by forces acting at its periphery (*passive grabens*); (3) the ones where the crust was stretched with magma intrusion at the base (*active grabens or rifts*).

The typical example of passive grabens is the series of grabens and small basins crossing the East of France to the Mediterranean: Rhine graben from Mayenz to Basel, Bresse and Lower Dauphiné basins, Roanne basin, Limagne, Lower Rhône basin south of Nîmes. They formed during the Oligocene. The Kaiserstühl volcanism, in the Alsace plain, is posterior (from the Miocene), and the still younger volcanism of the Chaîne des Puys, close to Limagne, dates from the Pleistocene. This magmatism may be a consequence, but cannot be the cause of the formation of these grabens.

These grabens are bounded by normal faults facing each other, affecting only the upper crust (down to 7 km in the Rhine graben). The lower crust stretched homogeneously, without breaking. Thus, the dip of normal faults decreases before they disappear (Figure 14.13). These faults are called *listric faults*.

The grabens in Eastern France are aligned in the N–S direction, and linked by NE–SW vertical faults. Unbiased geologists established the map of Figure 14.14; it shows that the grabens do not have the nice en-échelon pattern asserted by some theorists. The faults that moved during the Oligocene must have been old Hercynian faults. However, the E–W stretching cannot be denied.

Crustal stretching produced a Moho uplift. It is found at 25 km below the Rhine graben, instead of 31 km below the Vosges, the Black Forest and the Rhine Schistose Massif. The Moho uplift is similar below Limagne, the Lower Rhône, and suprisingly under the NE–SW fault which supposedly links the Vosges to Auvergne.

Crustal stretching does not require traction at the edges of the plate. In fact, during the Oligocene, the inner Western Alps were the location of an orogenic phase, showing a transverse NNE–SSE compression. At a moderate depth z, a principal stress is always vertical; this is the lithostatic pressure ρgz. The two main horizontal stresses were $\rho gz + P$ in the NNW–SSE direction, and ρgz in the perpendicular

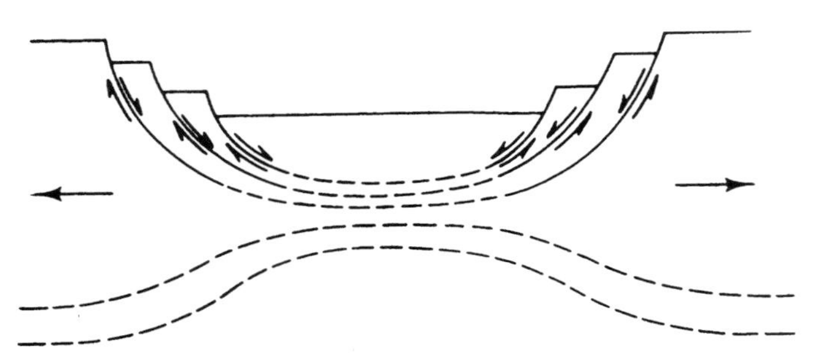

Figure 14.13. Faults accompanying the narrowing of stretched continental crust.

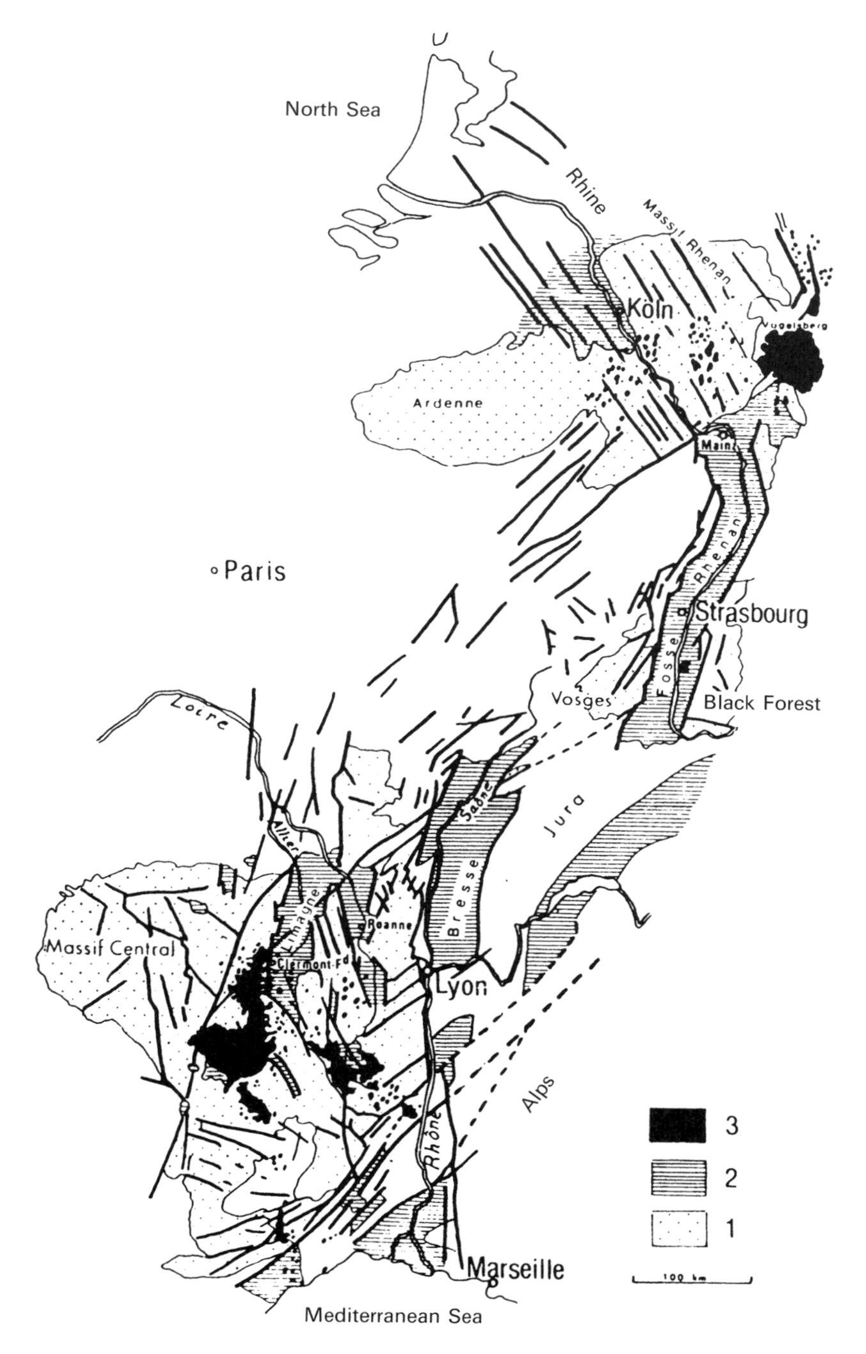

Figure 14.14. Oligocene basins and fractures of the Rhône–Rhine system. (1) Hercynian massifs; (2) Oligocene grabens; (3) Cenozoic volcanism. From Autran and Peterlongo (1980); © Dunod, 1980.

ENE–WSW direction. But deformations depend on the *deviatoric* stresses, which are a pressure $2P/3$ in the NNW–SSE direction and $-P/3$ (*traction*) in the ENE–WSW direction. There is also a maximal shear $P/(\sqrt{3})$ in the directions at 45° from the principal directions. This explains the sinistral strike-slip movements of older NE–SW faults.

Therefore, there is no reason to suggest that these grabens should not be genuine passive grabens, and that there was an incipient N–S rift in Western Europe during the Oligocene independently of the Alpine orogenesis.

Today, 30 Ma after these events, abnormally low seismic velocities are found below the Limagne area. They increase progressively from 7.2 to 8.4 km/s, between 24 and 45 km deep. It seems that melting has been reached, and that a mantle asthenosphere is currently in contact with the crust. But this is more likely the residue of the neighbouring Pleistocene volcanism.

14.16 ACTIVE GRABENS OR RIFTS

The big East African rifts are perfect examples of active grabens. The first one, in the North, starts from Afar (Republic of Djibouti) toward the SW and crosses Ethiopia. Another crosses Kenya from north (where it hosts Lake Rodolph) to south, and ends in Tanzania. A third one forms the eastern border of Zaire, from Lake Albert to Lake Tanganyika. The fourth one crosses Malawi, hosts the lake of the same name, and ends 3,200 km away from Afar. Like the previous one, it cuts at nearly right angle the Precambrian structural directions.

Figure 14.15 shows a model of the evolution of the Kenya rift since the Oligocene. It is assumed that an upwelling convection current in the mantle has:

(1) produced a large swell putting the plate under traction, of varying importance, along all horizontal directions;
(2) warmed and destroyed the mantle lithosphere;
(3) provided basic magmas which invaded the crust, before flowing at the surface and building volcanoes. This traction state allowed the intrusion of magmas as dykes; otherwise, underplating would have occurred.

Red Sea opening

Further north, past the triple junction of Afar with the Gulf of Aden mid-ocean ridge, the Africa rifts go on with the Red Sea ridge. Its opening has been well studied (Figure 14.16). The first opening phase occurred 40 to 34 Ma ago. A second phase of oceanic crust formation began at chron 5, around 10 Ma ago, but with limited extension. It did not reach Afar, where there is only transverse stretching and volcanism. The remaining relief between the two is the Danakil Horst.

Aulacogenes

A graben cutting a continent margin may end as a cul-de-sac, and become masked by later sediments. It is called an *aulacogene*.

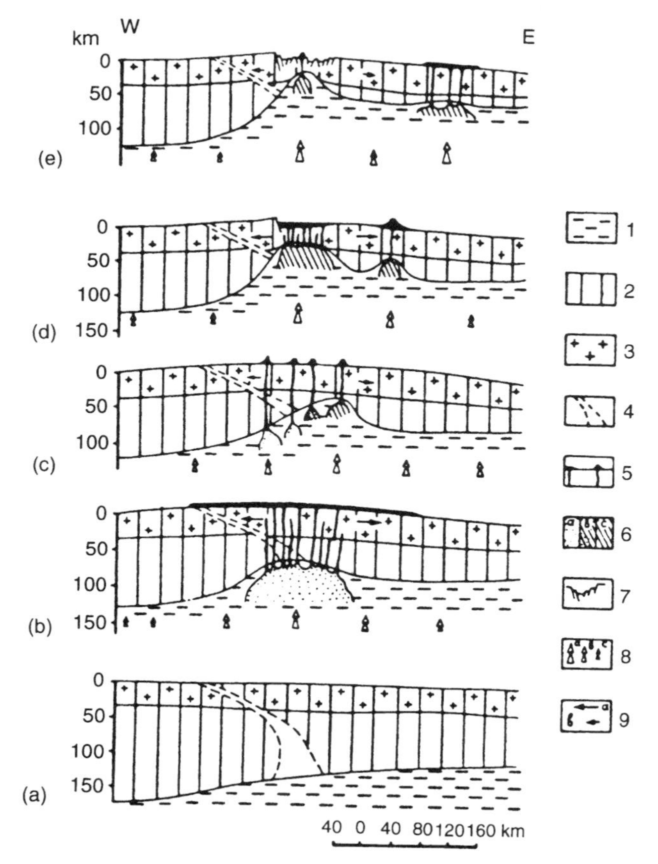

Figure 14.15. Hypothetical model of the evolution of deep structures and magma intrusion in the Kenya Rift (central segment). (1) asthenosphere; (2) lithosphere; (3) crust; (4) suture between the Precambrian Tanganyika shield and the folded Mozambique belt; (5) volcanoes and chimneys or fissures feeding them (*feeders*); (6) magma sources (*a*: highly alkaline; *b*: intermediate; *c*: lowly alkaline); (7) axial graben faults; (8) geothermal flux (with 3 degrees *a*, *b* and *c*); (9) extension. From bottom to top: (A) Oligocene; (B) Upper Miocene; (C) Lower Pliocene; (D) Upper Pliocene; (E) Pleistocene and Recent. From Logatchev, 1976.

An example is the Viking Trough, in the North Sea between Norway and Scotland. This aulacogene formed during the Cretaceous. Other aulacogenes exist on the Atlantic coastlines of South America and Africa, and are abortive branches of the large rift which gave birth to the South Atlantic in the Lower Cretaceous.

14.17 THINNED CRUST AND SUBSIDENCE OF PASSIVE CONTINENTAL MARGINS

The transversal stretching of the crust, which accompanies the formation of an ocean, must be much larger than during the formation of a graben. This crustal

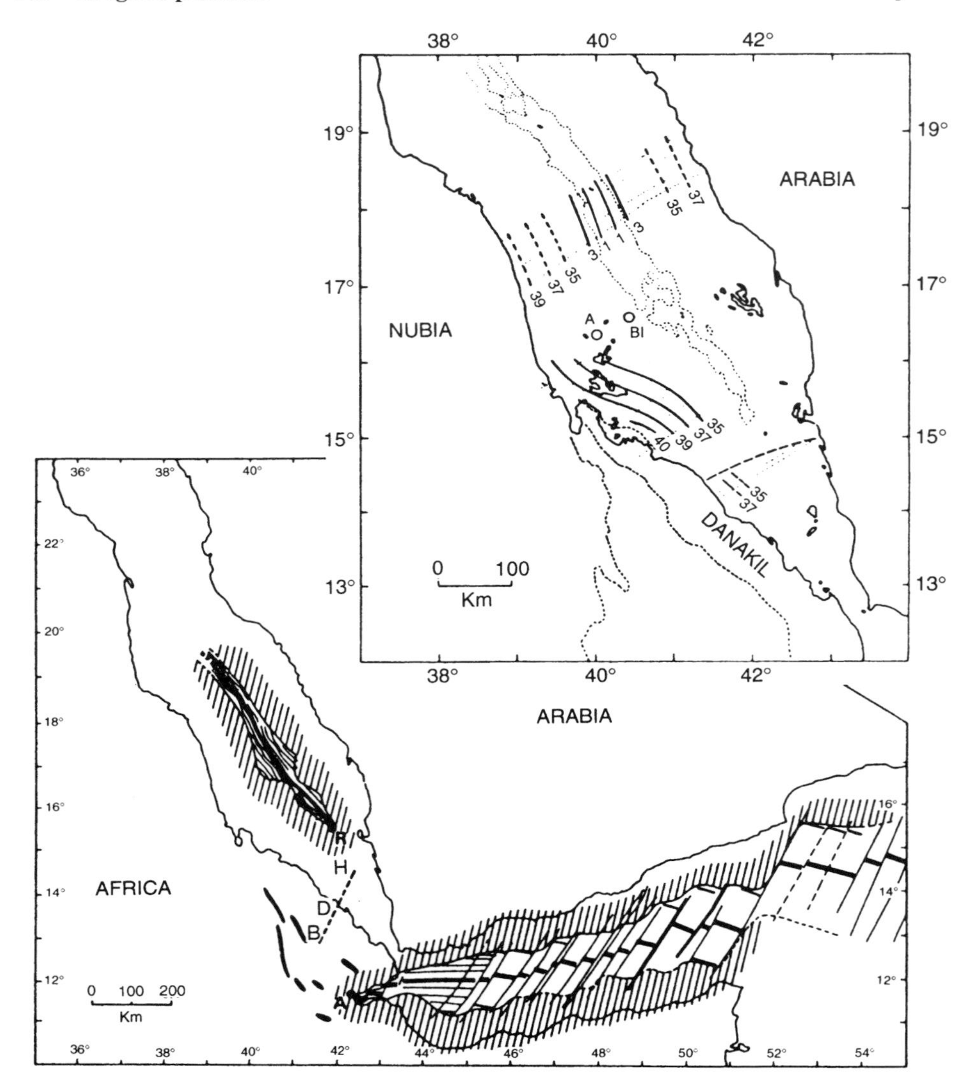

Figure 14.16. Top: isochrons (in Ma) for the Red Sea oceanic crust, inferred from magnetic anomaly stripes and a few K–Ar dates. They show the rotation of the Danakil Horst. From Girdler and Styles, 1976. Bottom: oceanic ridges from the Red Sea and the Gulf of Aden, and currently active volcanic chains of the Afar. The thin lines parallel to the ridges are the magnetic anomaly stripes. East of 45ºE, only anomaly 5 is represented. From Courtillot, 1982.

stretching was noted in Section 5.9, when discussing the Atlantic passive margins. But Figure 5.14 shows a continental crust thicker by 25 km, no more stretched than in the Rhine graben or in Limagne. And the submarine Rockall Plateau, part of a Precambrian continent which separated from Europe 140–150 Ma ago, and from North America 70 Ma and 54 Ma ago, still has a crust 33 km thick.

However, stretching and thinning are important at the continental margins of small oceanic crust basins, such as the South China Sea or the Western Mediterranean. The 'Provençal' basin between Provence, Languedoc, Catalonia, Corsica and Sardinia, contains a central third of typically oceanic crust, under 7 km of sediments deposited during the last 20 Ma. Apart from these recent sediments, the NW continental margin thins from 24 km south of the Cévennes Fault, or 20 km on the continental platform of the Gulf of the Lion to 6–7 km only. This thinning is accompanied by EW normal faults in the Provence Alps, subvertical at the outcrop but maybe listric at depth. This crustal thinning explains the great subsidence of continental margins (and not the weight of the sediments deposited over the continental shelf, as suggested in the past).

This subsidence is proved by the existence of *submarine canyons* deeply carved into the continental shelf, which have long intrigued geologists. They have often been eroded in hard rocks or have a sinuous path, which precludes erosion by turbidity currents. Free air erosion is the only explanation. But, during the glacial maximum, the level of oceans was only 140 m lower, at most. The Mediterranean dried out in the Messinian, indeed, but the same submarine canyons exist elsewhere, for example in Northern Alaska, and the Arctic Ocean never dried out.

The incomplete oceanization of a thinned crust, covered by a thick layer of sediments, and the formation of an *intermediate crust*, will be examined in the next chapter.

15

A short history of the Mesozoic and Cenozoic

15.1 FORMATION AND BREAKING UP OF PANGAEA

As in this whole book, the term 'continent' will designate a region with continental crust (normal or thinned), even if it is in fact a vast epicontinental sea.

In 1912, Wegener introduced the concept of continental drift. He thought that all current continents were the result of the breaking up of a super-continent, *Pangaea*, which grouped them together since the formation of the Earth. We know today that this grouping was only ephemeral in geological history, and never was complete. This is mainly inferred from geological studies, and also from palaeomagnetic poles, discussed in the next chapter. The breaking up stages of the Pangaea are well known, because of the magnetic anomaly stripes found in the oceans resulting from this dispersal.

In the Upper Devonian (360 Ma ago), there were two super-continents:

- *Gondwana*, grouping South America, Africa–Italy–Greece–Asia Minor–Arabia, East Antarctica, Madagascar, India (including Pakistan, Kashmir and South Tibet), Australia–New-Guinea–New–Zealand (Figure 15.1, with geological arguments).
- *Euramerica*, improperly called *Laurasia* by some (this is an anachronism, because the continents which formed Asia had not merged yet). Euramerica regrouped North America east of the Rocky Mountains and the MacKenzie River (*Laurentia*), Greenland and Europe (Italy and Greece excepted).

In the Upper Devonian and the Carboniferous, these two super-continents, separated by the *Hercynian Ocean,* collided. This collision resulted in the *Variscan* (or *Hercynian*) orogeny, and the formation of the *Hercynides*. Pangaea was the outcome of this merging (Figure 15.2).

In the Upper Carboniferous and in the Permian, the following blocks were added:

- The Kazakhstan continental block, to which the Siberian block should have already been attached. The Urals were born from this collision.

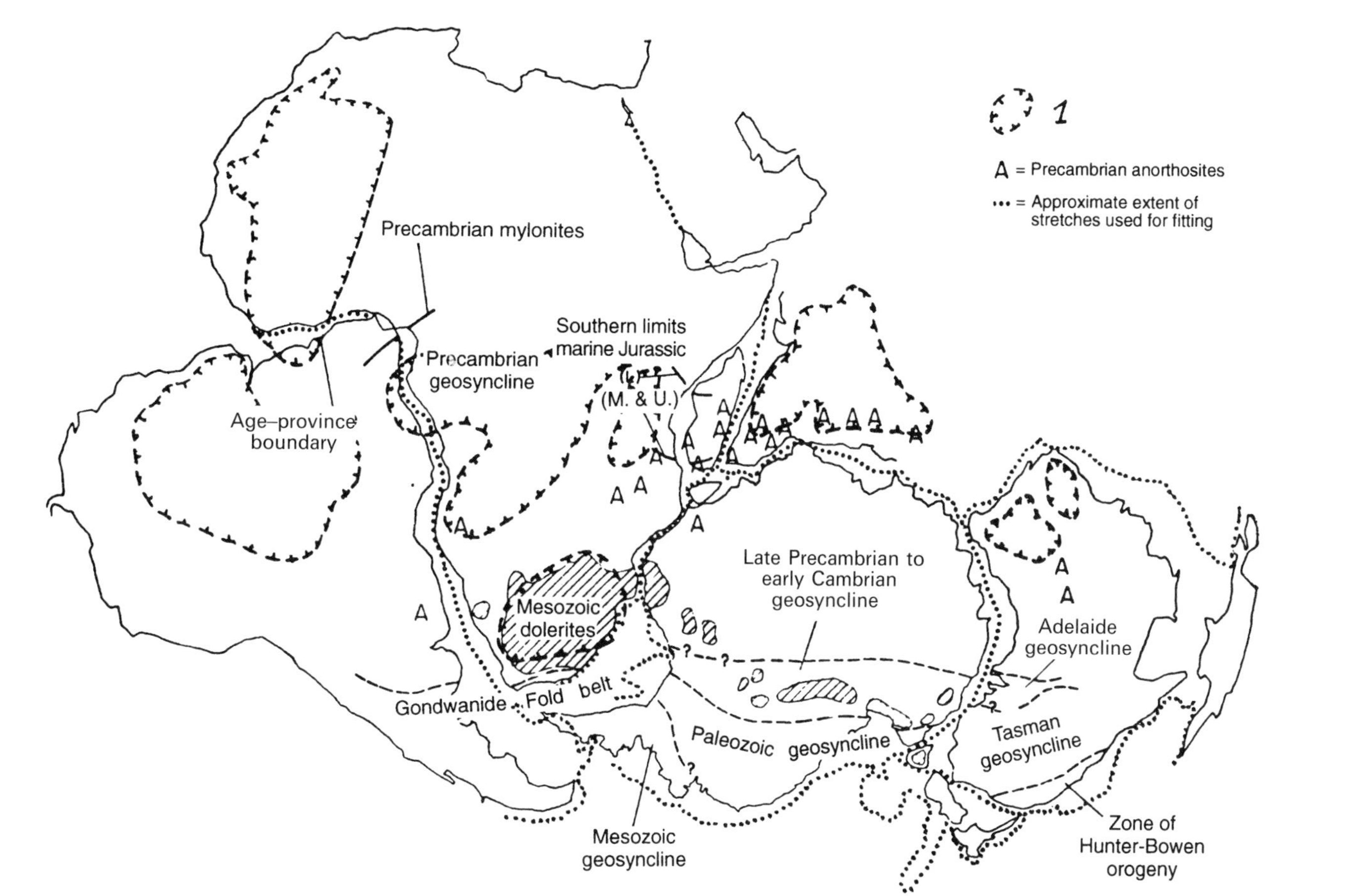

Figure 15.1. Geological proofs of the reality of Gondwana, put together in 1970 (some other proofs were added since). (1) Continental shields older than 2,000 Ma. (A = Precambrian anorthosites). Dotted lines = boundaries of continents used for fitting. From Smith and Hallam, 1970. © Macmillan Magazines Ltd.

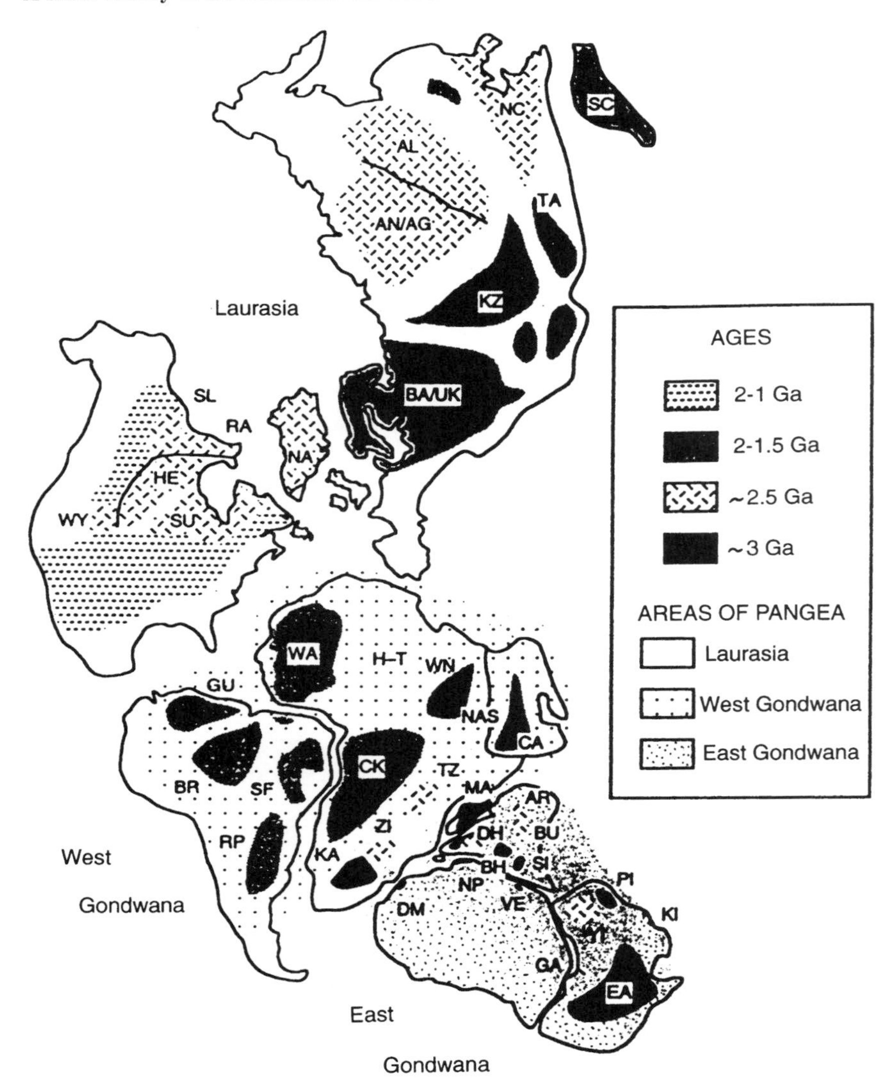

AL = Aldan; AR = Aravalli; AN/AG = Anabar/Angara; BA/UK = Baltic/Ukraine; BH = Bhandara (Bastar); BR = Brazil (Guaporé); BU = Bundelkhand; CA = Central Arabia; CK = Congo/Kasai; DH = Dharwar; DM = Dronning Maud Land (west); EA = Eastern Australia; GA = Gawler; GU = Guyana; HE = Hearne; HT = Hoggar/Tibesti; KA = Kaapvaal; KI = Kimberley Plateau (in the north of Australia); KZ = Kazakhstan; NA = North-Atlantic craton, including Greenland, Nain and Lewisian; NAS = Nubian-Arabian platform; NC = North China/Korea; NP = Napier; PI = Pilbara; RA = Rae; RP = Rio de la Plata; SC = South China (Yangtse); SF = São Francisco/Salvador de Bahia; SI = Singhbum; SL = Great Slave Lake; SU = Upper Great Lake; TA = Tarim; TZ = Tanzania; VE = Vestfold; WA = West Africa, including the small Brazilian craton of São Luis; WN = Western Nile (Nile/Uweinat); WY = Wyoming; YI = Yilgam; ZI = Zimbabwe

Figure 15.2. Approximate configuration of the Pangaea and the cratons it enclosed (see also Section 17.13). From Rogers (1996); © University of Chicago Press.

- The Tarim block, with formation of the Tien Shan cordillera between this block and the Kazakhstan block.

Most of the other continental blocks forming south Asia (India being the most notable exception) conglomerated with this 'Northern Asia' during the Triassic and Jurassic, so that Asia can be properly mentioned in the Cretaceous. But further west, back-arc basins were opening on the southern border of Eurasia, since the early Jurassic: Caucasian Basin (of which the south Caspian Sea and the Black Sea are remnants) and Carpathian Basin, extending westward with the narrow Valaisian Ocean. The Central Atlantic Ocean was also opening during the whole Jurassic, and Gondwana was beginning to break up. As early as 175 Ma, Africa had started separating from Antarctica. The South Atlantic Ocean had started opening from the south, 140 Ma ago (Jurassic/Cretaceous limit). Idem for the Indian Ocean.

Wegener's Pangaea therefore existed only at the end of the Palaeozoic, 250 Ma ago, and only contained the northern part of today's Asia.

15.2 FORMATION OF ASIA

Asian megatectonics is now much better understood, since a French mission to Afghanistan up to 1980, and next a long French–Chinese collaboration. A geological map of Central Asia is provided in Figure 15.6, and a NS cross-section of Afghanistan in Figure 15.4. Figure 15.3 shows the various continental blocks that may be recognized. They were formerly separated, as shown by the ophiolitic belts at their sutures, or they have moved respective to each other during the last 250 Ma. Let us describe these different blocks and their current topography, by successive W–E transects.

KAZ denotes the *Kazakhstan* block. It extends to the Arctic. Its suture with the *Siberian* block (SIB) (from which it was separated during the Lower Carboniferous) is imprecise. It should cross somewhere through the vast Siberian plain, extending from the Urals to the Ienisséi River. Its SE corner should perhaps be isolated from Kazakhstan; it is the Dzungarian Basin (JUN, for *Junggar Pendi*), a basin with altitudes ranging between 200 and 800 m.

Preceded by an island arc, this group merged with Europe during the Upper Carboniferous and the Permian, and formed a collision chain, the Urals. The Moho is 35–45 km deep below the Russian platform and the Siberian plain, but it is 60–65 km deep below the Urals. The Siberian crust seems underthrust below the island arc continental crust.

KAR denotes the *Karakum* Basin, in Turkmenistan. It is mistakenly called the Tadjik Basin on Figure 15.5. No relief separates it from the Tarun Basin around the Sea of Aral (Kazakhstan), and it is not proven that this block is indeed distinct from the previous one. It is bounded in the south by the *Hindu-Kush* and its western extensions, which formed in the Palaeozoic and were reactivated in the Cenozoic (with the formation of the Herat Fault).

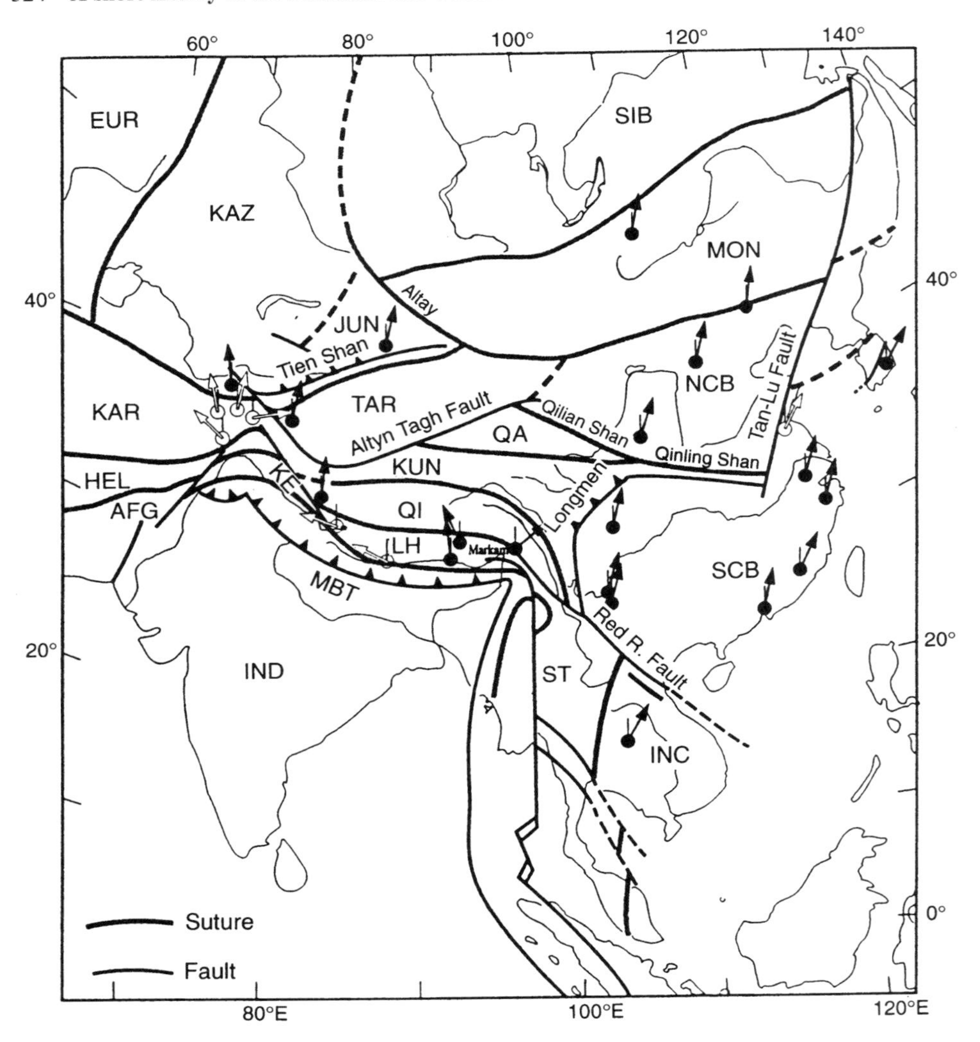

Figure 15.3. Schematic map of the different continental blocks forming Asia (see text for details). The angles between the arrows and the fine vertical lines indicate the palaeomagnetic declinations in the Cretaceous. From Chen *et al.* (1993); © AGU 1993.

Immediately south, the *Farah* block (not represented in Figure 15.3) merged in the Triassic (ophiolitic suture of Herat).

HEL denotes the *Helmand* block, around the river of the same name. It should not be confused with the large Helmand basin in southern Afghanistan. (In the cross-section of Figure 15.4, this block is called the Central Afghanistan block.) It merged with Eurasia during the Upper Jurassic, along the Panjab suture.

AFG denotes the *Kabul* block, between the Kabul and Khost sutures.

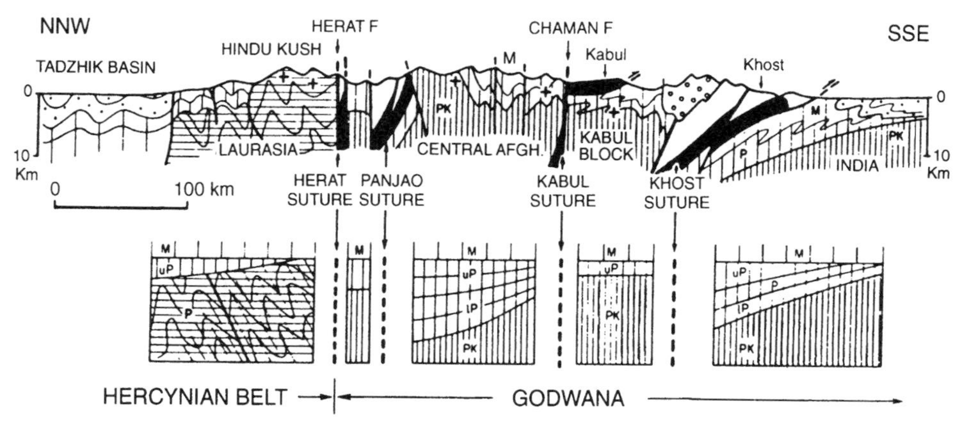

Figure 15.4. Simplified cross-section east of Afghanistan, showing the four continental blocks limited by ophiolitic sutures, which collided successively with the Hindu-Kush. From Tapponnier *et al.* (1981); © Elsevier.

Further east, the SW–NE sinistral fault of *Chaman* and the NW–SE dextral fault of *Karakoram* (KF) are linked by the Karakoram chain (between Kashmir and Pamir). This is manifestly where the Indian indenter penetrated the deepest in Central Asia.

TAR corresponds to the *Tarim* block. The desert basin of the Tarim River, the size of France, varies in altitude between 1,000 m and −154 m (in the depression of Turfan). It is separated from Dzungaria by the very high *Tien Shan* mountains, 1,400-km long and culminating at 7,439 m.

QA designates the *Qaidam* Basin block, between the Qilian and Kunlun cordilleras. 1,000- to 2,000-m high, this basin is separated from the Tarim Basin by the Altyn Tagh mountains, along a very long fault.

KUN denotes the *Kunlun Shan* block. This is a cordillera as long and high as the Tien Shan (it culminates at 7,723 m). It separates the low plains of Tarim and Qaidam from the high Tibetan plateau, which oscillates around 5,000 m.

From north to south in Tibet, one finds an Upper Triassic suture in the Hohxil range (also spelled Kokosili), the block of the *Qiangtang* plateau (QI), the Upper Jurassic BangongNujiang suture, the *Lhasa* block (LH), the Eocene Indus–Zangbo suture, and the north edge of the Indian continent (IND).

East of Tibet, very high mountain ranges are appearing, with a general WNW–ESE direction parallel to the Qilian range. They turn around the eastern extremity of Himalaya and Assam to become N–S west of Sichuan (where the culminating point is Mt. Gonga, at 7,556 m), and North Yunnan. This large mountainous region is not well known. It measures 1,300 km from north to south, and is 700-km wide, between the plains of Assam and Yunnan. According to Chen *et al.* (1993), these would be the eastern extremities of the Kunlun and Qiangtang blocks, rotated by a push from Southern China.

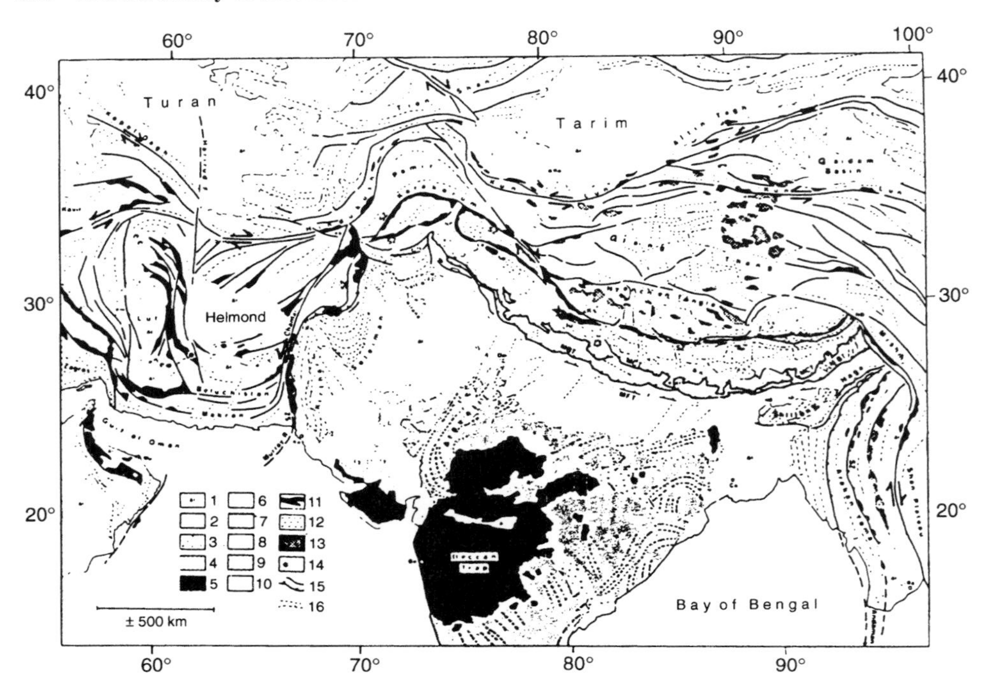

1 = Alluvial plains; 2 = Pre-Gondwana basement (Indian craton); 3 = Gondwanian sediments on the Indian craton; 4 = Mesozoic platform sediments on the Indian craton; 5 = Basaltic plateaus (traps); 6 = Siwalik molasse sediments (Sub-Himalaya); 7 = Lower Himalaya; 8 = High Himalayan crystalline; 9 = Thethysian platform sediments; 10 = Flysch; 11 = Ophiolites, mixtures and associated pelagic sediments; 12 = Trans-Himalayan plutons; 13 = Cenozoic volcanics; 15 = main faults and thrusts; 16 = lineations (axes of folds, faults, and secondary fracture zones)

Bo = Bombay; Ca = Calcutta; De = Delhi; He = Herat; Is = Islamabad; JL = Jolmo Lungma (Everest); Ju = Junghwa sheet; Ka = Karachi; Kb = Kabul; Ko = Khotan; Ks = Mt. Kailas; Kt = Kathmandu; La = Ladakh; Ls = Lhassa; MBT = Main Boundary Thrust; MCT = Main Central Thrust; MFT = Main Frontal Thrust; Mu = Muscat; NB = Namche Barwa; NP = Nanga Parbat; Qt = Quetta; Ra = Rangoon; Sp = Spongtang klippe; SZ = Indus–Zangbo suture zone.

Figure 15.5. Structural map of Himalaya and the surrounding regions. From Gansser, 1980. The half-arrows indicate the strike-slip directions, and are taken from Tapponnier and Molnar (1976).

Further east, one finds (from North to South):

- The blocks of *Mongolia* (MON) and *Northern China* (NCB), already merged in the Upper Permian.
- The *Southern China* block (SCB), separated from Northern China by the Qinling–Dabie–Shandong suture. This suture closed progressively from east to west, from the Upper Permian to the Middle Triassic or the Jurassic (there is some disagreement).
- The *Indochina* block, extending to Sumatra and Borneo (and therefore sometimes called *Sundaland*). This block was affected by the Indosinian orogenesis in the Upper Triassic.
- The *Shan-Thai* block (ST), comprising Malaysia and Thailand.

Enkin, Courtillot and others (1992) attempted reconstructing the formation of Asia from these different continental blocks (Figure 15.6). This reconstruction is based primarily on palaeomagnetism which, as explained in the next chapter, indicates the rotations of the continental blocks and their variations in latitude, but not in

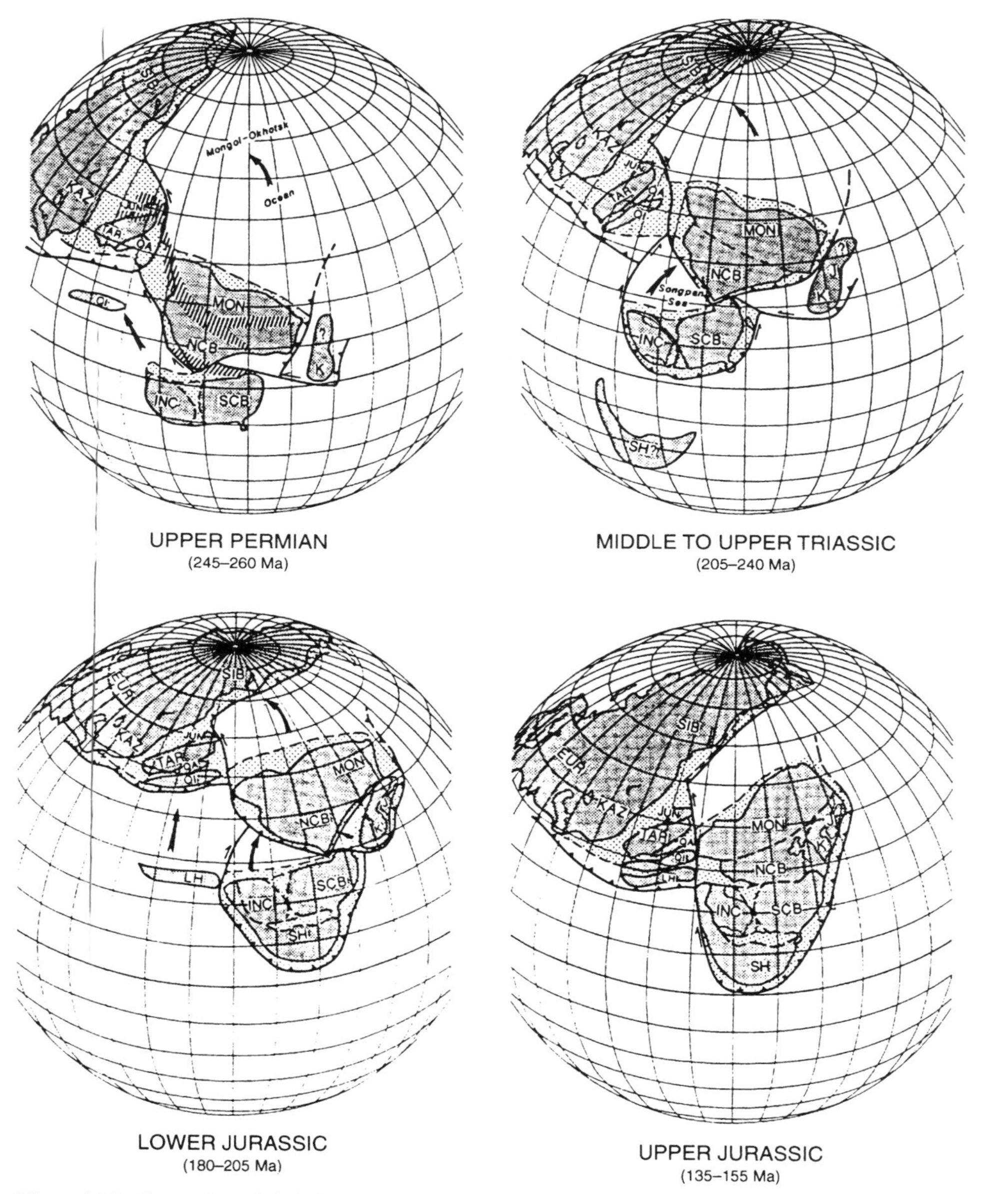

Figure 15.6. Formation of Asia by accretion of continental blocks. From Enkin *et al.* (1992); © AGU 1992.

longitude. Geologists have contested some chronology points. I can only point toward their fruitful discussion in two articles of the *Journal of Geophysical Research* (vol. 99, no. B-9, pages 18,035–18,048, 1994).

Since the beginning of the Carboniferous, Southern China and Indochina have been merged to form *Cathaysia*, and remained in an equatorial position. But in the Permian, the blocks Qiangtan and Shan were close to the Southern Pole. Enkin et al. (1992) do not think they were united, contrary to Metcalfe. In the Triassic and Lower Jurassic, Cathaysia turned clockwise. In the Upper Jurassic and Lower Cretaceous, according to Enkin *et al.* (1992), Cathaysia would have moved north by one thousand kilometres. Considering the extent of the mountains east of Tibet, I would have preferred a westward collision as well, against the four blocks of Central Asia.

15.3 PUNCHING BY INDIA AND ITS CONSEQUENCES

Magnetic anomalies in the Indian Ocean were used to determine the convergence rate between the Cretaceous Indian continent, which had separated from Gondwana, and Eurasia.

Today, according to the NUVEL-1 model, the convergence rate increases from 40 km/Ma at 65°E to 56 km/Ma at 97°E. This velocity difference made India rotate anticlockwise relative to Asia, when they moved closer. According to a study by Molnar and Tapponnier (1975), these velocities were twice as great between 38 and 55 Ma, and three times as great between 55 and 71 Ma (the recent NUVEL-1 model reduces these velocities, but the result remains).

Although geologists think they have detected signs of collision as early as 60 Ma ago, it did not complete until around 40 Ma ago. Giving India and Tibet their current extents, they are still found to be 2,200 km away. It is necessary to make a palinspastic reconstruction, as Westaway (1995). The distance between Tien Shan and the Indian craton is then increased by around 2,200 km at the time of contact (Figure 15.7). Since then, they moved closer through lateral extrusions and large overthrusts.

The south-eastward extrusion of Indochina and the Shan-Thai block should have stopped around 17 Ma ago. At this time, indeed, the opening of the China Sea ceased, and the sinistral strike-slip movement of the Red River fault stopped. Only a small EW lateral extrusion of Southern China remains.

Tibet was at sea level during the Cretaceous and, locally, up to the Eocene, around 50 Ma. With the Indian collision, it started rising and this uplift accelerated around 17 Ma. The uplift rate was 0.1 km/Ma in average for 50 Ma, to which must be added the erosion rate. The average erosion in the Indus and Brahmaputra basins should have been between 0.2 and 0.5 km/Ma before deforestation and anthropogenic degradation of the soils (which should have doubled it). Starting 11 Ma ago, N–S grabens appeared, indicating an E–W extension of Tibet. Around 8 ± 1 Ma, a

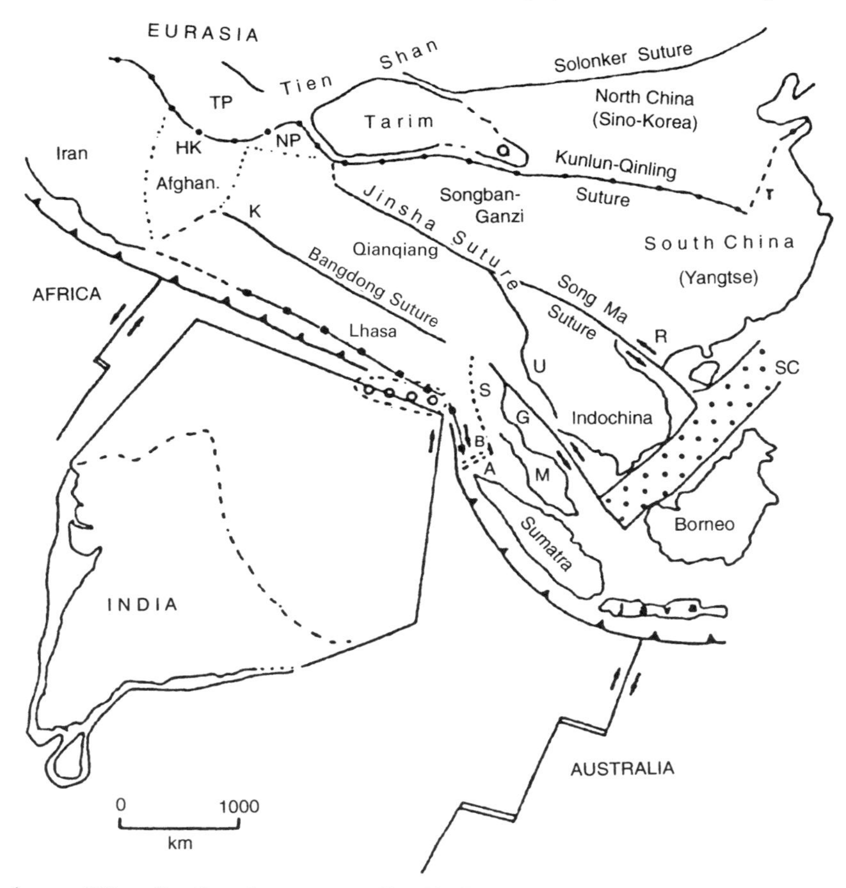

A = Andaman Ridge; B = Bangdong suture; G = Thailand Gulf Fault; HK = Hindu-Kush suture; K = Karakoram suture; M = Malaysia; Np = Nuristan-Pamir; Q = Qaidam Basin; R = Red River Fault; S = Sagaing Fault; SC = South China Sea continental zone; T = Tanlu Fault; TP = Turan Platform; U = Jinsha–Uttaradit suture.

Figure 15.7. Palinspastic reconstruction of India and Eurasia at the beginning of their collision, at chron 21 (48 Ma ago). From Westaway (1995); © AGU 1995.

higher aridity of the north of India and a stronger monsoon are attributed to the higher altitude now reached by the plateau.

The simplest explanation of this change in regime is defended by Chen, Courtillot and others. As lateral extrusion ceased little by little, the closing of India and Eurasia was now absorbed by the thickening of the Tibetan crust (not homogeneously, but by overthrusts; see Section 14.7).

Other authors, such as Westaway (1995), argue instead that the Tibetan crust thickened since the beginning of the collision. Their computation of the fluxes of matter accounts for the fluidity of the lower crust when not in old cratons (India, South China, and, hypothetically, Tarim). They choose for this crustal astheno-

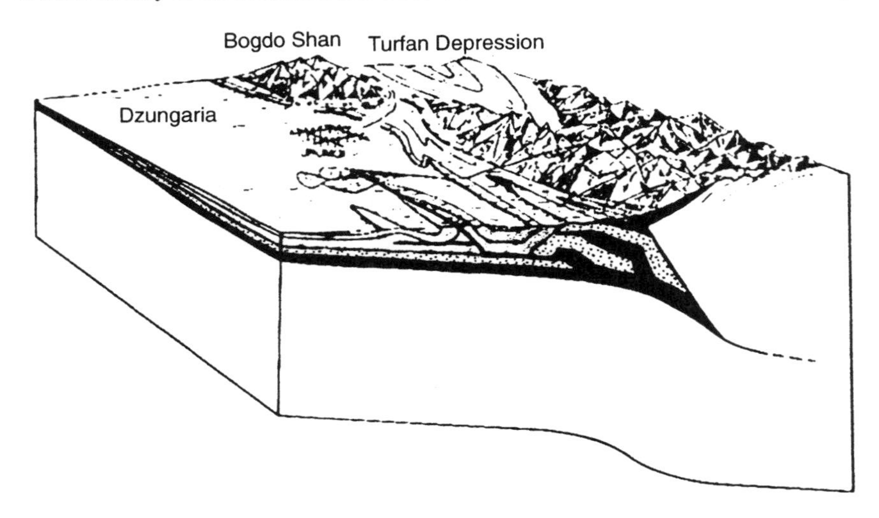

Figure 15.8. Eastern extremity of Tien Shan. From Avouac *et al.* (1993); © AGU 1993.

sphere the viscosity of 0.7 Ma MPa from Shackleton and Chang. This very low value comes from the fact that their model does not include the traditional mantle asthenosphere.[1] Westaway takes into account the erosion rate over Tibet, and the emplacement of cratons without crustal asthenosphere. But he does not manage to explain convincingly the change of uplift rhythm 17 Ma ago.

Less ambitious, but less hypothetical, studies were made during a French–Chinese collaboration in the highly seismic region of Urumchi (where earthquakes of magnitude 8.3 to 8.6 occurred in 1889, 1902, 1906 and 1911). The structure of the eastern extremity of Tien Shan was determined at this occasion. Apart from the two reverse faults bounding each side of Tien Shan, uplifting it, there are overthrusts on sub-horizontal faults, expressed in surface through kink folds (Figure 15.8). Cumulated throws of one century of earthquakes lead to a transverse compression of Tien Shan by 13 ± 7 mm/year.

15.4 TETHYS AND THE MEDITERRANEAN MOUNTAIN CHAINS

The word *Tethys* is source of confusion. When Suess introduced the word at the beginning of the 20th century, geologists were not yet making distinction between seas with a continental crust and seas with an oceanic crust. Any sea between Europe

[1] Their reasoning assumes a fixed and horizontal mantle lithosphere. A relief of average altitude h over a distance L would create in crustal asthenosphere a load of 0 'at the exit', an average load $\rho_c gh$ and therefore a load $2\rho_c gh$ 'at the entrance'. The load loss would be $2\rho_c gh/L$. With a thickness H, the volume flow per unit width would be: $q = (H^3/12h)(2\rho_c gh/L)$. This flow entering below Tibet from the South would create an uplift with the velocity $w = q/L$. Removing q and w, we get the estimate of η: $\eta = \rho_c ghH^3/6L^2 w$.

and Africa was called Tethys, or *Mesogea*. In this way, the Tethys already existed in the Triassic and closed only 10 Ma ago. But today, one uses Tethys for a genuine ocean, with oceanic crust. It is then necessary to distinguish:

- *A Palaeo-Tethys*, between Eastern Eurasia and Eastern Gondwana. This is the eastward extension of the Hercynian Ocean, whose closing did not reach beyond the Near East.
- *A Neo-Tethys*, formed in the Triassic between Gondwana and blocks which detached from it: Kunlun, Qiandang, Kabul, Iran, Northern Asia Minor (future Pontides).
- *A Para-Tethys*, series of back-arc basins formed in the Jurassic on the southern border of Eurasia, from Valais to the Caspian, and fully developed in the Cretaceous.

On the other hand, the word Tethys does not apply to the narrow *Maghreb* and *Ligurian* oceans, which linked the Central Atlantic to the Neo-Tethys in the Upper Jurassic. The ocean appearing in the Cretaceous between Turkey and Egypt, already forming the Eastern Mediterranean (Herodotus Basin close to Egypt, Levantine Basin close to Lebanon), is not called Tethys either, but *Mesogea*, even at the risk of increasing the confusion.

These definitions will become clearer when the palaeo-geographic maps of the Jurassic and Cretaceous (Figures 15.9, 15.10 and 15.11) will be examined. They were established in 1980–84 by a French–Russian group, and were published in *Tectonophysics* (vol. 123, 1986). We will only give the main results from this remarkable synthesis.

Relative movement of Africa and Eurasia

In 1972, Pitman and Talwani concluded to a sinistral E–W strike-slip movement of ca. 2,000 km between the African and Eurasiatic plates between 180 and 80 Ma, followed by dextral strike-slip over 850 km between 80 and 50 Ma. A better study of magnetic anomaly stripes in the Atlantic removes the reverse strike-slip (the stretching of continental crust before rifting is accounted for, but it is small: 20 to 50 km). From 80 Ma onwards, there is only a N–S closing of the two plates, increasing from west to east and reaching 550 km at the longitude of Suez (Figure 15.12).

The African and Arabic promontories

It has been long recognized that the thrust sheets carried northward in the Austrian Alps were still part of Africa, and that the Valais ophiolites marked the suture between the two continents. This northward African 'promontory' was named *Adria* (its centre being the Adriatic Sea, with continental crust) or *Apulia*. Apulia, the Roman name of the Puglia region, is the southernmost third of the Italian Adriatic coast. This is the only region of the peninsula not recovered by allochthonous nappes

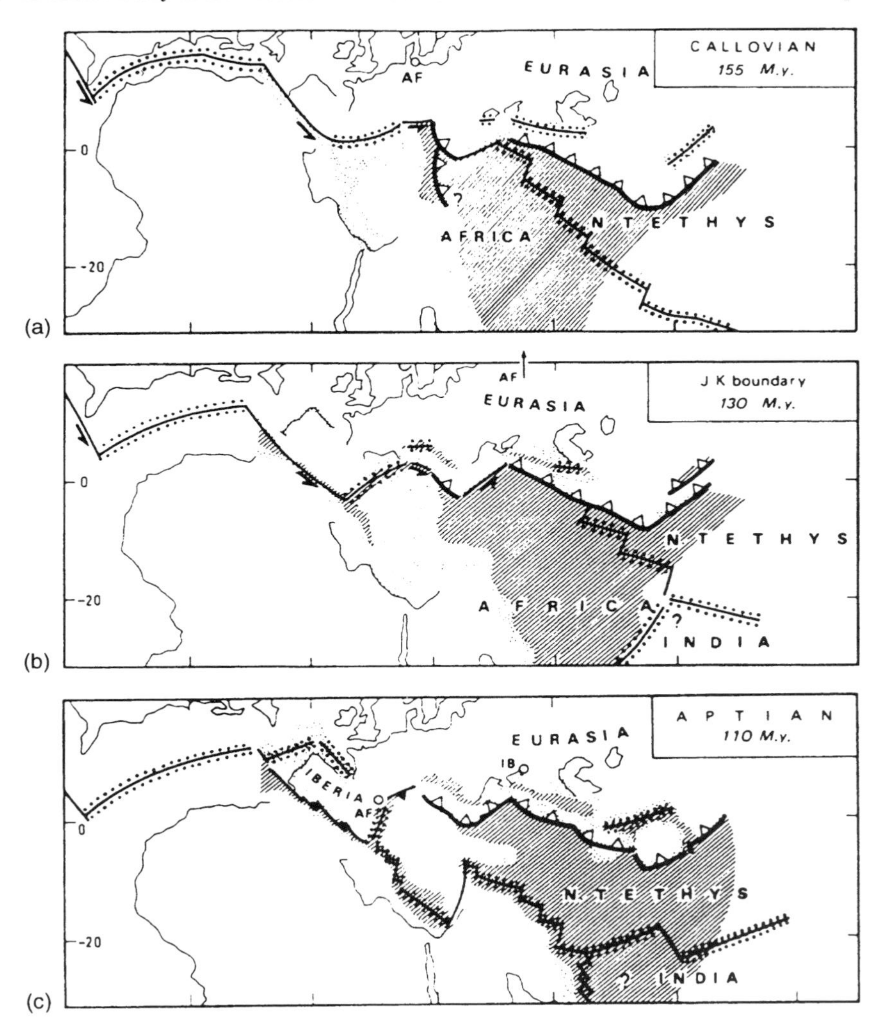

Figure 15.9. Simplified diagram of plate tectonics in the Tethys domain in the Jurassic (A, B, C) and in the Cretaceous (D, E). Each map corresponds to the geological period since the previous map. Note that the Jurassic ages have been revised since: the Callovian, at the end of the Middle Jurassic, or Dogger, dates back to 165 Ma and not 155 Ma; the Jurassic/Cretaceous boundary is 144 Ma and not 130. The Eulerian poles relative to Eurasia are indicated for the African (AF), Iberic (IB) and Indian (IND) plates. The hatched patterns correspond to oceanic crust (they were omitted for the Atlantic), the grey patterns to a stretched continental crust. The white triangles show the oceanic subduction, the black triangles the continental collision. From Dercourt *et al.* (1986); © Elsevier.

coming from the west. Adria contains most of the Dinarides basement (covered by sheets coming from the east, this time).

An Arabic promontory was also recognized, extending up to Armenia. The new fact is that there were no separate promontories. Until the end of the Jurassic, they were linked by Asia Minor (Pontides excluded).

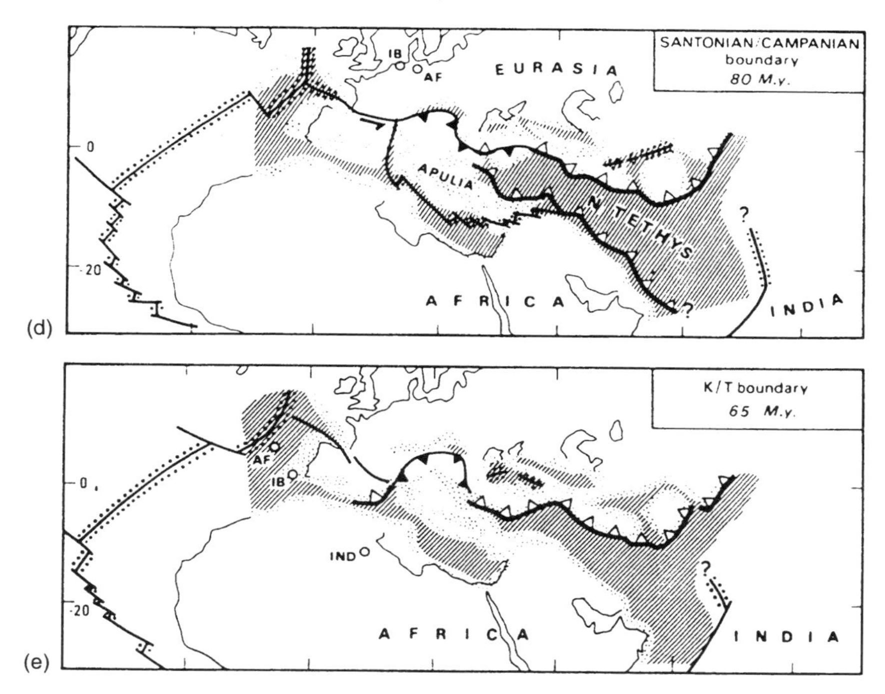

Figure 15.9 (*continued*)

Furthermore, until the Cretaceous, uninterrupted continental crust extended from Apulia to Iberia. This was the country of origin of the nappes which would later form the Apennines and Kabylia.

The rotation of Iberia and Italy in the Cretaceous

The rotation of the Iberian peninsula of ca. 30° anticlockwise in the Middle Cretaceous, between 120 Ma and 90 Ma, has been well established by palaeomagnetism (Figure 15.13). It was linked with the opening of the Bay of Biscay and the formation of the small thin-crust basins of Parentis and Adour-Mirande.

A similar rotation of the Italian peninsula was, however, subject to controversy. It seems there was indeed a 47° anticlockwise rotation between 130 Ma and 80 Ma. At this time, an Apulian plate, extending up to Asia Minor, detached from the African plate. The rotation of Apulia and its rectangular shape made one of its corners act as an indenter on Eurasia, inducing northward overthrusts in the Eastern Alps (first austro-alpine tectonic phases). Contrary to the old idea, no closing of the African and European plate was necessary; a sinistral strike-slip movement rotating Apulia was enough.

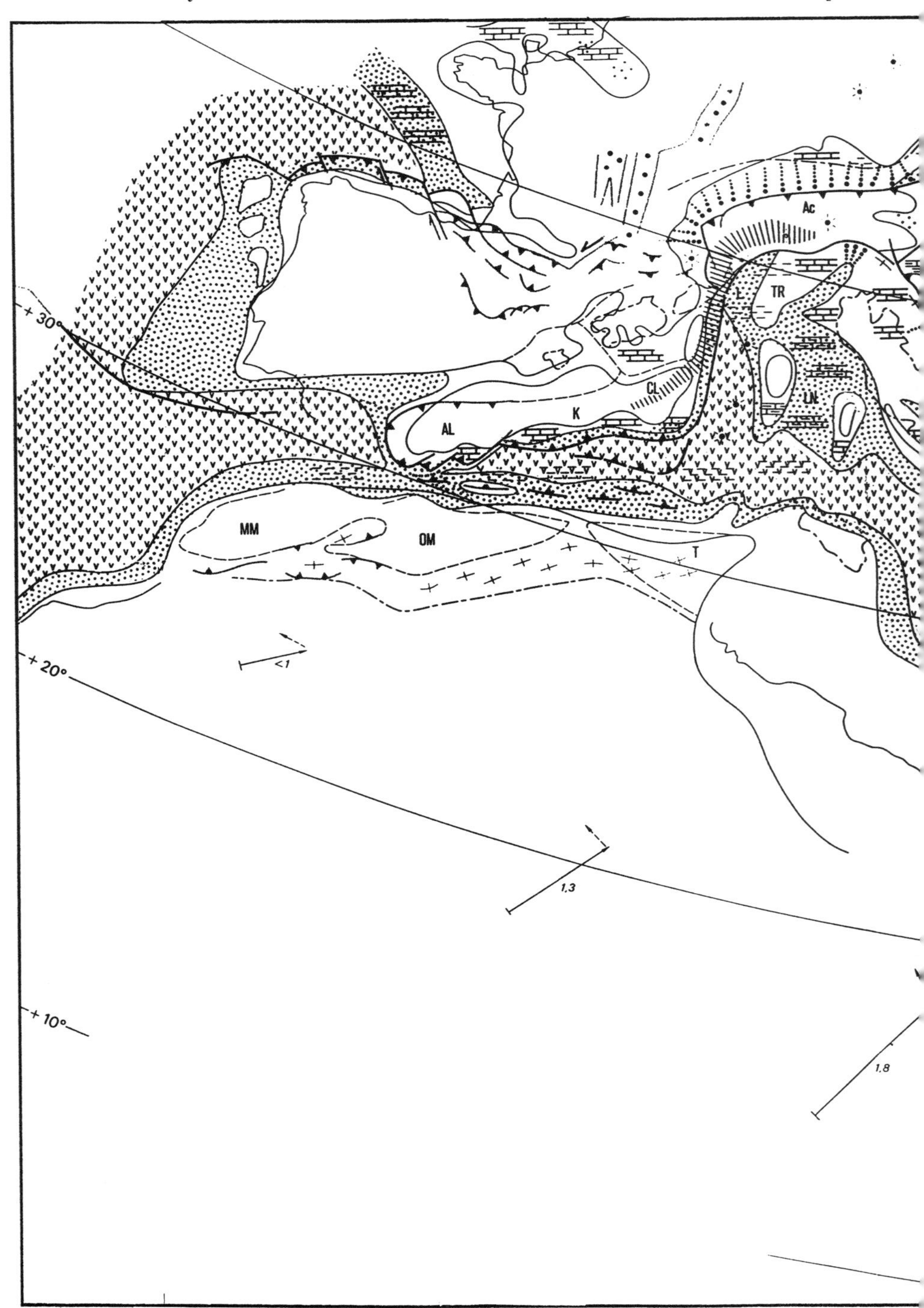
Ac
TR
CL
K
AL
LN
MM
OM
T
30°
20°
10°
<1
1,3
1,8

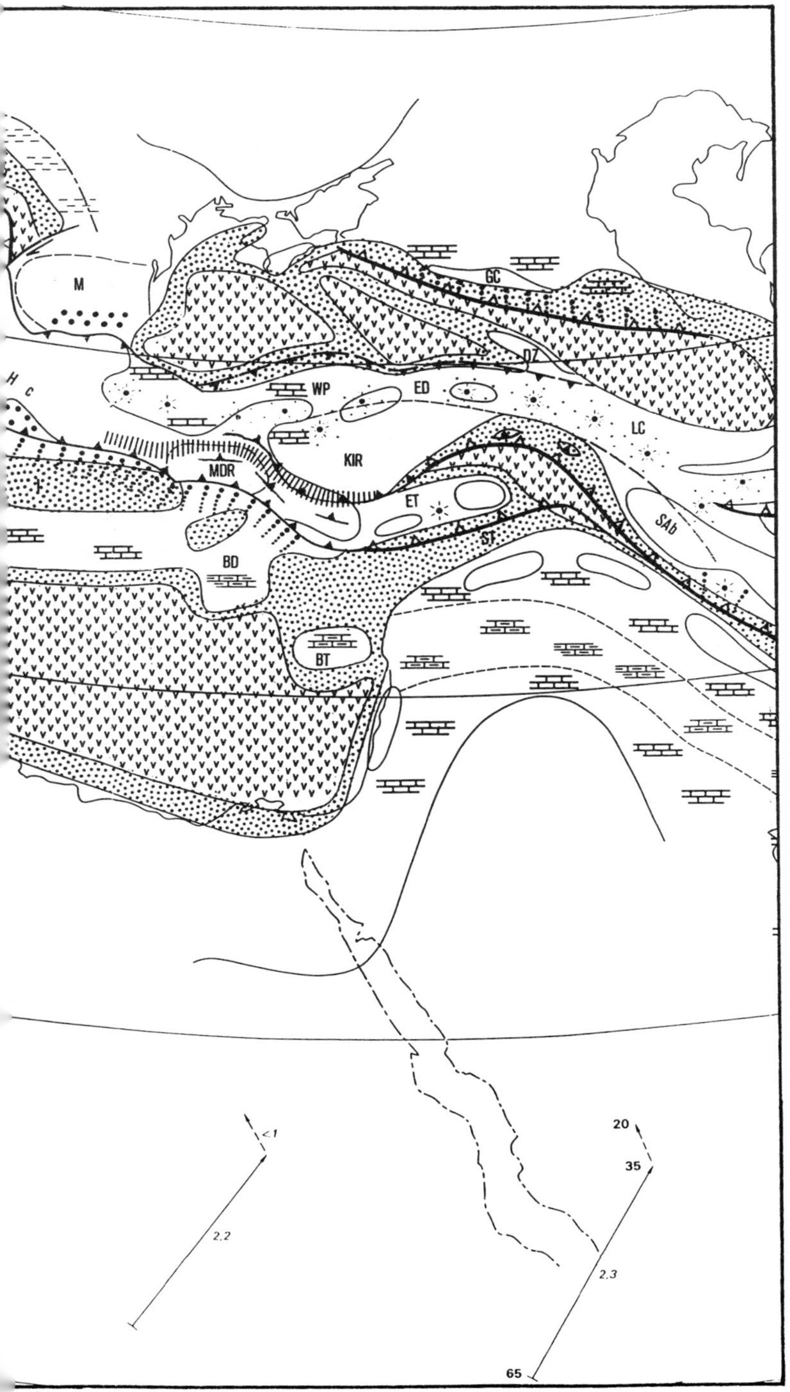

Ac = Alpine chain;
AL = Alboran;
BD = Bey Daglari;
BT = Troodos (Cyprus) basement;
Cfb = Carpathian flysch basin;
CL = Calabria;
DHc = Dinaro-Hellenic chain;
DZ = Dzirula;
ED = Eastern Dacides;
EP = Eastern Pontides;
ET = Eastern Taurides;
GC = Greater Caucasus;
I = Ionian trough;
K = Kabylia;
KIR = Kirschir;
L = Lombardy;
LC = Lower Caucasus;
LN = Lago Negro (Lucania);
M = Mesia;
MDR = Murat Dagi;
MM = Moroccan plateau;
OM = Oran plateau;
RH = Rhodope;
SAb = Southern Alborz;
ST = Southern Taurides;
T = Tunisia;
TL = Talish Mountains;
TR = Trentino;
V = Van (Turkish Kurdistan);
WD = Western Dacides;
WP = Western Pontides.

Figure 15.10. Palaeogeography of the Mediterranean Basin at the Eocene/Oligocene boundary (37 Ma). From Dercourt *et al.* (1986); © Elsevier.

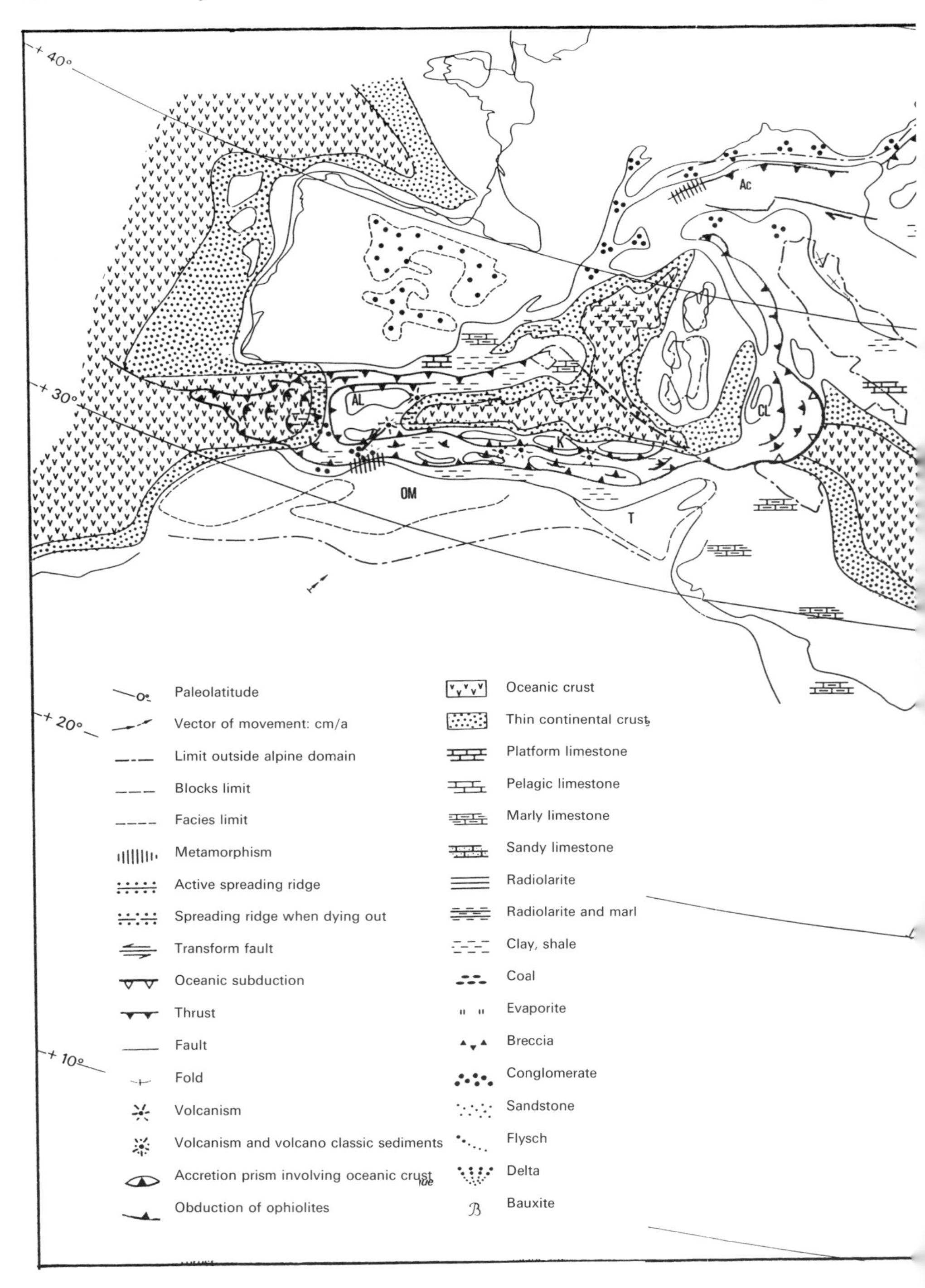

40°
30°
20°
10°
AL
OM
K
T
CL
Ac
Paleolatitude
Vector of movement: cm/a
Limit outside alpine domain
Blocks limit
Facies limit
Metamorphism
Active spreading ridge
Spreading ridge when dying out
Transform fault
Oceanic subduction
Thrust
Fault
Fold
Volcanism
Volcanism and volcano classic sediments
Accretion prism involving oceanic crust
Obduction of ophiolites
Oceanic crust
Thin continental crust
Platform limestone
Pelagic limestone
Marly limestone
Sandy limestone
Radiolarite
Radiolarite and marl
Clay, shale
Coal
Evaporite
Breccia
Conglomerate
Sandstone
Flysch
Delta
Bauxite

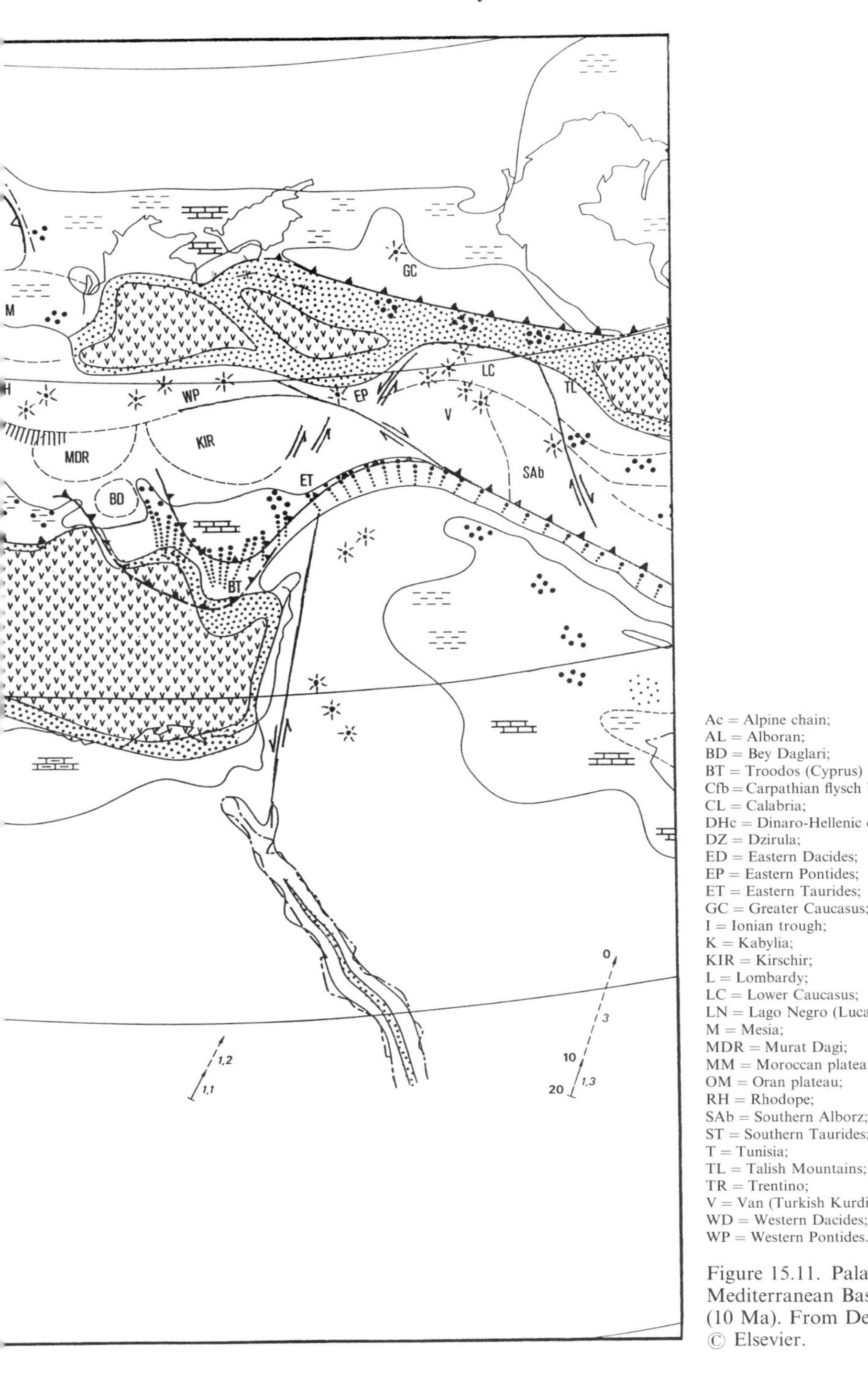

Ac = Alpine chain;
AL = Alboran;
BD = Bey Daglari;
BT = Troodos (Cyprus) basement;
Cfb = Carpathian flysch basin;
CL = Calabria;
DHc = Dinaro-Hellenic chain;
DZ = Dzirula;
ED = Eastern Dacides;
EP = Eastern Pontides;
ET = Eastern Taurides;
GC = Greater Caucasus;
I = Ionian trough;
K = Kabylia;
KIR = Kirschir;
L = Lombardy;
LC = Lower Caucasus;
LN = Lago Negro (Lucania);
M = Mesia;
MDR = Murat Dagi;
MM = Moroccan plateau;
OM = Oran plateau;
RH = Rhodope;
SAb = Southern Alborz;
ST = Southern Taurides;
T = Tunisia;
TL = Talish Mountains;
TR = Trentino;
V = Van (Turkish Kurdistan);
WD = Western Dacides;
WP = Western Pontides.

Figure 15.11. Palaeogeography of the Mediterranean Basin in the Tortonian (10 Ma). From Dercourt *et al.* (1986); © Elsevier.

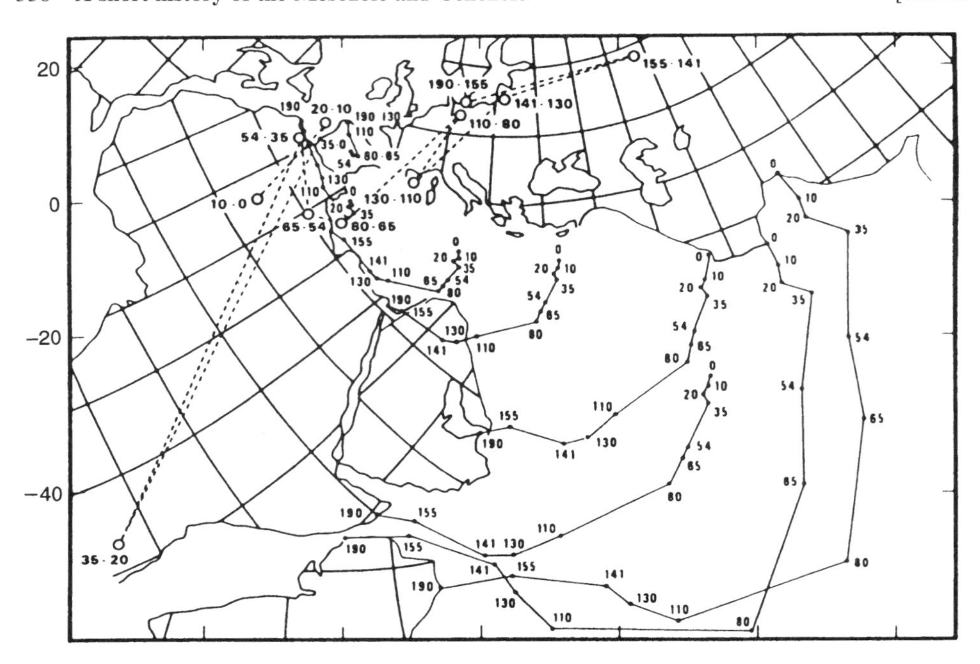

Figure 15.12. Movements of Africa and India relative to Eurasia. The numbers are dates in Ma B.P. The Eulerian pole for Africa is shown with small circles, joined by a dashed line which more or less forms the base of the rotation, as defined in Section 7.1. From Savostin *et al.* (1986); © Elsevier.

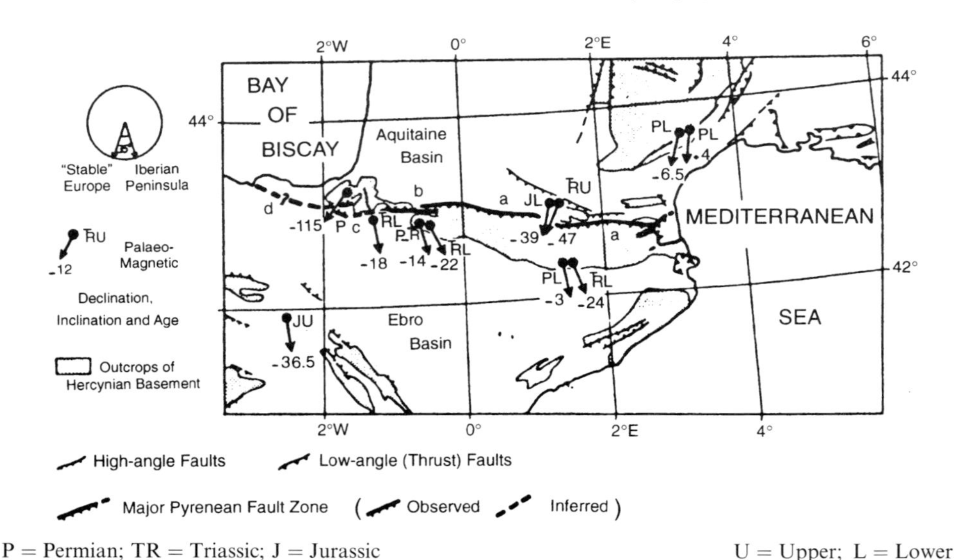

P = Permian; TR = Triassic; J = Jurassic U = Upper; L = Lower

Figure 5.13. Palaeomagnetic declinations (arrows) and inclinations (numbers) in the Pyrenees area. The Northern Pyrenees fault is shown: (a) lherzolite outcrops; subvertical en-échelon faults; (c) mylonitized zones. In (d), west of Bilbao, only pillow lavas are observed. From Van der Voo and Boessenkool (1973); © AGU 1973.

The 'Alpine cordillera'

Today's geography of the mountains in the Mediterranean Basin long argued in favour of two series of mountain chains, with a mysterious sinuous track. In the North, this was the Betic chain (Sierra Nevada of Andalusia), the Pyrenees, the Alps, the Carpathian and Balkan mountains, the Pontides and the Caucasus. In the South: Atlas, Peloritan Mts of Sicily, Calabria and Apennines, Dinarides, Hellenides, Taurides. In fact, most of these chains have nothing to do with each other and come from distinct megatectonic processes. Thus, the Carpathian arc comes from the subduction of a small external Carpathian ocean, the Pyrenees come from a beginning of subduction of Iberia along the trace of an old transform fault, the Atlas come from intra-continental folding, and the Apennines come from overthrusts and nappes coming from the west.

It is, however, possible to consider a series of WSW–ENE chains, forming what I call the 'Alpine cordillera', a rectilinear cordillera longer than 2,000 km. It manifested itself since the Cretaceous, through northward overthrusting and metamorphism (Figure 15.10). It enclosed the Eastern Dacides,[2] the Alps, Calabria, Kabylia, and the Betic chain. It was doubled in the north by the Briançonnais, Corsica, Sardinia and the Balearic Islands. This configuration has not changed since the Oligocene. Calabria and Kabylia only started migrating toward their current positions in the Lower Miocene (Aquitanian, ca. 20 Ma) (Figure 15.11).

Brevity of oceanic subductions in the Alpine domain

In 1984, this result ended a dispute (see Section 14.2, *in fine*). There had indeed been subductions in the Upper Jurassic and the Lower Cretaceous, but none during the Cenozoic. The discovery in Piedmont, in the Dora Maira massif, of crustal rocks having undergone ultra-high pressure metamorphism, 100 km or maybe 300 km deep (see Section 3.8) did not invalidate this affirmation. Considering the current subduction rates (see Section 7.1), one can expect the appearance of a 500- to 800-km long subducting plate in 10 Ma.

15.5 THE WESTERN ALPS

The building of the Alps is a very long story, starting 130 Ma ago. Ranges appeared, which were destroyed by erosion. The covering terrains contain limestones formed in epicontinental seas, schists coming from deep-water sedimentation, flysch sediments announcing the appearance of these mountains and molasse sediments resulting

[2] Geologists use the name of the old Roman provinces. *Docia* corresponds roughly to Transylvania, not yet to all modern Romania. The Dacides are therefore made of the Carpathian arc and the mountains inside. *Mesia* was between Dacia and Thrace, and corresponds to the Lower Danube basin, between the southern Carpathian and the Balkan mountains. *Pannonia* corresponds to the middle Danube basin, between Budapest and Belgrade (Hungary + Serbian Voivodina).

from their destruction. This geological history was not written on a blank sheet. There was here a Hercynian peneplain with its faults and structural lines. It is, for example, possible to discover in the Grandes Rousses and in Deux-Alpes the rest of a Hercynian N–S pinched fold. The outer crystalline massifs are plutons from this period, unroofed by erosion.

Three main periods can be seen in this history:

From 130 to 50 Ma: the austro-alpine phases of the Upper Cretaceous continue into the Eocene, with emplacement of Pennic nappes coming from Adria. These sheets end in France at the latitude of the Belledonne chain. Further south, in the Vocontienne trench (Diois and Baronnies, where Urgonian limestone, coming from littoral corals, is absent, whereas Tithonian limestone, coming from pelagic foraminifera, is abundant), folds roughly oriented E–W appear around 90 Ma (Turonian-Coniacian).

From 50 to 25 Ma: the Pyrenean–Provençal phase occurs in the Eocene, and the Alpine phase (strictly speaking) at the beginning of the Oligocene. There is compression and uplift of the Briançonnais area, between reverse faults and opposite overthrusts. And the outer crystalline massifs of the Northern Alps (Mont-Blanc, Belledonne) are carried westward. A sinistral strike-slip of the Alps, relative to the Lombardy–Trentino region, corresponds to these overthrusts. The consensus is that the alpine mountain chain, rectilinear until then, starts to take its arcuate shape at this time.

From 20 to 4 Ma: folding and uplift affect the outer zone, giving it its modern aspect (apart from the glacial erosion). The last phase, in the Pontian (Lower Pliocene, around 4 Ma) is paroxysmal in Jura. On the Italian side, the Po Basin subsides and is shortened by many reverse faults, now masked by several kilometres of recent sediments.

Figure 15.14 shows a rough geological map of the Western Alps. Figure 15.15 reproduces four geological cross-sections of the alpine arc, established in 1972 when the deep structures could still not be accessed. The French–Italian ECORS-CROP programme (1985–89), which disclosed them, did not change these conclusions (Figures 15.16–17).

Pennic nappes

The cross-section no. 1 shows four superposed Pennic nappes. From top to bottom:

- the Austro-alpine thrust sheet (A), whose last fragments are found south of Zermatt (Cervin, Dent Blanche).
- The lustrous shales of Piedmont, coming from the Liguro-Piedmontese Ocean closed 100 Ma ago after southward subduction.
- The thrust sheet of Grand Saint Bernard.
- The lustrous shales of Valais (V), coming from the Valais ocean. Its subduction southward took over the subduction of the Piedmont Ocean, enabling the continuation of the Africa–Europe closing.

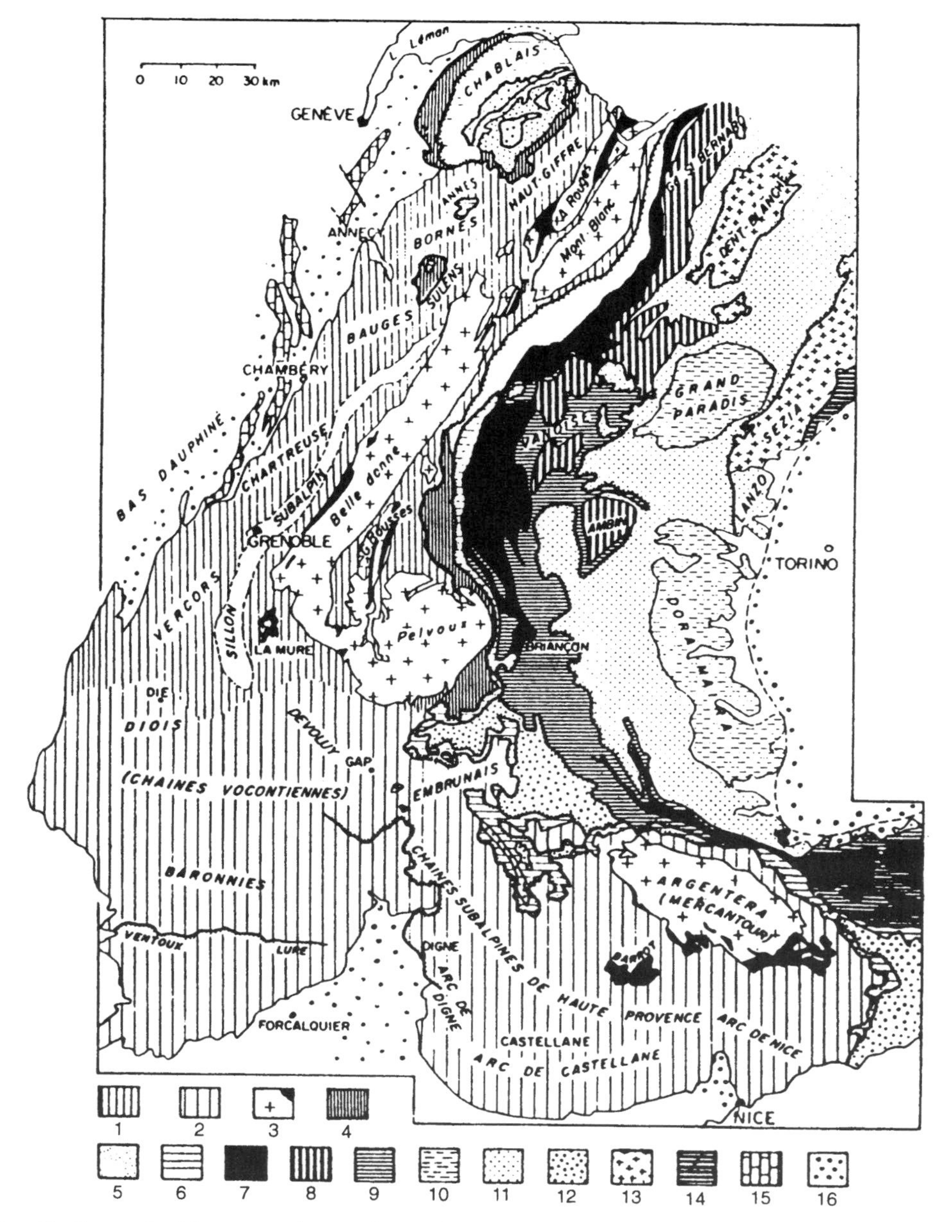

1 = Northern sub-alpine ranges; 2 = southern sub-alpine ranges; 3 = outer crystalline massifs and Permo-Carboniferous basins; 4 = ultra-dauphinese zone; 5 = Valais area; 6 = sub-Briançonnais area; 7 = Briançonnais coal-bearing zone; 8 = Vanoise/Mt. Pourri area (Permo-Carboniferous metamorphic Briançonnais); 9 = Briançonnais Mesozoic; 10 = inner crystalline massifs of Piedmont; 11 = Piedmont lustrous shale zone; 12 = neo-Cretaceous flysch zone; 13 = SesiaDent Blanche area; 14 = Canavese and Ivrea areas; 15 = Jura; 16 = peri-alpine molasse basins.

Figure 15.14. Simplified structural map of the French–Italian Alps. From Debelmas, 1974.

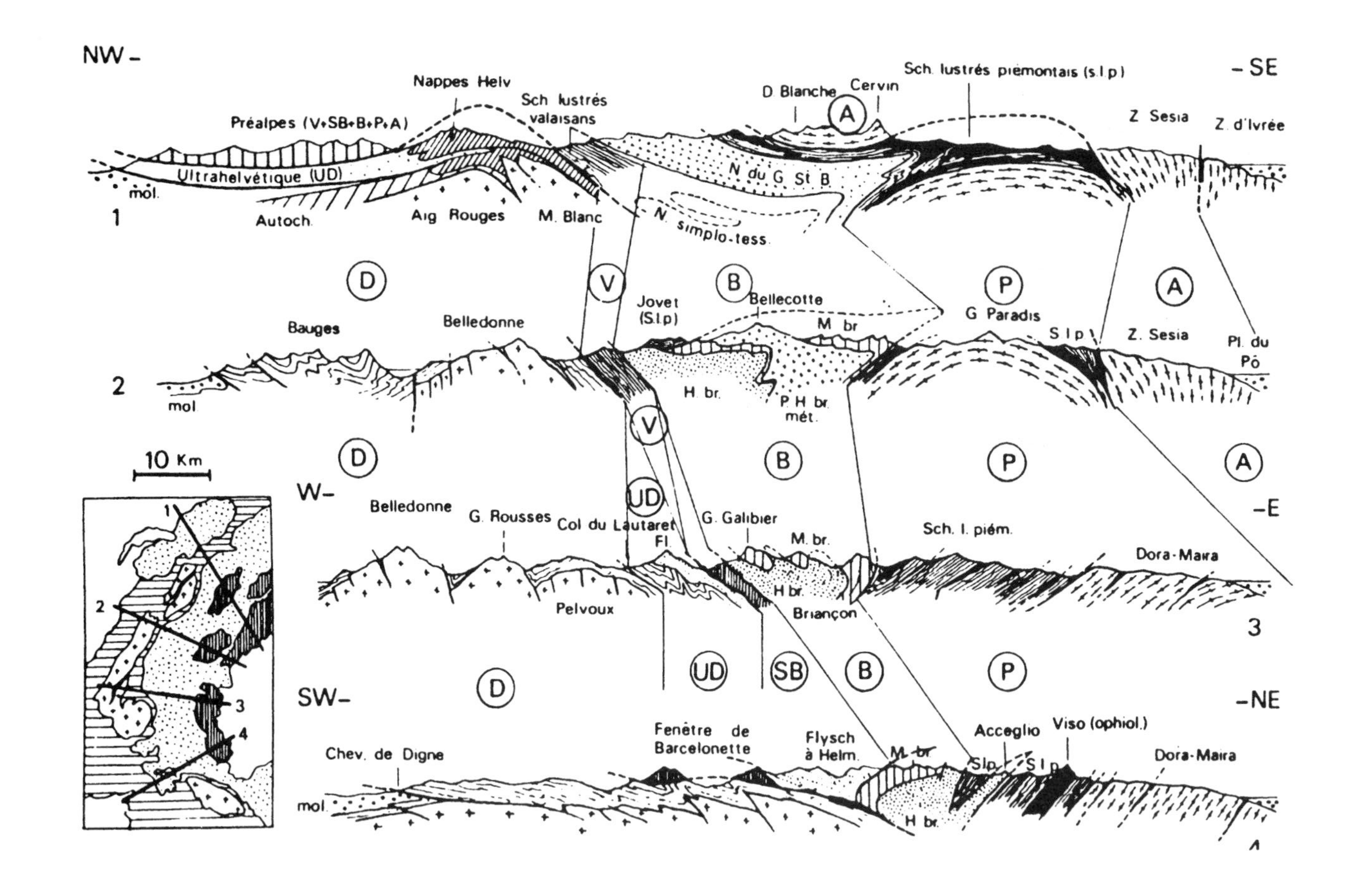

D = Dauphiné area; UD = Ultra-Dauphinese zone; V = Valais area; SB = sub-Briançonnais area; B = Briançonnais area; P = Piedmont area; A = austro-alpine; PHbr = Permo-Carboniferous Briançonnais; Slp = Piedmont lustrous shales.

Figure 15.15. Four cross-sections in the Western Alps arc. From Debelmas, 1974.

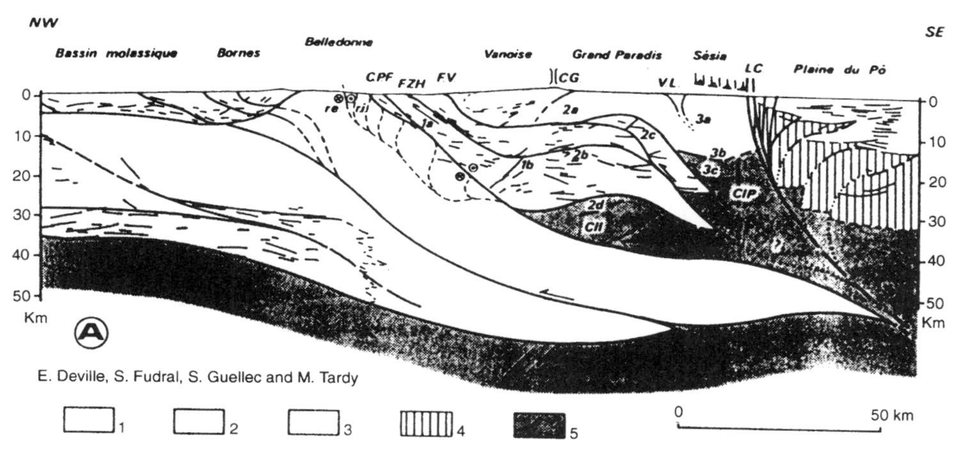

1 = sedimentary and meta-sedimentary covers; 2 = European continental crust, upper or undifferentiated; 3 = stratified lower crust; hashed lines = African continental crust (Apulian); grey patterns = upper mantle.

FZH = front of the coal-bearing area; CG = Galise Pass; VL = Viu-Locana scales; LC = Canavese line.

Figure 15.16. Interpretation of the ECORS-CROP seismic profiles in the Alps, proposed by Deville *et al.* The overthrusting of the outer crystalline massif of Belledonne is linked to deep crustal thrust. In this model, a lower Ivrea body (CII) is linked to the Pennic frontal thrust (CPF), and the main Ivrea body (CIP) is linked to the front of the Vanoise metamorphic units (FV). Another likely model was proposed by Ménard *et al.*: the late eastward overthrusts affecting the African crust would have extended further west and detached the main Ivrea unit. From Tardy *et al.* (in Roure *et al.*, 1990).

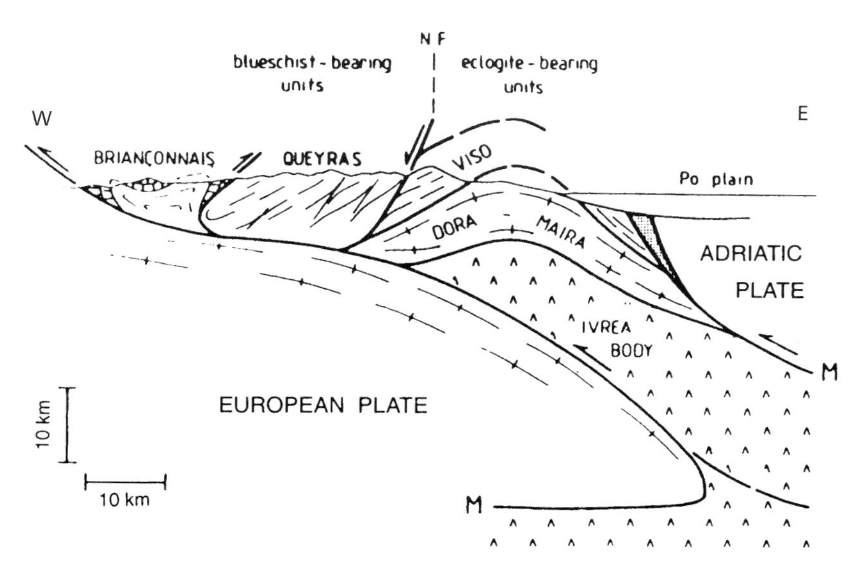

Figure 15.17. Hypothetical cross-section of the Western Alps, south of the cross-section shown in Figure 15.16. A normal fault (NF) would have enabled the exhumation of the Mt. Viso and Dora Maira units (see text for details). From Ballèvre et al., 1990.

The NNW limit of these nappes forms the *pennic front*. These sheets straighten up at their root, probably following later movements. This had formerly led to their interpretation as huge recumbent folds.

Briançonnais and outer nappes

The terrains forming the Grand Saint-Bernard nappe become in France the Briançonnais region, whose central part is autochthonous. It forms the Vanoise and the Briançon area. West, the lustrous shales of Valais are replaced by South-Briançonnais (SB) and Ultra-Dauphinese of Aiguille d'Arves (UD), both laid or thrust over the outer crystalline massifs. Between them, there is a layer of flysch with helminthoids (very sinuous gastropod tracks). In the Embrunais area, they form a covering nappe. East, the lustrous shales of Piedmont are thrust eastward.

The outer covering nappes form the limestone massifs of the Northern Pre-Alps (Bornes, Bauge, Chartreuse, Vercors) and most of the Alps of Upper and even Lower Provence (even if their exact extent is still discussed). As they set around 20 Ma ago, during which Europe and Africa only closed by one hundred kilometres, they can only be gravitational (see Section 14.6). Rather than imagining a sliding over a single huge dome, the 'Briançonnais Geanticline', or a progression over some mysterious 'orogenic waves', I suggest the progression on the slope bordering a plateau, during its post-orogenic subsidence and spreading.

The curvature of the Alpine arc

If one trusts the French–Russian palaeogeographic reconstructions, the alpine chain was rectilinear until the Oligocene. Its direction was not very different from the Pyrenees, and it is not surprising that the Pyrenean folds of the Eocene went on through Lower Languedoc and Provence. In the Oligocene, the Alps start curving and take a NW–SE direction in Upper Provence. But a rotation of 90° or more should be excluded, for two reasons:

(1) Palaeomagnetism does not detect such a rotation in the outer crystalline massif of Argentera (or Mercantour). It even seems that, at depth, a crystalline basement links this massif to the Maures massif.
(2) The previous set of folds approximately EW, formed earlier between the upper Cretaceous and the Eocene, roughly kept this direction in Provence. Palinspastic restorations of limestone beds in the Tithonian (Upper Jurassic) by Gratier *et al.* (1989) showed that there was indeed anticlockwise rotation, but of 15°–20° only. And the whole Southern Alps have not rotated as a block, because there was simultaneous strike-slip on SW–NE faults (parallel to the big Cévennes fault). Finally, it seems that only the upper crust rotated slightly.

Passing over recent speculations on the matter, I personally think this is a *false problem*. The only arcuate shape is in the topography. The higher areas have been eroded, and the arcuate shape is found in the map of geological outcrops. The outer sheets, gravitational, only obeyed topography to slide radially. But there was no rotation of the basement and its tectonic directions by 90° or more.

The only problem is now to explain an arcuate orography, which is a common occurrence on the Earth, in case of punching or basin subsidence.

The Ivrea body

The French–Italian project ECORS-CROP extended the geological cross-sections of Figure 15.15 into the upper mantle. Along with the vertical and 'wide-angle' data from seismic exploration (obtained with receiving stations far from the source), the gravity and magnetic anomalies were also used Nicolas *et al.*, 1990).

A cross-section through Ivrea, in the Aosta valley, is reproduced in Figure 15.16. It shows that the westward overthrusting of Belledonne concerned the whole crust down to the mantle, although the crust is up to 60 km thick. Furthermore, a large upper mantle slab went up to 15 km from the surface, creating the strong positive gravity anomaly of Ivrea.

At least two interpretations of the seismic soundings are possible. We reproduced the one in agreement with the interpretation of the Dora Maira massif, made by Ballèvre *et al.* (1989) (Figure 15.17). This inner crystalline massif, of moderate height, is not an old pluton, like the outer crystalline massifs, but an old portion of lower crust, metamorphosed at very high pressure, pushed close to the surface by the Ivrea unit and unroofed by erosion. There was some sort of huge *continental obduction*. The Mt. Viso ophiolites were pushed before Dora Maira.

The Insubrian Line

The Insubrian Line is a series of faulted zones marking in the Alps the southern limit of the European crust. Eastward, it is still in some places covered by the austro-alpine thrust sheet of African origin. It begins north of Torino by the SW–NE Canavese line. Its trajectory is rectilinear and W–E from Locarno to the Upper Adda, crossing the Sondrio valley and the Adamello massif, a later granitoid intrusion. After a trajectory SW–NE again, the Insubrian Line becomes W–E again in the Pusteria valley (northern limit of the Dolomiti) and the Austrian valley of the Gail, until Klagenfurt.

This major accident therefore dates from the formation of the Pangaea and the Hercynian orogenesis, but it was recently reactivated. In the Miocene, it enabled a sinistral strike-slip movement, like the Anatolian fault (it is tempting to see a common cause in a push from the Arabic Plate). Folding in the outer Alps could be associated to it.

Later, the Insubrian fault allowed the Lombard–Venetian region to subside. It became a reverse fault, allowing the Eastern Alps to overlie slightly the Po Basin. This was for example shown with the focal mechanism of the Friuli earthquake of May 6, 1976.

15.6 THE APENNINES AND THE TYRRHENIAN BASIN

The geology of the Italian peninsula is very well known, and may come as a surprise to students from other countries. Nearly everywhere, indeed, the peninsula consists in gravitational nappes having migrated east (southward in Sicily), in the course of events following the same path. By analogy with the French Alps (and according to the geosyncline theory), another high chain, origin of these sheets, should be found between the peninsula and the Corsica–Sardinia system. It should indeed have existed in the North, not in the South. The Tyrrhenian Basin is now found everywhere, going down to 3,620 m below sea level. South of the Arno River, in the Central Apennines and in the Calabrian arc, the autochthonous basement migrated relative to Apulia. It migrated west for a long way (800 km for Calabria), before the formation of the Tyrrhenian Basin. This displacement was similar in scale to the allochthonous Californian 'terranes', but with no large transform fault or mid-ocean ridge to explain it.

The reality of these displacements only came to light in the last 20 years. The geodynamical cause is still mysterious, as is the later collapse of the Tyrrhenian Basin. There was some similarity with the appearance of intra-arc basins in the Pacific. But the Tethysian domain did not experience any subduction of a large oceanic plate for a very long time after the Palaeozoic. Because this book is meant as an introduction to future research, and there are enigmas to investigate, we should examine the facts in greater detail.

Rotation of the Corsica–Sardinia block

Corsica is made up of two structural domains, divided by a narrow strip with ophiolites, along a line Balagne–Corte–Aléria. The stratigraphic series of Eastern Corsica has a facies similar to the alpine ones, and can therefore be called alpine. Western Corsica, however, is formed of Palaeozoic terrains affected by the Hercynian orogeny. So is Sardinia, with which it forms one single block. The same Hercynian veins are found in granites on each side of the Strait of Bonifacio, with no changes in orientation.

Palaeomagnetism has told that, since the Permian, Sardinia rotated anticlockwise by 60° relative to the European platform, whereas Corsica rotated by only 25°. One of these two results is evidently wrong. Geology suggests that the Corsica–Sardinia block was once aligned with Lower Provence, as part of a long rectilinear 'Alpine cordillera'. We should therefore keep the value of 60°.

More data comes from magnetic anomaly stripes corresponding to chron 6, found in the middle of the Provence Basin (see Section 14.17). Following French oceanographic cruises in 1979–82, it was discovered that two oceanic crust domains formed between 25 and 19 Ma ago. The first one was ranging from 10° to 15°, between Corsica and the Maures Massif. The second one was ranging from 25° to 30°, between Sardinia and the Bay of Biscay continental platform. But this does not imply that the rotations of Corsica and Sardinia were different. They both were 60°, but *the largest part was reached by crustal thinning*. Beyond a specific thinning, and therefore near the end of the rotation, a mid-ocean ridge appeared and generated some oceanic crust.

Simultaneously, in the Palaeogene, Lombardy and Trentino migrated NE, and a N–S range emerged east of Corsica.

The Northern Apennines (Figure 15.18)

Two superposed nappes made of Upper Cretaceous sedimentary rocks are found in the area of Genoa: the Mt. Antola sheet (flysch with helminthoids) and the Mt. Gottero sheet (sandstone). They lie over a series outcropping further east, in the area of Bracco Pass, and enclosing:

- clays with limestone interstratification from the Lower Cretaceous;
- pillow diabases and radiolarites from the Upper Jurassic;
- an ophiolitic base of gabbros and peridotites, which may be considered as the southern extension of the Insubrian line.

During the Lower and Middle Eocene (55–45 Ma), these sheets moved south from the Alpine chain (*Ligurian phase*). Much later, during the Tortonian (12–7 Ma), the region east of Corsica gained altitude, and they moved to the NE (*Tuscan phase*). There was no more Ligurian phase between the River Arno and the Ligurian zone, and only the NE overthrust of the thick Tuscany sheet.

After 7 Ma, in the Messinian and Pliocene, the nappes progressed eastward, over Umbria. Simultaneously, the mountain between Corsica and the continent subsided and the Ligurian Basin formed between Genoa and the Island of Elba. (Although normal faults appeared in the basin, 'subsidence' is better adapted than 'extension', because Corsica did not move away from Italy). This change of slopes means that gravity nappes in Alpine Corsica and in the Apennines now show a slope opposed to the slope which presided to their setting: the front of the Antola nappe now forms the Apennine crest!

Central Apennines

The Central Apennines are mainly formed of the Latium and Abruzzi carbonate platform. It is very irregular, and the detrital material is abundant. Indeed, grabens and horsts appeared during its slow formation during the Mesozoic and the Lower Miocene. There are no nappes, only folds, fold-faults and overthrusts with reverse faults and eastward overthrusting.

Some orogenic phases have only been accelerations in a continuous process, as elsewhere in the Apennines. This is the case of the *sub-Ligurian phase*, in the Oligocene–Aquitanian (32–28 Ma), the Tuscan phase in the Tortonian and two others in the Messinian and the Lower Pliocene. (Three distinct phases have been recognized between 12 and 4 Ma. As always in geology, the more recent the events the finer the analysis.)

Neotectonics is completely different, as the Assisi earthquakes of 25/26 September, 1997 (respective magnitudes of 5.6 and 5.8) correspond to an E–W

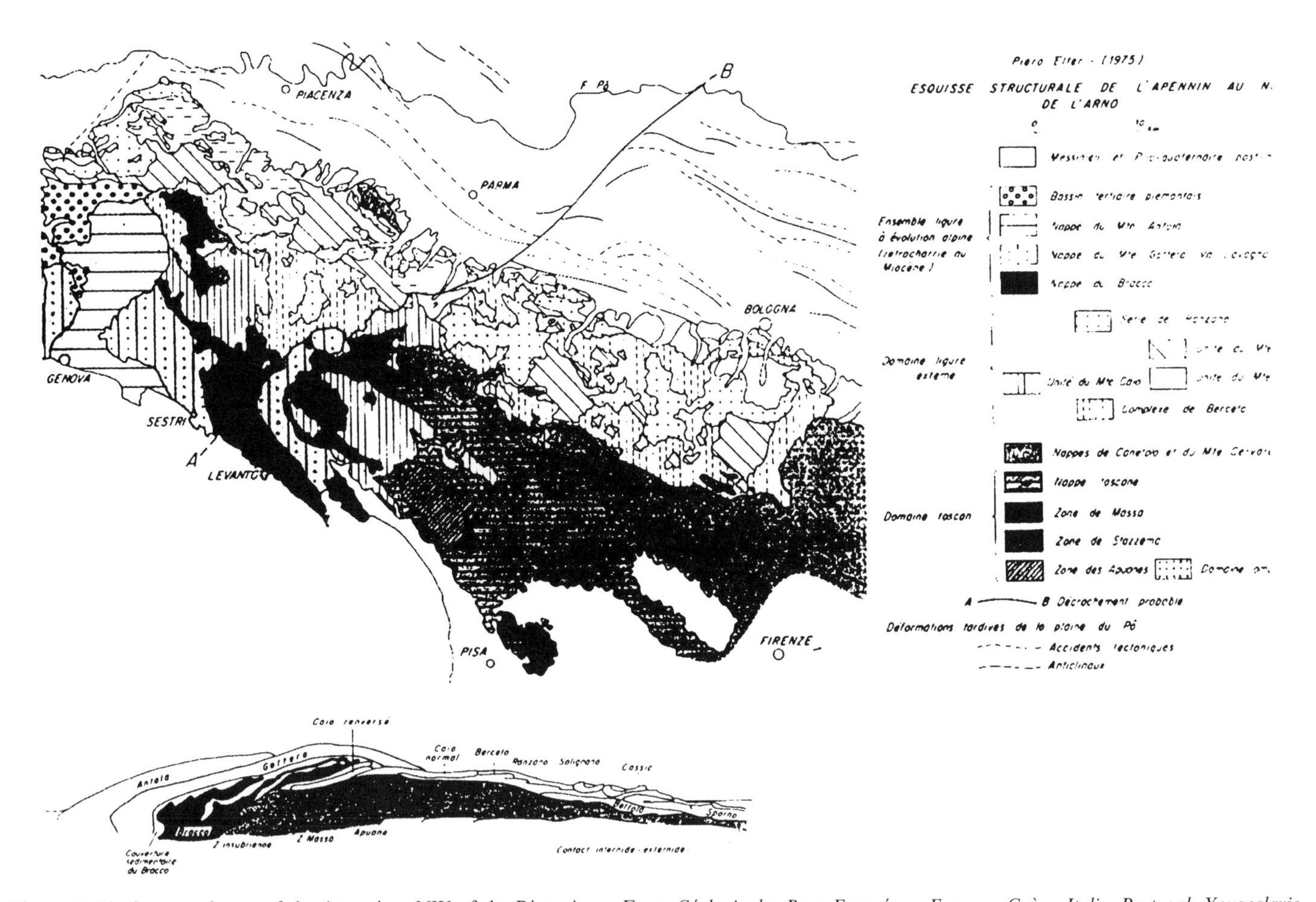

Figure 15.18. Structural map of the Apennines NW of the River Arno. From *Géologie des Pays Européens: Espagne, Grèce, Italie, Portugal, Yougoslavie*, Dunod, Paris, 1980.

extension, like all other earthquakes in the region. Previously reverse faults are now moving in the opposite direction, and have become normal.

All along the Adriatic Sea, the Palaeozoic basement of Apulia slopes down west below the Apennines. The Sicilian one outcrops on the southern coastline, near Sciacca and the Iblei mountains, and slopes down northwards. Completely filled with sediments from the Pliocene and Pleistocene, the long depression which surrounds the mountains is called the 'foredeep' by Italian geologists (a name inherited from the old geosyncline theory). Its depth varies from < 2 km to 7 km in central Sicily (Caltanisseta basin).

Southern Apennines and Sicily

Sicily was always fixed relative to Africa. The nappes covering its northern part, which have flowed southward, are extending into Tunisia. The Apulian basement has not moved either, since its Cretaceous rotation. However, at the edge of the Tyrrhenian basin, Calabria[3] and its Sicilian extension (Peloritan Mts.) migrated eastward by 800 km. It belonged to the 'Alpine cordillera' of the Upper Cretaceous-Palaeogene. This explains the alpine facies of these terrains.

Calabria was then separated from Apulia by the small Ligurian Ocean (which provided ophiolites), and by a large large epicontinental sea, with thinned continental crust, with alternating shallows and deep basins. This whole region provided material for the gravity nappes surrounding the Calabrian arc, with outgoing overthrusts. Figure 15.19 shows the simplified chronology of their setting.

Tyrrhenian Basin (Figure 15.20)

The periphery of the Tyrrhenian Basin is a thinned continental crust, 20-km thick. It thins progressively toward the centre of the basin, where there is oceanic crust. Below 600 m of sediments and a thin layer of basalts, the oceanic layer 3 is 5-km deep, with seismic velocities between 6.6 and 7.3 km/s. This crust lies on top of abnormal mantle ($v_P = 7.6$ km/s instead of the usual 7.9–8.2 km/s). The central oceanic crust is younger than 6 Ma, because it does not present the salt layer of the Messinian (see later for details), which exists nearer Sardinia and is very thick. Its eastern part, the *Marsili Basin*, is even younger (less than 1.9 Ma).

There are many magnetic anomaly lines in the Tyrrhenian Sea, coming from basaltic dykes, but they are usually less than 100-km long. And they are distributed along three azimuths (E–W, N–S and NE–SW), not a single one as with a mid-oceanic ridge.

Tholeiitic magmas, similar to mid-ocean ridge magmas, were output by the mantle in Ustica between 7.5 and 2.5 Ma (Ustica is the westernmost Aeolian Island, in front of Palermo). There were also eruptions of alkali basalts (e.g.

[3] Although the names of the old Roman provinces are used (e.g. Apulia), Calabria corresponds to the current province, close to Sicilia. The Romans called it Brutium, and called Calabria the region of Taranto and Brindisi.

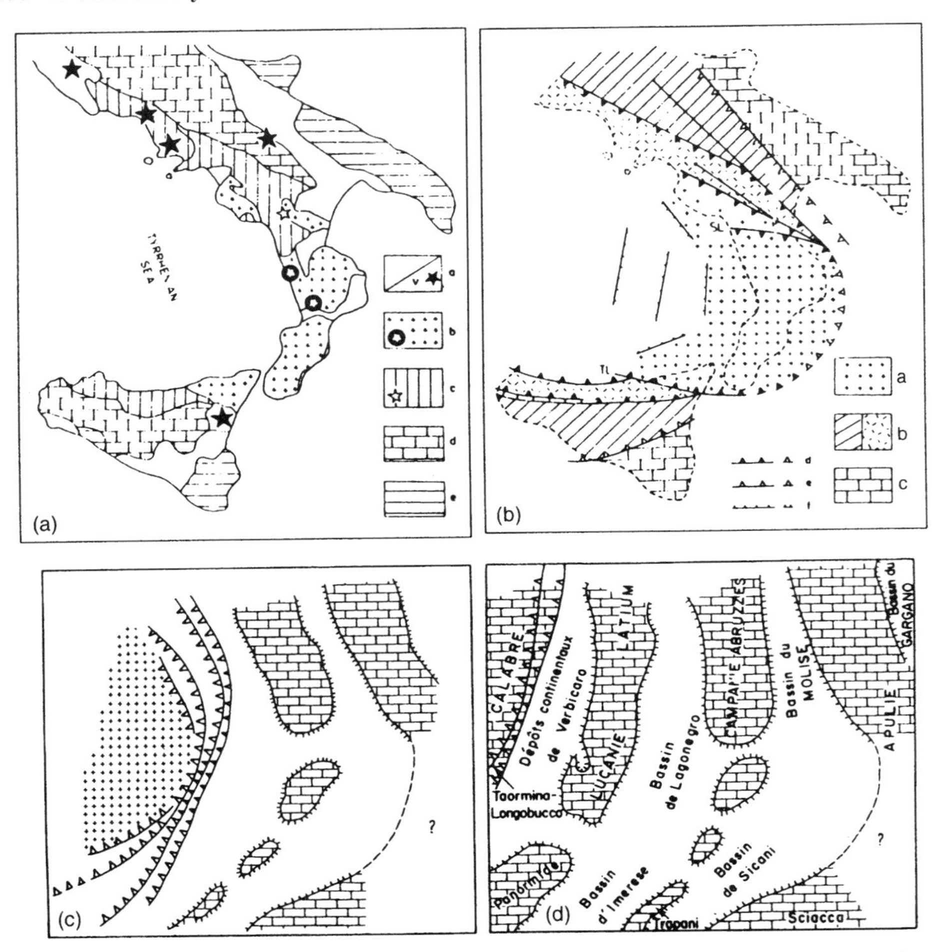

Figure 15.19. A: tectonic diagram of Southern Italy; (a) foredeep, main Pleistocene deposits (marine or continental, volcanics); (b) Calabrian arc and associated tectonic units; (c) Apennines and Sicily tectonic units deformed before the Tortonian; (d) Sicily and Apennines tectonic units, from the Tortonian to the early Pliocene; (e) foreland. B: Calabrian arc after the Lower Pliocene tectonic phase; (a) Calabrian massif; (b) units resulting from deformations before the Tortonian (right) or in the Tortonian and early Pliocene (left); (c) foreland; (d) pre-Tortonian overthrusting front; (e) later overthrusts; (f) normal faults; SL and TL represent the tectonic lines of Sangineto and Taormina. C and D: palinspastic restorations at the Middle Miocene and just before the Miocene. From Argenio *et al.*, 1980.

Etna). Between 1.3 and 0.2 Ma, volcanism moved away from the centre of the Tyrrhenian Basin and became calc-alkaline (lavas closer to andesites than proper basalts). Even more recently, lavas from the potassic series (shoshonite) appeared in the Aeolian Islands: 100,000 years ago in Vulcano, 30,000 years ago in Stromboli. This evolution of volcanism is completely different from the evolution observed in traditional intra-arc basins; it is attributed to magma contamination by continental sediments, more and more abundant with distance away from the centre.

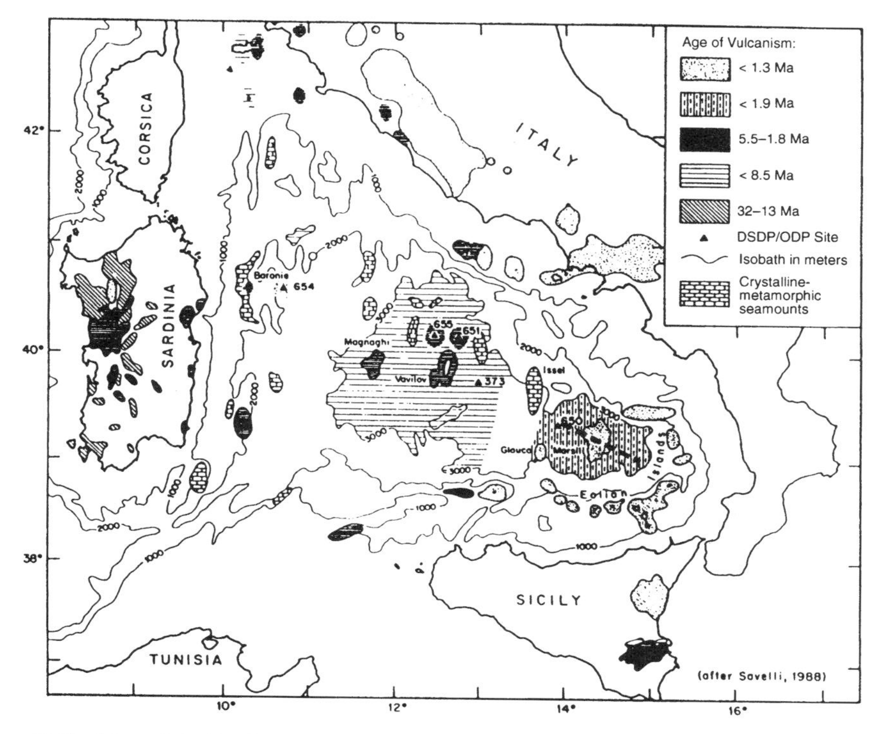

Figure 15.20. Map of the Tyrrhenian Basin and the age of its volcanism. The emplacement of deep-sea drillings ODP 373, 651 etc. and a seismic profile are also shown. From Wang *et al.*, 1991.

The earthquake sources are deeper and deeper as distance from the Calabrian arc increases to the NW. The depth is close to 500 km below the Vavilov submarine volcano, 300 km from the Messine Strait. But the Wadati–Benioff surface thus defined is very irregular, fluted, with narrow radial stripes of shallow earthquakes up to 150 km from the arc (Wezel, 1982). Let us remark immediately that the simple closing of Europe and Africa during post-Palaeozoic times was completely insufficient to create a 600-km long subducting plate.

Westaway (1993) infers from a study of raised beaches an uplift of all Calabria, starting 0.7 Ma ago, at the velocity of 1 mm per year. He suggests that the cause would be a breaking of this subducted plate, a totally unconvincing explanation. The uplift of Calabria in the Pleistocene is one more among the documented vertical movements in the area, that plate tectonics cannot predict.

15.7 OTHER ARCS AND BASINS IN THE MEDITERRANEAN

The other basins of the Western Mediterranean show some of the Tyrrhenian Basin characteristics.

The *Valencia Basin*, between Iberia and the Balearic Islands, has a 20-km thick crust above abnormal mantle ($\nu_P = 7.7$–7.8 km/s). Extension, fracturation, and intrusion of basaltic magmas in the continental crust (producing diorites and granites, through contamination, along with the gabbros) have transformed it into an *intermediate crust*, half-way between continental and oceanic crust.[4]

Below *Majorca*, the crust is 25 km thick. There, as in Ibiza and Formentera, the Palaeozoic does not surface. There are fold-faults and overthrusts toward NW.

Minorca (the NE island) is different. Its crust is 15 km thick, the Palaeozoic comes to the surface, and part of the Mesozoic underwent metamorphism. This crust is found in the whole *North-African Basin*, between the Balearic Islands and Algeria.

Further west, below the *Alboran Sea*, a thin continental crust ($\nu_P = 6.15 \pm 0.34$ km/s) overlays abnormal mantle ($\nu_P = 7.7$ km/s). (But, according to focal mechanisms, the crust is actually undergoing N–S compression.) The *Gibraltar arc* is all around, with centrifugal nappes, on the outside, in the Moroccan Rif and the Betic cordillera. There, as between Corsica and Italy, we assume the previous existence of an emerged micro-continent, that strongly subsided during the Pliocene. According to the palaeogeographic reconstructions of Figures 15.10 and 15.11, this Alboran block would have migrated west (relative to Europe and Africa) by 140 km between 35 and 10 Ma.

The uplift of the Gibraltar arc 10 Ma ago, and the uplift of the North of the Arabian Plate (Syria and Iraq), with maybe a lowering of the sea levels after a great Antarctic ice age (see Section 18.2) in the Messinian (6.5 to 6.1 Ma), made the Mediterranean an enclosed sea. It nearly dried out, as evidenced by a quasi-omnipresent layer of evaporites, of varying thickness.

Salts precipitate only when salinity is around 10 times larger than the salinity of sea water. Most of the salt was therefore deposited in the deep Messinian basins, the last ones to dry out. Even so, this makes up an evaporite layer of at most a few tens of metres, whereas Messinian evaporites are sometimes thicker than 500 m. There must have been repeated dryings, and during the Messinian, the Mediterranean behaved like a huge salt pan.

The *Ionian Basin*, between Sicily and Peloponnese, is not an oceanic basin subducted below the Calabrian arc, as believed by some. Its bottom is a thinned continental crust, 20 km thick, and the foreland of Africa.

A northward subduction of the Eastern Mediterranean oceanic crust happens indeed below Turkey, Crete, the Aegean Sea and the Peloponnese. A submarine accretion prism is formed, 180 km wide. It extends as an arc from the Levantine Basin to the Apulian platform, and is the *Mediterranean Ridge* of US oceanogra-

[4] One may talk of *oceanization* of the continental crust. But this term was completely discredited because its proponent was providing an absurd physico-chemical explanation of the actual process (as 'delamination' and, at its beginning, 'continental drift' were discredited by absurd mechanical explanations). Oceanization was introduced by Beloussov, the President of Soviet geologists, in 1967, to explain the presence of oceanic crust in the Okholtsk Sea. This long remained the official doctrine. Beloussov was a fierce opponent of plate tectonics until his death, and denied the creation of crust at mid-ocean ridges. He explained oceanization by a chemical transformation of continental crust into mantle, a transformation rejected by all specialists.

phers. Subduction should have started around 10–13 Ma, at the same time than the second phase of opening of the Red Sea. Movement reorganization should have happened around 3 Ma, and a row of intra-arc volcanoes appeared in the Cyclades 2.7 Ma ago. This set forms the *Hellenic Arc*.

Today's relative movements in this region have been well determined with spatial geodesy (LePichon *et al.*, 1995). Relative to Europe, the Anatolian Plate (including the Aegean Sea) moves laterally, pushed by the Arabian Plate. It rotates by 1.7 microdegree per year, anticlockwise, around a very close Eulerian pole, between Cyprus and Egypt (Figure 13.16). This rotation produces a sinistral strike-slip of 28 mm/year on the North-Anatolian Fault. It adds to a clockwise rotation of Northern Greece and Albania, 2.8 microdegrees per year, around a pole in the north of the Adriatic Sea (this explains the change in azimuth of the coastline between Dalmatia and Albania). These two different rotations are adjusted by a deformation of Central Greece, in particular with the widening of the Gulf of Corinth, reaching 23 mm/year on the west side. This explains the very high seismicity of Greece.

These two rotations add to the closing of Europe and Africa, making the Hellenic Arc move closer to Africa by 23 mm/year. According to an old estimation by LePichon and Angelier (1979), they would have moved closer by 300 km during the last 13 Ma, and only 90 km would be due to the movement of Europe toward Africa.

15.8 KRIKOGENIC MECHANISMS IN THE WESTERN MEDITERRANEAN

The recent evolution of the Mediterranean shows the formation of arcs and intra-arc basins, and these processes are sometimes called *krikogenic* processes (from the Greek word *krikos* = ring, creek). They are different from the processes in the Western Pacific. Western Mediterranean processes are irrelevant to plate theory, and cannot be explained either by the existence of a crustal asthenosphere. Let us sum up their characteristics:

(1) Below all basins, a thinned continental crust lies directly on top of a mantle with a low seismic velocity (an LVZ in some sort). Geothermal fluxes are high. But there are no active mid-ocean ridges, only *mantle diapirism*, i.e. intrusion of magmas coming from the mantle. All these facts show that the mantle directly beneath the crust is at the melting point and forms the asthenosphere. According to the seismologists Calcagnile and Panza (in Udias and Channel, 1980), the lithosphere is in average only 30-km thick below the Tyrrhenian Basin, compared to more than 100 km below the Alps, the Po Basin, the Adriatic or Calabria.
(2) Around the Tyrrhenian and Alboran basins, one finds overthrusts and gravity nappes with centrifugal direction, formed before the recent subsidence of these basins.

(3) Calabria drifted relative to Apulia and Sicily, pushing in front of it 600 km of thinned crust, found as nappes and overthrusts. But what befell the corresponding mantle lithosphere? The same question can be asked for the Alboran block.
(4) Although there was no important subduction of the Ionian Basin below the Calabrian Arc, there is a 600-km long Wadati–Benioff surface below the basin.

The last two facts can be explained by assuming that the subduction below the Tyrrhenian Basin is due to the displacement relative to Apulia (and Europe) of the small Calabrian block, and not of Africa. As the thinned crust preceding Calabria is pushed and compressed, with overthrusts, the mantle lithosphere was progressively delaminated and getting deeper and deeper in the asthenosphere.

The displacement of Calabria is linked to the opening of the Valencia Basin and the Provence Basin. The thinned crust behind must have lifted and then collapsed, giving birth to the Tyrrhenian Basin. The centrifugal nappes around it are indeed gravitational in origin. This process has been agreed on in the north. Only local asthenosphere currents, with 'ebb' and 'flow', can explain this original krikogenic process. But a thermo-mechanical explanation and a geodynamic model should still be set up.

15.9 WEST INDIES ARC

North and South America were contiguous in the first reconstructions of Pangaea, and Central America was difficult to place. It was since understood that, in the Jurassic, it was west of Mexico. Palaeogeography of the Caribbean domain was clarified by the Thetys research group, grouping scientists of BRGM (*Bureau de Recherches Géologiques et Minières*, the French Geological Survey), the *Institut Français du Pétrole* (French Oil Institute) and the 4 main oil companies (*Bull. Soc. Géol. De France*, **8**(6), 1990).

The North Atlantic opened from south to north, around 180 Ma ago (Lower Jurassic). After repeated lagoon drying, creating evaporite deposits between 175 and 160 Ma, the opening of the Gulf of Mexico began around 160 Ma ago (Callovian–Oxfordian, in the Middle Jurassic). Oceanization of the *Cuicatec Basin* occurred around 140 Ma (Lower Cretaceous). Its westward subduction induced the formation of the *West Indies palaeo-arc*, which remained active throughout the Cretaceous. The Equatorial Atlantic opened around 111 Ma ago (Aptian), and around 98 Ma (Upper Albian) it extended to the coast of Venezuela (Figure 15.21a).

Subduction of the Farallon Plate (see Section 6.7), west of the West Indies palaeo-arc, singled out a Caribbean plate. It formed on the edge of this plate the Columbian Andes, and pushed the Caribbean plate eastward. Around 70 Ma ago (Campanian), the West Indies palaeo-arc collided with the North American passive margins (Cuba, Hispaniola = Haiti + Dominican Republic) and the South American ones (Venezuela) (Figure 15.21b). The volcanism of the Greater West Indies became calc-alkaline.

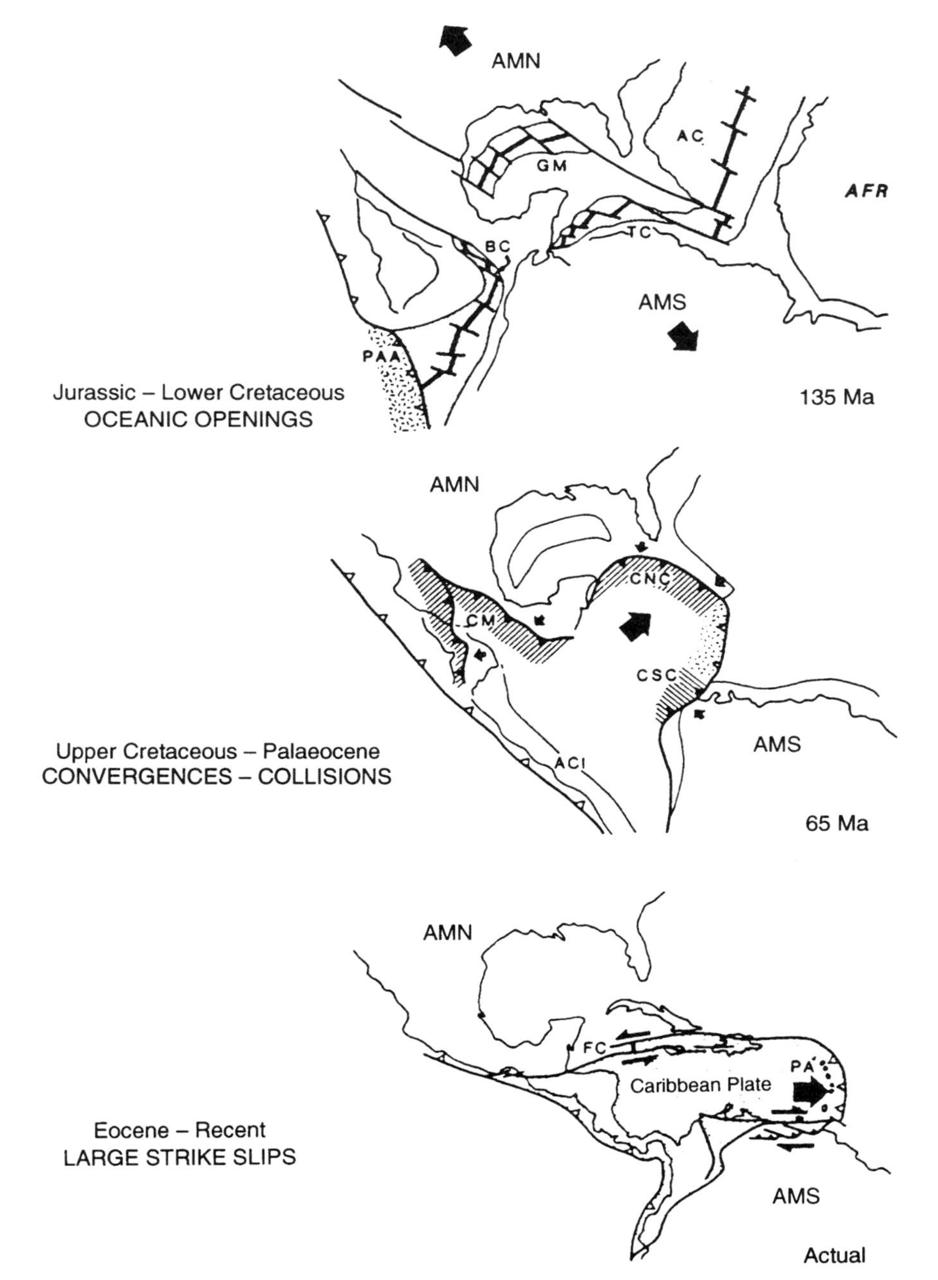

AC = Central Atlantic; GM = Gulf of Mexico; TC = Caribbean Tethys; BC = Cuicatec Basin; PAA = West Indies palaeo-arc; CM = Mexican sierras; CNC = North Caribbean sierra; CSC = South Caribbean sierra; FC = Cayman Trench; PA = Lower West Indies. From Calais *et al.*, 1989.

Figure 15.21. Synthesis of the geodynamic evolution of the Caribbean domain, from Lias to Recent.

In a third stage, the Caribbean plate, separated from Columbia, continues drifting east. South of it, dextral strike-slip movements occur along the Venezuela coast, east of Caracas. The northern strike-slip movement happens through two long transform faults, linked by a short active ridge in the Cayman Trench (Figure 15.21c). The southern fault follows the River Motagua (in Guatemala), crosses Jamaica and South Hispaniola. The northern fault starts in the Cayman Islands, follows the Sierra Maestra south of Cuba, the northern coast of Hispaniola and the PuertoRico Trench. (These two trenches are very deep: 7,119 m for the Cayman Trench, 8,648 m for the Puerto–Rico Trench.)

The West Indies arc has drifted this way, east relative to North America, without any interruption in 140 Ma, at an average velocity of 30 km/Ma. According to the NUVEL-1A model, the Caribbean/North America strike-slip velocity was only 11.3 km/Ma during the last Ma, and the Caribbean/South America strike-slip velocity was 18.2 km/Ma. It seems that the largest east-drifting velocity was transmitted to the Cocos Plate, which, in an absolute frame of reference, took the place formerly occupied by the Caribbean Plate.

15.10 PROGRESSIVE ISOLATION OF ANTARCTICA

A similar drift has been recognized for the *Scotia Arc* (Figure 15.22) between South America and the Antarctic Peninsula. South Georgia, 1,600 km east of Tierra del Fuego, is a micro-continent which formerly belonged to the Andes. According to Cunningham *et al.* (1995), the Scotia Plate drifted east with north and south strike-slip displacements of 890 km (NSR and SSR in the figure), and after opposite rotations of the extremities of Patagonia (SSA) and the Antarctic Peninsula (AP), contributing to 500 km. There also appeared a NS spreading ridge, isolating a small

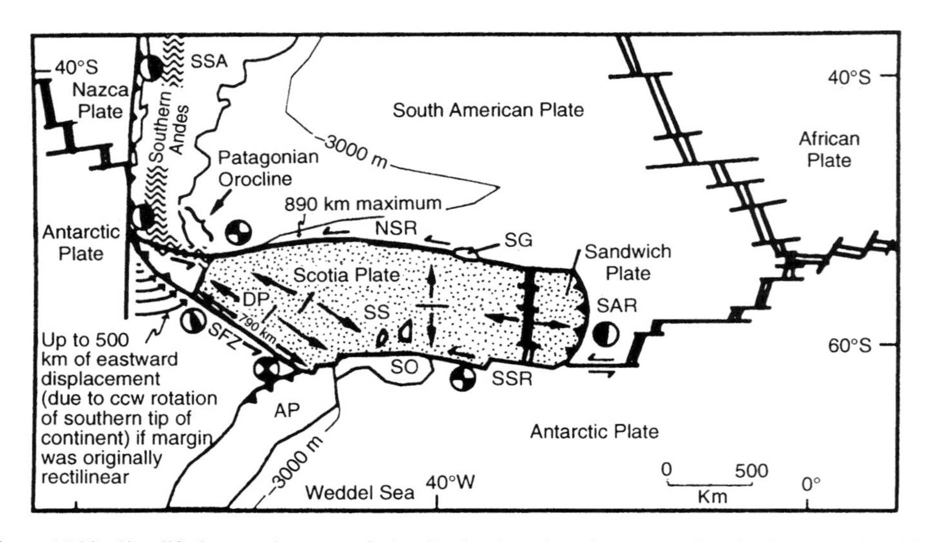

Figure 15.22. Simplified tectonic map of the Scotia Arc domain. From Cunningham *et al.*, 1995.

South Sandwich Plate, which pushed further east the eastern part of the Scotia Arc (South Sandwich Arc, SAR). Simultaneous with this drift, started 84 Ma ago, South America and Antarctica moved 1,400 km apart.

According to recent palaeomagnetic studies (Grunow, 1993), the first fission of Gondwana through the appearance of a spreading ridge occurred as soon as 175 Ma ago in the Weddell Sea basin, between the Antarctic Peninsula and the Ellsworth mountains. (The Ellsworth mountains are a large massif lying besides Eastern Antarctica, and only its northern part currently emerges from the ice, with the Sentinel Range mountains culminating at 5,140 m). But this ridge was only active for ca. 20 Ma, and as early as 130 Ma ago, subduction below the Antarctic Peninsula had practically re-established the initial situation.

An eastward overthrust of the Palaeozoic basement has been recognized in the south of the Patagonian Andes, over the Lower Cretaceous and Jurassic folded layers. The overthrusting slab forms a high N–S range emerging from the broad ice field, the Cordón Mariano Moreno (3,536-m high). It is highly likely that this overthrusting happened during the closing phase above, the tip of the Antarctic Peninsula colliding with Patagonia. To this, we can add the persisting subduction of the Phoenix oceanic plate (south of the Farallon Plate), below Patagonia and Western Antarctica. The Antarctic Peninsula and the Pacific coastline of Antarctica are extensions of the Andes, and the volcanism from the last 10 Ma only reinforced their similarity.

Australia split off from East Antarctica and New Zealand split off from Mary Byrd Land since 130 Ma, but this separation only accelerated after 45 Ma. In the meantime, between 80 and 50 Ma, the Tasmania Sea basin was appearing, separating Australia and its SE continental platform (200,000 km^2, improperly called *Tasmania Ridge*) from New Zealand and its large continental platforms (*Lord Howe Rise* and *Norfolk Ridge* in the northwest, *Chatham Rise* in the east, *Campbell Plateau* in the south).

Around 40 Ma ago, the Tasmania continental platform and the Campbell plateau became deep enough below sea level for the circum-Antarctic marine current to flow south of Australia without having to detour to the north. This must have induced important climate changes.

16

Polar wander and continental drift during the Palaeozoic

16.1 ASTRONOMICAL CLIMATE FACTORS

The term *climate* is generally restricted to averages over 30 years or more of mean air temperatures and rainfall, monthly and yearly, despite the importance of the statistical dispersion of values and of the sequences. These averages depend on the distribution of sea and land, on topography, on marine currents, cloud coverage, the amount of greenhouse gases and dust in the atmosphere, etc. They also depend on a single factor outside the Earth: the heat received from the Sun, per unit area on a spherical Earth, out of the atmosphere. This solar energy flux depends on latitude, the time of year and the *solar 'constant'*, the energy flux received by the Earth perpendicularly to the light rays (around 1400 W/m^2). We still do not know whether the flux emitted by the Sun is variable, or slightly absorbed by cosmic dust during its journey, and whether this is significant. Excluding these possible factors, the other factors controlling the solar energy flux outside the atmosphere are the *astronomical climatic factors*.

At a specific latitude and date in the year, the fluctuations of this solar flux are very small. They only affect climate because the atmosphere–oceans–land–sea-ice system amplifies enormously some perturbations. This is to the point where large climatic oscillations like the ice ages would surely exist without any external periodic factor. The astronomical factors would only synchronize these random oscillations on their own frequencies, in the way that the slight electric impulses of a pacemaker regulate heart contractions without needing to provide the energy that these contractions require.

But *one should not believe that the astronomical climatic factors have very precise frequencies, permanent on the geological time scale*. Let us insist on the mathematical aspect of the question, because we will find it again in Chapter 19, when talking about convection in the Earth, not about palaeo-climatology.

The oscillating variables are noted $x_i (i = 1, \ldots N)$. The system of equations regulating their evolution with time (t) is of the form $dx_i/dt = f_i(x_i, \cdots, x_N)$. If all f_i were linear functions, the astronomical system would have N intrinsic frequencies. But this is not the case, as soon as more than 2 bodies are involved: the Sun, the Earth, the Moon and the other planets, or even as soon as these bodies are liable to deformation. Non-linear terms appear in the system of equations, and there is no more exact analytical solution. Furthermore, a minute variation in the initial conditions can, at any time, totally change the solution, so that it cannot be predicted accurately with numerical models. Henri Poincaré studied this behaviour a century ago. This came back as *deterministic chaos*, and can now be studied with powerful computers.

The parameters of the Earth's orbit and rotation axis are very stable, because the Earth–Moon system has a large angular momentum. Nevertheless, the numerical solution of the system of equations, starting from the current values and going back through time, could not be completed beyond 5 Ma, because the inaccuracy was becoming too large (Berger, 1988).

Numerical computation expresses each variable with a few sine functions of decreasing amplitudes. For the *terrestrial orbit eccentricity*, the amplitudes of the successive terms decrease very slowly and their periods are very different:

Term #	Amplitude	Period (years)
1	0.0110	412,900
2	−0.0087	94,950
3	−0.0075	123,300
4	0.0067	99,600
5	0.0058	131,250
6	−0.0047	2,035,400

If one tries adjusting the eccentricity variations of the last 5 Ma onto a single sine, the best period is 100,000 years. This should not be used to establish some astronomical chronology from the rhythms of sedimentation beyond 5 Ma (see Section 6.3). To stress this point, it is best talking of a 100,000-year *quasi-period* (even if a quasi-periodic function is a different thing for mathematicians).

Today, the eccentricity c/a of the Earth's orbit is 0.016. The Earth–Sun distance is minimal and equals $(a - c)$ at *perihelion*, and it is maximal and equal to $(a + c)$ at *aphelion*. The solar 'constant' varies during the year, with the ratio $(a + c)^2/(a - c)^2 = 1.066$.

The Earth is now at perihelion on January 3, in the middle of winter in the northern hemisphere. This decreases the temperature difference between seasons in this hemisphere, and increases it in the southern hemisphere. (The differences are in fact stronger in the northern hemisphere because it is more continental.) But the position of perihelion moves around the orbit, in the direct sense (anticlockwise) when looking from the North Pole. According to numerical computation, it makes one revolution in approximately 140,000 years.

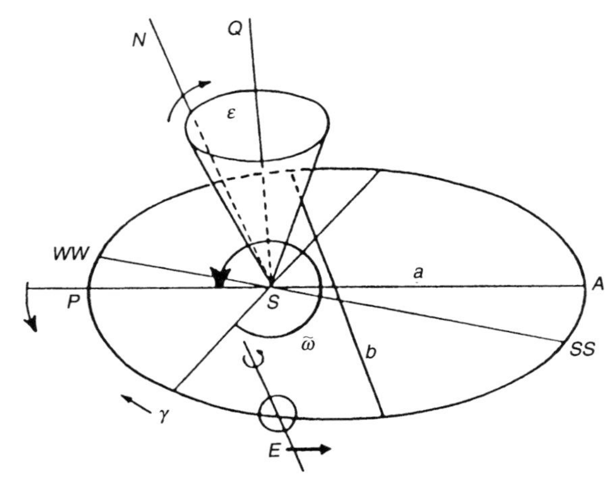

Figure 16.1. Elements of the Earth's orbit. In a Galilean frame of reference, the Earth (E) rotates around the Sun (S), in the direction of the thick arrow, and describes an ellipse of half-axes a and b, with S as one focus. The aphelion (A) and perihelion (P) rotate much more slowly. The Earth revolves around the axis of the poles, parallel to SN. The plane perpendicular to SN in S crosses the Earth's orbit in γ and γ'. The Earth is at the spring equinox in γ, at the summer solstice in SS, at the winter solstice in WW. SN rotates around the perpendicular to the orbit SQ (and therefore γ, γ', SS and WW rotate around the orbit), in the directions of the arrows. From Berger, 1988.

The Earth's rotation axis currently makes an angle of 23°27′ with the perpendicular to the orbit (Figure 16.1). This is the *inclination of the ecliptic*. (For a terrestrial observer, the ecliptic is the Sun's trajectory on the sphere of fixed stars. It is its inclination on the equatorial plane.) This inclination oscillated between 21.8° and 24.4° during the last 5 Ma, with a quasi-period of 41,000 years.

Because of the Earth's equatorial bulge, the Sun's attraction, besides keeping the Earth–Moon system on its orbit, exerts a torque on this system. There results a slow rotation of the Earth's rotation axis around the perpendicular to the orbit. Thus, the *nodes* of the orbit (the points γ and γ' where the equatorial plane crosses the orbit) are moving around the orbit in the retrograde sense (clockwise). This is the luni-solar precession, or *equinox precession*. In a Galilean system, the nodes describe the orbit with a quasi-period of 25,700 years. The astronomical factor of importance for climate is the precession relative to perihelion; a node crosses the perihelion every 21,700 years. (For a terrestrial observer, the ecliptic is always fixed among the stars. The world's axis motion makes the celestial North pole move relative to the stars).

In a Galilean system, the terrestrial rotation axis also describes a small cone, of half-angle at the summit 9.2″, with a period of 18.6 years, the same period than the node precession for the Lunar orbit. This is the *astronomical nutation*

16.2 WANDER OF THE GEOGRAPHICAL POLE

Until now, we only considered the movement of the centre of mass of the Earth–Moon system, and the direction of the instantaneous rotation axis in a Galilean

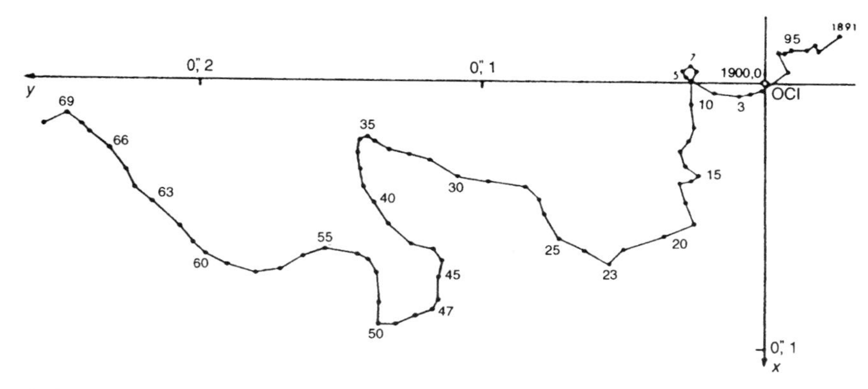

Figure 16.2. Secular movement of the instantaneous rotation pole, from the *Service International des Longitudes* (now SIMP), analysed by Mrs Stoyko. From Guinot (in: Coulomb and Jobert, 1973).

system of reference. The movement of the Earth relative to this axis is a completely different problem.

The diameter parallel to this rotation axis (not the axis itself, which is excentred because of the Moon, see Section 8.10) crosses the Earth's surface at the *geographic poles*. The question is now the path of the geographical North Pole over the Earth. It is split up into a fast movement around a mean point rather arbitrarily chosen (*nutation*) and a slow movement of this mean point (*polar wander*).

Beats are observed in the nutation, between a yearly period, due to seasonal displacements in the atmospheric mass (e.g. monsoon) and a 436-day period called the Chandler period. The latter comes from the fact that the Earth does not rotate exactly around its principal axis of inertia; and the period does not correspond to the one calculated by Euler for a non-deforming body (for details, see Lliboutry, 1992, Section 6.11).

The radius of the nutation varies between 1.2 and 7.5 m. This movement precludes a precise determination of polar wander. Between 1891, the date of the first determinations, and now, it would have been 0.13 m/year in average, at meridian 85°W, but would have been very sinuous in detail (Figure 16.2).

The relative displacements of the observatories are a more fundamental source of inaccuracy. The path of the Pole is determined by the *Service International du Mouvement Polaire* (SIMP), from data provided by 5 observatories all located at 39°08′N: Carloforte (Sardinia), Kitab (Adzerbaidjan), Mizusawa (Japan), Ukiah (California) and Gaithersburg (Maryland). With the accuracy available today from spatial geodesy, a non-deforming world system of reference should be established, and observations from the SIMP should be continually referenced to it. Observatories also provide the positions of the Pole relative to the local vertical. But it can vary slowly because of geological processes; none of the 5 observatories is on an old craton, in the middle of a large lithospheric plate with a known movement.

For all these reasons, and because the very sinuous aspect of the polar wander observed for a century may also exist on a geological time scale, the latter cannot be deduced from the former. (With the mean velocity mentioned, Indonesia would have

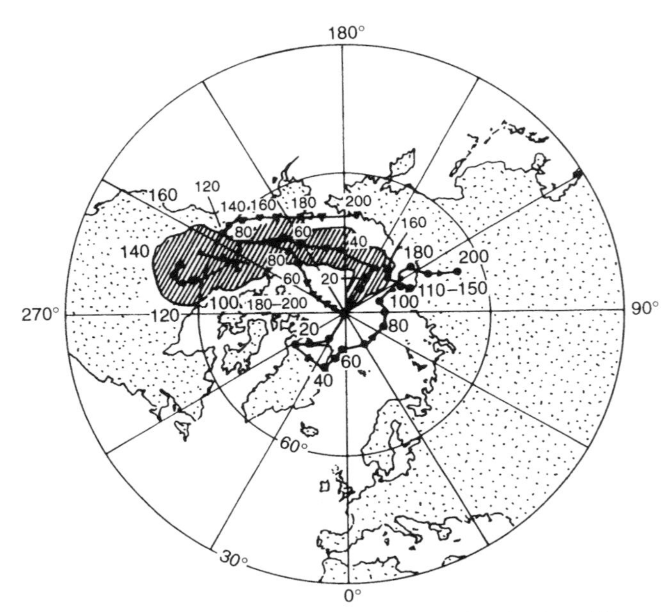

Figure 16.3. Comparison of the apparent wander of the geomagnetic pole of the African plate (black dots; the hatched zone corresponds to the error margin) and the movement of the hot-spot system relative to this plate (triangles). The difference, i.e. the drift of the hot-spots relative to the geomagnetic pole, is indicated by the black squares. The ages are given in Ma. From Courtillot and Besse, 1987.

been on the North Pole 77 Ma ago, and this is completely excluded by palaeontology.)

It will be shown that the thermoremanent magnetization of rocks yields the position, to a few degrees in the best case, of the *mean geomagnetic pole* (MGP) in the geological past. 'Mean' corresponds to a mean position over 10 or 20 Ma, and excluding the inversions of the geomagnetic field. It may be a south geomagnetic pole, or a north geomagnetic pole, to ensure continuity of the path. We assume (and this is discussed later) that this MGP coincides with the geographical North Pole (NP). With this assumption, the MGP wander is identical to the NP wander. (Let us remark that palaeomagnetism experts talk of 'polar' wander, without any other precision, which may be a source of confusion.)

On this time scale, we may use a frame of reference, which is linked to most of the mantle (let us say, to be short, the 'lower mantle', the hot-spot reference frame (Section 6.6). We know the movement of a large plate with several hot spots and a continent large enough for sampling and determination of the geomagnetic palaeopoles: the African Plate. Its movement in the hot-spot system (HS absolute velocities) is well known for the last 110 Ma, and with less accuracy for the last 170 Ma. Courtillot and Besse (1987) determined the *absolute HS wander of the Mean Geomagnetic Pole (MGP)* during the last 170 Ma.

Their results are reproduced in Figure 16.3. Between 150 and 110 Ma, the MGP has not moved in the HS system, and was at the current coordinates of 78°N and

120°E. Since then, it described a hairpin curve which brought it, between 40 and 20 Ma, at the point of present coordinates 78°N, 30°W.

16.3 CAUSES OF THE HS ABSOLUTE WANDERS OF NP AND MGP

The wandering of the geographical North Pole relative to the HS reference frame of the lower mantle is the vectorial sum of 4 terms:

(1) the wandering of the North Pole relative to the Mean Geomagnetic Pole;
(2) the wandering of the Mean Geomagnetic Pole relative to the African Plate;
(3) the wandering of this plate relative to its hot spots;
(4) the wandering of these hot spots relative to the HS frame.

The third term can only be computed for the last 110–170 Ma, and during this period the fourth term is negligible compared to the inaccuracy with which the second term can be determined. There are two possibilities for the first term: this term equals 0, as assumed, and the wandering of Courtillot and Besse (1987) is the wandering of the Earth's rotation axis relative to the lower mantle. Or this term is different from 0, and the NP wander is unknown, maybe null.

The cause of the NP wander relative to the lower mantle can only be a change of the orientation of the principal axes of inertia, more precisely of the axis around which the moment of inertia is maximal. The position of this axis indeed depends only of the location of the equatorial bulge. But this bulge is due to the Earth's rotation, and will move in function of the location of the North Pole (but with some delay, depending on the mantle viscosity). The rotation axis and the equatorial bulge are linked, and both depend on mass distribution inside the Earth. Because of isostasy, the position of continents and oceanic basins is not important. The location of convection currents and the corresponding lateral density variations should be the main cause of change.

On a short term, the displacement of surface masses causes a North Pole wander, until the equatorial bulge has moved or isostatic equilibrium is reached. The strong NP wander measured by astronomers seems related to the melting of the last ice caps of the northern hemisphere, the resulting isostatic readjustments being far from finished.

16.4 CORE AND TERRESTRIAL DYNAMO

The question of the origin of the Earth's magnetic field is dealt with in Section 4.4. We review here the considerable progress made in recent years because, even if core processes may seem marginal in the present book, they intervene in some recent geodynamic speculations.

The first progress comes from new techniques to study materials under extremely high pressures (see Section 17.1). Very small samples (diameters of ca. 0.1 mm) could be statically compressed with diamond 'anvils', to pressures higher than in the

Table 16.1. Core temperatures and pressures.

Limit	CMB	ICB	Centre of the Earth
Depth	2894 km	5150 km	6370 km
Pressure	135 GPa	329 GPa	364 GPa
Temperature	3500°C–3600°C	4500°C–4800°C	5500°C

Earth's centre (364 GPa, i.e. 37.1 tonnes per square millimetre). Among many results, the phase diagram of the allotropic varieties of iron and its melting temperature were determined. This greatly reduced the uncertainty about densities and temperatures in the core. Table 16.1 shows the pressures and temperatures at the core-mantle boundary (CMB), at the inner core boundary (ICB) and at the centre of the Earth.

The inner core is formed of iron and 5–6% nickel, in a hexagonal compact phase (ε-iron) which is not ferromagnetic at any temperature. It must contain a bit of potassium and uranium. Their radioactivity is a heat source which, with the stored latent heat, warms the base of the liquid outer core (a necessary condition for maintaining convection currents).

Convection can appear in a fluid only if the vertical temperature gradient is larger than the adiabatic gradient (resulting from simple decompression with altitude, as observed in a 'neutral' atmosphere). For a liquid adiabatic core, conductivity only would emit heat estimated to 5 TW. More than 5 TW must therefore come out of the core, and the heat sources mentioned do not seem sufficient.

Gravitational energy is another energy source for convection (Gubbins and Masters, 1979). The liquid iron core contains, apart from 5% nickel, 10% of a light element, oxygen or sulphur. (At the local pressures, O and S are metallic and soluble in liquid Fe). The core is cooling down and slowly growing, catching pure ferronickel from the liquid core. The interface liquid, enriched in light elements and therefore less heavy, comes up and provides some mechanical energy. The gravitational energy so liberated is used at 100% to run the terrestrial dynamo. With heat as the principal source, the efficiency would be limited by the Carnot theorem to 22%.

Let us remark that the solidification at $p = 329$ GPa of a liquid of density 13.55 Mg/m^3 to produce a solid of density 14.1 Mg/m^3 absorbs $pV = 0.95$ MJ/kg. This is close to the latent heat liberated (which is the enthalpy variation: $H = U + pV$). Under this high pressure, pV becomes much larger than the variation of internal energy U. At the inner core boundary, additional gravitational energy is therefore liberated, and appears as heat, which is transformed into work with an efficiency of less than 22%.

Other progress was made recently in the numerical solution of the MHD equations governing velocity in the liquid core, and the electrical intensities and magnetic field in the liquid core and the inner core. These equations are not linear and the amplitudes of variations are such that any solution of the linearized system is fallacious. Solutions where the velocities in the liquid core are pre-established do not prove anything. The velocities, like the electric intensities, must come out from the equations and the imposed boundary conditions. In the jargon, this solution must be 'self-consistent'. Glatzmaier and Roberts (1995) obtained such a solution. The computer had to compute the solution of a system of non-linear equations, by successive approximations, at each of the successive times. For the method to converge, the time steps must be very short (1 week). These authors simulated numerically the system evolution during 40,000 years, i.e. more than 2,000,000 time steps. This required 2,000 hours of CPU on a Cray-90 supercomputer!

Numerical solution showed that after 10,000 years of adjustment, the geomagnetic field is stable and statistically independent from the initial conditions. But, 4,000 years before stopping the simulation, there was a polarity reversal which took around 1,000 years. Glatzmaier and Roberts (1995) insist in their article on the field modifications during the inversion, but other results are more unexpected and should be remembered.

A theorem, established by Cowling in 1934, states that to get a poloidal magnetic field, the velocity field must not be axisymmetrical. It was therefore thought that the current lines were looking like helices, left- or right-handed. Numerical simulation showed that the helicity of current lines and the polarity of the magnetic field were inverse on each side of a particular level in the liquid core. Helicity and polarity of the outer part would not be stable, and would reverse very frequently, at unrealistic pace if there were no electric currents in the inner core, stabilizing the magnetism of the centre. From time to time, the inner core does not limit enough the continual fluctuations of the outer part, and a reversal occurs.

This result seems to predict that, contrary to some recent speculations, the conditions at the core–mantle boundary have little influence on the changeable frequency of reversals (recall that there were very long periods with no inversions: the superchrons.) These conditions are not well reproduced in the model; the topography of the coremantle boundary was not reproduced, although it is not exactly a sphere, and the mantle electrical conductivity was neglected, although it is not negligible close to this boundary. To deduce from the simulation that these factors are not important would be a circular reasoning, but this seems highly likely.

Conversely, the conditions at the inner core boundary must be better reproduced, as recognised by Glatzmaier and Roberts (1995). They did not introduce the liberation of a light element, and contented themselves with increasing the energy provided to the system, by giving a high value to the heat flow from the mantle (one of the parameters of the model). The segregation process at the surface of the inner core may however not be regular (e.g. Ito *et al.*, 1995), and may intervene in the frequency of reversals.

Another result of numerical simulation is that the inner core would rotate slightly faster than the mantle, and that the phase shift would reach one revolution in around 500 years. This difference of the rotation velocities was confirmed by seismologists (Song and Richards, 1996). The transit time of the seismic waves crossing the inner core (PKIKP waves) had shown that their velocity depended on direction. It is higher along a direction at approximately 10° from the rotation axis (common to core and mantle). This is due to some uniform fabric of hexagonal ε-iron in the inner core of unknown origin. During earthquakes with epicentres in the South Sandwich Islands (25°W) recorded at College, Alaska (148°W), the PKIKP waves were 0.3 s faster in the 1990s than in the 1960s. But the PKIKP waves emitted at the Kermadec Islands (180°E) and recorded in Norway (11°E) were slower. In 30 years, the fast axis of the inner core has therefore moved away from the latter path of seismic waves, and become closer to the former.

16.5 GEOMAGNETIC PALAEOPOLES

A *virtual geomagnetic pole* is the location of the geomagnetic pole if the local magnetic field were exclusively coming from a dipolar field (see Section 4.1). It is located in the diametrical plane of the Earth containing this field, at an angular distance θ linked to the measured inclination i by the relation: $\tan i = 2 \operatorname{cotan} \theta$.

Assume that we know the direction of remanent magnetization in many rocks from the same period, collected in the same region, and therefore we know many virtual geomagnetic palaeopoles. They will not coincide because, even if the non-dipolar components of the regular planetary field are most often negligible, there are some local magnetic anomalies. Taking an average location gives an estimate of the Mean Geomagnetic Pole (MGP) at this time. The following precautions must of course be observed:

- The rocks must have kept the remanent magnetization acquired at that time, with no modification from later warming episodes or chemical alterations. A large part of the magnetic parasites affecting the past field memory can be eliminated through rigorous magnetic cleaning (see Section 4.11).
- One must be sure that the sites sampled belonged to the same continental block during their magnetization, and remained so (there may have been fission of the block, then re-merging, but in a different relative position). This uncertainty casts doubts over the MGP wanders of Africa in the Precambrian which have been published so far, and they will not be given here.
- The terrain where the samples were collected must not have changed orientation during later tectonic events. Only good geologists can collect suitable oriented samples.

When using the depositional remanent magnetization of sediments, it must be remembered that their compaction modified the inclination, and only the declination can therefore be used. Instead of a cluster of virtual geomagnetic palaeopoles, one gets large circles. If the region sampled is large, the MGP will still be determinable.

Whether one uses DRM or TRM, one always finds two probable opposite MGPs, because during the period considered, at least a few Ma, there were many reversals of the magnetic field. (There may also be completely impossible virtual palaepoles, corresponding to a time of polarity reversal; they should be removed before computing the MGP.) Choosing between the two antipodal MGPs is possible because they are determined at successive geological ages: the MGP wander must be continuous, without sudden jumps of nearly 180°.

It is assumed that *MGP and NP always remained close*. The angular distance θ between the site considered and the MGP is, more or less, the *colatitude* of the site at the time. The angle of the lineation of the region investigated with the large circle crossing the site and the MGP, provides the *azimuth* of the lineation at this particular time. Palaeomagnetism therefore gives the past latitude and orientation of a continental block, but *does not give any information about the longitude*. This keeps some arbitrary in the palaeographic reconstructions of old ages, with no seafloor with magnetic anomaly stripes left. Geological considerations are then of great help. The proposed continental drift must agree with the formation of collision chains in particular places and times.

A qualitative test of proposed palaeolatitudes comes from fossils, and particularly some geographically restricted species, and some deposits characteristic of the climate: coal, evaporite, bauxite, kaolinite, glacial tillites, etc.

NP determination from a large number of virtual geomagnetic palaeopoles requires statistical processing, to output the best possible estimate of its location, and the *confidence circle* where it has 95% chances to be. We must therefore know the probability distribution *dp* for the angular interval between the true NP position and its estimate to range between φ *and* $\varphi + d\varphi$. *The Fisher distribution* is used. Writing that $dS = 2\pi \sin\varphi \, d\varphi$ is the elementary area of unit radius on the hemisphere, it reads:

$$\frac{dp}{dS} = \frac{K}{2\pi(e^k - 1)} \exp(K\cos\varphi) \qquad (16.1)$$

When K varies from 30 to 500, the angular radius of the confidence circle varies from 25.8° to 6.3°. As K is very high, 1 is neglected relative to e^K:

$$\frac{dp}{dS} \approx \frac{K}{2\pi} \exp\left(-2K\sin^2\frac{\varphi}{2}\right) \qquad (16.2)$$

When the dispersion is small, this tends toward the Gauss distribution, or normal law, which has a serious theoretical base (the Moivre–Laplace theorem, also called the central limit theorem).

16.6 MGP WANDERING FROM ORDOVICIAN TO PERMIAN

When two continents were joined for some time on the geological scale, the MGP trajectories during this period, determined separately for the two continents, should be superposable through a simple rotation on the sphere: this is the relative drift of the two continents since their separation.

Because of the imprecision in determining the MGPs, and the probable differences with NPs, this theoretical deduction could be applied until now to the case of Laurentia and Europe only (Figure 16.4). The MGP trajectories can coincide from the Ordovician/Silurian limit (420 Ma ago) to the Lower Jurassic, through a rotation of 38° about the North Pole. These two continents have therefore merged around

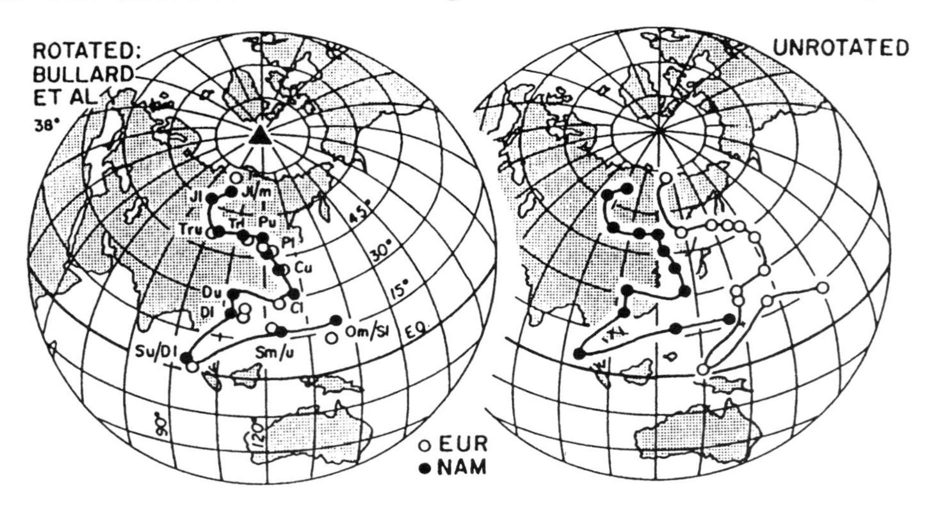

Figure 16.4. Right: apparent drift of the geomagnetic pole, deduced from the palaeomagnetism of European rocks (white circles) and North American rocks (black circles), from the Middle Ordovician (Om) to the Middle Jurassic (Jm) (S = Silurian; D = Devonian; C = Carboniferous; P = Permian; l = lower; u = upper). The continents are represented in their modern positions. Left: the two trajectories coincide after rotating Europe of 38° around the North Pole (black triangle). From Van der Voo, 1990.

420 Ma ago, and separated again in the Lower Jurassic. The drift of one compared to the other was indeed a rotation of 38° around the North Pole. This remarkable success justifies the assumptions made to determine geographic palaeopoles.

If we consider now that the geographical pole is fixed, Euramerica has slowly drifted north, starting at the Silurian/Devonian limit (ca. 400 Ma ago). This drift accelerated from the Lower Carboniferous (340 Ma ago). Until then, Euramerica was in the intertropical region, which explains the red colouring of Devonian sandstone. Euramerica was often called 'the old red continent'.

The Palaeozoic drift of Gondwana is less well known. Tillites are indicators of glaciations in the Ordovician (450 Ma ago). It is certain that, at this time, Australia and Eastern Antarctica were straddling the equator, but the South Pole was in Gondwana. It is also certain that, in the Devonian, Gondwana migrated rapidly north, catching up with Euramerica and colliding with it (beginning of the formation of the Pangaea). But the first trajectories proposed for Gondwana were erroneous.

The studies by Ziegler and Crowell (1981) are already more reliable. The Palaeozoic MGP trajectory for Gondwana is much more complicated than in the previous publications, and fits the traces of the Ordovician glaciation (Figure 16.5). In terms of continental drift, Gondwana would have rotated anticlockwise by 180° in the Silurian. Before, in the Cambrian and Ordovician, Australia, Eastern Antarctica, South Africa and Patagonia were at the equator, whereas North Africa was at a high southern latitude. Australian scientists have since shown that the rotation occurred in two stages: 90° in the Silurian, 90° in the Lower Carboniferous, between 340 Ma and 320–310 Ma (Chen *et al.*, 1994).

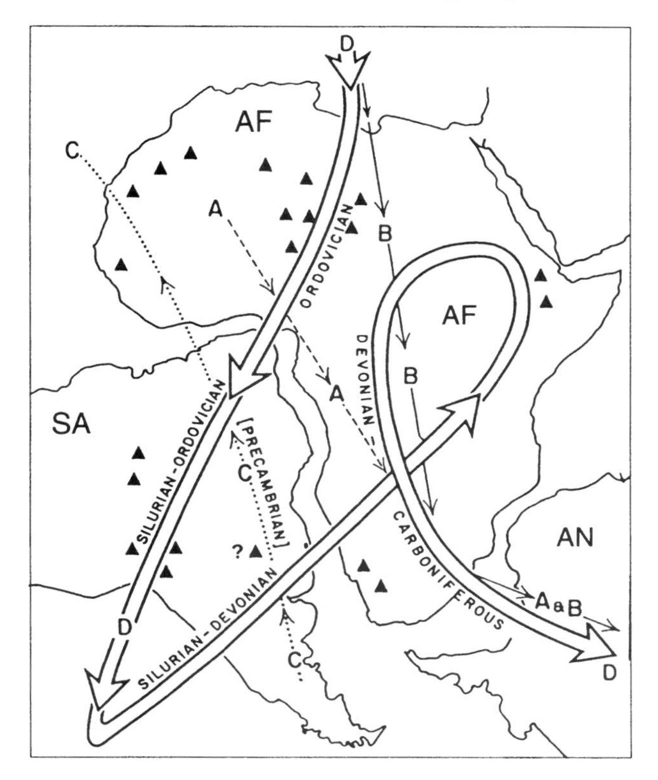

Figure 16.5. Apparent geomagnetic polar wander relative to Gondwana. C = Precambrian (from McElhinny *et al.*, 1974); A, B, D = Palaeozoic (A = from McElhinny, 1973; B = from McElhinny and Embleton, 1974; D = from Morel and Irving, 1978). Remnants of the Ordovician glaciations (black triangles) agree with it. From Crowell, in McElhinny and Valencio (1981).

16.7 OROGENIC PHASES IN THE PALAEOZOIC

The latitudes and orientations of continents since the end of the Cambrian, during the last 500 Ma, are now becoming well known, but their relative longitudes are still discussed. Palaeogeographic reconstructions use geology: continental collision chains between continents must be accounted for. One uncertainty comes from the fact that some collisions may have happened not between two large continents, but between a continent and separated continental fragments.

For example, during the Middle Ordovician, a collision with allochthonous terrains created the Sierra de Famatina in NW Argentina. Dalziel (1992) sees this as evidence of a collision between North and South America, and uses it as a basis for his reconstruction. But Jurdy *et al.* (1995) think that it was only a fragment detached from North America which later collided with South America. They could place South America 90° further west from North America, 500 Ma ago, than Dalziel (1992). This gives more reasonable relative continent velocities.

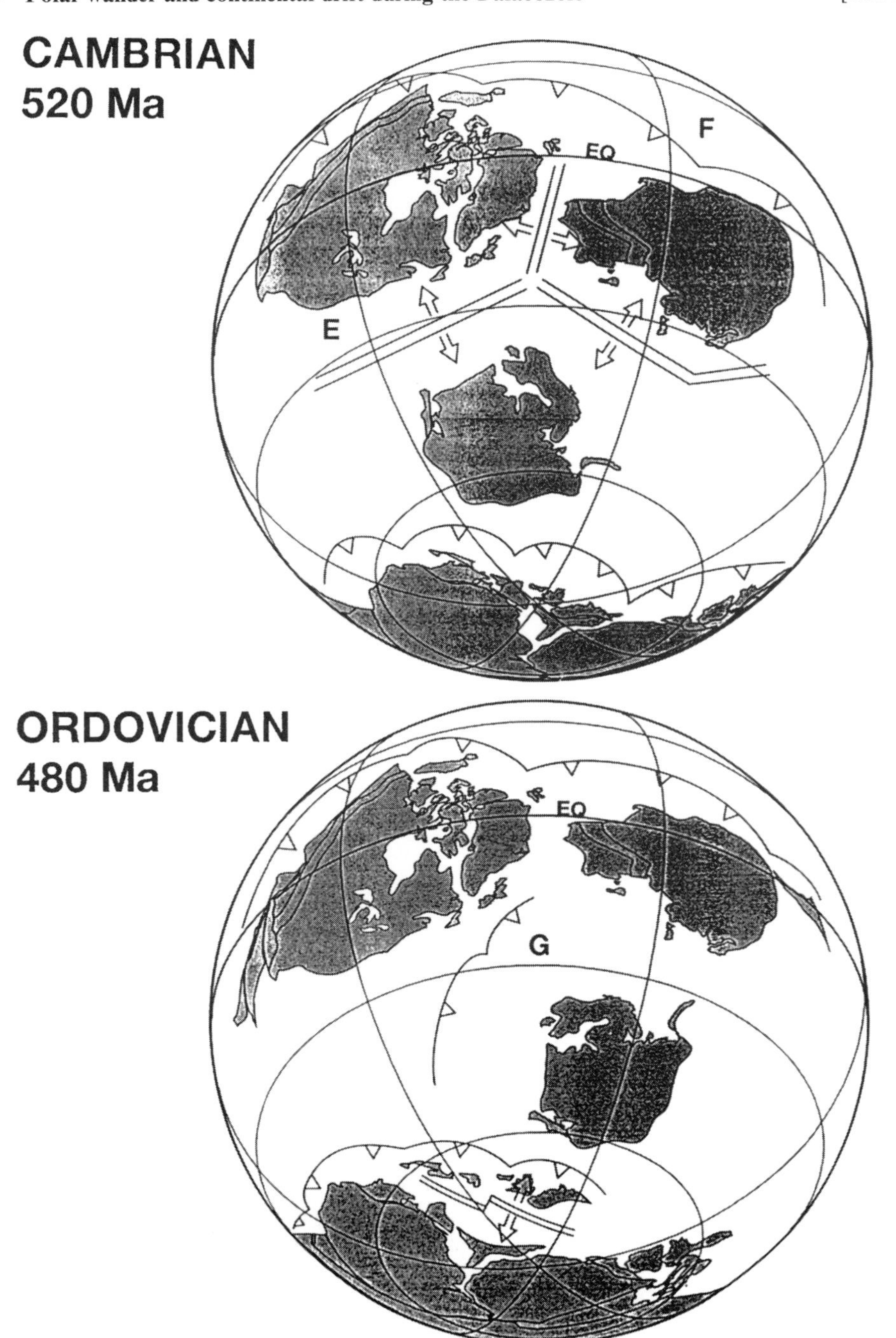

Figure 16.6. Continental drift in the Palaeozoic. EQ = equator, E = Iapetus Ocean; F = Panthalassa Ocean; G = Taconic island arc H = Rheic Ocean; I = Acadian collision. From Jurdy *et al.*, 1995.

SILURIAN
440 Ma

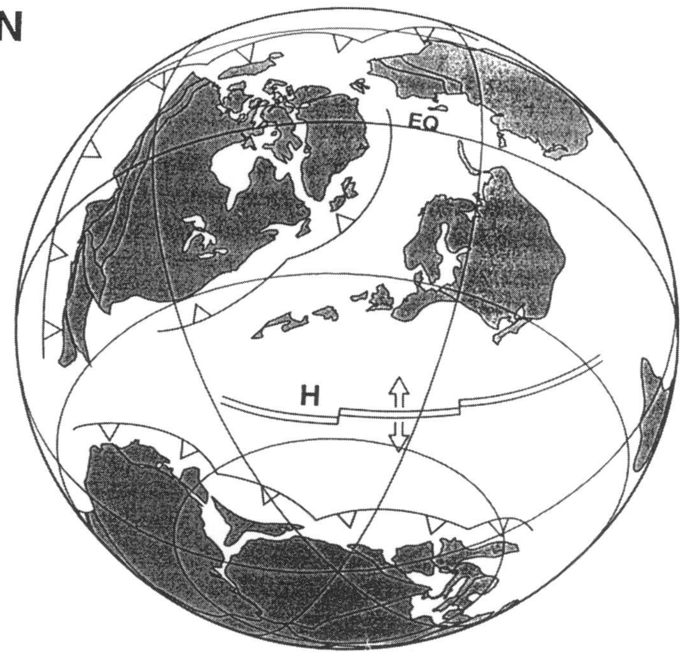

DEVONIAN
400 Ma

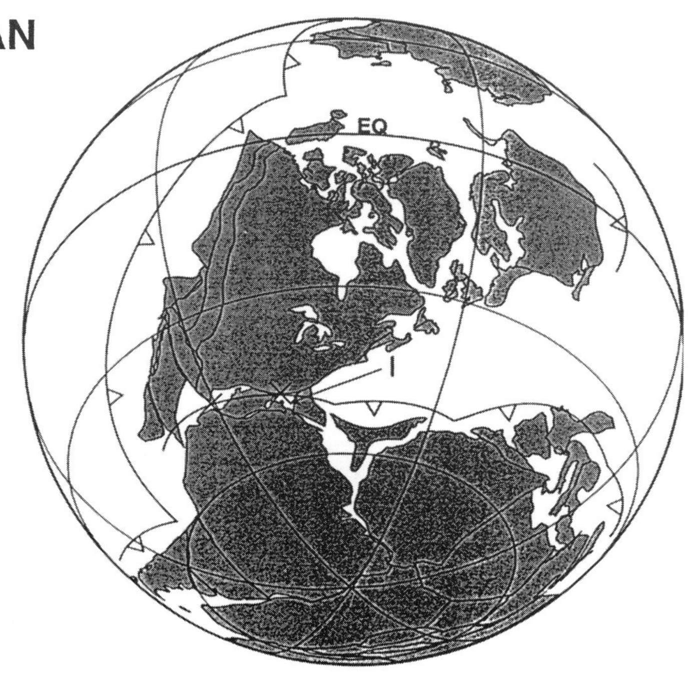

This difference is insisted upon, to show the relatively fragile bases of palaeogeography, for periods where the magnetic anomaly stripes of modern oceans are no help. We will present here the Palaeozoic history according to Jurdy *et al.* (1995), illustrated by the maps of Figure 16.6. Although this history cannot be considered final, it is unavoidable in any geological history of the Palaeozoic. But it does not figure (yet) in any geology textbook.

Let us note by the way that the reconstruction of Irving (1977) was completely disproved since, as new palaeomagnetic data lead to a more complex vision of Hercynian orogeneses (Molina Garza and Zijderveld, 1996).

At the end of the Cambrian, there were on the Earth one supercontinent (Gondwana), and three main continents: *Siberia*, *Laurentia* and the *Baltica* (or *Fenno-Sarmatia*). The latter was composed of the Baltic and Ukrainian cratons, and of the Russian platform, i.e. Eastern Europe beyond a line Tornquist–Teisseyre, from Denmark to the Black Sea.

During the Ordovician and the Silurian, the *Iapetus* ocean separating Laurentia from the Baltica narrowed and closed through subduction. One island arc it contained collided with Laurentia around 460 Ma ago (*taconic* orogenic phase). Around 420 Ma (end of the Silurian), the collision of Laurentia and Baltica caused the Caledonian orogenesis strictly speaking, or *Scandinavian* phase, giving rise to the *Caledonides*.

During the Ordovician, a Gondwana margin called *Avalonia* was detached around 480 Ma ago, to merge with Baltica around 440 Ma (*Erian* phase). Avalonia contained the New England coast, Nova Scotia, Newfoundland, SE Ireland, southern England and the Ardennes.

During the Silurian, another Gondwana margin (*Armorica*), comprising all the Hercynian massifs of continental Europe, from Portugal to Bohemia (Ardennes excluded) detached and merged with the Baltica–Avalonia block. The accretion took place in the Lower Devonian (400 or 380 Ma, depending on the author), during the *Eo-Hercynian* phase. The suture, with Armorican overthrust over the terrains further north, forms the *Hercynian front*.

According to Tait *et al.* (1994), Armorica would have rotated anticlockwise by 140° during its journey: at the beginning, Portugal was in the east, and Bohemia in the west.

The Laurentia–Baltica collision created a *Laurussia*, and it was only 20 or 40 Ma later that the pre-alpine Europe began, and *Euroamerica* replaces Laurussia.

The ocean existing in the Silurian between Gondwana (in the south) and Baltica, Avalonia, and Laurentia (in the north) is the *Rheic Ocean* (or Proto-Hercynian Ocean). It closed in the west, around 400 Ma ago, after a collision between North and South America (part of Gondwana). This is the *Acadian* orogenic phase, synchronous or earlier than the Eo-Hercynian phase.

The Taconic, Erian, Scandinavian and Acadian phases are traditionally grouped in a '*Caledonian orogeny*'. These phases have very diverse causes, and this grouping is arbitrary: why should the Erian phase be included, and not the Eo-Hercynian one ?

The ocean opening in the Silurian between Armorica and Gondwana, while the Rheic Ocean is closing, is the *Hercynian Ocean*. Since the Acadian collision, it is

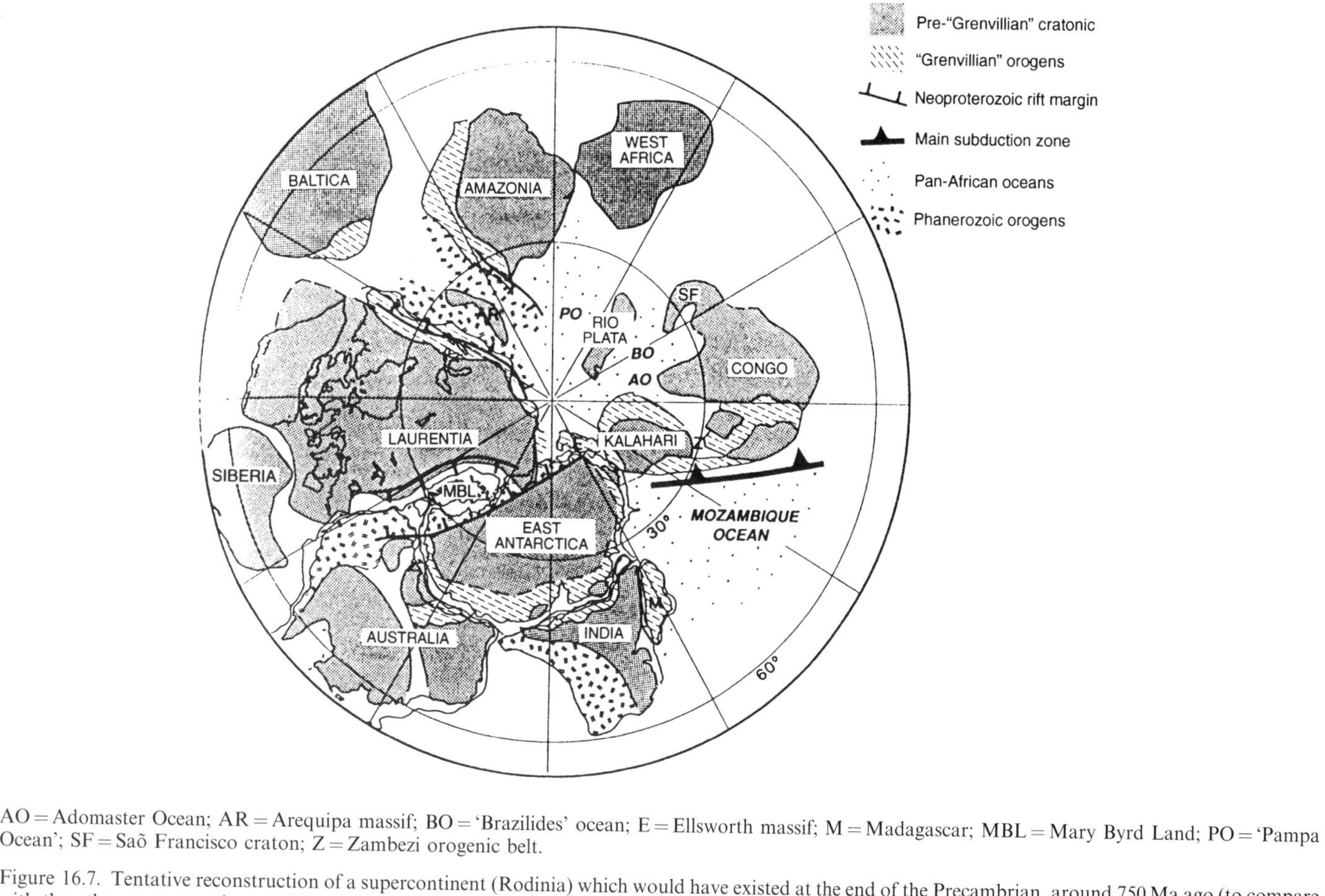

AO = Adomaster Ocean; AR = Arequipa massif; BO = 'Brazilides' ocean; E = Ellsworth massif; M = Madagascar; MBL = Mary Byrd Land; PO = 'Pampa Ocean'; SF = Saõ Francisco craton; Z = Zambezi orogenic belt.

Figure 16.7. Tentative reconstruction of a supercontinent (Rodinia) which would have existed at the end of the Precambrian, around 750 Ma ago (to compare with the other reconstruction, on Figure 16.8). From Dalziel, 1992.

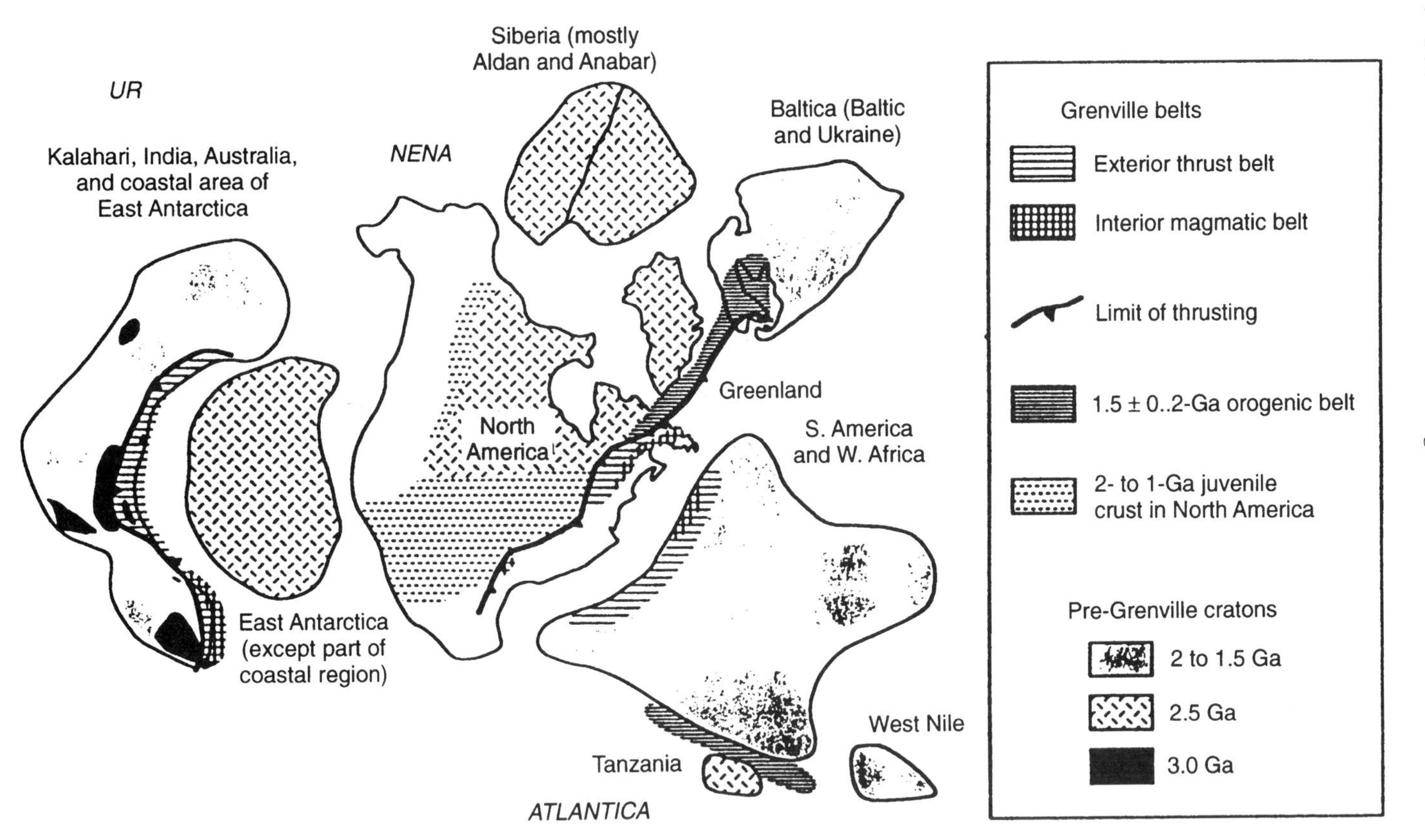

Figure 16.8. Another tentative reconstruction of Rodinia, which would have formed during the Grenville orogenesis, around 1 Ga ago. From Rogers, 1996.

more or less closed in the west, between the two Americas. The second anticlockwise rotation of Gondwana, in the Lower Carboniferous, will close it completely. Mauritania will collide with the United States east coast, producing the Appalachian Mountains. North Africa and Apulia will collide with Europe and produce the *Variscan chains (Hercynides)*. The complete collision took at least 50 Ma, and presented several orogenic phases, two of which at least concern Western Europe: the *Briton* phase since 350 Ma (Upper Devonian) and the *Asturian* phase around 310 Ma.

16.8 CONTINENTAL DRIFT BETWEEN 1,000 AND 500 MA B.P.

Around 500 Ma ago, at the limit between Cambrian and Ordovician, Gondwana was the setting of a poorly known orogenic activity, called *Brazilian* phase in South America and *Pan-African* phase in Africa. The orogenesis affected the Eastern Antarctica coastline, and the building of the *Gondwanides* probably began then.

It seems now quite certain that Gondwana formed at this time, by merging of an *Eastern Gondwana* (Australia, East Antarctica, India and East Africa), and a *Western Gondwana* (most of Africa and South America). The collision gave birth to mountain ranges in East and South Africa, called 'intracratonic' at a time where the different African cratons had not been well recognized (see Figure 15.2).

In the last years, several authors have attempted the reconstruction of continents in the Precambrian. All agree that all continents were merged around 1 Ga ago. At this time, the *Grenville orogenesis* (precisely dated in North America) must have resulted from the collision of Laurentia with the Amazonia craton. Around the same time, there must have been another collision on the other side of Laurentia, where the Pacific coast of North America is now lying, attributed to East Antarctica. Laurentia was already united to Baltica and Siberia, in a hypothetical continent called *Arctica* by Rogers (1996). Amazonia was part of a Western Gondwana (*Atlantica*, for Rogers), and East Antarctica was part of an Eastern Gondwana. Another Pangaea would have formed this way, in the Upper Precambrian. It was baptized *Rodinia* (from the Russian *rodina*, fatherland), and would have subsisted between 1 Ga and 700 Ma approximately. Figure 16.7 shows the Rodinia proposed by Dalziel (1992) and Figure 16.8 shows the Rodinia proposed by Rogers (1996). The differences between the two remain sizeable: In the former the cratons of Kalahari and Congo-Kasai are as close as today, whereas in the latter they are very distant from each other.

Rodinia would have dislocated around 700 Ma ago. Two rifts appeared, dividing none of the agglomerated cratons, but running along the two orogenic belts on each side of the Canadian cratons. These rifts became mid-oceanic ridges and generated large oceans: Iapetus, which roughly prefigured the Atlantic Ocean, and *Panthalassa* which roughly prefigured the Pacific Ocean. A complete hemisphere of oceanic crust or more must have disappeared simultaneously on the other side of the Earth, between 700 and 500 Ma, in one or several subductions. The mean subduction rate (total, if there were several subductions) would have been 130 km/Ma in Dalziel's reconstruction, and 100 km/Ma in Rogers's. This is not impossible, but all this still remains highly hypothetical.

17

Mantle chemistry and continent formation

17.1 VERY HIGH PRESSURE STUDIES

Our knowledge of matter at the pressures and temperatures existing inside the Earth has been obtained with two complementary techniques:

(1) The compression of a volume of around $1\,cm^3$ by a *shock wave*, for less than a microsecond. The advantage of this technique is that pressure and density are directly and unambiguously obtained, by a simple application of the laws of mechanics.
(2) Static compression, i.e. compression maintained as long as necessary, of a sample of $0.001\,cm^3$ between *diamond anvils*. This allows reaching equilibrium when there are phase changes.

Compression by a shock wave

When a projectile with a speed of a few km/s strikes a fixed target, two supersonic shock waves start from the contact surface. One moves inside the target, forward, with the velocity U, and the other moves in the projectile, backward (Figure 17.1). Between the two fronts, the medium is compressed adiabatically and behaves like a fluid.

Per unit time, over a length U, the speed of matter inside the target goes from 0 to u and this length becomes $U - u$. Noting ρ_0 and ρ_1 the densities at rest and after the shock, the conservation of the mass M (per unit area, perpendicularly to the velocities) means that:

$$M = \rho_0 U = \rho_1 (U - u) \tag{17.1}$$

The momentum of this mass M goes from 0 to Mu in one time unit, under the effect of the pressure p exerted by the projectile. Therefore:

$$p = Mu = \rho_0 Uu \tag{17.2}$$

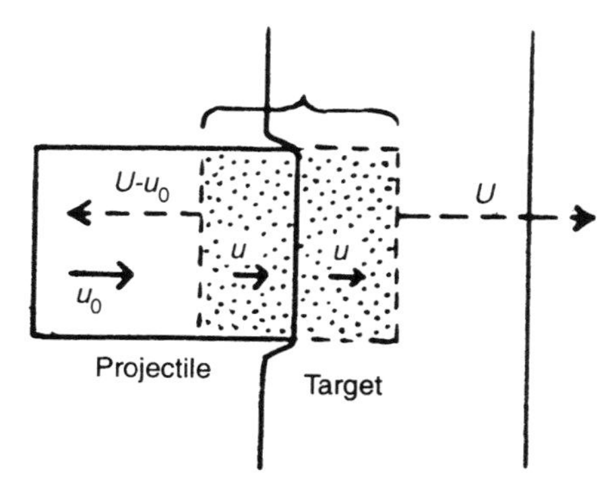

Figure 17.1. Demonstration of the Hugoniot relations. See text.

This pressure exerts a work, per unit area and unit time, equal to $pu = \rho_0 U u^2$. One half becomes kinetic energy (which rises from 0 to $1/2\ Mu^2$), the other half is expressed as an increase ΔE of internal energy. The latter manifests itself as a temperature increase, and maybe fusion:

$$\Delta E = \rho_0 \frac{Uu^2}{2} \tag{17.3}$$

The equations (17.1–3) are the *Hugoniot relations*. A fourth equation gives the speed u_0 of the projectile after the shock. The speed of matter in the compressed area of the projectile went from u_0 to u. If the projectile and the target are of the same material, there is perfect symmetry at the interface and $u - u_0 = -u$. Therefore $u = u_0/2$.

The experiments consist in measuring U and u_0 during this first phase of the shock. It ends when a shock wave has crossed the target or the projectile. It is then reflected on the free surface and becomes a dilatation wave which, once at the target–projectile interface, causes their separation. As U is of the order of 10 km/s, and the target is thinner than 1 cm, the first phase lasts less than 1 microsecond. Fortunately, fast electronics allows to measure time with a nanosecond precision. Each shot provides one point in the graph (p, ρ), called the Hugoniot of the substance and representing its state equation (Figure 17.2).

A recent improvement of the technique is the simultaneous measurement by radiometry of the temperature reached, which is used to determine the specific heats and melting points.

Static compression

Static compression uses a hydraulic press with anvils of tungsten carbide, capable of exerting pressures of 1 tonne over 10 mm^2, i.e. an axial compression of about 1 GPa. This pressure is amplified by a factor 100 with two diamond anvils, concentrating the force between their large face (the table[1]) and the opposite face (the culet). A great

[1] These are the jewellery terms. The diamonds must be without defects – of 'gem' quality.

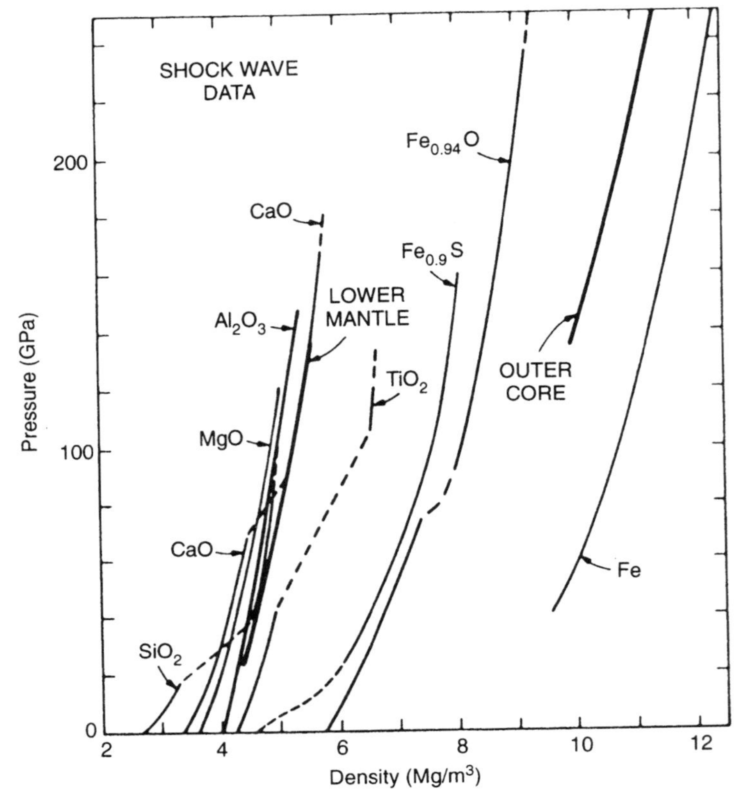

Figure 17.2. State equations obtained with shock waves (hugoniots) for iron, several oxides and an iron sulphide (the latter, as the iron oxide, did not have the stoichiometric composition). In bold lines: temperature–pressure profiles in the lower mantle and the inner core, from seismology measurements. From Jeanloz, 1989.

experimental difficulty is the keeping of these two small faces, pressing the sample, exactly parallel. The sample (10 μm thick, 20–300 μm in diameter) is enclosed in a highly ductile solid, surrounded by a metal gasket crushed under the pressure (Figure 17.3).

The pressure on the sample is not completely homogeneous. It cannot be deduced from the force exerted by the press and the system geometry. This is why microscopic ruby grains are placed on the sample and illuminated with a laser. By fluorescence, they emit a red light, whose wavelength depends on pressure (calibrating was performed with shock waves). In 1986, Xu *et al.* reached a static pressure of 550 GPa, larger by one-half than the pressure at the Earth's centre.

The transparency of the diamond enables these measurements, and also the determination of the crystalline structure of the phases appearing, by X-ray diffraction.

The complete set (diamond anvils, joint, sample) can be heated for hours or days to 500–1,000°C at most (otherwise, the diamonds are damaged). However, in a small

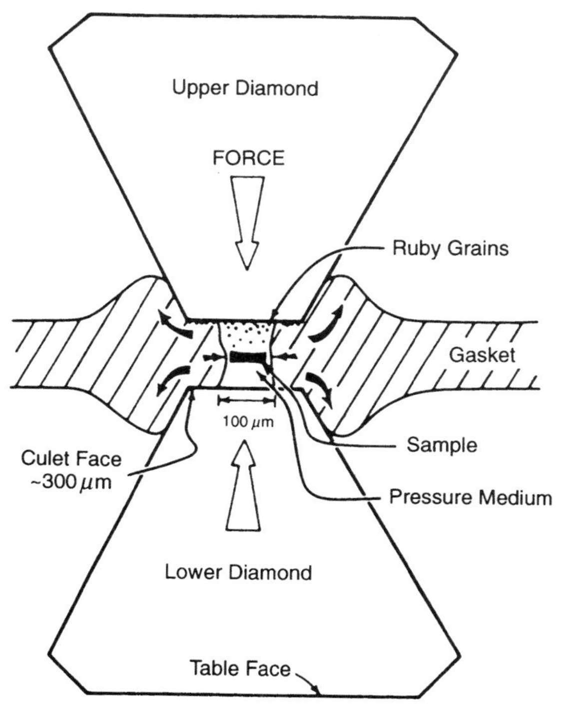

Figure 17.3. Cross-section of the diamond-anvil cell used to reach very high pressures. From Jeanloz, 1989.

part of the sample only, and not homogeneously, laser heating can produce temperatures of 5,000–5,700°C. In this case, the resulting products have been examined only at much lower temperatures, after quenching of the sample.

17.2 MANTLE DENSITY AND PHASE CHANGES

If we neglect the lateral variations of the density ρ inside the mantle (they are important in the first 200 km only), i.e. if we assume that ρ is uniform for a given distance r to the Earth's centre, the gravity $g(r)$ is deduced from the Gauss theorem (Annex A.1). The flux of the gravity vector entering a sphere concentric with the Earth and of radius r is:

$$\begin{cases} (4\pi r^2)g = 4\pi G \int_0^r dm \\ dm = \rho(r') \times 4\pi r'^2 dr' \end{cases}$$

$$g(r) = \frac{4\pi G}{r^2} \int_0^r \rho(r') r'^2 dr' \tag{17.4}$$

At surface ($r = R$), g is known. We also know, from the precession of equinoxes, the Earth's moment of inertia:

$$C = \tfrac{2}{3}\int_0^R \rho(r) \times 4\pi r^4 dr \tag{17.5}$$

Deviatoric stresses are negligible, and the pressure p varies according to the equation:

$$\frac{dp}{dr} = -\rho(r)g(r) \tag{17.6}$$

The hugoniots of the materials forming the different layers have been determined; they provide the third relation necessary to calculate g, ρ and p as functions of r. Seismic wave velocities were previously used:

$$v_p^2 - v_S^2 = \frac{k_S}{\rho} \tag{17.7}$$

If there is no change in phase or in global chemical composition, then:

$$\frac{1}{k_S} = \frac{1}{\rho}\frac{d\rho}{dp} \tag{17.8}$$

It seems today established that *the mantle discontinuities located by seismologists are almost entirely due to phase changes of the minerals*, and not, as defended obstinately by Anderson and Bass (1986), to major modifications of the chemical composition. The discordant hypothesis of Javoy (1995) (Section 17.12) is, however, worth examining seriously.

At 410 ± 10 km in depth, the α-olivine with orthorhombic structure transforms into β-olivine with modified spinel structure (see Section 3.1), along with a density increase of 7.8%. This reversible transformation is *exothermic* (in the direction light phase $\rightarrow$ dense phase).

At 520 km, a slight discontinuity corresponds to the transformation of β-olivine into γ-olivine with exact spinel structure, along with a slight density increase (2.3%).

At 660 ± 10 km, the γ-olivine (global formula $SiO_4Mg_{1,8}Fe_{0,2}$) does not transform into a new 'post-spinel' phase. It rather decomposes *endothermally* into two minerals of very compact structure: cubic *perovskite* SiO_3Mg, unknown at the surface of the Earth, and *magnesio-wustite* $Mg_{0,8}Fe_{0,2}O$. (This was the first discovery from diamond anvils, by Liu and Ringwood (1975)). The transition is very sudden, 5 km at most. This suddenness was a strong argument in favour of a chemical discontinuity, as for the Moho. This argument is lessened by the fact there is also transformation at 660 km depth of the second component of the upper and intermediate mantle, pyroxenite. It takes a perovskite structure, along with a density increase of 10%.

Garnets can cross the 660-km discontinuity without transformation (O'Neill and Jeanloz, 1994), but they should take at a lower level the very high-pressure form of *majorite*.

The chemical composition in major elements of the upper and lower mantle does not seem perfectly identical. Recent measurements of the thermo-elastic properties of magnesian perovskite and magnesio-wustite lead to a lower mantle richer in Si and Fe than the intermediate and upper mantle (Bina and Silver, 1990; Stixrude *et al.*, 1992). The resulting density difference at the interface would be a few per cent.

17.3 PENETRATION OF SUBDUCTED PLATES INTO THE LOWER MANTLE

Around 60 km deep, the basalt of subducting plates transforms into eclogite, denser by 17% (see Section 3.8). In the intermediate mantle, the pyroxenes of eclogite should transform into *stishovite*, even denser, but the relatively high Ca and Al content of eclogite should prevent its transformation into perovskite at 660 km. These transformations facilitate subduction in the upper and intermediate mantle, but they hamper the penetration of the subducted plate into the lower mantle.

Let us now consider the mantle lithosphere, which forms nine tenths of a subducted plate. By crossing a discontinuity, the light phase of olivine transforms into a dense phase, and does not seem to subsist as a metastable state. The phase change occurs at pressure p, and at the temperature T. The Clapeyron formula shows that dp/dT is of the same sign than the transformation heat, i.e. it is positive if the transformation from light phase to dense phase is exothermal. (The fact that olivine is a solid solution modifies the value of dp/dT, actually $2.9\,\mathrm{MPa\,K^{-1}}$ (Bina and Helffrich, 1994).)

This is the case of the 410-km transformation. In a plate colder than the surrounding mantle, this transformation will occur at lower pressure, at a higher level. The plate becomes denser, and the warming by heat emission will not compensate this effect. The phase change at 410 km must therefore increase traction at the trench, and facilitate subduction and convection.

The 660-km transformation is endothermic, and thus if the plate penetrates into the lower mantle the opposite occurs. In the colder subducted plate, the spinel/post-spinel transformation happens lower down. According to Bina and Helffrich (1994), $dp/dT = -2\,\mathrm{MPa\,K^{-1}}$.

The delayed move of the oceanic crust and mantle lithosphere to dense phases, and the small difference in chemical composition of the lower mantle, are as many factors that prevent the convection currents from crossing the 660-km discontinuity. To cross it, a subducted plate must be much colder than the surrounding mantle, and an upwelling plume must be much warmer; this is not easy. The controversy on this question has long been obscured by confusion between 4 questions completely distinct:

(1) *Are earthquake foci observed below 670 km?*

This is rarely the case (Figure 17.4). But this does not exclude penetration of the subducted plate into the lower mantle. If it penetrates, the deep transformation of the crystalline structure must allow a complete relaxation of deviatoric stresses transmitted by the plate.

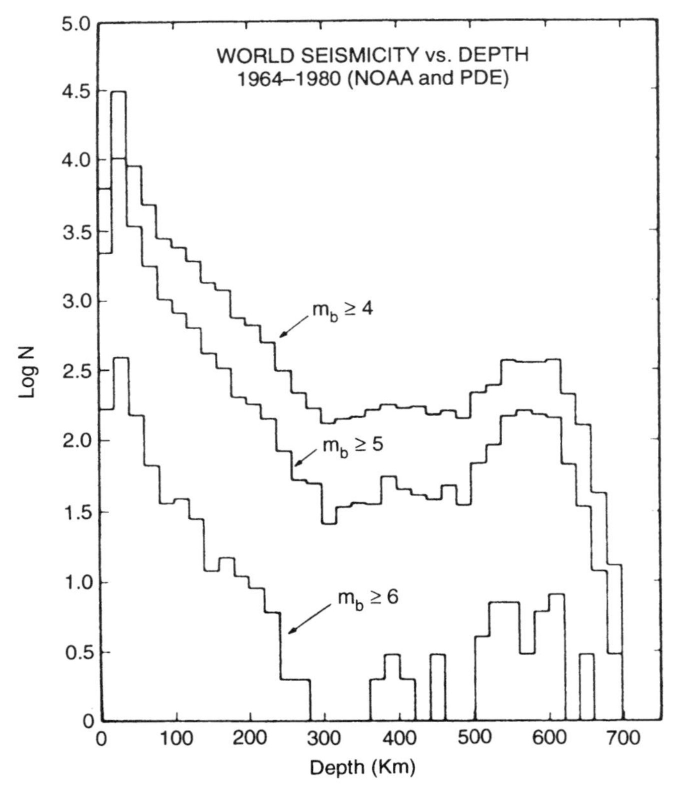

Figure 17.4. Total number of earthquakes recorded worldwide in 1964–80, by 20-km depth intervals. From Vassiliou *et al.* (1984).

(2) *Are there convection currents in the lower mantle or not?*

There must be some, to carry the heat away, because the upper mantle is not enough of a heat source to explain the geothermal flux measured at the surface. Contrary to former theories, the thermal conductivity of the lower mantle is insufficient and heat transfer must be ensured mainly by advection, through convection currents. But, as we shall see in the last chapter, a very moderate convection is enough.

(3) *Do the velocities in the lower and upper mantle have the same magnitudes?*

Deviatoric stresses are of the same order (because of the equilibrium equations), the velocities would have the same magnitudes if the viscosities were of the same order. Some authors (like Peltier in Toronto) thought this could be inferred from the analysis of glacio-isostatic uplifts. But the relative stability of the hot-spots, linked to the lower mantle, demonstrates that the velocities there should be lower by at least one order of magnitude to those in the upper mantle. The analysis of geoid undulations leads to the same conclusion. Today, everybody (including Peltier)

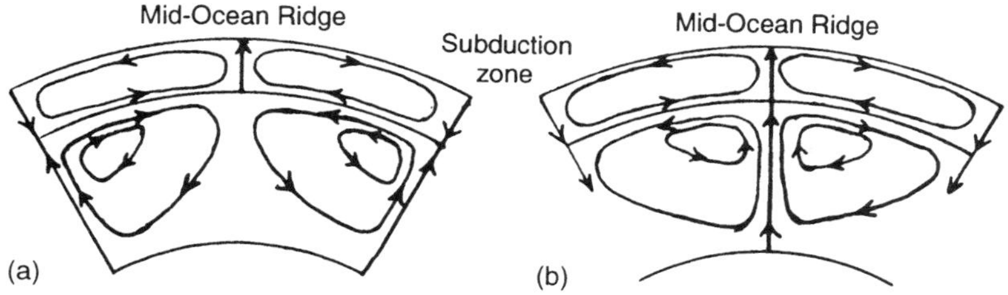

Figure 17.5. Two-layer convection: (a) mechanical coupling; (b) thermal coupling.

agrees that the lower mantle viscosity is 12 to 40 times higher than the upper mantle viscosity.

(4) *Is there a 'one-layer' or 'two-layers' convection?*

If the 660-km discontinuity cannot be crossed, there will be two superposed layers of convection cells or rolls. They may be coupled mechanically (Figure 17.5(a)) or thermally (Figure 17.5(b)). This is at least the intuitive prediction in the case of a steady regime, with fixed current lines, but we shall see that the regime is not steady.

Deep foci were only located precisely after the installation of the first global seismic networks. Jordan (1977) could then affirm that in the subducted plate north of the Kuriles there had been a 1000 km deep focus, below the Okholtsk Sea. Similar deep earthquakes, requiring a penetration of the subducted plate into the lower mantle, were observed since in the subduction zones of the Mariannas and Sunda Islands. But, in other places, the deep foci show that the subducted plate does not cross the 660 km discontinuity, is deviated and stays just above for hundreds of kilometres. This is the case for subductions in the South Kuriles, Honshu, the Izu-Bonin Islands, the Banda Sea, the Tonga and Fiji (Lundgren and Giardini, 1994).

Lower mantle penetration or not, the two cases are currently co-existing. This does not tell us whether these are permanent convections, here with one layer and there with two layers, or whether convection is constantly evolving and the two cases may succeed in time, in a same place.

17.4 SEISMIC TOMOGRAPHY

The study of seismic velocity lateral variations in the Earth started 20 years ago at Harvard University, under the impulse of Dziewonski. It preceded the installation of worldwide seismographic networks covering a large spectrum of frequencies, essential for the analysis of any wave and the collection of reliable results. These are the Global Seismograph Network (GSN) run by the US Geological Survey and IRIS (Incorporated Research Institutions for Seismology), and the French network GEOSCOPE. Other networks, not covering the whole world, have been set up by Italy (MEDNET, in the Mediterranean), Japan, China, Germany and Canada.

The main reason brought forward to defend these costly projects is that a better knowledge of propagation times would allow to better locate seismic sources, and would be essential for seismic hazard studies and the detection of underground nuclear tests. But the main motivation of scientists is probably pure science. The lateral variations of seismic velocities are correlated to the lateral temperature variations. Roughly, we have:

$$\frac{d(\ln v_P)}{dT} = -5 \times 10^{-5}\,K^{-1}; \qquad \frac{d(\ln v_S)}{dT} = -7.5 \times 10^{-5}\,K^{-1} \tag{17.9}$$

We shall see in the next chapter that lateral temperature variations are the driving mechanism of convection, and that the latter can be inferred from the former, at least qualitatively.

The values of the reference seismic velocities v_P and v_S are taken from the Preliminary Reference Earth Model (PREM), published by Dziewonski and Anderson (1981). This model has a spherical symmetry, with velocity jumps at 670, 400, 220 and 80 km (Figure 17.6). The 220-km one is at the base of the LVZ (see Section 5.4), and the 80-km one is at the base of the lithosphere. The PREM model has been corrected to account for the flattening of the Earth.

The variations of v_P, v_S and ρ, say δv_P, δv_S and $\delta\rho$, may be represented by a limited series of spherical harmonics, whose coefficients are functions of the radius.

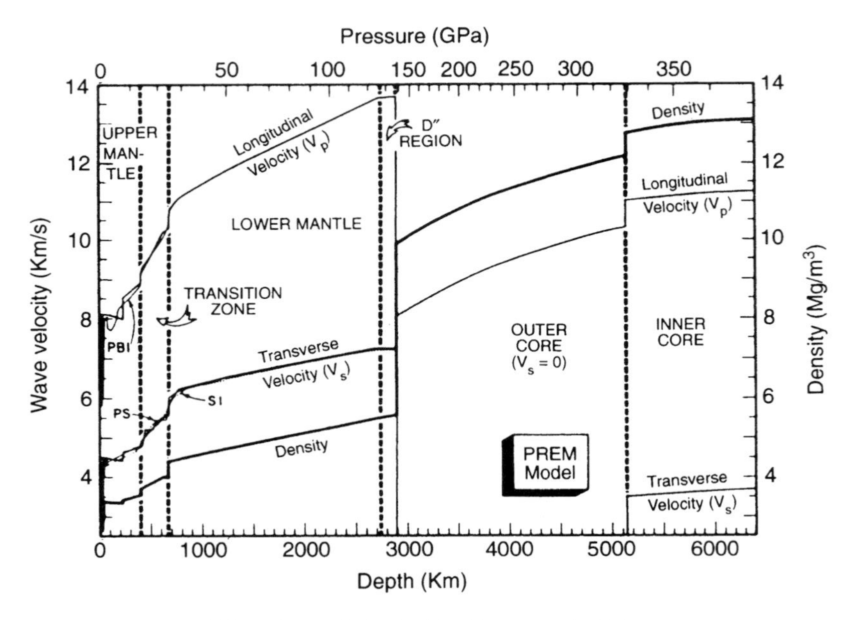

SI = Hart, 1975; PBI = Burdick and Powell, 1975; PS = Fukao *et al.*, 1982

Figure 17.6. Seismic velocities and densities in the Earth, from the PREM model of Dziewonski and Anderson (1981). From Jeanloz, 1989.

These functions are represented by Legendre or Tchebychev polynomials with few parameters. In the most recent studies, the wealth of data allowed to use spherical harmonics to degree 12 for the upper mantle, and to degree 6 for the lower mantle. With this representation, the anomalies that extend over only a few thousands of kilometres (e.g. below a hot-spot or in a subduction zone) are heavily smoothed and reduced, and the smaller anomalies disappear.

Variations of v_S are of 1% at most; this corresponds to a lateral temperature variation of 130°, although it should be larger but more localized. But whether the corresponding Archimedes thrust spreads over 300 or 3,000 km has a secondary influence on convection.

The *S*-wave velocity is best known, in the upper mantle by surface waves, in the lower mantle by the Earth's eigen-frequencies (see Figure 2.9) and by transit times. For *P*-waves, we can only access the *P*, *PKP* and *PKIKP* transit times. The main source of error with transit times comes from the corrections made to account for the exact nature of the crust near the receiving stations.

The ratio of the relative variations of δv_P and δv_S is usually assumed, to have fewer unknown variables. It is assumed equal to the ratio $d(\ln v_P)/d(\ln v_S)$ when the radius varies (PREM model) or when temperature varies (laboratory experiments). This ratio equals 1.25 in both cases. But when models were established separately with only v_S data and only v_P data, the relative variations ratio was 2.0 ± 0.4. The assumption therefore seems unsound.

A critical examination of 4 global v_S models, performed by different teams with partly different datasets and mathematical techniques, was recently published (Ritzwoller and Lavely, 1995). There is a good agreement for the location of anomalies (less for the amplitudes), at all levels except in the higher layer of the upper mantle, for which the data used provide the least information.

- *In the upper mantle,* the δv_S fit the predictions from plate theory: positive in subduction zones, where matter is colder, negative below mid-ocean ridges, where it is warmer.
- *In the intermediate mantle,* the positive anomalies of subduction zones are continuing. (But they sometimes correspond to the subducted part of a plate now disappeared.) The negative anomalies no longer correspond to mid-ocean ridges, or to negative anomalies in the lower mantle. This shows that *the currents of matter feeding mid-ocean ridges remain in the upper mantle, whereas subduction reaches the lower mantle.* This (temporary) conclusion is not in contradiction to the conservation of matter fluxes. If a cold convection current locally descends in the lower mantle, the return current may be spread over the whole Earth and its velocity become insignificant. For example, if half the flux of subducted matter (130 km^3/year) penetrates into the lower mantle, the vertical return current would only have an average velocity of 1 mm/year.
- *In the lower mantle,* the anomalies δv_S are smaller. The predominant factor is a high-speed, cold ring around the Pacific. The low velocities are not located below mid-ocean ridges. They are diffuse (which can be attributed to the low

resolution of tomography) and correspond roughly to high altitudes at the surface of the Earth, or to 'bumps' in the geoid, or to hot spots.

17.5 NUCLEOSYNTHESIS AND NATURAL RADIOACTIVITY

Before talking about dating and other isotope-based techniques, let us recall the laws of natural radioactive decay.

Different radioactivities

The nucleus of a radioactive isotope may free a helium nucleus He–4. This *α-radioactivity* decreases its atomic number Z (number of protons) by 2, its number N of neutrons by 2, and its atomic mass $A = Z + N$ by 4.

In β^- radioactivity, one neutron from the nucleus transforms into a proton. There is emission of an electron, a γ photon and an undetectable antineutrino. Z increases by 1, N decreases by 1, A does not vary.

In *K capture*, a proton from the nucleus captures an electron from the inner atomic layer (K layer) and transforms into a neutron, with emission of a photon and a neutrino. Z decreases by 1, N increases by 1, A does not vary.

Therefore, the elements of a same radioactive family have equal mass numbers, or their mass numbers differ by a multiple of 4.

Radioactive constant

The decay rate of a radioactive nucleus is independent from the chemical body it belongs to, from the fluid or solid state of this body, from temperature or pressure. In an isolated system where the isotope P is not created (or is not created any more), its concentration, here denoted (P) will decrease according to the law:

$$\frac{d(P)}{dt} = -(P)\lambda \implies (P) = (P)_0 e^{-\lambda t} \tag{17.10}$$

The (half-disintegration) *period* of P is the time interval T after which, in these conditions, (P) has decreased by half:

$$T = \frac{\ln 2}{\lambda} = \frac{0.693\ldots}{\lambda} \tag{17.11}$$

Secular equilibrium

Consider a closed system (i.e. not exchanging any matter with the outside world) containing a radioactive isotope P not created (or not created any more), and all its descendants: $P \rightarrow F_1 \rightarrow F_2 \rightarrow \cdots \rightarrow F_n \rightarrow S$, stable. The evolution of concentrations, in *atoms*, is given by the system of equations:

$$\begin{cases} \dfrac{d(P)}{dt} = -(P)\lambda \\ \dfrac{d(F_1)}{dt} = (P)\lambda - (F_1)\lambda_1 \\ \dfrac{d(F_n)}{dt} = (F_{n-1})\lambda_{n-1} - (F_n)\lambda_n \\ \dfrac{d(S)}{dt} = (F_n)\lambda_n \end{cases} \tag{17.12}$$

If the period of P is very large compared to the periods of all its radioactive descendants ($T \gg T_1, T_2, \ldots T_n$, i.e. $\lambda \ll \lambda_1, \lambda_2, \ldots, \lambda_n$), after some time, all descendants present in the system will be coming from the initial P. A steady regime is established:

$$\begin{cases} (P)\lambda = (F_1)\lambda_1 = \ldots = (F_n)\lambda_n \\ (P)/T = (F_1)/T_1 = \ldots = (F_n)/T_n \end{cases} \tag{17.13}$$

Even if this has taken more than a century, the so-called *secular equilibrium* has been reached.

A first application of (17.13) is the determination of extremely long radioactive periods, when the much shorter period of a descendant is known. If there is certainty that the system remained closed, the ratio of periods equals the ratio of atomic concentrations.

Origin of heavy elements

All astrophysics and nuclear physics studies lead to the idea that heavy natural elements come from fusion reactions in the heart of stars, when all their hydrogen had been converted into helium. During this terminal phase in the life of a star, the temperature of its core increases more and more, and successive nuclear fusions, from helium to carbon etc., create heavier and heavier elements. If the mass of the star exceeds some critical value, the evolution ends up with a final explosion, the formation of a supernova and the dispersion of most of its matter around the Universe. Dust disks transforming into stars with possible planets therefore already contain heavy elements.

The list of natural radioactive isotopes was given in Section 9.12. Apart from the ones continuously forming in the atmosphere, they either belong to the heavy elements existing before the formation of the solar system (4.5 billion years ago), or they are descendants of these radioactive isotopes with very long lives. In the latter case, they are only present as minute traces. Let us look at some figures:

- Th–232 (giving after 10 successive disintegrations, Pb–208, stable) has a period of 13.9 Ga. Its radioactive descendant with the longest life is Ra–228, with a period of 5.77 years only. Therefore, at secular equilibrium, it is 4.1 10^{10} times less abundant than thorium–232.

- U–238 (giving, after 14 successive disintegrations, Pb–206, stable) has a period of 4.51 Ga. Its most abundant radioactive descendants are U–234 ($T = 245,000$ years) and Th–230 ($T = 75,200$ years). Therefore, at secular equilibrium, (U–234)/(U–238) = 5.43 10^{-5}.
- U–235 (giving, after 11 successive disintegrations, Pb–207, stable) has a period of 0.71 Ga. Its most abundant radioactive descendant is Pa–231, with a period of 32,500 years. At secular equilibrium, (Pa–231)/(U–235) = 4.6 10^{-5}.
- Np–237 (neptunium) has a period of 2.20 Ma. It completely transformed into Bi–209, the stable isotope with the largest mass number and the largest atomic number ($Z = 83$).

Similarly, I–129 (iodine), of period 16.4 Ma, was completely transformed into Xe–129. Assuming that the nucleosynthesis of I–129 happened in a very short time, the Xe–129 content of chondrites should make it possible to estimate the time between nucleosynthesis and their formation, and with them the formation of the first planetary bodies in the solar system (see Albarède and Condomines, 1976). The time found was 60 Ma only. But, in fact, I–129 appeared continuously by spontaneous uranium fission and the hypothesis above is not valid (Caffee and Hudson, 1987).

17.6 RADIOCHRONOLOGY UP TO 300,000 YEARS

Organic products (wood, bones, shells) can be dated with the carbon-14 technique. In living matter, the (C–14)/(C–12) ratio differs from the atmospheric value by a constant factor, close to 1, as carbon is constantly renewed. At the death of the tissue ($t = 0$), it is no longer renewed, and the C–14 content decreases with $\exp(-t/T)$. C–14 contents are minute indeed, and the individual disintegrations of atoms are counted in specific rooms, protected as much as possible from cosmic rays, built with materials exempt from any radioactivity, with air filtering etc.

Selecting for the initial C–14 content the value measured today in similar living tissues, and $T = 5568$ years, ages were obtained and are now noted 'BP C–14' (BP = before present = before 1950). We now know that $T = 5,730$ years, and that the atmospheric C–14 content varied with time. Its production rate depends on the flux of solar particles reaching the atmosphere, which varies with the intensity of the Earth's magnetic field. (Currently, this value does not depend on latitude, showing a rapid homogenization.) The isotopic equilibrium between atmosphere and ocean also depends on temperature.

C–14 chronology was calibrated until 9,000 years BP, using the annual rings of very old sequoias (*dendrochronology*). The correction to BP C–14 ages reaches 10% for 7,000 years. Bard *et al.* (1990) were able to extend the calibration with corals from Barbados, whose aragonite skeleton could be safely dated by the (U–234)/(Th–230) ratio. They reached $18,200 \pm 400$ BP C–14, which is equivalent to

21,930 ± 150 BP. Between 9,000 BP C–14 and this date, the correction needed varies irregularly between +11% and +24%. Datings were performed up to 50,000 BP C–14 and earlier, but they cannot be corrected.

The measurement of (U–234)/(Th–230) ratios nowadays makes dating up to 300,000 years possible. This method is sometimes called the *ionium method*, ionium being the old name for Th–230.

17.7 DATING OF ROCK MATERIALS

Beyond 1 Ma, dating is performed by using the radioactivity of rubidium–87, potassium–40, or U–238 and U–235 simultaneously.

Rb/Sr method

Around 27.8% of natural rubidium is made of its slightly radioactive isotope Rb–87, and the rest of Rb–85. Through β^- radioactivity, with a period of 47 Ga, Rb–87 transforms into Sr–87, stable (Figure 17.9). Strontium has 4 stable isotopes, of which only Sr–87 is at least partially issued from radioactivity (this fraction of Sr–87 is called *radiogenic*). Respective contents in natural strontium are approximately: 0.5% of Sr–84, 9.9% of Sr–86, 7.0% of Sr–87 and 82.6% of Sr–88. The ratios (Rb–87)/(Sr–86) and (Sr–87)/(Sr–86) can be directly measured with a mass spectrograph. Sr–88 is not used as a stable reference isotope because it is much more abundant, and the direct measurement of ratios would be much less accurate.

If the rock crystallized at $t = 0$ from some magma or heavy metamorphism, the global ratios (Rb)/(Sr) took different values in each mineral (see later), but the isotopic ratios inside each chemical element, (Rb–87)/(Rb–85), (Sr–84)/(Sr–86), (Sr–87)/(Sr–86) etc. took the same value in all minerals. There was *isotopic homogenization*. If there were no later isotopic exchanges between minerals, the *atomic* concentrations became:

$$\begin{cases} (\text{Rb–87}) = (\text{Rb–87})_0 e^{-\lambda t} \\ (\text{Sr–87}) = (\text{Sr–87})_0 + [(\text{Rb–87})_0 - (\text{Rb–87})] \\ \qquad\qquad = (\text{Sr–87})_0 + (\text{Rb–87})(e^{\lambda t} - 1) \end{cases} \tag{17.14}$$

Isotopic ratios measured in the different minerals are therefore linked by the *chronometric equation*:

$$\frac{(\text{Sr–87})}{(\text{Sr–86})} = \frac{(\text{Sr–87})_0}{(\text{Sr–86})} + \frac{(\text{Rb–87})}{(\text{Sr–86})}(e^{\lambda t} - 1) \tag{17.15}$$

By plotting (Rb–87)/(Sr–86) as abscissa and (Sr–87)/(Sr–86) as ordinate, the points relative to the different minerals are aligned on a line with slope $(e^{\lambda t} - 1)$, called the *isochron*. This slope gives the age. If the points are not aligned, this is because of an insufficient isotopic homogenization at the beginning, or because of isotopic exchanges between minerals after their formation, and dating is no longer possible.

K/Ar method

Potassium contains 1/10,000th of the radioactive isotope K–40, which disintegrates in Ar–40 by *K* capture and in Ca–40 by β^- radioactivity (Figure 17.9). In a closed system:

$$\begin{cases} \dfrac{d(\text{Ar–40})}{dt} = (\text{K–40})\lambda_K; \qquad \dfrac{d(\text{Ca–40})}{dt} = (\text{K–40})\lambda_\beta \\ \dfrac{d(\text{K–40})}{dt} = -(\text{K–40})(\lambda_K + \lambda_\beta) \\ \lambda_K = 0.0585\,\text{Ga}^{-1}; \qquad \lambda_\beta = 0.472\,\text{Ga}^{-1} \\ T = \ln 2/(\lambda_K + \lambda_\beta) = 1.31\,\text{Ga} \end{cases} \tag{17.16}$$

Ca–40 is the calcium isotope most abundant in rocks, and Ar–40 is the argon isotope most abundant in the atmosphere. Recognizing the very small radiogenic fractions is impossible. But, in the micas appeared during the solidification of a magma or during important metamorphism, it can usually be admitted that all Ar–40 is radiogenic, because:

- during the formation of mica, there was complete degassing and any other Ar–40 left;
- once formed, mica constitutes a closed system, whereas other minerals are degassing during later non-destructive warming (e.g. feldspars) or admit atmospheric argon in. Ar–36 dosing is used to check for this contamination, and perform corrections if necessary.

Ar–40 and K–40 dosing in a single mica are then enough to give the age. The chronometric equation becomes:

$$(\text{Ar–40}) = \frac{\lambda_K}{\lambda_K + \lambda_\beta}(\text{K–40})[e^{(\lambda_K+\lambda_\beta)t} - 1] \tag{17.17}$$

Ar–39/Ar–40 method

This method is more complicated, but only requires measurements of relative and not absolute contents, and it can also provide some indications about the thermal history of the mineral.

The samples to analyse are irradiated in a nuclear reactor with a flux of fast neutrons, along with a reference sample *R* of known age. Some argon 39, a β^+ radioactive isotope with a period of 270 years, is produced by the reaction:

$$\text{K–39} + n \rightarrow \text{Ar–39} + p \tag{17.18}$$

Each sample is then heated by stages. At each stage, the ratio (Ar–40)/(Ar–39) is measured in all the freed argon, since the beginning of heating. The values may be different. It is then likely that radiogenic Ar-40 diffused out of the formation sites during thermal events. If such perturbing event did not happen, the age t is deduced from the age t_R of R by the relation:

$$\frac{(\text{Ar–40})/\text{Ar–39})}{(\text{Ar–40})_R/(\text{Ar–39})_R} = \frac{\exp[(\lambda_K + \lambda_\beta)t] - 1}{\exp[(\lambda_K + \lambda_\beta)t_R] - 1} \tag{17.19}$$

17.8 LEAD DATING

Lead ($Z = 82$) has 4 stable isotopes: Pb–204 (not radiogenic) and the three isotopes already seen: Pb–206, Pb–207, Pb–208, respectively coming from U–238, U–235 and Th–232. For a closed system, when secular equilibrium is reached, it is possible to write three chronometric equations similar to (17.15). Only the first two are generally used:

$$(\text{Pb–206}) = (\text{Pb–206})_0 + (\text{U–238})(e^{\lambda t} - 1)$$

$$(\text{Pb–207}) = (\text{Pb–207})_0 + (\text{U–235})(e^{\lambda' t} - 1)$$

$$\lambda = 0.154\,\text{Ga}^{-1}; \qquad \lambda' = 0.971\,\text{Ga}^{-1} \tag{17.20}$$

We will only consider '*rich systems*' (rich in U and Th), where the lead isotopes 206, 207 and 208 are nearly entirely radiogenic. This is the case of *zircons* (see Section 9.13), which are abundant in granites and are the most used to date their formation. $(\text{Pb–206})_0$ and $(\text{Pb–207})_0$ can then be neglected in equation (17.20). And, as in all minerals today, we have:

$$(\text{U–238})/(\text{U–235}) = 137.8 \tag{17.21}$$

(This is due to isotopic homogenization of uranium during the Earth's formation. As U–235 disintegrates faster than U–238, this ratio increased with time.) If the system has remained closed since a time t, the isotopic ratios satisfy the equation:

$$\frac{(\text{Pb–207})}{(\text{Pb–204})} = \frac{(\text{Pb–206})}{(\text{Pb–204})} \times \frac{1}{137.8} \times \frac{e^{\lambda' t} - 1}{e^{\lambda t} - 1} \tag{17.22}$$

For different samples with the same geological history, for example different zircons from a same granitic massif, the points of abscissae (Pb–206)/(Pb–204) and ordinates (Pb–207)/(Pb–204) are located on a line, called the *isochron*, whose slope gives t.

In fact, the above model of a system closed since the rock formation is often inexact: U and Th may have escaped in part, during a later metamorphism episode. To date it, the uranium isotopes are also dosed, and we used for abscissa $x =$ (Pb–207)/(U–235) and for ordinate $y =$ (Pb–206)/(U–238). If the system had remained closed, the points (x, y) would be on the *Concordia curve*, of equation:

$$1 + y = (1 + x)^{\lambda/\lambda'} \tag{17.23}$$

When they are not on this curve, they are usually aligned on a line D. The age of the perturbing metamorphism is given by the intersection point of D and the Concordia

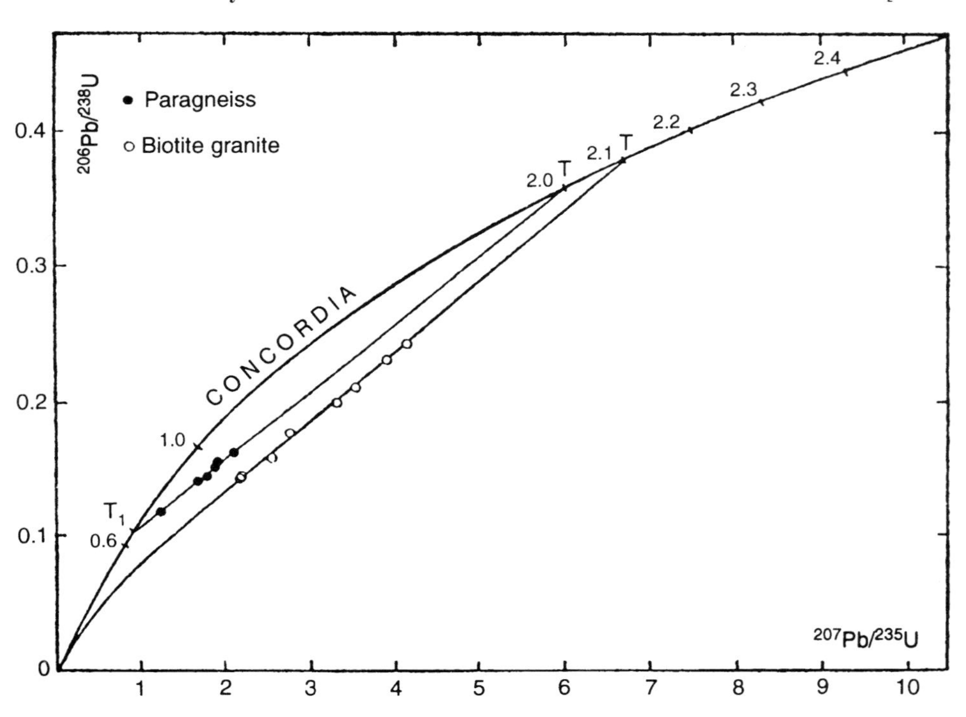

Figure 17.7. Concordia diagram for two zircon populations. The paragneiss zircons were formed in T, 2 Ga ago, and underwent an opening event 0.62 Ga ago, during the metamorphism which created the paragneiss. The biotite granite zircons crystallized 2.1 Ga ago, and continuously lost lead until now. From Lancelot, 1975 (reproduced in Albarède and Condomines, 1976).

curve (Figure 17.7). For more details (and 3 models for explanation), the reader may consult Allègre and Michard (1973).

17.9 AGE OF THE EARTH

By age of the Earth, we mean the time since a relatively short period where it underwent a complete isotopic homogenization, following important melting and intense mixing. Let us note $-t_0$ this time, using the present as the origin.

The first precise determinations of the age of the Earth were performed in 1942–46, and were exclusively using the isotopic ratios of lead in galenas (S Pb), which do not contain any uranium. We assume the system remained closed until their formation at the time $-t_1$. At this moment, uranium left the galena and the isotopic ratios have not varied since (the system became closed again). Let us note (U–238) and (U–235) the fictive modern concentrations in galena if the lead had not migrated elsewhere. We must have: (U–238)/(U–235) = 137.8. The indices 0 and 1 corresponding to the atomic concentrations at the times $-t_0$ and $-t_1$ respectively, the evolution of the closed system means that:

$$(\text{U–238}) = (\text{U–238})_0 e^{-\lambda t_0} = (\text{U–238})_1 e^{-\lambda t_1}$$

$$(\text{Pb–206}) = (\text{Pb–206})_1 = (\text{Pb–206})_0 + [(\text{U–238})_0 - (\text{U–238})_1]$$

$$= (\text{Pb–206})_0 + (\text{U–238})[e^{\lambda t_0} - e^{\lambda t_1}] \quad (17.24)$$

Similarly, for the family U–235 → Pb–207, we have:

$$((\text{Pb–207}) = (\text{Pb–207})_0) + (\text{U–235})[e^{\lambda' t_0} - e^{\lambda' t_1}] \quad (17.25)$$

Let us write:

$$\frac{(\text{Pb–206})}{(\text{Pb–204})} = X; \qquad \frac{(\text{Pb–207})}{(\text{Pb–204})} = Y \quad (17.26)$$

We must therefore have:

$$\frac{Y - Y_0}{X - X_0} = \frac{1}{137.8} \times \frac{e^{\lambda t_0} - e^{\lambda t_1}}{e^{\lambda' t_0} - e^{\lambda' t_1}} \quad (17.27)$$

In the first calculations, the ages of galenas were determined independently, by dating accessory minerals. For galenas of the same age, the points (X, Y) must be aligned, their slope giving the age of the Earth. The lines corresponding to different ages must converge on the point (X_0, Y_0), which must be close to the origin, because these are rich systems. The imprecision was coming from the dating of galenas.

In 1955, Patterson determined the age of the Earth with more precision by inferring the X_0 and Y_0 from the isotopic ratios in the troilite (FeS) of siderites (ferrous meteorites, supposed to be as old as the Earth) and by using only recent galenas, for which $t_1 \approx 0$. Later studies have also used silicate meteorites, which proved to have been closed until now ($t_1 = 0$), with varying initial uranium contents. As shown in Figure 17.8, all the points (X, Y) are perfectly aligned, on a line whose slope corresponds to: $t_0 = 4.55 \pm 0.07$ Ga.

17.10 MARKING BY TRACE ELEMENTS OR ISOTOPIC RATIOS

The elements of the silicate Earth (whole crust and mantle) can be classified into three groups:

- *The major elements* (more than 2% in weight). By decreasing order of concentration in the crust (this order is different in the mantle): O, Si, Al, Fe, Ca, Na, K, Mg.
- *The minor elements* (0.44 to 0.06% of the crust, in weight): Ti, H, P, Mn, F, Ba, Sr, S, C.
- *The trace elements*. By decreasing order of concentration in the crust: V (0.0135% = 135 p.p.m.), Cl, Cr, Rb, Ni, Zn, Ce (60 p.p.m.), Cu, Y, La (30 p.p.m.), Nd, Co, Sc, Li (20 p.p.m.), N, Nb, Ga, Pb (13 p.p.m.), etc.

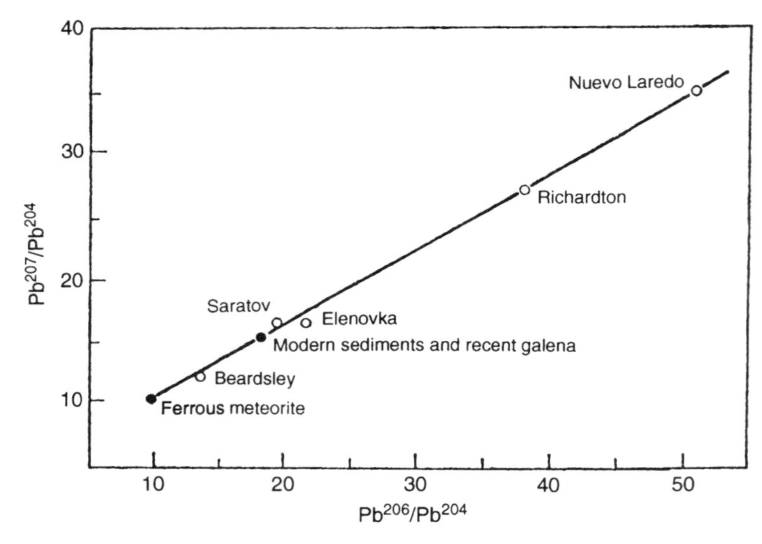

Figure 17.8. 4.55-Ga isochron defined by the analysis of meteorites. From Murthy and Patterson (1962), reproduced in Albarède and Condomines (1976).

Siderophile elements are amongst the last ones; they have a strong affinity with iron and were nearly entirely captured by the core (Ni, Co, etc.). So are the *chalcophile* elements, with a strong affinity with sulphur, probably trapped with it in the inner core: Zn, Cu, Ga, Pb, etc.

Amongst the *lithophile* elements concentrating in silicates, the *lanthanides* ($Z = 57$ to 71) are very interesting because of their potential as markers. They were formerly called *rare earth elements* (REE). Their names and their different isotopes are given in Figure 17.9. They all have the same chemical properties and are always trivalent, except cerium (passing by oxidation from Ce^{3+} to Ce^{4+}) and europium (passing by reduction from Eu^{3+} to Eu^{2+}). Thus, the anomalies in Ce and Eu contents provide indications of the redox potential in the medium during the formation of the mineral in which they are included.

The contents in trace element Rb, like the contents in different lanthanides, distinguish markedly from tholeiitic basalts forming at mid-oceanic ridges (MORB) the alkali basalts, with olivine, of intraplate volcanism (see Section 3.3). This is shown in Figure 17.10.

Other tracers, more interesting and easy to measure, are the isotopic ratios between a stable isotope, more or less radiogenic, and another stable isotope, non-radiogenic, of the same element. The ratios between different chemical elements vary with each episode of melting or metamorphism, but these isotopic ratios usually do not vary.

These studies began with the discovery by Gast (1964) that the ratio (Sr–87)/(Sr–86) is relatively constant in mid-ocean ridge tholeiites (0.7028 ± 0.0005), and it is high and variable in oceanic islands (0.703 to 0.706). The ratio (Nd–143)/(Nd–144),

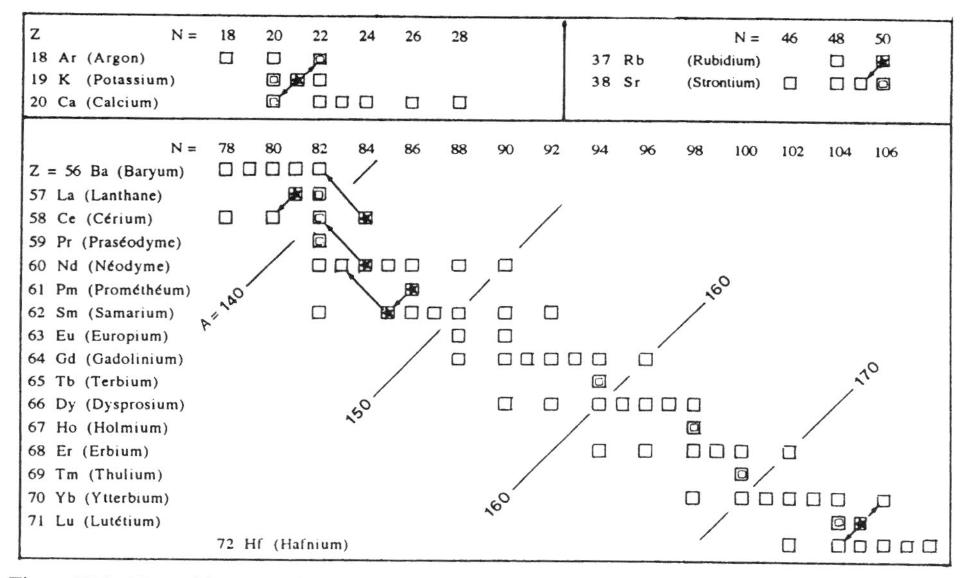

Figure 17.9. Natural isotopes of Ar–K–Ca, Rb–Sr and lanthanides (rare earth elements). The latter were completed with prometheum, an artificial product. A double square means that the isotope forms more than 80% of the natural chemical element. A star means it is radioactive, and the arrow(s) point to the products of its disintegration.

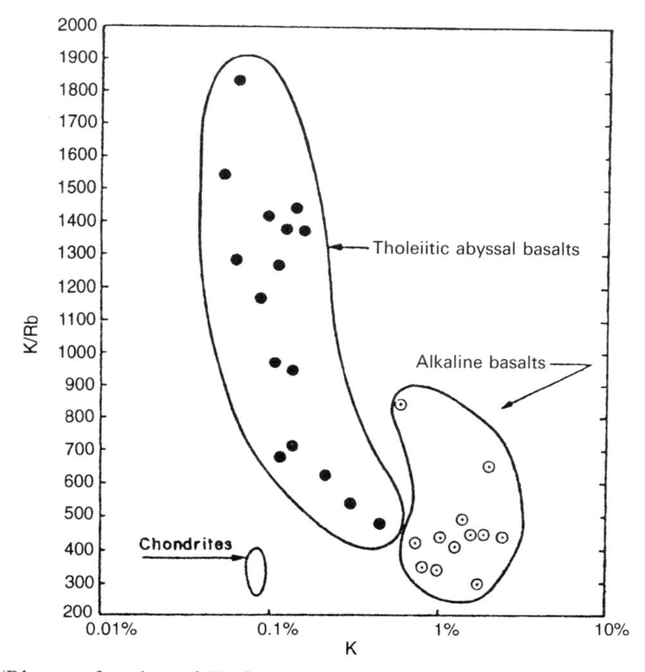

Figure 17.10. K/Rb as a function of K, for the two large basalt families and for chondrites. From Albarède and Condomines, 1976.

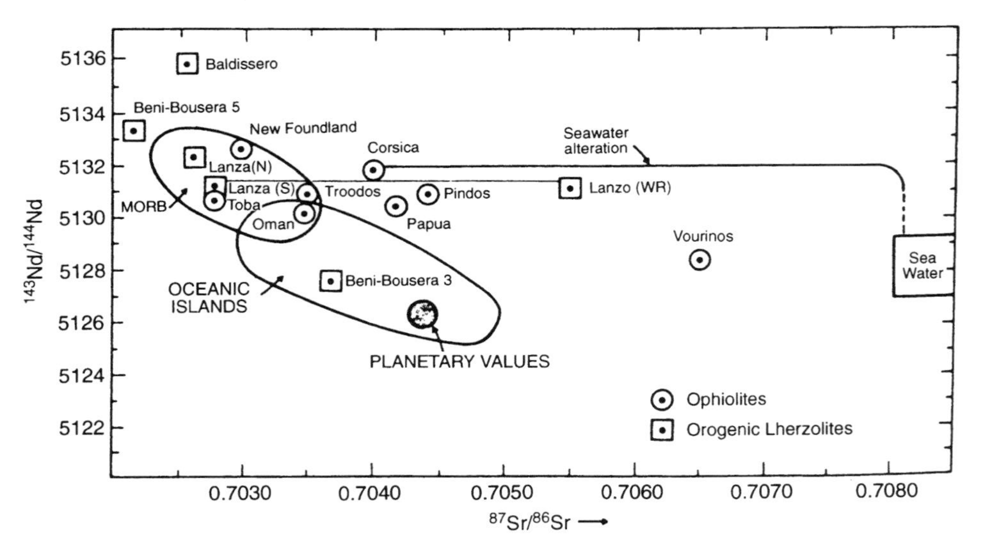

Figure 17.11. Isotopic Nd–Sr diagram for orogenic ophiolites and lherzolites. The domains corresponding to mid-ocean ridge tholeiites (MORB), to oceanic islands, to chondrites ('planetary values') and to sea-water are indicated for comparison. From Richard and Allègre, 1980.

which was introduced in 1976 by Richard, Shimizu and Allègre (1976) and is more interesting. It is negatively correlated with the previous ratio in unaltered rocks, but at the difference of (Sr–87)/(Sr–86), it is nearly insensitive to alteration and metamorphism. Strontium is an alkali metal, producing highly soluble salts. The Sr–87 atom also has a diameter clearly different from its parent (Rb–87), and it tends to migrate away from the crystallographic site where it was born. This is clearly visible in Figure 17.11, taken from a study of orogenic ophiolites and lherzolites; the alteration of these rocks has strongly modified (Sr–87)/(Sr–86) but not (Nd–143)/(Nd–144).

The ratios for meteorites and lunar rocks are also reported in this figure: (Sr–87)/(Sr–86) $= 0.7043$ and (Nd–143)/(Nd–144) $= 0.51264$. We see that the values for oceanic islands are between the MORB and 'planetary' values.

At this time, geochemists were aiming to explain the isotopic ratios with a *model of boxes*, a few homogeneous reservoirs, communicating more or less with each other, without destroying the homogeneity of each reservoir. Dupré and Allègre (1980) were advocating two reservoirs in the mantle: an 'undepleted' lower mantle, having kept the initial chondritic composition, and a 'depleted' mantle from which the continental crust came.

The isotopic ratios of lead and helium were also determined. They are rarely correlated with the previous ratios. As the studies were developing, heterogeneities were discovered away from the mid-ocean ridges, at all scales, from 10 m to 10,000 km. The isotopic ratios were fluctuating, the more so as samples were coming from sites more distant from each other. Solid-state diffusion only destroys heterogeneities at the metre scale, in a time of approximately 100,000 years.

Multiplying the number of reservoirs is therefore not enough, and this naïve concept of a reservoir at a precise location should be abandoned. It is better, at least for the moment, to talk of 'components', in the sense of the *principal components* drawn from the statistical analysis of large number of related random variables. The possibility of isolating some components, present or absent depending on the location, is empirical, and does not require any rational justification.

Zindler and Hart (1986) recognize the following components:

(1) *The most depleted mantle* (because of the continental crust exudation). It provides most of the mid-ocean ridge basalts, by melting 10 to 23% of the rock at depths of less than 100 km. Some xenoliths emitted by continental volcanoes also show this component.

(2) *A mantle enriched* in U and Th, with a (Pb–206)/(Pb–204) ratio very high and a (Sr–87)/(Sr–86) very low. It produces the basalts of St. Helena and the Tubuai. These authors attribute it to the presence of recycled oceanic crust in the magmas.

(3) and (4) *Two types of mantles enriched* by the mixing of marine sediments of continental origin or of continental crust, with or without metasomatism. This gives the basalts of Walvis, Kerguelen, the Samoas, and it contaminates the mid-ocean ridge basalts from the Indian Ocean.

(5) *A mantle with very weak (He–4)/(He–3) ratio*, below Iceland and Hawaii. It was therefore not well degassed at the origin, but still not primitive according to the isotopic ratios of Sr, Nd and Pb.

(6) *The predominant mantle*, which is no longer the primitive mantle, with the same isotopic ratios as chondrites, and is already half-depleted. This component could be a mixing of the previous enriched and depleted components, but this would require a mixing always with the same proportions, and this cannot be explained. In weight, the predominant mantle forms less than 60% of the whole mantle. This contradicts the old idea that it forms the entire lower mantle (73.4% of the mantle).

This geochemical argument is not the only one to argue against the idea of separate convections in the lower and upper mantle. But geochemistry cannot contradict the hypothesis, advanced by some, that mixing of the entire mantle would have happened a long time ago and stopped today.

17.11 FORMATION OF THE EARTH

By analogy with nebulae observed in our galaxy, it is thought that a disk of gas (mainly hydrogen) and dust surrounded our Sun at its formation. This disk was containing most of the angular momentum. Dust was at the origin of planets. Temperature decreasing with distance from the Sun, H_2O and CO_2 were only condensed as particles at distances larger than the current distance of Jupiter. The *terrestrial planets* (Mercury, Venus, Earth, Mars), their satellites and the asteroids between Mars and Jupiter formed by agglomeration of iron and silicate dust into larger and larger planetary bodies.

Asteroids with diameters of around 10 km, on average 20,000 km distant from each other, could have formed in 1 Ma. Opinions are diverging about the existence of a very rapid growth phase (Stevenson, 1989). In any case, the number of planetary bodies reduced, and their encounters became widely spaced. Some of the numerical simulations performed by Wetherhill in 1980–85 lead to the relative distances and masses of modern terrestrial planets. According to these simulations, 90% of the Earth's mass would have gathered in 40 Ma; 40% of the Earth would result from giant impacts by planetary bodies with diameters between 500 and 3,000 km.

We have already seen in Section 8.13 that the late collision, around 4.2 Ga ago, of a small planet with the Earth could have formed the entirely silicate Moon on one part, and enlarged the terrestrial core on the other. Another large collision should explain why Mercury lost most of its silicate mantle. (Although its mass is 18 times smaller than the Earth's, and therefore the elastic or phase change compression inside is much smaller, Mercury has a nearly similar density: 5.4 Mg/m^3 instead of 5.5 Mg/m^3.)

Opinions are also diverging about the atmosphere around the primitive Earth. According to some, the disk around the Sun would have lost all its gases in 3 Ma at most. The primitive Earth's atmosphere would have mainly consisted in H_2O and CO_2 present in the minerals of particles or asteroids. These minerals would have dissociated during the accretion. For others, the loss of the gases would have been much slower, and they would have been part of the primitive atmosphere. For yet others, more water was captured later, from comets.

According to Chang *et al.* (in Schopf, 1983), partial pressures in the early Earth's atmosphere would have been: 35 bar CO_2, 4 bar CO, 1.5 bar N_2, 0.1 bar CH_4, traces of NH_3 and HCN, plus water vapour (the Earth would have then enclosed twice more H_2O). This question is of interest for the scientists interested in the emergence of life on Earth. It seems that bacteria (cells without nuclei, but already reproducing identically) appeared 3.5 Ga ago and that photosynthesis appeared 3.2 Ga ago, with the first cyanophyceae (blue algae). The near totality of the atmospheric oxygen appeared by photosynthesis, at the expense of carbonic gas. The fixation of CO_2 in carbonates must have started 2.5 Ga ago with the development on the coastlines of continents of *stromatolites*, colonies of cyanophiceae with a calcareous skeleton. The greenhouse effect of CO_2, which increased the Earth's surface temperature a lot, must have decreased then. The first beings based on cells with nuclei (eukaryotes) seem to date from 1.3 Ga ago.

At the beginning, the small asteroids in the solar nebula had circular orbits around the early Sun, and similar velocities during their encounters. The random perturbations from the mutual attractions have progressively made many orbits quite elliptic, with two consequences:

- The chemical composition of the solar nebula being a function of the distance to the Sun, the primitive asteroids did not form by accretion of matter with a single composition;

- The impact velocities were more or less increased, and thus the energy liberated as heat during the impacts. For an Earth with half of its modern mass, an asteroid with no relative velocity at high distance would have acquired at impact time a speed of 9 km/s. (Reversing time, we see this is the liberation speed for this Earth). With a soft shock, without rebound, the ensuing energy liberated as heat would have reached 4×10^7 J/kg. With some relative speed at high distance, the energy liberated could have been 2 to 4 times larger.

As 3×10^6 J/kg are enough to induce melting, it surely happened during impacts. In the absence of atmosphere, the cooling of the surface by infrared radiation would have been very fast. It would have prevented any important warming of the early Earth at depth. This occurred, however. But as the statistical distribution of asteroids sizes and their penetration depths are unknown, any estimate of temperatures in the early Earth would be very speculative.

The prevailing opinion is that the Earth never was completely liquid, but that large oceans of silicate magma over a liquid iron layer (and siderophile and chalcophile elements) would have formed many times. The undepleted interior of the primitive Earth, always solid, would have warmed up. And after acquiring some fluidity, pockets, large 'blobs' of liquid iron would have gone down by gravity and progressively formed the current iron core.

But this scenario comes up against a non-solved difficulty. The solidification of a large silicate magma ocean (mainly producing olivine and garnets in their high pressure phases) should lead to fractionation of trace elements. Laboratory experiments demonstrated this. But in all samples of the upper mantle, the ratios (Sm)/(Hf), (Sc)/Sm), (Ir/Au), (Co)/(Ni) have the same values, similar to the ones measured in chondrites. After its solidification the silicate Earth would then need to have been intensely mixed by convection currents.

Furthermore, this scenario should be compatible with the composition of chondrites, which is not evident.

17.12 CHONDRITES

The meteorites made entirely of silicates, or chondrites (see Section 9.12) result from the relatively recent fragmentation of one or several asteroids which never reached the size necessary for generalized mixing and chemical fractionation. The segregation of iron in a core had already taken place, however. This core is at the origin of *siderites*, meteorites nearly entirely made of iron (and thus much easier to pick out in the field). It does not seem that the formation of these proto-planetary bodies, sources of both chondrites and siderites, has been modelled. More rarely (1% of cases), one also finds meteorites formed of half silicates and half iron (*lithosiderites*). They must come from the silicatesiron core interface, and have not been studied much.

It is often written that the silicate Earth, crust + mantle (bulk silicate earth, BSE), has a chondritic composition. This affirmation should be more detailed, because there are in fact important differences, and chondrites do not have a uniform composition.

A chondrite contains a *porous matrix*, with grains smaller than 10 μm, undoubtedly coming from dust aggregation with no melting. It also contains 30–70% of *chondrules*, and often some *refractory inclusions* of angular shape and all sizes, richer in Al, Ca, Ti, forming as much as 10% of the meteorite. The spheroidal form of the chondrules seems to indicate a melting episode, but, since their composition is not in equilibrium with the matrix composition, this episode must have been very short (around 1 hour, for some). Many events leading to local chondrule formation would have happened in the primitive dust ring. The nature of these events and the origin of the heat inducing the temporary melting of chondrules remains mysterious (lightning, maybe?).

Chondrites are classified according to their chemical composition in 7 groups: CI, CM, CO, CV, E, H, L, LL. All classes show the same ratios (Ca)/(Al) = 1.09 and (Sm)/(Nd) = 0.325. In all classes except CM, CO and CV, the ratio (Nd)/(Ca) is the same. Each meteorite is also attributed a number indicating its alteration and its amount of metamorphism: 1 for hydric alteration at low temperature (transformation of some minerals into clays); 2 for the same alteration to a lesser degree; 3 to 6 for increasing metamorphism in a dry environment. For example, the meteorite that fell at Allende (Mexico) in 1969, subject of many studies, is a CV3.

The isotopic ratio (O–18)/(O–16) varies within chondrites, and differs from its terrestrial value: $0.200 \pm 0.001\%$. These differences seem to require that the protoplanetary disk was not homogeneous, and that the ratio (O)/(H) was 5 to 20 times larger than its cosmic value of $7\ 10^{-4}$ (Wood, 1988).

After a very long and complex discussion, Zindler and Hart (1986) came to the following conclusion. The global composition of the silicate Earth is the composition of a CI chondrite very depleted in volatile elements, less depleted in Si (−19%) and Mg (−8%), and highly enriched in the most refractory trace elements like lanthanides (2.5 times more abundant than in CI chondrites).

Javoy (1995) suggests however a different global composition for the Earth. It would come from E chondrites, mainly formed of enstatite (SiO_3Mg), free silica, iron and troilite (FeS). Their iron content corresponds often to the Earth's in its entirety, this iron is not oxidized as in the other chondrite classes, and the ratios (O18)/(O16) and (O17)/(O16) are exactly those of the Earth and the Moon, to 0.02‰, which cannot be by chance. E chondrites, whose minerals are refractory and have high temperature forms, would have been predominant in the part of the original dust disk closer to the Sun.

The composition of E chondrites is the one that differs most from the composition of the upper mantle. Javoy's hypothesis requires a high proportion of Si in the core (in addition to S and O), which remains to be proved, and a high difference in the chemical composition of the lower and upper mantle. To find again the lower mantle density from the PREM model, we must assume a rapid temperature increase of ca. 500°C around 660 km. This would fit with a two-layer convection (become compulsory) and the formation of limit thermal layers at this level.

17.13 PROGRESSIVE APPEARANCE OF THE CONTINENTS

The distribution in trace elements differs a lot in continental crust and in the mantle from which this crust comes (Carlson, 1944; Taylor and McLennan, 1995). Whereas continental crust only represents 0.35% of the Earth's mass, it contains more than 30% of some lithophile elements called *incompatible* because they concentrate in magma during partial melting: Cs, Rb, K, U, Th, La, ... But it seems that, at no time, there was complete melting of the silicate Earth. Amongst other arguments, if this had been the case, the mantle olivine would have captured all the nickel from magma, and no cobalt at all. But the (Ni)/(Co) ratio is nearly the same in olivines and in chondrites.

Scientists now think that *the segregation of the continental crust was progressive*, that it happened during many local melting episodes, while solid-state convection was continuously mixing the mantle. Around 1980, Allègre *et al.* used a naïve model with 3 reservoirs (primitive lower mantle, depleted upper mantle and continental crust). From isotopic ratios, they deduced that the continental crust formation had started very slowly, was maximal 2.5 Ga ago and slowed down to become negligible during the Phanerozoic. This conclusion remains with the more realistic and more complex modern models. It may well be that only 60% of this segregation happened before 2.7 Ga.

Let us define the terminology. The *Precambrian* extends from the Earth's formation (4.55 Ga) to the Phanerozoic (last 0.58 Ga, for which fossil-based stratigraphic chronology is possible). The Precambrian is divided into *Hadean*, from which no continental crust remains, *Archaean* (until 2.5 Ga), and *Proterozoic* (2.5 to 0.58 Ga).

Detrital zircons aged 4.1 to 4.3 Ga were found in Western Australia. The oldest dated rock is a gneiss of clearly continental composition (granite–tonalite) from the Slave Province of Canada, and is 3.96 Ga old. But the existence of continental crust does not mean that there were continents, in the usual sense. A continent comprises one or several cratons, blocks stable enough to serve as basement for volcanic and sedimentary series, formed in shallow water or in free air. These cratons are surrounded by orogenic belts, containing sediments from deep water, continental margins and island arcs (with sometimes, locally, fragments of older cratons).

Depending on whether the Hadean is defined as the period with no continental crust of the period from which no continental crust subsists in a craton, it ends 4 Ga or 3.2 Ga ago.

The oldest cratons are around 3 Ga old. They are found in South Africa, Madagascar, Eastern India and Western Australia (see the map of Figure 15.2). There may be some in East Antarctica, hidden by the ice cap. This proves that the first large continent to appear was East Gondwana, deprived from part or whole of East Antarctica. This first continent formed 3 Ga ago, and was given the name of *Urgondwana*,[2] or, more simply, *Ur*.

[2] The German prefix *ur* means very old, primitive. It happens that *Ur* was one of the main towns of the Sumerian civilization, the first one to create writing. (According to the Bible, Abraham would have come from Ur.) But the site of this town, south of Baghdad, is not in Urgondwana.

Another group of cratons is around 2.5 Ga old. They are found in Canada, Greenland, Central Siberia (Anabar–Angara craton), and in Yakutia (Aldan craton). According to Rogers (1996), a large continent would have appeared then. He named it *Arctica*, and these cratons would have been its cores. But, according to a well documented article by Hoffman (1988), the different provinces of the Canadian craton would have merged only 1.91 to 1.63 Ga ago.

The other cratons of Figure 16.2 are around 2 Ga old. According to Rogers, two other large continents would have formed at this time: Baltica and Western Gondwana (Atlantica).

According to geochemists, this progressive apparition of the continents is due not only to the collision and merging of many small and dispersed blocks of continental crust, but mainly to the increasing total volume of this crust.

In the Archaean, in a hotter upper mantle, convection must have been much more intense, with formation of many small convection cells modifying rapidly. The lithospheric plates, between mid-oceanic ridges and subduction zones, should already exist and their relative velocities should have been much higher.

17.14 CONTINENTAL CRUST EVOLUTION AND RECYCLING

It should be expected that the composition of new, emerging continental crust will not remain the same from the beginning to the end of segregation. Figure 17.12 sums up the differences between modern and Archaean magmatism, according to Taylor and McLennan (1995).

The oldest cratons show lavas globally very rich in MgO (24 to 32%), the *komatiites*, which were not created since. They can be found in Zimbabwe (Rhodesia), in NE Ontario and in Gorgona, an island on the Pacific coast of Colombia (a fragment of very old craton, incorporated into the Andes orogenic belt). Experimental work shows that a magma with 30% MgO has a liquidus of 1,600°C at atmospheric pressure, 1,820°C under a pressure of 3 GPa (i.e. around 100 km deep). This leads to the idea that the upper mantle was hotter than now by 300°C approximately. It was also observed that komatiites were lighter than their magma of origin, an exception to the general rule of which ice is another example. Large pockets may have therefore subsisted during a long time, true seas of magma under a solid komatiite lid, 30 to 50 km thick (Nisbet and Walker, 1982).

Other differences appear when looking at trace elements. In crust formed at the Archean, the ratios of lanthanide concentrations vary a lot from one rock to the other, while they are highly uniform in crust formed later. The Archaean crust is less enriched in light lanthanides (La to Sm), and presents no europium anomaly. (However, according to Taylor and McLennan (1995), these differences could attenuate if the whole continental crust was considered, not only its superficial layer.)

The average thickness of continental crust increased with time, because of underplating or intrusions of magmas from the upper mantle. These magmas were probably andesitic in the oldest periods, and progressively became basaltic. According to Jordan *et al.* (1989), the mantle so depleted would still be found

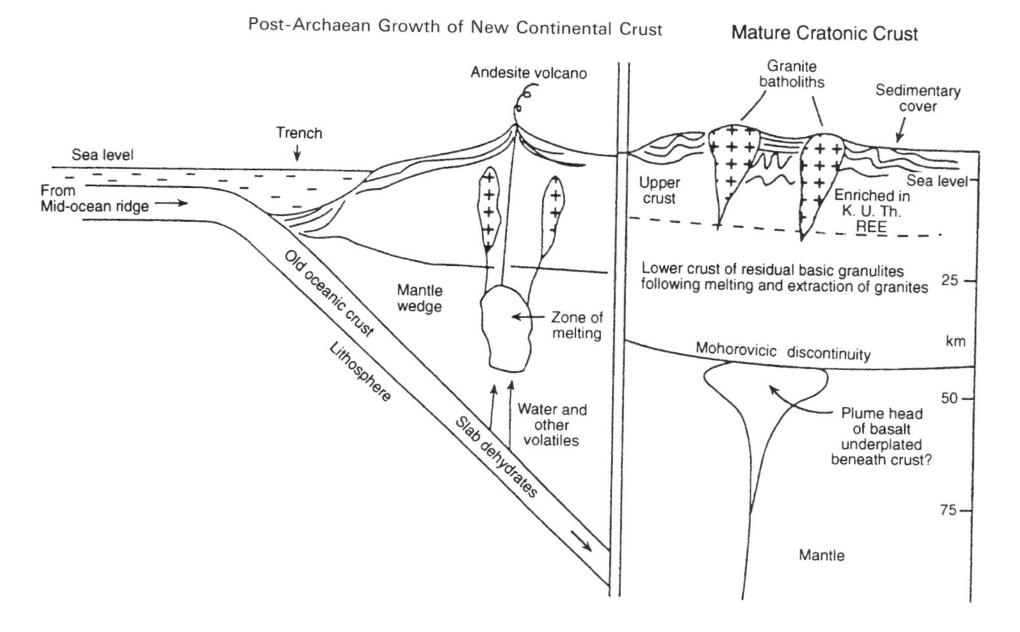

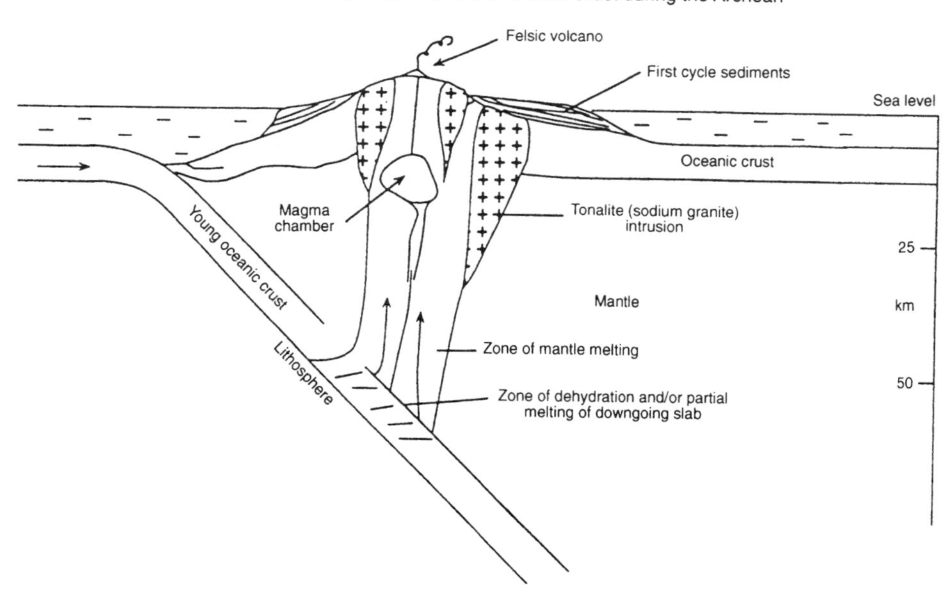

Figure 17.12. Sketch showing the difference in the continental crust created in a subduction zone during the Archaean and after. During the Archaean (bottom), the subducted oceanic crust was younger and hotter, and the temperature/pressure conditions necessary for dehydration melting in the subducted plate and the mantle above were very different. From Taylor and McLennan, 1995.

below the continents, even in the case of old cratons, because it would be part of the lithosphere.

This is of course the lithosphere of geodynamicians, the one of plate theory, called *tectosphere* by Elsasser and Jordan. Below oceans, it stops at the LVZ, where the solidus of mantle lherzolites is reached. Below continents, seismological and geochemical arguments show that it would be more than 200 km thick, and could reach 400 km under the oldest cratons.

The area occupied by emerged lands is currently 150×10^6 km^2, and their volume above sea level 135×10^6 km^3 (hence a mean altitude of 875 m). Rivers carry in the oceans 13 km^3 of sediments per year, coming from the alteration and erosion of the upper crust. (At this rate, the emerged lands will not have emerged any more in 75 Ma, taking only isostasy into account: see Section 8.9.) Rich in silica, these sediments thicken the continental platforms. Sediments carried far from the shores are brought back by plate movements. They can form accretion prisms, or be subducted and regurgitated in active margins, as andesitic magmas or granitic plutons. The continental crust that appeared at the surface of the Earth during the Precambrian is therefore following a cycle of destruction and recycling without ever going deeply into the mantle.

In addition to following this erosion–accretion cycle, the continental crust thickens considerably during orogeneses, mainly through overthrusting (Section 14.6). But it also undergoes thinning (Section 14.16).

17.15 RECYCLING OF THE OCEANIC CRUST

17 km^3 of basaltic crust are now appearing each year at mid-ocean ridges. It is completely sunk in the mantle at subduction zones. Only around 1 km^3/year reappears at the surface, in island arc and active continental margin volcanism. If 16 km^3/year of old oceanic crust were accumulating at 660 km deep, without crossing the discontinuity, they would form a layer with a particular chemistry (the density passing from 2.7 to 4.0) thickening at a rate of 27 km/Ga. If the subducted oceanic crust were accumulating at the core-mantle boundary, forming the D$''$ layer (density 5.5), it would thicken at a rate of 77 km/Ga. Although both hypotheses were suggested, it is currently thought that the oceanic crust sunk does not stay in a particular place, but mixes with the mantle and is recycled.

We shall see in Chapter 19 how this dissemination and mantle mixing can happen. We will only mention here the most recent study on the subject, by Christensen and Hoffman (1994). It is based on the numerical modelling of 2-D convection, with a temperature-dependent viscosity. The temperatures at the bottom of the mantle and the surface plate velocities are specified, instead of being determined by the convection itself. This simplifies the problem. According to this model, around 5/6 of the oceanic crust would remain in the upper mantle and spread there. One-sixth would reach the D$''$ layer, from which it would come up, at hot spots. It would acquire the isotopic signature characteristic of hot-spot basalts during its long time near the liquid core (which is progressively solidifying, and exuding light elements) (see Section 16.4).

18

Glaciations, glacio-isostasy and sea level

18.1 SEA ICE, ICE SHELVES AND ICE SHEETS

Sea ice may form a continuous layer (*pack ice*). It may be along a coast, and fixed to it (*fast ice*) or drifting (*drift ice*). In drift ice, channels (*leads*) and areas of free water (*polynyas*), pressure ridges and hummocks appear and disappear continuously.

The surface of a permanent sea ice field is an ablation zone; its annual melting is more important than the annual accumulation of snow. Conversely, at its bottom, sea ice is accreting, as the cold crosses the whole ice layer, some metres thick. This is a stable regime, because when thickness increases the accretion at the base decreases. In permanent ice fields, the ice takes around 10 years to go up from the bottom, where it forms, to the surface, where it melts.

If the surface becomes a zone of accumulation (annual snow accumulation larger than the melting), the ice pack thickens by 100 to 400 m and becomes an *ice shelf*. The amount of atmospheric cold that can reach the bottom of the ice shelf is now negligible, and in general its base melts, according to the heat that oceanic waters afford. This situation is the reverse of the former. In shallow seas, it would be unstable, because thinning increases the water thickness below the shelf and therefore the melting by ocean currents. But the ice shelf is also fed by ice streams coming from the continent. There are 1.5×10^6 km^2 of ice shelves around the 12.3×10^6 km^2 of the Antarctic ice cap.

Without outside alimentation, an ice shelf must disappear or thicken until it lies on the seabed. It becomes a *marine ice sheet* or *marine ice cap*. This is currently the case for a large part of West Antarctica, Mary Byrd Land. When an ice stream feeds an ice shelf, mechanical equilibrium requires that the grounding line be not on any slope reversal. Therefore, following progressive climatic variations, the extension of a marine ice cap at the expense of an ice shelf, or, conversely, its partial transformation into an ice shelf, may include one or several rapid and irreversible changes.

Faster sliding of a glacier on its bed can occur irreversibly after the modification of the hydraulic or thermal regime at its bottom, unbalancing the mechanical equilibrium. The word *surge* should be exclusively reserved to these catastrophes (in the mathematical sense), not to the jumps of the grounding line.

We will not consider the hypothesis advanced by some, that the Arctic Ocean may have become an ice shelf during Ice Ages. With such an area, a permanent anticyclone would have set up, and without precipitation, the ice shelf would have turned into mere sea ice again.

Following isostatic subsidence, a plain supporting an ice cap may find itself below sea level. This is currently the case of the Wilkes sub-glacial basin, between the Transantarctic chain and Adélie Land. The ice cap is not called marine, however.

Danish call *Indlandsis* the Greenland ice cap, the 'inland ice'. Some French geographers have used 'inlandsis' (with a mistaken spelling) for any ice cap, and I have used it in my 1965 glaciology textbook. But this should be proscribed. Indeed, the ice cap often ends in an uninterrupted cliff of ice (on the NW Greenland coast, for example). But, usually, the ice flows out through rapid ice streams (a few hundred metres to a few kilometres per year) ending at the bottom of fjords, or as ice tongues lying on the continental platform, or in ice shelves.

Today, isostatic readjustment is finished far from the formerly ice-covered areas. But tide gauges show a slow rise of sea level in the last century; 1.28 ± 0.10 mm/year in Marseilles (1886–1987 average), 1.35 ± 0.12 mm/year in Brest (1880–1987 average), 1.47 ± 0.13 mm/year in San Francisco (1880–1988 average). The average of these tide gauges and 8 others is 1.37 ± 0.04 mm/year. This rise of sea level is due to human-induced climate warming (burning of fossil carbon). The sea-level rise due to the thermal dilatation of oceans is estimated to 0.4 mm/year, the part due to the Indlandsis melting to 0.25 mm/year, the part due to the melting of mountain glaciers to 0.4 mm/year. There remains a rise of 0.32 ± 0.2 mm/year, attributable to a melting of the Antarctic ice cap. But it would be wrong to believe that such a large ice sheet shows a uniform behaviour. Climate warming increases the local precipitations, which remain snowfalls. The ice cap thickness should increase until a new equilibrium between accumulation and outflowing.

Stratigraphy of the firn in East Antarctica and in the Antarctic Peninsula show this increase, which should cause a negative rise of the ocean level (-0.6 ± 0.5 mm/year). But the Filchner and Ross Barriers (vertical fronts of the homonymic ice shelves) are strongly retreating, freeing tabular icebergs sometimes larger than Corsica (especially the former). At first glance, the melting of icebergs and ice shelves does not provide a sea-level rise, because they are floating. But warmer oceanic waters create a stronger melting below the shelves. This increases the continental ice streams feeding the ice shelf. Consequently, the part of the ice cap along Filchner Shelf and the West Antarctic ice cap along Ross Shelf are thought to be thinning (James and Ivins, 1995).

18.2 GLACIATIONS PRIOR TO THE PLEISTOCENE

The traces of a glaciation are morphological, and therefore likely to disappear along geological history. Simple grooves on stones may not be of glacial origin. If they are in a conglomerate of angular rock debris, not sorted according to their size (as it would be in talus of mudflows), they are indeed part of a moraine, but this can be due to some small local glacier on a high summit. *Tillites*, which are compressed and cemented clays, are surer clues about a thick ice layer.

It was established in this way that an ice cap existed between 350 and 250 Ma over part of Gondwana, migrating with its drifting, and that no ice cap existed on Earth between 225 and 38 Ma. For an ice cap to form, the continent must be at high latitudes, and the mean temperature at the surface of the Earth must be low enough. Temperature varied widely during geological times, and this is thought to be mainly because of the slow variations of the atmospheric CO_2, and of the corresponding greenhouse effect. The heat dissipated by the Sun may have progressively increased since its formation, roughly balancing the decrease of the greenhouse effect, but irregular variations of solar energy on a 100-Ma time scale do not seem possible.

Our knowledge of the most recent glaciations comes from the study of oceanic sediment cores in deep water. They provide a continuous stratigraphy and, with detrital remanent magnetization, the chronology can be calibrated against the inversions of the Earth's magnetic field. Changes in microfossil species make possible a finer correlation between cores.

High latitude oceanic cores show moraine debris transported in the bottom layer of icebergs (*ice-rafted debris*).[1] This was used to infer that at the Eocene–Oligocene limit (34 Ma), some East Antarctic glaciers must have reached the sea, and that there were more to reach it from 25 Ma onwards. But the glacierization of East Antarctica would have only been completed in the Upper Miocene (15–14 Ma). (Let us remark that the large Trans-Antarctic range bounding this half-continent on the west only gained its modern altitude 19 Ma ago.) If confirmed, the results from atoll coring noted in Section 9.8, would show that the glaciation of East Antarctica had already been complete three times in the Oligocene.

In the Oligocene and Lower Miocene, West Antarctica possessed a relatively mild climate. Sea-reaching glaciers were only found from 11.3 Ma ago, once the Drake Passage south of America was widely open.

Glaciation was the wider in the southern hemisphere during the Messinian (6.5–5.2 Ma). The entire Antarctica, East and West, was then covered by an ice cap which reached the edges of the continental shelf. Atmospheric circulation had then considerably changed, isolating the Antarctic continent. Corrected to sea level, the difference in mean temperatures between the Equator and the South Pole was of only 10°C in the Mesozoic, and reached its current value of 27–28°C at this time (Flöhn, in Waddington and Walder, 1987).

This glaciation around 6 Ma is called *Ross Ice Age* in New Zealand and Tasmania, and was also found by Mercer and Sutter (1982) in Patagonia. A glacial deposit from this period might have been found in British Columbia

[1] In the nineteenth century, the adversaries of the existence of large glaciations in the northern hemisphere were explaining this way the presence of sometimes large erratic blocks, far from their place of origin. Hence the name of *drift* given in English to any ground carried by glaciers, which survived after the giving up of the *Drift theory*.

(Canada). Another glaciation would have occurred in Patagonia 3.5 Ma ago (Mercer, 1976). The oldest glacial deposits in the Russian Arctic are 3.3 Ma old. During the long interglacial period between 6 and 3.3 Ma, the transformation of the West Antarctic marine ice cap into a shelf is not excluded.

18.3 PLEISTOCENE STRATIGRAPHY

Deposits left by the large glaciations form the superficial layers and the current topography of the Netherlands, Denmark, Northern Germany and Poland. They were therefore intensely studied by geologists from these countries (e.g. Ehlers, 1983). The existence of several glaciations cannot be deduced from terminal moraines, because the last glaciation has destroyed most of the earlier moraines. Research thus focused on finding in sections the interglacial periods, during which the bare ground was much altered. In Europe, the interglacial character of deposits was confirmed by the study of the fossil pollens they enclose. Let us remark that there is no strict synchronicity between the mild climate indicated by pollens and the absence of an ice cap, because of the time it takes to disappear and reappear.

The last glaciation in northern Eurasia, called the *Weichsel* glaciation and in Russia the *Valdai* or *Sartan* glaciation, had its maximal extent around 20 ka ago. It formed the terminal moraines of Brandenburg, south of Berlin, and in the centre of Jutland (continental Denmark). Figure 18.1 shows its extension, known by the highly documented study of Grosswald (1980). A continuous ice cap was extending over 5600 km, from Ireland to the Putorana Plateau (Ienissei right bank, culminating today at 1701 m). It covered the continental shelves of the Barents and Kara seas, up to Spitsbergen, the Franz-Josef Archipelago and Severnaia Zemlia. Its total area was 8,370,000 km^2, half lying on now-submerged continental shelves.

The hydrographic network of North Eurasia was totally modified, with the formation of 1,500,000 km^2 of shallow periglacial lakes. The Angara River, outlet of Lake Baikal, was flowing into the largest of these lakes, Lake Mansi. Its waters were flowing into the Mediterranean (120–140 m lower then, as were all oceans), through the Aral Sea, the Caspian Sea, the Black Sea and the Sea of Marmara, all transformed into freshwater lakes.

The glaciation before last, called the *Warthe* glaciation, must have culminated around 130 ka BP. The one before, called the *Saale* glaciation, around 240 ka BP, reached down to the current location of Amsterdam and Nijmegen in the Netherlands. There was at least one older glaciation, the *Elster* glaciation, around 4,000 ka BP. It dug the largest part of the network of U-shaped valleys, now filled with later deposits, found in Germany close to the Baltic Sea (*Rinnentäler*, or *Urströmtäler*).

In the Alps, a book published in 1909 by Penck and Brückner has long been the reference. They attributed 4 terraces found on the valley slopes along the Bavarian tributaries of the Danube River to 4 glaciations: *Würm*, *Riss*, *Mindel* and *Günz* (from the younger to the older). It would seem that the Riss glaciation corresponds both to

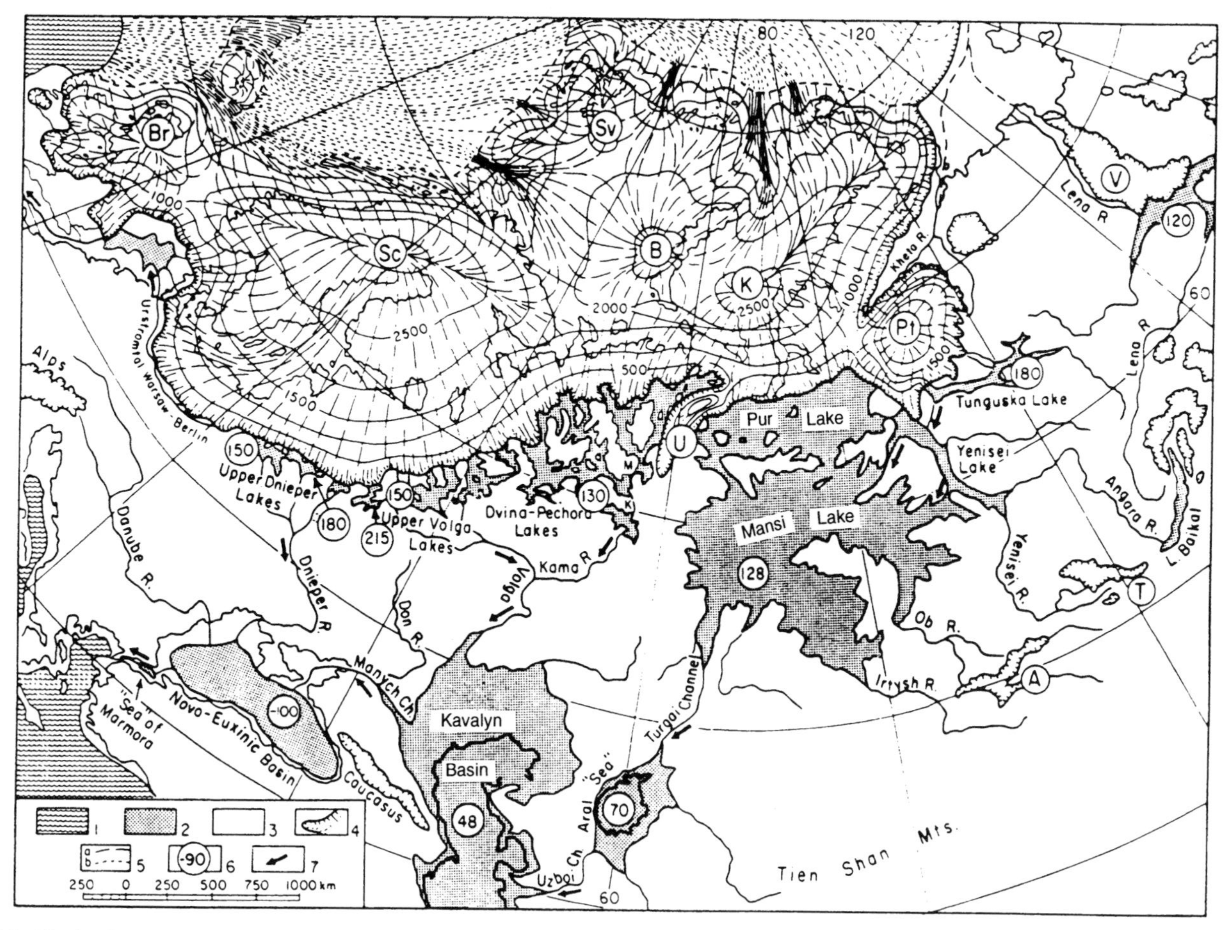

Figure 18.1. Maximal extent of the Weichsel glaciation over northern Eurasia. From Grosswald, 1980.

the Warthe and Saale glaciations, and the Günz glaciation corresponds to the Elster glaciation (Kukla, 1978).

Four ice ages were also recognized in North America; *Wisconsin*, *Illinois*, *Kansas* and *Nebraska*. The ice cap of these periods (*Laurentide ice-sheet*) was as long and larger than the Eurasian ice cap. It extended north through the *Innuitian ice sheet* on the Canadian archipelago. The *Cordilleran ice sheet* was spreading on the western mountains. (An ice-free corridor between this ice complex and the Laurentide ice sheet should have allowed the peopling of America through the emerged Bering Strait.)

Except for the last one, the correlations between ice ages in these three regions of the Earth are not known for certain, because *the stratigraphic series used are far from complete*. In general, a moving ice sheet has carried away most of the earlier deposits. And these classical studies do not clarify whether an interglacial deposit corresponds to a total disappearance of the ice sheet, or only to a temporary reduction of its extent.

Fortunately, continuous stratigraphic series are available; mainly the oceanic sediment cores from the last 2.7 Ma.

The precipitation content in stable isotopes O–18 and D (H–2 deuterium) is linked to the temperature at their site of formation from atmospheric water vapour. Since 1953 this fact has been the main basis of palaeoclimatology from polar ice cores (Dansgaard, 1961). It was believed for some time that, similarly, the O–18 content of Foraminifera in oceanic cores was reflecting the temperature of the surface water where they lived (Emiliani, 1958). In fact, O–18 variations mainly express the total volume of ice on the Earth, because the heavy-isotope enrichment of the ocean by evaporation was not balanced by the return to the sea of rain or melting water depleted in this isotope. This determination becomes relatively accurate if using benthic Foraminifera (see Section 3.5), because the deep ocean temperatures have not varied much. (To obtain palaeotemperatures of surface water, the abundance and nature of Globigerinas must be used.)

$\delta^{18}O$ is always given; it equals 1,000 times the difference between the concentration ratio (O–18)/(O–16) and a standard seawater sample. In graphs of $\delta^{18}O$ as a function of time, the ordinate axis is inverted to facilitate comparisons with palaeotemperatures or with sea levels.

Ice-rafted debris show that in the northern hemisphere, sea-reaching glaciers have been abundant for only 2.5 Ma. Nothing leads us to think that the Antarctic ice sheet varied much during the last 2.5 Ma. In Patagonia, the largest glaciation occurred approximately 1.1 Ma ago, but it did not reach the current Atlantic coastline (Mercer, 1976). It is therefore thought that *benthic oceanic* $\delta^{18}O$ *reflects quite faithfully the total volume of ice sheets in the northern hemisphere*. This forms the basis of a modern stratigraphy of the Pleistocene, with more than one hundred successive *stages* around the main peaks of the curve. They were numbered from the present with Arabic numerals, odd for the 'interglacials' corresponding to a minimum of $\delta^{18}O$, even for the 'glacials' corresponding to a maximum. Only a few interglacials saw the ice volume at the modern level or below. Their beginning is a *termination*, noted with a Roman numeral: I = beginning of stage 1 (11.5 ka BP);

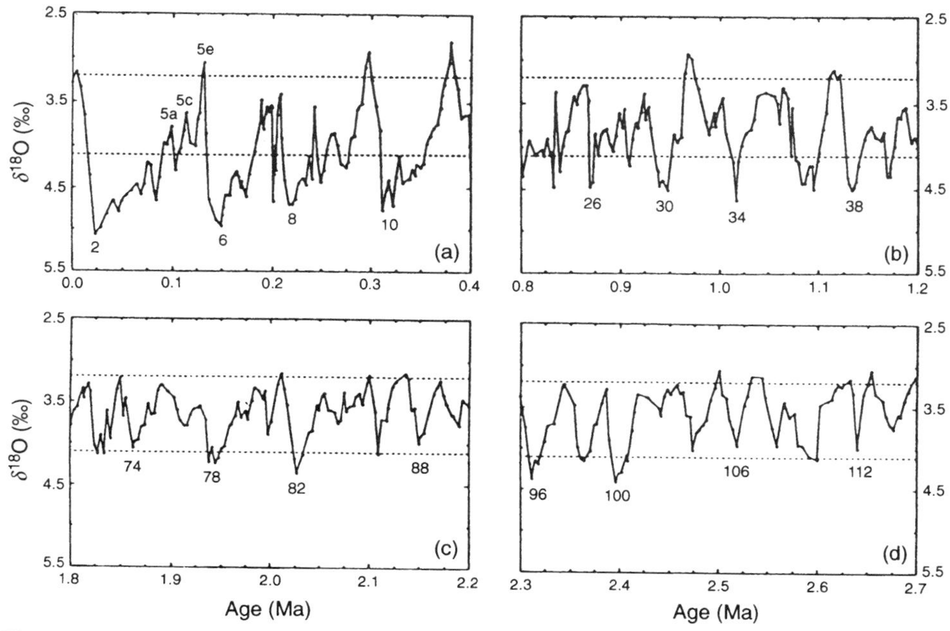

Figure 18.2. Deviation of the O–18 content of microfossils in oceanic cores, as a function of sediment ages (determined by palaeomagnetism). This $\delta^{18}O$ is correlated with surface water palaeotemperature. The numbering of isotopic peaks is the numbering of Imbrie *et al.* (1984), extended. From Raymo (in Kukla and Went, 1992).

II = beginning of stage 2 (128 ka BP); III = beginning of stage 7 (ca. 243 ka BP); etc. The time interval between two terminations is called a *cycle*.

The graph of oceanic $\delta^{18}O$ has roughly the same shape for the 11 cycles between terminations I and XII. Each is 60–116 ka long, with an average of 78 ka, even if a 100-ka peak appears during harmonic analysis of the signal (this is hastily attributed to an astronomical factor; see Section 16.1). According to the definition of terminations, the beginning of each cycle saw complete deglaciation of Northern Eurasia and North America. Only part of the Indlandsis and mountain glaciers remain. Along a cycle, the ice volume oscillates several times with a period of about 40 ka, each maximum being larger than the previous one.

The pattern of oceanic $\delta^{18}O$ changes between 1.2 and 2.7 Ma BP (Figure 18.2). The ice volume increases approximately every 41 ka, more moderately than in stages 2 or 6, and always with close amplitudes (this is why terminations and cycles were not extended to this period).

The 41-ka period corresponds to the variations in the inclination of the ecliptic, changing the contrast between seasons. Milder winters and cooler summers favour the appearance of ice sheets. The complex mechanisms amplifying this very small astronomical factor cannot be treated in this book, and are still much discussed.

(Note that the beginning of the Pleistocene, fixed by palaeontologists at 1.9 Ma BP, has nothing to do with the Ice Ages.)

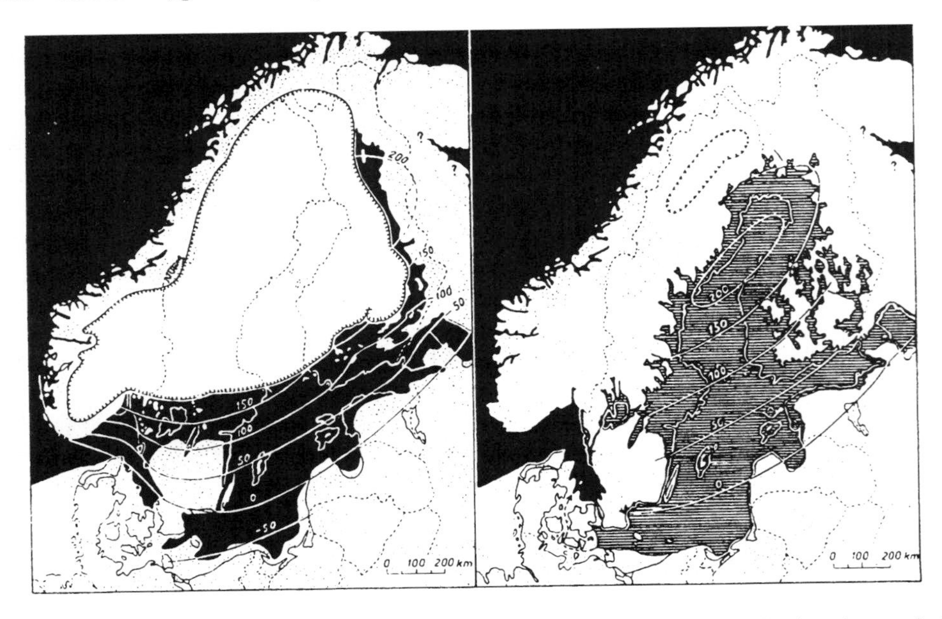

Figure 18.3. The Yoldia sea (7,650 BC) and the Ancylus lake (6,050 BC). Beach uplifts since these periods are in metres (isobases). From Granlund, 1949 (reproduced in Charlesworth, 1957).

18.4 STAGES OF THE LAST DEGLACIATION

The withdrawal periods of the last glaciation in Northern Europe are very well known and dated.

Withdrawal started in a climate only allowing tundra vegetation, and therefore still very cold but very dry (*Dryas I* of palynologists, or *oldest Dryas*). This withdrawal was interrupted by at least one temporary advance around 16,000 BP, which formed the Pomerania moraines. Around 13,300 BP, the limit of the ice sheet was following the south coastline of the modern Baltic Sea.

The appearance of birch pollen in peat bogs shows a less cold climate is then starting. This is the *Bölling*, which should correspond in France with the Lascaux period, during the Magdalenian. Tundra and dry cold start again during the *Dryas II*, between 12,400 BP and 11,900 BP approximately, and the birch and pine forest starts around 10,900 BP (*Alleröd*). During the 2000 years that *Bölling* and *Alleröd* lasted, the ice sheet moved back by 500 km, evacuating the whole south of the Baltic Sea up to the Gulf of Finland included. The Baltic of this time is a marginal lake, most often frozen, whose surface is 74 m above modern sea level. Between 10,900 BP and 10,000 BP approximately, the last offensive of the cold and tundra (*Dryas III* = *young Dryas*) sees the building of the three parallel moraines of *Salpausselkä*, in the south of Finland, ca. 60 km away from the Gulf of Finland in 225, 183 and 100 years respectively. In Sweden, the corresponding moraines are around 58.5°N, crossing the lakes Vättern and Vänern. This last episode of returning cold does not seem to have extended to North America. It would be due to a cooling of surface waters in the North Atlantic, icebergs reaching then the Bay of Biscay.

Next, between 10,000 and 9,000 BP, the ice sheet withdrawal is very fast (250 to 500 km, depending on places). For the course of a few centuries, the Baltic lake communicates with the North Sea, becoming the *Yoldia* sea, and reforms as a great lake, larger and deeper than the actual Baltic, the *Ancylus* lake (Figure 18.3).

Withdrawals so rapid can only be explained by assuming that during the Dryas periods, the continental side of the ice sheet was starved, became stagnant and thinned considerably. I think that it must have become an ice shelf and calved large icebergs. The ice sheet thickness can be estimated by comparison with the current Greenland ice sheet only at the beginning of the Salpausselkä stage, after a fast withdrawal, and not at the end.

Once exclusively on firm land, the ice sheet shrank on all sides. Coarse deposits consecutive to the rupture of a large glacial dam lake in Jämtland (Sweden, east of Trondheim) show a partition into two ice sheets around 8,750 BP. They had entirely disappeared around 8,000 BP.

In North America, the chronology of the Laurentide ice sheet withdrawal (after correcting dates still made with C–14) proves to be different. At the maximum of the Wisconsin glaciation, the ice sheet extended from the NE coasts of Labrador and Baffin Islands to 40°N (Tazewell moraine, 22,000 BP). The Cary moraine, on the south shore of Lake Michigan, dates back to 19,000 BP, and the Valders moraine south of the Upper and Ontario Lakes dates back to 12,000 BP (Dryas II). The last moraine, Cochrane (200 km south of St. James Bay) dates from 9,300 BP and is therefore younger than the European Dryas III.

18.5 GLACIO-ISOSTASY

Ground uplift along the Baltic coast is drawn from old beaches now become terraces along the shoreline. In the years 1910–28, De Geer and his school dated them and the last moraines with a 1-year accuracy, relative to the year of the ice sheet partition, by counting *varves*. Varves are alternate sediments deposited by glacial rivers: light sands in summer and dark fine silt the rest of the year. The beginning of a series must be matched with the end of another, as is done in dendrochronology with the annual rings of trees. Varve chronology extends from 1,000 BP to 12,000 BP (the part between 9,000 and 12,000 BP has been reviewed and modified recently). The ages measured directly from C–14 in bog cores, rafted wood or shells are much less precise (see Section 17.6).

Up to 28 raised beaches in parallel can be counted. All have been well dated. Toward the mouth of the Ångerma, the last point of the Swedish coastline to have been freed from ice, the highest beach is at 270 m and dates from 9,000 BP. A correction must be made to account for ocean level variations. It raised irregularly by 20 metres between this date and 6,000 BP (end of the melting of the North American ice sheet), and then oscillated by a few metres only.

Simultaneously, the water rose along the south coasts of the North Sea, the Channel and the Atlantic down to Biarritz. Until 6,000 BP, this was mainly a general rise of sea level (called *eustatic*), but a later sea level rise of the order of 10 m shows there is also ground subsidence.

The raised beaches show that, in detail, the uplift happened by jolts, and that the upper crust behaved like a mosaic of blocks with various sizes. According to Sauramo (1944), the beaches were built during stopping phases lasting up to several centuries. Ångermanland raised of 71 m in the Finiglacial, in 550 years, but because of three stops, these were in fact only 150 years. This must have been during strong and frequent earthquakes. The region is still seismic.

The smoothed profiles become rectilinear and show a global tilt of the region about a hinge line. The first tilt was about the *gothiglacial hinge line* (along isobase 0 on the map in Figure 18.3a), the second one was about the *finiglacial hinge line* (along isobase 100 on the map in Figure 18.3b).

According to old but well-documented studies by Sauramo, the tilting of a region about a hinge line happened only once this entire region was freed from ice. During the withdrawal, the region would have raised without tilting. This is shown in Figure 18.4 in the case of the finiglacial withdrawal. (To better understand the successive stages of ground uplift, the diagram (6) should be flipped, and the broken line BG III should be considered as the ground profile when the beach BG III was built.)

Hinge lines had long been recognized in North America, in the raised beaches of the large Agassiz periglacial lake. But Walcott (1972) exposes facts differently in his review. In the central zone (Hudson Bay and around), there was continuous emergence relative to the ocean. Quite far from the glaciated zone (US Atlantic coast, from Delaware to Florida), there was continuous subsidence. In a transition zone along the ice sheet during its maximal extension (Prince Edward Island, Maine, down to Boston), there was block emergence, and, after 12,000 BP, subsidence (and therefore tilting relative to the central region).

Let us consider only the Ångermanland uplift. If $h(t)$ is the modern altitude of the beach of age t, and h_0 is the future uplift until complete isostasy, the depression has been $Z = h + h_0$. The uplift rate $V = dZ/dt$ can be drawn. The analysis of data published by Lidèn (Figure 18.5) shows that we can write:

$$V = kZ^m \tag{18.1}$$

Imposing $m = 3$, one must take $h_0 = 164$ m and $k = 1.66 \times 10^{-9}\,\mathrm{m}^{-2}\,\mathrm{a}^{-1}$. But $m = 5$, with $h_0 = 302$ m and $k = 2.84 \times 10^{-15}\,\mathrm{m}^{-4}\,\mathrm{a}^{-1}$ is better. *A depression decreasing exponentially with time* ($m = 1$) *is completely excluded.* The reason is that the uplift in a place does not depend on the depression only in this place, but on depressions everywhere, because there are horizontal transfers of matter in the asthenosphere.

18.6 FIRST MODELS OF THE GLACIO-ISOSTATIC REBOUND

The first models of ground uplift after complete deglaciation date back to 1934. They were very unrealistic, because they only considered an isoviscous layer with a free

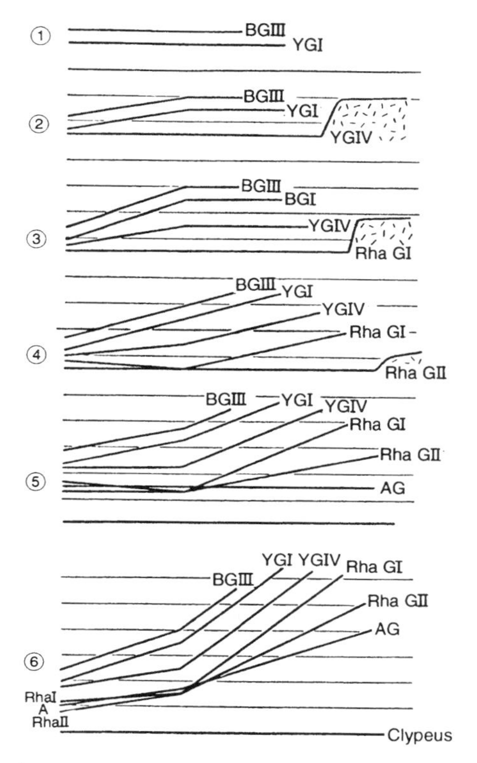

Figure 18.4. Beach uplift on the Baltic coasts, between the formation of the stage III of the Baltic Lake (around 10,900 BP) and the formation of the *Clypeus* beach (around 7,000 BP). From Sauramo (1944).

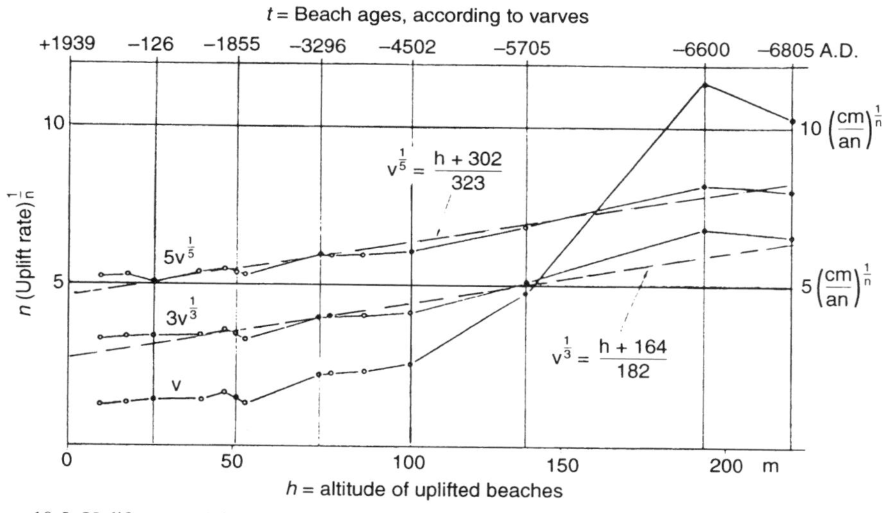

Figure 18.5. Uplift rate ν (obtained by parabolic interpolation between three consecutive pairs (h, t) as a function of the uplift $h.n\nu^{1/n}$ was used in ordinate, to try getting a linear law. From Lliboutry, 1965.

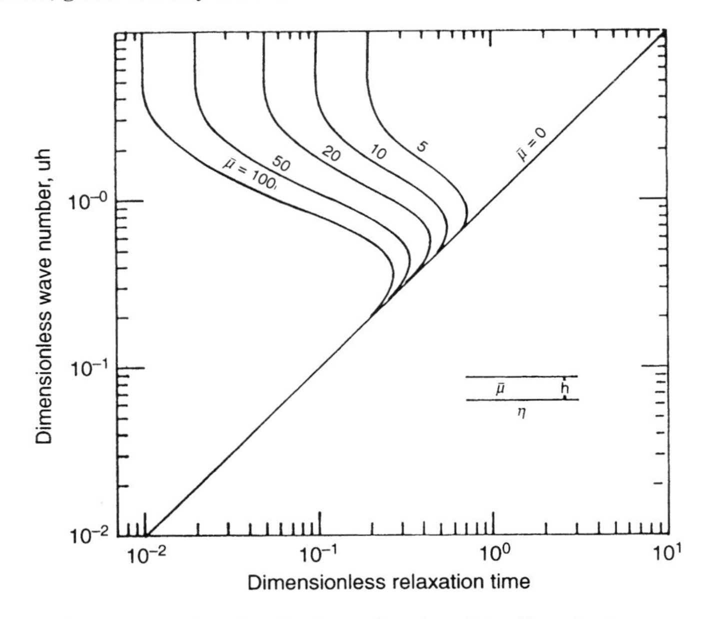

Figure 18.6. Dimensionless relaxation time Tm/h as a function of the dimensionless wave number h/l, for an isoviscous half-space of viscosity h, overlain by an elastic layer of thickness h, Coulomb modulus m and Poisson modulus $n = 0.25$. From McConnell Jr. (1965).

surface, without superimposed elastic lithosphere. This model was treated in Section 10.10, and we will use the same notations. It was next believed that the lithosphere could be simulated with a very supple membrane with no flexural rigidity, and that the null shear stress surface condition could be replaced by a null horizontal velocity. But this does not change the qualitative result: when ω varies from $-\infty$ to $+\infty$, $T(\omega)$, still varies from $+\infty$ to $+\infty$ through a minimum.

In fact, as demonstrated by McConnell in 1965, the results are completely different with a superimposed elastic layer: $T(\omega)$ varies from 0 to a finite value, through a maximum (Figure 18.6).

McConnell (1965) showed how to deal with a stratified medium formed by several superimposed plane layers, elastic or isoviscous. The Thomson–Haskell vector θ, defined and calculated in Section 10.8, is continuous from layer n to layer $(n{+}1)$. One must first consider a fixed level for the interface, although it moved from U. In the nth layer, σ_z must be increased by $\rho_n g U_z$, in the $(n+1)$th layer it must be increased by $\rho_{n+1} g U_z$. No correction will be needed if $\rho_n = \rho_{n+1}$. We will only look at this simple case: the layers all have the same density. Then, θ's continuity is written as:

$$[\mathbf{M_{n+1}}(\omega, 0)]\mathbf{c_{n+1}}(\omega) = [\mathbf{M_n}(\omega, H_n)]\mathbf{c_n}(\omega) \tag{18.2}$$

We shall stop writing ω, which is the same for all layers. Multiplying on the left-hand side by the inverse of matrix $[\mathbf{M_{n+1}}(0)]$, we get:

$$\mathbf{c}_{n+1} = [\mathbf{M_{n+1}}(0)]^{-1}[\mathbf{M_n}(H_n)]\mathbf{c_n} = [\mathbf{T_n}]\mathbf{c_n} \tag{18.3}$$

T_n is the *transmission matrix*. In these calculations, the operator $\partial/\partial t$ must be handled as an algebraic quantity. It is removed from the transition matrix between two viscous layers. We can indeed write:

$$\mathbf{M} = \mathbf{DM}'; \qquad \mathbf{D} = \begin{bmatrix} \partial/\partial t & 0 & 0 & 0 \\ 0 & \partial/\partial t & 0 & 0 \\ 0 & 0 & 1 & 0 \\ 0 & 0 & 0 & 1 \end{bmatrix}$$

$$\mathbf{M}' = \begin{bmatrix} 2\eta\omega\, e^{\omega z} & 2\eta\omega^2\, e^{\omega z} & -2\eta\omega\, e^{-\omega z} & -2\eta\omega^2\, e^{-\omega z} \\ 2\eta\omega\, e^{\omega z} & 2\eta\omega(1+\omega z)e^{\omega z} & 2\eta\omega\, e^{-\omega z} & -2\eta\omega(1-\omega z)e^{-\omega z} \\ e^{\omega z} & (1+\omega z)e^{\omega z} & -e^{-\omega z} & (1-\omega z)e^{-\omega z} \\ e^{\omega z} & \omega z\, e^{\omega z} & e^{-\omega z} & \omega z\, e^{-\omega z} \end{bmatrix} \tag{18.4}$$

Therefore, using the rule for the inversion of a product of matrices, we get:

$$\begin{aligned} \mathbf{T}_n &= [\mathbf{DM}'_{n+1}(0)]^{-1}[\mathbf{DM}'_n(H_n)] \\ &= [\mathbf{M}'_{n+1}(0)]^{-1}\mathbf{D}^{-1}\mathbf{D}[\mathbf{M}'_n(H_n)] = [\mathbf{M}'_{n+1}(0)]^{-1}[\mathbf{M}'_n(H_n)] \end{aligned} \tag{18.5}$$

The boundary conditions of the stratified medium concern θ at the surface, worth $[\mathbf{M}_1(0)]\mathbf{c}_1$, and θ at the base, worth $[\mathbf{M}_N(H_N)]\mathbf{c}_N$ if there are N layers. Now, $\mathbf{c}_1$ and $\mathbf{c}_N$ are linked by the relation:

$$\mathbf{c}_n = [\mathbf{T}_1 \quad \mathbf{T}_2 \quad \dots \quad \mathbf{T}_{N-1}] \tag{18.6}$$

These matrix computations are only feasible with computers, particularly because several models must be tested before finding one that fits the field data.

McConnell (1965) used the profiles from 6 raised beaches from the Baltic, of ages between 9,800 BP to 6,000 BP. He assumed that deformation was axisymmetrical. The Fourier transforms, which use $\cos\omega x$ and $\sin\omega x$ must then be replaced by Fourier-Bessel transforms, which use the Bessel functions $J_0(\omega r)$ and $J_1(\omega r)$, but the calculations are similar. He computed numerically the Fourier–Bessel transforms of these profiles, $\zeta(\omega, t)$. They can be determined precisely only for $\lambda = 2\pi/\omega$ ranging from 80 km to 1,000 km. McConnell calculated this way the relaxation time $T(\omega)$ in this interval of ω (small circles in Figure 18.7). He could reproduce very exactly these values with his model 62-12 (Table 18.1).

Table 18.1. Parameters of model 62-12 (McConnell, 1965).

Layer 1:	elastic $H_1 = 120$ km; $\mu = 65$ GPa; $\nu = 0.25$								
Layers 2–5:	isoviscous								
Depth:	120		200		400		800		1200 km
Viscosity:		3		2		10		25×10^{20} Pa s	

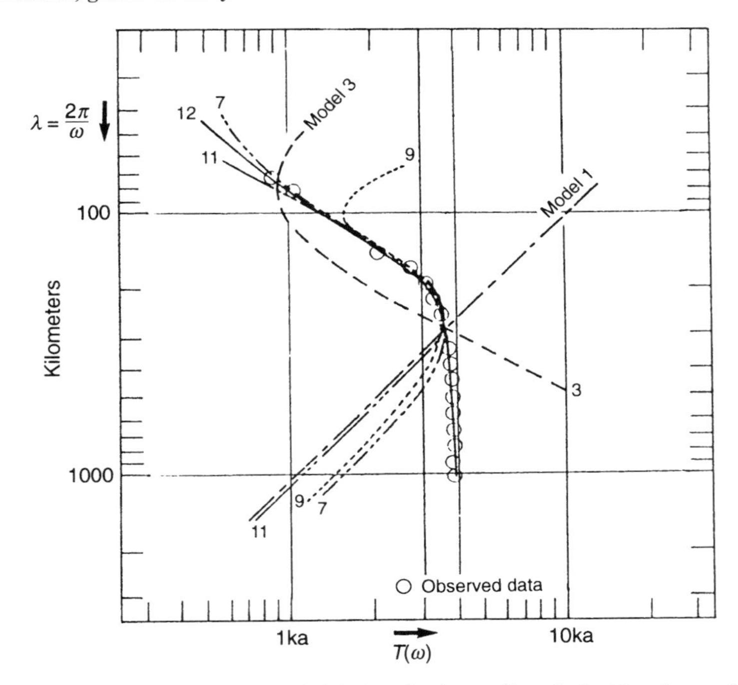

Figure 18.7. Comparison of the function $T(1/\lambda)$ given by the profiles of raised beaches, and the functions given by different models. From McConnell Jr., 1965.

The viscosity selected for the layer 6, extending down to the inner core, is not provided because its influence on $T(\omega)$ is negligible. In 1968, McConnell tried to determine this deep viscosity from the acceleration of the Earth's rotation, corrected from the effect of tides (see Section 8.12). But his model was too regional to handle this worldwide process.

18.7 MODELLING WITH VISCO-ELASITIC SPHERICAL LAYERS

A more exact calculation of glacio-isostatic rebound must account not only for the adjustment to the disappearance of the ice load, but also for the adjustment to the overload of the seafloor by melt water, and the variations of gravity and the geoid resulting from these processes. We must consider the whole Earth. Instead of using the Fourier transform of the variables, functions of ω_x, ω_y, z and t, we will use spherical harmonics, functions of two integers m and n, of r (distance to the centre of the sphere) and of t. It is necessary to go to $n = 180$, which means there are 180×361 coefficients for each variable. (As the Fourier transform is done numerically, by discretizing the continuous variables ω_x and ω_y, the difference between a plane stratified medium of infinite extent and a closed spherical medium is less than it seems.)

The selected model is a spherical globe whose rheological properties depend on r only, without lateral variations. This was acceptable on the scale of the old Fennoscandian basement, or the Canadian shield. But this is not valid on the scale of the Earth (we will come back to this in Section 18.10). But the problem can only be solved with this assumption. The variation of density with r requires, to consider it as uniform in each of the concentric layers, to consider a high number of layers. But rheologically, only a very limited number will be considered. Until recently, these were only an elastic lithosphere, and visco-elastic lower and upper mantles.

This is visco-elasticity in the Maxwellian sense (see Section 10.2). For a given stress state, we add an elastic strain and an isoviscous strain, respectively noted with the exponents e and v. We use the Laplace transforms of the variables (see Appendix A.4). When there is no initial deformation, the linear operator $\partial/\partial t$ is replaced by the algebraic variable s. To simplify, the law of the Maxwell body will be expressed in Cartesian coordinates; the transformation into spherical coordinates is very easy.

From equations (10.22) and (10.41), we have:

$$\begin{aligned} \tilde{\sigma}_x &= 2\mu\tilde{\varepsilon}^e_{xx} + \lambda(\tilde{\varepsilon}^e_{xx} + \tilde{\varepsilon}^e_{yy} + \tilde{\varepsilon}^e_{zz}) \\ \tilde{\tau}_{yz} &= 2\mu\tilde{\varepsilon}^e_{yz} \\ \tilde{\sigma}_x - \frac{\tilde{\sigma}_x + \tilde{\sigma}_y + \tilde{\sigma}_z}{3} &= 2\eta s\tilde{\varepsilon}^e_{xx} \\ \tilde{\tau}_{yz} &= 2\eta s\tilde{\varepsilon}^\nu_{yz} \end{aligned} \tag{18.7}$$

The other equations are not written, but they are obtained by cyclic permutation of x, y and z. The average pressure may be removed from the third equation because:

$$\tilde{\sigma}_x + \tilde{\sigma}_y + \tilde{\sigma}_z = 3k(\tilde{\varepsilon}_{xx} + \tilde{\varepsilon}_{yy} + \tilde{\varepsilon}_{zz}) \tag{18.8}$$

We multiply the elastic equations by s, and the viscous equations by η/μ. Adding them together gives:

$$\begin{cases} \tilde{\sigma}_x = \dfrac{2\mu s}{s + \mu/\eta}\tilde{\varepsilon}_{xx} + \dfrac{\lambda s + k\mu/\eta}{s + \mu/\eta}(\tilde{\varepsilon}_{xx} + \tilde{\varepsilon}_{yy} + \tilde{\varepsilon}_{zz}) \\ \tilde{\tau}_{yz} = \dfrac{2\mu s}{s + \mu/\eta}\tilde{\varepsilon}_{yz} \end{cases} \tag{18.9}$$

We therefore move from a purely elastic body, where the only elastic parameters are λ and μ (ν, in the solution of (10.36) can be written $\nu = \lambda/(2(\lambda + \mu))$ to the visco-elastic body by the simple change:

$$\mu \rightarrow \frac{\mu s}{s + \mu/\eta}; \qquad \lambda \rightarrow \frac{\lambda s + k\mu/\eta}{s + \mu/\eta} \tag{18.10}$$

All authors took for μ and λ the 'seismic' values coming from the velocity of S and P waves, and, in general, the ones provided by the PREM model (Section 17.4).

In the **direct problem**, the viscosity for each layer and a history of the ice load are given. And we look for the resulting ground response. The problem can be treated separately for each spherical harmonic. The following variables (in spherical coordinates r, θ, φ) must be continuous when passing from one layer to the next:

- the displacements U_r, U_θ, U_φ;
- the shear stresses $\tau_{r\theta}$ and $\tau_{r\varphi}$;
- the traction σ_r. But, because we need fixed levels and because density depends on depth, superficial masses must be introduced at each level.
- the three components of gravity, linked by a relation expressing this is a gradient.

We must therefore consider eight coefficients function of r and s (the analogue of the components of the Thomson–Haskell vector defined in Section 10.7). They obey a system of linear differential equations of the first order in r (more details can be found in Lliboutry, 1987). Separately, for each spherical harmonic, the system is solved for a unit load imposed on the Earth's surface from $t = 0$. The resulting impulse response is convolved with the load, function of time. (As we work with Laplace transforms, this is a simple multiplication.) The contributions of the different spherical harmonics are added. And we get at last to the real movements of the ground by inverting the Laplace transform we just computed, by numerical integration.

The reader may not have followed everything. May his consolation be that when Peltier (1974) published this method, most glacio-isostasy scientists were out of the running for at least ten years!

18.8 INVERSE PROBLEM SOLUTION FOR A MAXWELL EARTH

Peltier (1974) was at first trying to determine the viscosities of the upper mantle (η_s) and the lower mantle (η_i) only from the uplifts or subsidences in several locations during the last thousands of years. His first results confirmed those found independently by Cathles (1975). The best fit is obtained with $\eta_s = \eta_i = 10^{21}$ Pa s.

This result was very disconcerting for those explaining the relative fixity of hot spots by hot plumes coming from a poorly convective mantle. This result was also conflicting with the analysis of the geoid undulations, better and better known from spatial geodesy (Lago and Rabinowicz, 1984; Hager, 1984). The modelling of the convection currents causing them (glacio-isostasy contributes only to the regions formerly covered by ice, and can be neglected elsewhere) gives $\eta_i/\eta_s = 10$ to 50.

It was later realized that the ground uplift does not yield any worthy information about viscosity deeper than 1,400 km. If it is assumed very different, the resulting uplift rates are still nearly the same. For example, Mitrovica (1996) recognises that with $\eta_s = 3 \times 10^{20}$ Pa s and $\eta_i = 75 \times 10^{20}$ Pa s, the uplifts of Ångermanland and Oslo are as well found than if $\eta_i = \eta_s = 10^{21}$ Pa s.

On the other hand, from solid state physics, Poirier and Libermann (1984) conclude to a viscosity increase of a factor 100 between the depths of 700 and 1,700 km, viscosity deeper becoming constant.

To find the lower mantle viscosities, we must use other data, sensitive to movements in the deepest part of the mantle: the secular drift of the Pole (Section 16.3) and the acceleration of the Earth's rotation, once corrected for the slowing by tides (Section 8.12). The latter is linked to dJ_2/dt, now directly accessible from spatial geodesy (see Sections 8.2 and 8.4 for the definition and the determination of J_2). Using these data, Peltier and Jiang (1996) find that $\eta = 10^{21}$ Pa s until 1,400 km deep, and (12 to 15) $\times 10^{21}$ Pa s below.

Today, mantle viscosities are no longer assessed without the data relative to thermal convection. This is why the viscosity model published by Pari and Peltier (1995) will only be presented in the next chapter.

Because we work with Laplace transforms, we can use older sea level data, from the time where the northern hemisphere ice sheets still existed. The problem is to estimate correctly their thickness as a function of time.

This approach was made by Peltier since 1982, and again by Lambeck and Nakada (Lambeck *et al.*, 1990). They used a method exposed by Farrell and Clark (1976). Sea level variation is the sum of three terms:

- A variation due to ice melting, supposing the Earth is rigid but accounting for the gravity variations induced by the disappearance of ice and the adjunction of a water layer on the oceans. (The ocean surface must remain an equipotential of the gravity field.)
- A ground uplift due to the melting of ice.
- A subsidence due to the load of the added water layer. These last two terms account for the Earth's visco-elasticity.

These three terms are calculated in sequence, and iterations are performed until there is convergence. Lambeck and Nakada tried different models, with different elastic lithosphere thickness, and different values of η_s and η_i. The best fits for variations of the sea levels on the coasts of NW Europe and in Ångermanland are obtained with:

- An elastic lithosphere 100 to 150 km thick;
- Viscosities $\eta_s = (3.2 \text{ to } 4.1) \times 10^{20}$ Pa s and $\eta_i = (25 \text{ to } 70) \times 10^{20}$ Pa s.

This is in good agreement with model 62.12 of McConnell (1965), obtained 25 years earlier with a much more sparing use of mathematical tools. But in another study, Nakada and Lambeck use data from Australia and some Pacific islands, and this time obtain: $\eta_s = (0.5 \text{ to } 10) \times 10^{20}$ Pa s and $\eta_i \approx 100 \times 10^{20}$ Pa s. This is a confirmation of the existence of lateral variations on the Earth's scale.

18.9 ICE CAP THICKNESS

Like all scientists trying to explain the large glaciations, the authors referenced above have used an ice cap model recognized wrong for a long time. It must be said that this model was used in 1976 by the CLIMAP working group to compute the thickness of old ice caps, and that geophysicists and climatologists were trusting these so-called experts. Peltier (1994) recently saw this error, and remarked that the thicknesses given by CLIMAP should be reduced by 35% to find proper uplift rates. But, as others persist in this mistake, an explanation is necessary.

Everything started with an article by Nye (1951), a few years before the ice creep law was established. He proposed to model its behaviour by an ideal plastic body (Section 11.3), with a plasticity threshold τ_c close to 1 bar (= 0.1 MPa). He assumed that the friction on the bed always had this value τ_c. For a 2-D flow on a horizontal and plane bed, the thickness h must obey the relation $\tau_c = \rho g h \, dh/dx$. If τ_c is constant, then:

$$h = \left[\frac{2\tau_c}{\rho g}\right]^{1/2} \approx (23\ \mathrm{m} \times x)^{1/2} \tag{18.11}$$

In fact, this *parabolic profile* is valid only for an active ice sheet, over a rocky bed which has not been smoothed on the metre scale by loose sediments or water pockets. In the last case, the friction of the bed of an active glacier may be much smaller; this is the case in outlet streams from ice caps. Furthermore, the flow velocities and the friction on the bed become very small in the central regions of polar ice caps. If the model of the ideal plastic body is kept, these regions must be considered as motionless areas, where the plasticity threshold is not reached. A rule of thumb was presented in Section 9.7 to estimate the thickness of a cold ice cap. We cannot escape from the estimation of past precipitations and temperatures.

Peltier and his co-workers assumed that the northern hemisphere ice caps thickened regularly, as a linear function of time, during 90,000 years approximately, and then disintegrated in 10,000 years if they were lying on firm ground, and nearly instantaneously if they were lying on the seafloor (e.g. Barents and Kara Seas ice caps). Let us remark that this distinction is illusory: what is the difference between these seas over continental shelves and the Baltic or the very large Siberian periglacial lakes? In fact, we saw that inside a 100-ka cycle, there are several stronger and stronger glacial stages, separated by interglacial episodes.

Also, the North Atlantic sediments indicate intense iceberg calving every 5,100 years approximately in the heart of glacial stages. The study of the (O18)/(O16) ratio in deep ice cores from Greenland and Vostok (Russian station near the centre of East Antarctica) confirmed these events. They are called *Heinrich events*, and are indeed temporary important warming episodes in the middle of a glacial period (Yiou *et al.*, 1995). The growth of an ice cap must have therefore been as irregular as its decrease.

18.10 OBJECTIONS TO THE TRADITIONAL MODEL OF A VISCO–ELASTIC EARTH

The first defect of the preceding model is that *it ignores the lateral variations in the thickness of each layer and in its rheological parameters.*

The lithosphere thickness is far from uniform, whether this is the elastic lithosphere of isostasy or the tectosphere of plate theory and convection (Section 5.4).

Elastic thickness is inferred from the deformation of oceanic plates (Section 14.2) or, in continents, from the correlation between Bouguer anomaly and topography (Sections 14.3 and 14.4). But, in both cases, it assumes that isostatic equilibrium is achieved. The elastic thickness is less than 50 km below oceans and of the order of 100 km below old continents. As for the tectosphere, it extends down to the seismic LVZ, i.e. 100 km below old oceans and 250 km below Scandinavia (Sacks *et al.*, 1979). In continents, there also seems to exist a thin crustal asthenosphere (Sections 14.8 and 14.9). The lateral variations can probably be ignored when only looking for the lower mantle viscosities. But, in glacio-isostasy, not accounting for makes illusory the accuracy expected from the passing of the plane model to the spherical model, and the taking into account gravity field modifications.

In the asthenosphere, the model ignores the very high viscosity variations coming from lateral temperature variations between upwelling hot plumes and downwelling cold plumes. We will come back to this point in the next chapter.

Another arguable point is whether *the phase changes in the mantle are at fixed levels or at material levels*. According to Peltier, the allotropic changes causing most of the 660-km discontinuity would occur with a time constant of around 500,000 years. The discontinuity would therefore move with the medium during isostatic readjustments. But if the phase changes are very rapid (as argued by Christensen), the discontinuity must be placed at the level corresponding to a specific pressure, whatever the vertical movements.

The most discussed point is the *viscosity law*. It has been pointed out that (1) laboratory experiments on rocks never show a Newtonian viscosity, but a power law with $n \approx 3$; (2) during the isostatic readjustment, the stage of transient creep may never be crossed, at least far away from the glaciation centre.

The discussion is biased because it is based on laboratory results with strain rates of the order of $10^{-6}\,s^{-1}$, compared to 10^{-13} to $10^{-16}\,s^{-1}$ in the Earth, and on concepts established for metals. Even if the creep also occurs through dislocation movements, the microscopic processes in metals are different (Section 11.4). These questions are exposed in detail in Chapter 11. Let us sum up for the mantle:

- The interface energy tends to grow rock grains, the better as temperature is higher (Section 11.7). Syntectonic recrystallization tends to diminish them, the more so as the deformation is faster (Section 11.6). It can therefore be predicted that in the mantle, and especially the lower mantle, the grains will be very big, at least decimetric. This excludes creep by diffusion of vacancies or interstitials, the process always advanced by the proponents of Newtonian viscosity (Section 11.5).
- The anisotropy of seismic wave velocities implies the existence of a fabric, but this does not imply recrystallization. It may have been acquired during crust formation at mid-ocean ridges. And a material with an anisotropic fabric may have an isotropic rheological behaviour (e.g. the ice of a temperate glacier). Seismic anisotropy is therefore not a decisive argument.
- The deformations required by isostatic readjustments are below 10^{-3}, minute compared to those from convection currents below the tectosphere, lasting

10 Ma or more. The latter are therefore controlling the density of dislocations. The rheological law concerning isostasy is not the relation stress/strain rate, but the relation between stress increments and strain rate increments. It is the same only in the case of a Newtonian viscosity.

- The deformation by displacement of dislocations may produce a Newtonian viscosity, when their density is very low and constant. This is *Harper–Dorn creep*. Observed by these authors on very pure aluminium in 1957, it was observed since for non-metallic substances: CaO (Dixon *et al.*, 1982), olivine (Justice *et al.*, 1982), $KZnF_3$, an analogue of (Mg, Fe)SiO_3 with a perovskite structure (Poirier *et al.*, 1983) and NaCl (Banerdt and Sammis, 1985). According to them, Harper–Dorn creep has the following characteristics:
  (1) The strain rate is proportional to deviatoric stress.
  (2) This rate does not depend on the grain size, and is of the same order for a monocrystal or a polycrystal.
  (3) On loading, there is indeed transient creep, but it ends for a strain smaller than 0.1%, whereas permanent creep could be continued beyond 1% deformation.
  (4) The activation enthalpy H^* is approximately the one of vacancy or interstitial diffusion, as in the normal dislocation creep.
  (5) The dislocation density is low: 1,000 to 30,000 cm^{-2} for aluminium, 4×10^6 cm^{-2} for NaCl. If the initial density is too high, Harper–Dorn creep is not observed anymore.
- Transient creep, suggested by Weertman (Section 11.5) was the irreversible creep of Andrade, the only one well known at the time. It comes from the progressive mutual blocking of most original dislocations. It is therefore incompatible with Harper–Dorn creep. This is not the same for reversible, logarithmic, transient creep which seems due to limited sliding on the grain joints. We suggest that it may occur just below the elastic lithosphere, in the tectosphere layer which is no longer elastic, but does not participate in convection.

In conclusion, *nothing seems to contradict the hypothesis of a Newtonian viscous behaviour, with no noticeable transient creep, of the mantle below the lithosphere. The same viscosity would then intervene in glacio-isostasy and in convection currents.*

Wu (1993) looked if a power-law mantle viscosity could account for glacio-isostatic rebound. The Earth's sphericity and the modifications of gravity only play a secondary role, and he therefore limited himself to a model with superimposed plane layers. He assumes a generalized Maxwell body behaviour, with $n \neq 1$ for the viscous component. The traditional solving method is then useless, but it is possible to use the finite element method with commercial software. To simplify, he adopted $\nu = 1/2$ for the elastic component; this incompressibility removes the term in λ of equation (18.6).

Wu can account for the rebound in the central region, but not in the transition zone around. He thinks that the reason would be a lack of knowledge of the exact evolution of the ice cap thickness (and, indeed, he still uses the parabolic profile). The modelling of glacio-isostasy remains therefore an open problem.

19

Thermal convection in the mantle

19.1 VELOCITY AND TEMPERATURE EQUATIONS IN AN ISOVISCOUS MEDIUM

Navier–Stokes equation, Boussinesq approximation

For an elastic body, the Navier equation established in Section 10.7 reads (keeping the inertia force $-\rho\gamma$):

$$\mu\nabla^2\mathbf{U} + \left(k + \frac{\mu}{3}\right)\mathbf{grad}(\operatorname{div}\mathbf{U}) + \rho(\mathbf{g} - \boldsymbol{\gamma}) = 0 \tag{19.1}$$

Since $k \operatorname{div}\mathbf{U} = \boldsymbol{\sigma}_0 = -p$, we can rewrite this equation as:

$$\mathbf{grad}\, p = \mu\left[\nabla^2\mathbf{U} + \tfrac{1}{3}\,\mathbf{grad}(\operatorname{div}\mathbf{U})\right] + \rho(\mathbf{g} - \boldsymbol{\gamma}) = 0 \tag{19.2}$$

For an isoviscous fluid, the same calculations lead to the *Navier–Stokes equation*, with $\eta\partial/\partial t$ replacing μ. The components of the velocity $\boldsymbol{u} = \partial\boldsymbol{U}/\partial t$ must verify:

$$\mathbf{grad}\, p = \eta\left[\nabla^2\mathbf{u} + \tfrac{1}{3}\,\mathbf{grad}(\operatorname{div}\mathbf{u})\right] + \rho(\mathbf{g} - \boldsymbol{\gamma}) = 0 \tag{19.3}$$

In a terrestrial system of reference, the Coriolis force should be added, but it is generally negligible, except when dealing with atmospheric or oceanic movements. The acceleration $\boldsymbol{\gamma}$ is not the partial derivative $\partial\boldsymbol{u}/\partial t$, with x, y, z fixed, but the material derivative, following a particle with time (see Section 9.4). The material derivatives are written D/Dt. They can be expressed with the vector gradient matrix (see Appendix A.1). In particular:

$$\frac{D\mathbf{u}}{Dt} = \frac{\partial\mathbf{u}}{\partial t} + (\nabla\mathbf{u})\mathbf{u} \tag{19.4}$$

The Navier–Stokes equation is usually written for a homogeneous and uncompressible fluid, for which $\operatorname{div}\mathbf{u} = 0$:

$$\mathbf{grad}\, p = \eta\nabla^2\mathbf{u} + \rho\left(\mathbf{g} - \frac{D\mathbf{u}}{Dt}\right) \tag{19.5}$$

For a medium not strictly incompressible, or with a non-uniform density induced by thermal dilatations, $\operatorname{div}\mathbf{u}$ does not equal 0 exactly. To admit it, and write equation (19.5) instead of equation (19.3) is the *Boussinesq approximation*.

Rotation equation

The pressure p can be removed from equation (19.5) by writing that $\mathbf{curl}(\mathbf{grad}\, p) = 0$. We write that $\mathbf{curl}\,\mathbf{u} = \boldsymbol{\omega}$ (*rotation vector*, twice the instantaneous rotation vector defined in Section 10.5). Since $\mathbf{g}$ is the gradient of the gravity potential, its curl is zero, and the *rotation equation* is obtained:

$$\rho\frac{D\mathbf{u}}{Dt} = \eta\nabla^2\mathbf{u} + (\mathbf{grad}\,\rho) \wedge \left(\mathbf{g} - \frac{D\mathbf{u}}{Dt}\right) \tag{19.6}$$

The density ρ depends on the temperature T and the elastic compressibility modulus k:

$$\rho = \rho_0\left(1 - \alpha T + \frac{p}{k}\right) \tag{19.7}$$

Elastic compressibility is neglected, and $\rho_0\alpha\nabla T$ is assimilated to $\rho\alpha\nabla T$. For slow flows, $D\boldsymbol{u}/Dt$ can be neglected compared to $\mathbf{g}$. Writing $\eta/\rho = \nu$ (*kinematic viscosity*), the rotation equation becomes:

$$\frac{D\mathbf{u}}{Dt} = \nu\nabla^2\mathbf{u} - \alpha(\mathbf{grad}\, T) \wedge \mathbf{g} \tag{19.8}$$

A vertical z-axis, pointing upward, is adopted. The components of $\nabla T \wedge \mathbf{g}$ are then $(-g\,\partial T/\partial y,\ g\,\partial T/\partial x,\ 0)$. We see that:

(1) *Lateral temperature variations induce a rotation inside the medium*. Otherwise, it would only appear at contact with the inner walls, where the velocity usually vanishes.
(2) *The fluid in movement carries the rotation with it*, because of the material derivative. In the next paragraphs, the rotations will vary only slowly from one point to another, and the velocities will be small; the material derivative will thus be assimilated to the partial derivative.
(3) *Rotation diffuses in the medium*, as well as it is transported, because we find the heat equation again. The kinematic viscosity ν plays the role of the diffusivity coefficient.

Plane flow

For a plane flow, with $u_y = 0$ and null derivatives relative to y, the only non-zero component of ω is ω_y, which will be abbreviated by ω. The three scalar equations of (19.8) are reduced to a single equation:

$$\frac{\partial \omega}{\partial t} = \nu\left(\frac{\partial^2 \omega}{\partial x^2} + \frac{\partial^2 \omega}{\partial z^2}\right) - g\alpha\frac{\partial T}{\partial x} \tag{19.9}$$

The Boussinesq approximation allows us to introduce a *stream function* $q(x, z, t)$, defined to a constant:

$$\begin{cases} u_x = \dfrac{\partial q}{\partial z}; \qquad u_z = -\dfrac{\partial q}{\partial x}; \qquad \omega = \dfrac{\partial^2 q}{\partial x^2} + \dfrac{\partial^2 q}{\partial z^2} = \nabla^2 q \\ \nabla^2 \omega = \nabla^4 q = \dfrac{\partial^4 q}{\partial x^4} + 2\dfrac{\partial^4 q}{\partial x^2 \partial z^2} + \dfrac{\partial^4 q}{\partial z^4} \end{cases} \tag{19.10}$$

The material area (volume per unit width parallely to the y-axis) crossing a fixed infinitesimal segment (dx, dz) during a time unit equals $u_x dz - u_z dx = dq$. At any moment, the curves $q =$ constant are therefore tangent to the velocity vector. These are *stream lines*, functions of time. When the movement is steady, they merge with trajectories of material points.

Temperature equation

It was written in Section 9.4. It takes into account the heat appeared in the medium, per unit time. There are four possible sources of internal heat:

- The viscous dissipation of energy, W, which is given by the equations (10.36) to (10.38): $W = \eta\dot{\gamma}^{-2}$.
- The heat dissipation from radioactivity, Q_r, per unit volume and time.
- The latent phase change heats (L), which appear at discrete levels. At these levels, the heat flux increases suddenly by $-\rho L u_z$, and therefore $\partial T/\partial z$ is not continuous. This case will not be considered here.
- The heat emitted by elastic compression of the medium. If it were not diffusing (adiabatic compression), the corresponding temperature increase would be the one indicated in Section 11.6. Denoting by T_a the absolute temperature:

$$dT = \frac{\alpha T_a}{\rho C} dp \tag{19.11}$$

As $dp/dt = -\rho g u_z$, the resulting temperature increase would be $dT/dt = -G_a u_z$, where G_a is the *adiabatic gradient*:

$$G_a = g\alpha\frac{T_a}{C} \tag{19.12}$$

The heat emitted per unit volume and time is $-\rho C G_a u_z$. In the case of the plane problem, if K and ρC are the same everywhere, the heat equation becomes:

$$\frac{\partial T}{\partial t} = \kappa \nabla^2 T - \frac{\partial T}{\partial x}\frac{\partial q}{\partial z} + \left(\frac{\partial T}{\partial z} + G_a\right)\frac{\partial q}{\partial x} + \frac{Q_r + \eta\dot{\gamma}^2}{\rho C} \tag{19.13}$$

If G_a is independent from z, by writing $T = T_0 - G_a z + \theta(x, z, t)$, the *super-adiabatic temperature* θ follows the same equation, but without the term $G_a \partial q/\partial x$.

19.2 BÉNARD–BOUSSINESQ CONVECTION

Heat transfer by advection of matter has been known for a long time (it was named convection by Prout in 1834), but it was necessary to wait for Bénard's experiments in 1900 to discover two facts:

- When the bottom of a layer of viscous fluid is uniformly heated, convection appears only if the temperature difference between the bottom and the surface reaches a critical value.
- The fluid circulation gets itself organized in regular convection cells.

Boussinesq published the theory of this process for an isoviscous fluid in his 2-volume treatise *Théorie Analytique de la Chaleur* (Gauthier-Villars, 1903). Lord Rayleigh published the same in 1916 only, in the *Philosophical Magazine*. This is the reason why the convection of a fluid layer whose bottom and surface are maintained at fixed temperatures, called *Rayleigh convection* by British scientists and, more generally, Bénard–Rayleigh convection, should in fact be called *Bénard–Boussinesq convection*.

Consider an isoviscous layer of thickness H, and assume that all variables are independent from y. Any internal heat source other than adiabatic warming is neglected. We assume C, K and G_a are uniform, and we make the Boussinesq approximation. The stream function q and the super-adiabatic temperature θ must obey the set of equations:

$$\left\{\begin{aligned} \frac{\partial(\nabla^2 q)}{\partial t} &= \nu \nabla^4 q - \alpha \frac{\partial \theta}{\partial x} \\ \frac{\partial \theta}{\partial t} &= \kappa \nabla^2 \theta - \frac{\partial \theta}{\partial x}\frac{\partial q}{\partial z} + \frac{\partial \theta}{\partial z}\frac{\partial q}{\partial x} \end{aligned}\right. \tag{19.14}$$

The set uses fourth-order derivatives of q and second-order derivatives of θ. To have a single solution, we must select two conditions for q and one condition for θ over the whole boundary of the domain. These conditions are:

(1) All along the boundary of the domain, $q =$ arbitrary constant $= 0$.

(2) At the surface and at the bottom, if they are free, the shear strain rate $\partial^2 q/\partial z^2 - \partial^2 q/\partial x^2 = 0$, i.e., because q is constant, $\partial^2 q/\partial z^2 = 0$. (In fact, the free surface slightly moves from the horizontal plane but the correction to be made is negligible.) A free bottom is not unrealistic if the viscous layer floats

over a very dense and very fluid liquid (mantle convection above the liquid outer core, or, in the laboratory, oil above mercury). If the upper or lower limit is rigid, the condition changes to $u_x = \partial q/\partial z = 0$. But we will restrict ourselves here to the simplest case, where the surface and the bottom are free.

(3) The temperatures are fixed at the surface and at the bottom, or, indifferently, the over-adiabatic temperatures. We admit that $\theta = 0$ at the surface and $\theta = \theta_b$ at the bottom.

(4) On vertical sides, we must have a second condition for velocities, and one condition for temperatures. We admit that the vertical shear strain rate is 0 $(\partial^2 q/\partial x^2 = 0)$ and that no heat flux crosses the sides $(\partial\theta/\partial x = 0)$ ('mirror' conditions). This is true only with virtual sides placed in the axis of upwelling or downwelling streams, very far from the actual sides.

It is worth introducing dimensionless variables: x', z', t', Q, Θ such that:

$$x = Hx'; \qquad z = Hz'; \qquad t = \frac{H^2}{K}t'$$

$$q = \kappa Q; \qquad \theta = \theta_b\Theta \tag{19.15}$$

With the notation:

$$\begin{cases} Ra = \dfrac{g\alpha\theta_b H^3}{\kappa\nu} \; (Rayleigh\ number) \\ Pr = \dfrac{\nu}{\kappa} \; (Prandtl\ number) \end{cases} \tag{19.16}$$

The set becomes:

$$\frac{1}{Pr}\frac{\partial(\nabla^2 Q)}{\partial t} = \nabla^4 Q - Ra\frac{\partial\Theta}{\partial x'} \tag{19.17a}$$

$$\frac{\partial\Theta}{\partial t} = \nabla^2\Theta - \frac{\partial\Theta}{\partial x'}\frac{\partial Q}{\partial z'} + \frac{\partial\Theta}{\partial z'}\frac{\partial Q}{\partial x'} \tag{19.17b}$$

In the Earth's mantle, κ is of the order of $10^{-6}\,\mathrm{m^2\,s^{-1}}$, ν of 10^{16} to $10^{18}\,\mathrm{m^2\,s^{-1}}$. The Prandtl number is practically infinite, and the left term of equation (19.17a) becomes zero (this expresses the fact that inertia forces are completely negligible). But, in laboratory experiments, it is very difficult to get high Pr values. For water, $Pr = 3$. The largest values are $Pr \approx 100$ for some silicon oils at room temperature.

With the mentioned boundary conditions, the set of equations (19.17) has one trivial solution: no convection and vertical and uniform super-adiabatic gradient:

$$\left.\begin{aligned} Q &= 0 \\ -\frac{\partial\Theta}{\partial z'} &= 1 \end{aligned}\right\} \implies -\frac{\partial\theta}{\partial z} = \beta = \frac{\theta_b}{H} \tag{19.18}$$

But this solution still needs to be stable when submitted to small perturbations. Note that the considered perturbations only induce 2-D convection, whereas Bénard's experiments induced a 3-D convection.

The study of stability is traditional and simple, because we start from a state where the temperature gradient is uniform and non-zero. Its value will be kept in the terms on the right-hand side of (19.17b), to make this equation linear. Taking the origin in a lower corner of the domain, a perturbation respecting the boundary conditions will be expressed by the following double Fourier series:

$$\begin{cases} Q = \sum_{m,n} Q_{mn}(t) \times \sin\dfrac{m\pi x'}{c} \times \sin n\pi z' \\ \Theta = 1 - z' + \sum_{m,n} \Theta_{mn}(t) \times \cos\dfrac{m\pi x'}{c} \times \sin n\pi z' \end{cases} \tag{19.19}$$

The set (19.17) is obeyed if, for any pair (m, n):

$$\begin{cases} \left(\dfrac{\pi^2 k^2}{Pr}\dfrac{\partial}{\partial t'} + \pi^4 k^4\right) Q_{mn} + Ra\dfrac{m\pi}{c}\Theta_{mn} = 0 \\ \dfrac{m\pi}{c} Q_{mn} + \left(\dfrac{\partial}{\partial t'} + \pi^2 k^2\right)\Theta_{mn} = 0 \\ k^2 = \dfrac{m^2}{c^2} + n^2 \end{cases} \tag{19.20}$$

Q_{mn} and Θ_{mn} must be of the form $c_1 \exp(s_1 t) + c_2 \exp(s_2 t)$, s_1 and s_2 being the roots of the second-degree equation:

$$\frac{k^2}{Pr}s^2 + \pi^2 k^4\left(1 + \frac{1}{Pr}\right)s + \left(\pi^4 k^6 - Ra\frac{m^2}{c^2}\right) = 0 \tag{19.21}$$

(The values c_1 and c_2 for Q_{mn} are also linked to the values for Θ_{mn}.) The sum of roots is negative; if the roots are imaginary conjugates, their real parts are negative: Q_{mn} and Θ_{mn} oscillate with exponentially decreasing amplitudes. If the roots are real, one is negative. Q_{mn} and Θ_{mn} increase exponentially if the product of roots is negative, i.e. if:

$$Ra > \pi^4 k^6 c^2 / m^2 \tag{19.22}$$

Instability will appear if $Ra > Ra_c$. The critical threshold Ra_c is the smallest value attainable by $k^6 c^2/m^2$. Noting $m^2/c^2 = \zeta$, it can be demonstrated that this factor is minimal for $n = 1$ (1-layer convection, from the bottom to the surface) and $c/m = \sqrt{2}$ (convection rolls $\sqrt{2}$ times larger than they are high). Therefore:

$$Ra_c = 27\pi^4/4 = 657.5 \tag{19.23}$$

With other boundary conditions, the values of c/m and Ra are different (Table 19.1). For a spherical layer, the study of stability is similar, and the double Fourier series is replaced by a double series of spherical harmonics. The results of Table 19.1 correspond to a bottom with radius half the radius of the surface. (This is nearly the case for the Earth's mantle, but as it is not isoviscous and not only heated from

Table 19.1. Numerical values of the convection coefficients.

Limit		Plane layer		Spherical layer		
Outer	Inner	$\lambda/H = 2c/m$	Ra_c	Degree	$\lambda/H = 3\pi/n$	Ra_c
Free	Free	2.83	657.5	3	3.14	887
Free	Rigid	2.34	1101	4	2.36	1330
Rigid	Free	2.34	1101	4	2.26	1840
Rigid	Rigid	2.016	1708	5	1.89	2380

below, the exact numerical values are of only limited interest). The wavelength λ has been chosen at the half-height of the convective layer.

19.3 LORENZ EQUATIONS

We found a convection increasing indefinitely and exponentially because we had linearized the equations of (19.17). The non-linear terms limit its amplitude in the end. To see that, it is only necessary to keep in the double series of (19.19) the terms Q_{11} and Θ_{11}, and the term in Θ_{02} (to simplify, they will be written as Q_1, Θ_1 and Θ_2). Giving c/m the value $\sqrt{2}$, the most favourable to convection, we get:

$$\begin{cases} Q = Q_1 \sin\dfrac{\pi x'}{\sqrt{2}} \sin \pi z' \\ \Theta = 1 - z' + \Theta_1 \cos\dfrac{\pi x'}{\sqrt{2}} \sin \pi z' + \Theta_2 \sin 2\pi z' \end{cases} \tag{19.24}$$

The non-linearized equations can be satisfied by these heavily truncated series to the term $\cos(\pi x'/\sqrt{2}) \sin \pi z' \cos^2 \pi z'$, which remains small in all the convective domain. For that, we need:

$$\begin{cases} \dfrac{1}{Pr}\dfrac{dQ_1}{dt'} = -\dfrac{3\pi^2}{2} Q_1 - Ra\dfrac{\sqrt{2}}{3\pi}\Theta_1 \\ \dfrac{d\Theta_1}{dt'} = -\dfrac{3\pi^2}{2}\Theta_1 - \dfrac{\pi}{\sqrt{2}} Q_1 - \pi^2\sqrt{2}\, Q_1\Theta_2 \\ \dfrac{d\Theta_2}{dt'} = -4\pi^2\Theta_2 + \dfrac{\pi^2}{2\sqrt{2}} Q_1\Theta_1 \end{cases} \tag{19.25}$$

We make the following variable changes:

$$\begin{aligned} Q_1 &= \frac{3}{\sqrt{2}} X; \qquad \Theta_1 = -\frac{1}{\pi r} Y; \qquad \Theta_2 = -\frac{1}{2\pi r} Z \\ t' &= \frac{2}{3\pi^2} t''; \qquad \left(r = \frac{Ra}{Ra_c} = \frac{4Ra}{27\pi^4} \right) \end{aligned} \tag{19.26}$$

And we get the *Lorenz equations*. $\dot{X}$, $\dot{Y}$, $\dot{Z}$ denote the derivatives of X, Y, Z relative to t'':

$$\begin{cases} \dot{X} = Pr(Y - X) \\ \dot{Y} = -Y + rX - XZ \\ \dot{Z} = -8Z/3 + XY \end{cases} \tag{19.27}$$

In mathematical physics, X, Y and Z are called the *phases* of the system. The set of equations (19.27) gives the trajectory of the point (X, Y, Z) in the *phase space*, describing a *flow*.

Let us first study the case where the Prandtl number is infinite. At each time, $Y = X$. There are only two independent phases, which may be Y and Z. The phase space reduces to a plane, the flow reduces to:

$$\begin{cases} \dot{Y} = (r - 1 - Z)Y \\ \dot{Z} = Y^2 - 8Z/3 \end{cases} \tag{19.28}$$

Without solving this system of equations (only feasible numerically), we can get an idea of what the trajectories look like, from the signs of:

$$\begin{cases} \dfrac{dZ}{dY} = \dfrac{\dot{Z}}{\dot{Y}} = \dfrac{Y^2 - 8Z/3}{(r - 1 - Z)Y} \\ \dfrac{1}{2}\dfrac{d}{dt''}(Y^2 + Z^2) = Y\dot{Y} + Z\dot{Z} = (r - 1)Y^2 - \dfrac{8}{3}Z^2 \end{cases} \tag{19.29}$$

For $r < 1$ (Figure 19.1a), the trajectories converge toward the origin, corresponding to a stable state (absence of convection). The origin is an *attractor*. For $r > 1$ (Figure 19.1b), there are two attractors A and A', whose (attraction) *basins* are the half-planes $Y > 0$ and $Y < 0$. (They correspond respectively to convection rolls rotating in one direction, and in the opposite direction.) The representative point (Y, Z) moves toward the attractor on a spiral.

The attractor corresponds to the permanent, stable regime, with finite amplitudes:

$$\begin{cases} Z = Z_A = r - 1 \\ X = Y = Y_A = \pm\left[\frac{8}{3}(r - 1)\right]^{1/2} \end{cases} \tag{19.30}$$

Coming back to the original dimensional variables, this gives:

$$\begin{cases} (q_1)_A = \pm 2\sqrt{3}\,\kappa\,\sqrt{r - 1} \\ (\theta_1)_A = \mp \dfrac{\theta_b}{\pi r}\sqrt{\frac{8}{3}(r - 1)} \\ (\theta_2)_A = \dfrac{\theta_b}{2\pi}\left(1 - \dfrac{1}{r}\right) \end{cases} \tag{19.31}$$

The stream function q_1 increases with $r = Ra/Ra_c$, and the velocities as well. The

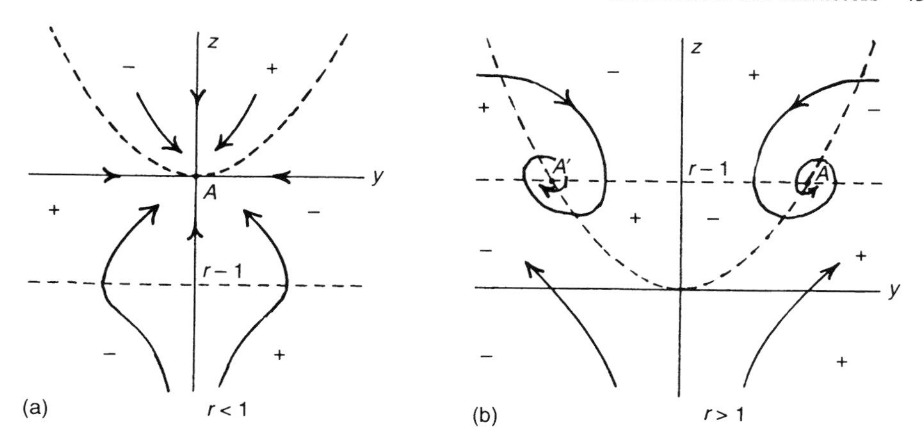

Figure 19.1. Lorenz flow in the phase space, when $Pr = \infty$. The sign of the trajectory's slope is indicated for the different domains. This allows us to approximately draw the trajectories without needing to integrate numerically the Lorenz equations.

super-adiabatic temperature tends toward:

$$\theta = \beta(H - z) - \frac{\beta H}{2\pi} \sin \frac{2\pi z}{H} \tag{19.32}$$

The super-adiabatic gradient $-\partial\theta/\partial z$ vanishes around $z = H/2$. The vertical temperature variation concentrates close to the surface and close to the bottom, in *thermal boundary layers*. They should not be confused with the boundary layers of fluid mechanics, forming at contact with the walls for high velocities, and characterized by very high shear strain rates and rotations.

19.4 BIFURCATIONS AND ATTRACTORS

The Lorenz flow has no physical basis when r is high, but its study allows us to introduce essential modern notions, useful for mantle convection and many other domains of physics, chemistry, ecology or economy. We will therefore detail them further.

We are not interested in transient regimes, where the flow in the phase space moves toward the attractor, but to the trajectories and modifications of the attractor when one of the system parameters (r, in this case) varies continuously.

Bifurcation

In the above case $Pr^{-1} = 0$, the phase space is reduced to two dimensions. The attractor has the coordinates (Y_A, Z_A) and, if Y_A is known, Z_A can be deduced from it. Therefore, we only need to study $Y_A(r)$, plotted in Figure 19.2a. When r increases, $Y_A = 0$ until $r = 1$, where a bifurcation occurs. In the present case, where the point

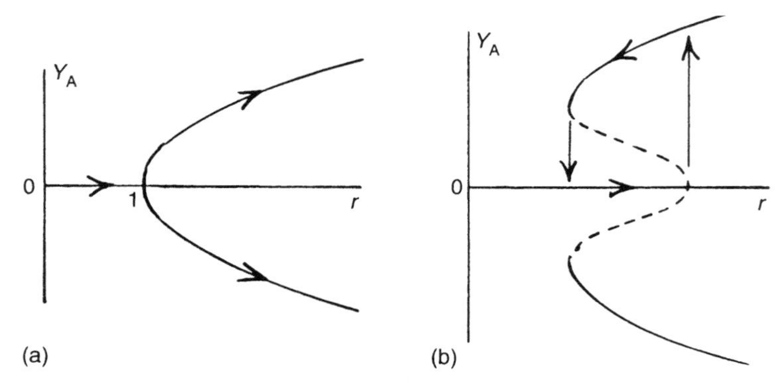

Figure 19.2. Fork bifurcation during the continuous variation of a parameter. (a) Fork bifurcation of the Lorenz system for $r = 1$. (b) Reverse bifurcation.

attractor remains a point attractor, this is a *fork bifurcation*. Moreover, it is said to be *normal* (or *over-critical*).

For other flows, the fork bifurcation can be *reverse* (or *sub-critical*), as represented in Figure 19.2b. When r varies monotonously, the attractor jumps irreversibly from one position in the phase space to another position far away. This is a *catastrophe*, in the mathematical sense. The attractor may then be in a metastable state. When the parameter r oscillates, it may describe a hysteresis cycle, as shown with the arrows in Figure 19.2.

Periodic attractors

In a 2-D phase space (Y, Z), the attractor may not be a single point (Y_A, Z_A), and may instead be a close curve, always described by the flow (Y, Z) in the same sense, with some period. This is a *periodic attractor*. For example, the fluid velocities will vary periodically with a slower rhythm than a fluid particle following the convection roll.

The investigation of the Lorenz flow has been made with computers (mathematics may sometimes also be an experimental science). The following results were obtained with the fixed value $Pr = 10$.

A periodic attractor exists for $r > 214.364$. Although the Bénard–Boussinesq convection is very badly represented by the Lorenz system, it allows us to intuitively grasp the process. Let us remember that Y_A (not much different from X_A if Pr is high), measures the amplitude of the convection, and Z_A measure the super-adiabaticity of the medium (boundary layers excluded). We have the circular causal chain: strong convection → low super-adiabaticity → low convection → high super-adiabaticity → strong convection.

There is a time delay between causes and effects, because Pr is finite; i.e. the forces of inertia are not negligible. When Pr is infinite, such a time delay is possible if the lower limit condition is a fixed heat flux instead of a fixed temperature, because with a stronger convection, the mean temperature of the entire medium must decrease.

With $Pr = 10$, there are two 'periodicity windows': $99.524 < r < 100.795$ and $145 < r < 166$.

In a phase space with three independent dimensions, the attractor may be a surface with the topology of a torus. The closed trajectory of the point (X, Y, Z) on this attractor may be shifted continuously on the surface of the torus, to form one or several circles, passing or not passing through the hole in the torus. These two topologically distinct circuits correspond to two fundamental, distinct periods of phase variation with time.

Any periodic function of time can be expressed as the sum of a fundamental and its harmonics. A linear expression of functions with two fundamental periods and their harmonics does not present other periods. But this is no longer the case with the flow studied, where the time derivatives are given by non-linear expressions. If f_1 and f_2 are the two fundamental frequencies of the attractor, frequencies $n_1 f_1 + n_2 f_2 (n_1$ and n_2 are positive or negative integers) appear with diverse amplitudes. If one investigates with Fourier analysis the frequencies of the sine oscillations present in a phase $X(t)$, one finds a high number of peaks emerging from a continuous-looking background. Indeed, theoretically, except in the unlikely case where f_1 and f_2 are multiples of the same frequency f_0, any frequency f can be approximated with a combination $n_1 f_1 + n_2 f_2$.

Aperiodic or strange attractors

There are natural processes of oscillatory nature, but with no dominant frequencies because the number of phases is very high, allowing for a very high number of fundamental frequencies. White light emission by an incandescent solid is the best example. These are sometimes called *aperiodic oscillations*, or the system is called *chaotic*. It is sometimes said that there is turbulence in the phase space. Beware: this turbulence may exist in thermal convection (*Péclet turbulence*) but the fluid movement remains laminar.

Lorenz and others have discovered that a chaotic regime may appear as soon as the number of phases is 3 or more. The attractor fills entirely a closed part of the phase space (i.e. it does not extend to infinity). In the course of time, the phase trajectory can be as close as possible to any point in this domain. With a 3-D phase space, the attractor is no longer a 2-D surface, but the limit of a 2-D surface infinitely folded onto itself (a *fractal*). (A ball of thin paper, extremely crumpled and tending to fill a volume, may give an idea of it.) A 'fractal dimension' can be defined, which ranges between 2 and 3 in the present case.

Such an attractor is called an *aperiodic attractor*, or, to avoid confusion with aperiodic oscillations (which do not require such an attractor) a *strange attractor*.

In the case of the Lorenz system, with $Pr = 10$, a strange attractor was found at $r = 24.06$, with a small fractal dimension (2.06). It coexists with the two point attractors appeared at $r = 1$ until $r = 24.74$. It remains alone then, until $r = 214.364$ where it is replaced by the periodic attractor mentioned above. (There were, however, two periodicity windows before.) The bifurcation at $r = 24.06$, where this attractor appears, is reverse. Thus, by decreasing r, its existence may be extended beyond 24.06, until $r = 13.926$.

Intermittences

The list of characteristics of the Lorenz system is far from exhausted, despite its simplicity. For $r > 30.1$, when the attractor is strange, it may become periodic sporadically, in a random and unpredictable way. The duration of these periodicity episodes increases when closer to the periodicity windows (for r around 100 and 155). The unknown process which creates these intermittences of periodicity becomes more efficient when r is closer to these values, and can even establish a permanent periodicity.

19.5 EXPERIMENTAL STUDIES OF BÉNARD–BOUSSINESQ CONVECTION

Laboratory studies investigated 3-D convection of a nearly isoviscous fluid, but not for very high Prandtl numbers. It seems however that the values of Pr for which it becomes unimportant have been reached. The pattern is of most interest. Busse, who worked on this theme since 1962, has reviewed the results, sometimes discordant and not always reproducible (Busse, 1989).

When Ra is above a certain threshold, which must be smaller to the one given by Boussinesq theory, an *l-hexagonal pattern* appears, where the fluid goes up along the axes and descends along the side faces.

For larger values of Ra, the rolls of the theory appear. With a layer much wider than it is thick, for $Ra/Ra_c = r$ between 9 and 20 (Ra_c is the theoretical value of 657.5), Krishnamurti (1970) observed a *chessboard pattern*, the 'white' squares with rolls in one direction and the 'black' squares with rolls in the perpendicular direction.

For $r > 20$ approximately, the hexagonal pattern reappears but the fluid goes up along the side faces and goes down along the axes (*g-hexagonal pattern*). There is hysteresis during the crossing from one pattern to the next.

For $r > 30$ approximately, this pattern is replaced by a *spoke pattern*. The hexagonal paving is still there, but the fluid goes down along radial planes joining the axes, and goes up along the prismatic edges. (This is not the l-hexagonal pattern, because there are 6 concurrent descending planes instead of 3). Less frequently, it is also possible to observe *bimodal convection* (on two scales). It derives from the roll pattern (with one single mode) by twisting the circular stream lines, so that they form helices at the surface of a torus.

Finally, for $r > 150$, a stable *square pattern* has sometimes been observed. Its periodicity is much shorter than the spoke and bimodal patterns. The fluid goes up in the centre of a square and descends at its four corners (the inverse is possible: this amounts to shifting the pattern).

Beside these invariable patterns and convection velocities, other behaviours can be observed where the cell size and the velocities oscillate periodically. The oscillations appear where the pattern is the less regular (it is never perfect) and progressively invade the whole domain. Krishnamurti (1970) observed them for $48 < r < 136$. For $r > 136$, the pattern was modifying incoherently. Busse (1989) does not signal this possibility, but few experiments have been done for very high values of r. And if intermittences appeared, they were not recognized as such.

It is not clear how to extend the description in a single phase space to these processes, where the patterns and not only some parameters are changing. Anyway, with three space dimensions, the same sudden changes of the convection type (sometimes non-reproducible and unpredictable) are found as in 2-D convection.

19.6 HEAT TRANSFER BY CONVECTION: NUSSELT NUMBER

Let us go back to the stable 2-D convection, in a steady regime, in an isoviscous medium, i.e. convection rolls with a height H, a width LH, and with two perpendicular symmetry axes. For very high values of Ra/Ra_c, we can assume that the super-adiabatic temperature θ equals zero, except in thermal boundary layers of thickness Δ_s at the surface and at the bottom, and Δ_h in vertical plumes, upwelling or downwelling (Figure 19.3). (The thickness of a plume, which contains the limit layers of two adjacent rolls, is $2\delta_h$.) Beware: these boundary layers are not mechanical boundary layers; $\theta = 0$ out of these layers does not imply $q =$ constant.

The heat flux due to the only adiabatic gradient is neglected. By symmetry, the cooling by decompression of an upwelling hot plume is exactly balanced by the compression heating of a downwelling cold plume.

Heat transfer from the bottom to the surface only uses the super-adiabatic temperatures θ. In the absence of convection, the heat flux transferred per unit area and time is $K\theta_b/H$. In the presence of convection, it is multiplied by a factor Nu, the *Nusselt number*.

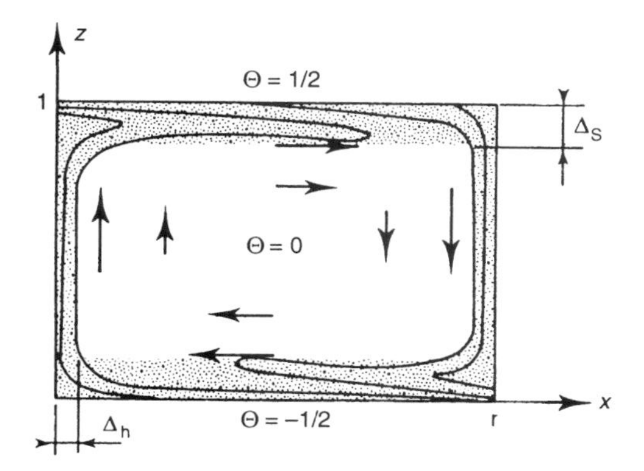

Figure 19.3. Boundary layer theory.

Nu can be calculated with the *boundary layer theory*, which may be found in my book '*Very slow flows of solids*' (Lliboutry, 1987). In this theory, it is assumed that the heat transfer between the bottom boundary layer and the surface boundary layer is exclusively occurring through plumes (by advection), whereas the cooling of the former and the heating of the second are exclusively by conduction. For a free surface and bottom, it is found that Nu is proportional to $Ra^{1/3}$. And for a rigid surface and bottom, Nu is proportional to $Ra^{1/5}$, with numerical prefactors functions of L. As the theory includes several unwarranted approximations, numerical checks were necessary. A synthesis was made by Jarvis and Peltier (1989).

The power law becomes exact only for $Ra/Ra_c = r > 50$. For free lower and upper surfaces and $L = 1$, the theory predicts: $Nu = 0.93r^{1/3}$, $\Delta_h = 0.465r^{-1/3}$, $U_h = 10.81r^{2/3}$. The numerical experiments yield: $Nu = 2.24r^{0.318}$, $\Delta_h = 0.300r^{-0.278}$, $U_h = 8.465r^{0.695}$.

Except for Nu's prefactor, the agreement is good. Nevertheless, the theory becomes wrong for a fluid with temperature-dependent viscosity, and this makes its application to mantle convection unreliable. As a rigid plate (the lithosphere) appears at the surface, rigid boundaries seem more appropriate. The theory predicts: $Nu = 0.494r^{1/5}$. The numerical simulations of Fleitout and Yuen (1984) yielded: $Nu = 1.13r^{0.15}$.

19.7 SUPER-ADIABATIC TEMPERATURES AND RAYLEIGH NUMBER IN THE MANTLE

Although the thermal convection studied in the first part of this chapter is very different from convection in the silicate Earth, it is interesting to apply its results to better understand the problems and the unknowns. The presence of rigid lithosphere plates being excluded from the isoviscous model, we assume the presence of a rigid upper limit at the depth of 100 km. The upper limit conditions are not well reproduced, but this is better than introducing the very large temperature differences present in the lithosphere, because they do not contribute to convection.

The densities are reasonably well known. From them, we deduce that the gravity g does not vary much from the surface to the core–mantle boundary; $g \approx 10\,\text{m/s}^2$. The calorific capacity is $C = 1,260\,\text{J/kg}$ for upper mantle peridotites (assuming that melting is not reached below 100 km). This value can be kept for the whole mantle without risk of large errors. The uncertainties concern the viscosity, the thermal conductivity K, the volume dilatation coefficient α and the temperature profile. The last two are essential to calculate the adiabatic gradient $G_a = g\alpha T_a/C$ and the super-adiabatic temperatures.

In the following, the indices s and i stand for the upper mantle (down to 660 km) or the lower mantle.

Thermal conductivity

Thermal conductivity is normally due to the diffusion of atomic vibration energy quanta (*phonons*). It increases with pressure more than it decreases with temperature. In the mantle, it must remain between 3 and 20 $W m^{-1} K^{-1}$. At high temperature, there is also transmission of infrared light energy quanta (*photon conduction*). It was suggested in the 1950s that the post-spinel phase in the lower mantle would be highly transparent to infrareds, but it does not seem to be the case, because of the presence of ions Fe^{++}. At most, the conduction by photons equals the conduction by phonons. Another high temperature process, of which we do not know much, is however possible: the migration of electron excitations (*exciton conduction*). The extrapolation of values obtained with MgO led Gilvarry to assume the exciton conduction was 10 times higher than phonon conduction, but no one else comforted this hypothesis.

I shall take for the upper mantle the value adopted by McKenzie in 1974: $K_S = 6.66\, W m^{-1} K^{-1}$. For the lower mantle, I shall take a minimal value: $K_i = 20\, W m^{-1} K^{-1}$ (Jarvis and Peltier (1982) adopted for the entire mantle a value of $14.8\, W m^{-1} K^{-1}$). The resulting diffusivities are: $\kappa_S = 1.4 \times 10^{-6} m^2 s^{-1}$ and $\kappa_i = 3.4 \times 10^{-6} m^2 s^{-1}$.

Temperatures

The temperature at 100 km depth is well known because of the metamorphosed xenoliths coming from this depth. The temperatures in the core are well known too (see Section 17.1). But, between the two, there are two possibilities (Figure 19.4): either there are two separate convections in the upper and lower mantle, which leads to the formation of two thermal boundary layers on each side of the 660-km discontinuity ('hot' geotherm) or this is not the case ('cold' geotherm).

I modified the temperatures from Figure 19.4, so that the geothermal gradient in the bulk of the lower mantle be roughly equal to the adiabatic gradient and not much smaller. These modified values are noted with an asterisk in Table 19.2.

According to Jeanloz (1989), the volume dilatation coefficient equals $(2.7 \pm 0.6) \times 10^{-5} K^{-1}$. It must be less in the lower mantle, with more compact phases. I chose $\alpha_S = 3.0 \times 10^{-5} K^{-1}$, and, always to ensure that the adiabatic gradient is never larger than the geothermal gradient, $\alpha_i = 2.1 \times 10^{-5} K^{-1}$. The adiabatic gradient then becomes:

- Upper mantle (mean $T_a = 1710\,K$): $G_a = 0.408\, K\, km^{-1}$
- 'Cold' lower mantle (mean $T_a = 2456\,K$): $G_a = 0.409\, K\, km^{-1}$
- 'Hot' lower mantle (mean $T_a = 3132\,K$): $G_a = 0.522\, K\, km^{-1}$

Table 19.2. Comparison of temperatures at depth, for the 'hot' and 'cold' geotherm hypotheses.

Depth (km)	100	120	400	600	660	800	2700	2900
'Cold' geotherm (K)	1360	1560	1665	1920	1956	2040	2760*	3830
'Hot' geotherm (K)	1360	1560	1665	1920	2130*	2620*	3630*	3830

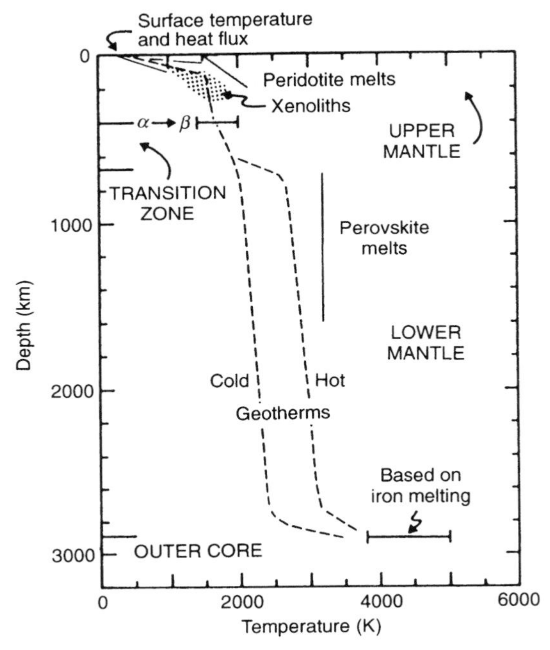

Figure 19.4. Experimental constraints for actual temperatures, as horizontal averages in the mantle. From Jeanloz and Morris, 1986.

Rayleigh number

Let us adopt the values of Peltier and Jiang (1996): $\eta_S = 10^{21}$ Pa s and $\eta_i = 15 \times 10^{21}$ Pa s. With $\rho_s \approx 3700\,\mathrm{kg\,m^{-3}}$ and $\rho_i \approx 4700\,\mathrm{kg\,m^{-3}}$, the resulting kinematic viscosities are: $\nu_s = 2.7 \times 10^{17}\,\mathrm{m^2\,s^{-1}}$ and $\nu_i = 3.2 \times 10^{18}\,\mathrm{m^2\,s^{-1}}$.

We can now estimate the Rayleigh number, θ_b in (19.15) being replaced by $\Delta\theta$, the difference of super-adiabatic temperatures above and below the layer. For a two-layer convection (hot geotherm):

- Upper mantle: $\Delta\theta = 540°\mathrm{C}$, $Ra = 75,000$, i.e., if $Ra_c = 1,101$, $r = 68$.
- Lower mantle: $\Delta\theta = 531°\mathrm{C}$, $Ra = 120,000$, i.e. if $Ra_c = 887$, $r = 135$.

In the case of a one-layer convection in the entire mantle (cold geotherm), $\Delta\theta$ is calculated by adding its value for the upper mantle (392°C) and the one for the lower mantle (933°C). We choose for $\alpha/\kappa\nu$ a weighted average for the two mantles ($1.74 \times 10^{-17}\,\mathrm{m^{-4}\,s^2\,K^{-1}}$). We find that $Ra = 5.0 \times 10^6$, i.e. if $Ra_c = 1,840$, $r = 2,700$.

Although the model used is very rough, these estimates seem to show that if the 660 km discontinuity results only from a phase change, it cannot resist a one-layer convection. If the nature of the discontinuity is also chemical, making two-layer convection the only possibility, r remains of the same order between the two layers, and remains high enough to form well-marked thermal limit layers.

If we trust the formula of Jarvis and Peltier giving the Nusselt number, with the parameters adopted, we find that in the upper mantle $Nu = 9.6$ and it would transmit 35 TW. The lower mantle, with $Nu = 10.7$, would transmit 43 TW. This value should be lessened because of internal heating (quite a lot, if K was under-estimated). These values are acceptable. But, with a single-layer convection, this formula gives an approximate value of $Nu = 27.6$, leading to a geothermal flux much too high and unrealistic.

Setting time for a permanent regime

With the dimensionless units of equation (19.16), this time is of the order of 1. But this time unit H^2/κ is huge: 7 Ga for the upper mantle and 47 Ga for the lower mantle. *It is therefore impossible that a steady state did set up in the mantle*, a fact long missed by geophysicists.

The study of a steady state caused by internal heat only does not have any meaning. Numerical computations are not necessary to affirm that without core heat to evacuate, a thermal boundary layer would not form at its contact, and localized upwelling hot plumes would not form either. To treat the problem directly and realistically, we must consider the Earth's thermal history since its creation, with decreasing radioactivity, a sizeable excess of heat stored since the beginning and a later cooling, etc. This endeavour seems premature.

19.8 INFLUENCE OF PLATES AND VARIABLE VISCOSITY ON CONVECTION

The convection studied until now leads to convection rolls or cells, with horizontal dimensions similar to the thickness of the convective layer. With the 1970s hypothesis that each lithospheric plate forms the upper part of a convection roll, this is manifestly not the case in the Earth: the Pacific plate measures 14,000 km along its largest diameter. Convection limited to the upper mantle would *a fortiori* be unthinkable.

This argument is in fact invalidated by the presence in the upper boundary layer of a rigid plate moving as a block. In one of the first numerical convection experiments, Houston and De Bremaecker (1975) imposed this condition to 2-D convection in a rectangle 700 km high and 6,000 km wide. In this kind of calculation, one assumes an initial stream function, modified by successive iterations until all equations and boundary conditions are obeyed. The initial two symmetrical convection rolls were replaced by a single roll during the iterations.

Later, Jaupart and Parsons (1985) followed the stability study by Boussinesq, assuming that viscosity was decreasing with depth. (This would be the case for a hot fluid suddenly cooled at the surface.) A rigid 'lid', a plate, forms at the surface. Two modes of convection can appear below: elongated rolls if the viscosity contrast is low, and convection with small isodiametrical convection rolls if the contrast is high.

The relevance of these studies to mantle convection is, however, very dubious, because two primordial processes were not introduced into the models (or in later studies):

- the formation of basaltic magmas, migrating, warming and thinning the plate, lessening its resistance to traction and enabling the formation of a new mid-ocean ridge;
- the existence for the plate of a plasticity threshold, enabling the appearance of new subduction zones, even in the middle of oceans.

The proper way to study the influence of plates on convection below seems to be the one advocated by Davies (1988), with a more modest ambition. The aim is not predicting plate movements, and they are imposed, this condition replacing the zero shear strain rate condition at the surface. But Davies (1988) assumes the heating is internal only, by radioactivity, and that the flow is steady. His conclusion that 'only single-layer convection can account for the geothermal fluxes measured' is therefore highly arguable.

Furthermore, these old studies only deal with the appearance of convection. When it has developed, it should be imperative to account for the influence of lateral temperature variations on viscosity, which is huge. The activation enthalpy of dry dunite is $H^* = 465\,\text{kJ/mol}$ (see Section 11.3). With this value, at the same depth, the viscosities in a hot plume at 1,800 K and a cold plume at 1,400 K would differ by a factor of 700. We must therefore introduce in the program a temperature-dependent viscosity, which will be recomputed at each iteration, to get a self-consistent solution.

Even if, as most recent authors, we only account for temperature, and not for pressure (which increases viscosity), the results cannot be applied to the entire mantle. Viscosity would become very low in the lower thermal limit layer. It would continuously emit plumes, but randomly and chaotically, completely differently from the hot spots issued from the bottom of the mantle.

Christensen (1984) has accounted for the augmentation of viscosity with pressure. He studied 2-D convection in a square, with a mid-ocean ridge in one corner and a subduction zone in the other. The 'trick' to suppress a singularity where the plate folds is not mentioned. The stream lines get closer below the mid-ocean ridge, showing higher velocities in an upwelling plume. They move further from each other when away from the ridge, showing a diffuse return current. Christensen also studied the influence of a non-Newtonian viscosity (power 3 law). It has the same effects than a diminution of temperature and pressure influences (H^* and V^* divided by 2). But he does not find the subducted plates, unless the effect of pressure is nearly removed.

All these studies are more *numerical experiments*, to study the influence of one factor, than *numerical simulations* aiming at solving the direct problem of convection in the mantle. They were performed with 2-D models. Convection with one or two spherical layers was simulated by Glatzmaier and Schubert (1993). It shows the formation of very complex patterns. But, even with modern super-computers, it seems that 3-D convection cannot be simulated with *Ra* high enough to be realistic.

19.9 'HARD' TURBULENCE AND MANTLE MIXING

Fluid mechanics scientists have used the possibilities of modern super-computers to study convection with very high Rayleigh numbers, but *assuming the medium to be isoviscous*, and in the 2-D case only (equations 19.17). Their results cannot be applied directly to the Earth, but they open interesting possibilities (Solomon and Gollub, 1991; Yuen *et al.*, 1993).

At very high Rayleigh numbers, there is 'soft' turbulence. Convection evolves continuously, without periodicity; it is chaotic. But hot plumes reach the surface, and cold plumes reach the lower limit layer. At a given time, the entire 2-D convection is organized into convection rolls.

For $Ra > 10^8$, however, many plumes are continuously forming, which are not the boundaries of convection rolls. Along their progression, they subdivide and some branches stop on the way. These plumes are deviated by general convection, still ensuring most of the heat transfer (Figure 19.5(a)). If Ra increases again, the general

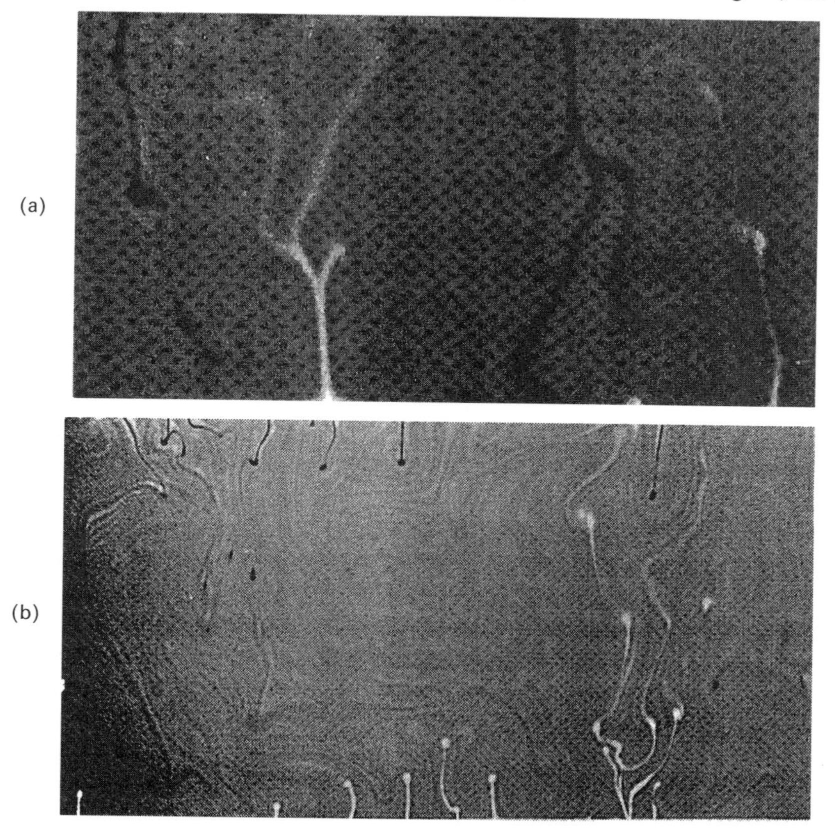

Figure 19.5. Hard turbulence snapshots in Bénard–Boussinesq convection: (a) for $Ra = 10^8$; (b) for $Ra = 10^{10}$. The hot plumes are in white, the cold plumes in black. Very weak thermal traces are visible, remaining from old plumes. From Yuen *et al.*, in (Stone and Runcorn, 1993).

convection disappears progressively, the plumes become very thin, and most of them do not cross the layer, forming simple diapirs (Figure 19.5(b)). This is 'hard' turbulence.

Yuen *et al.* (1993) studied the case of a power-3 viscosity law, always independent from temperature. Convection intensity is better characterized then by the Nusselt number *Nu*. The transition to chaotic regime and hard turbulence occurs for much smaller values of *Nu* than for an isoviscous fluid. If the heating is internal, overheating occurs in stagnant regions in the middle of the medium, and next plumes leave them to evacuate the excess heat.

Similar processes may be present in the upper mantle. The chaotic character of convection solves one enigma: what happens to the basalt of subducted oceanic crust, or rather the eclogite it became at depth? The constant composition of mantle basalts through the geological ages implies the subducted basalt is recycled, and *mixing of the upper mantle*. With traditional convection rolls, the oceanic crust could be laminated, but it was not mixing with the mantle. The same is not true with chaotic convection, as shown in Figure 19.6.

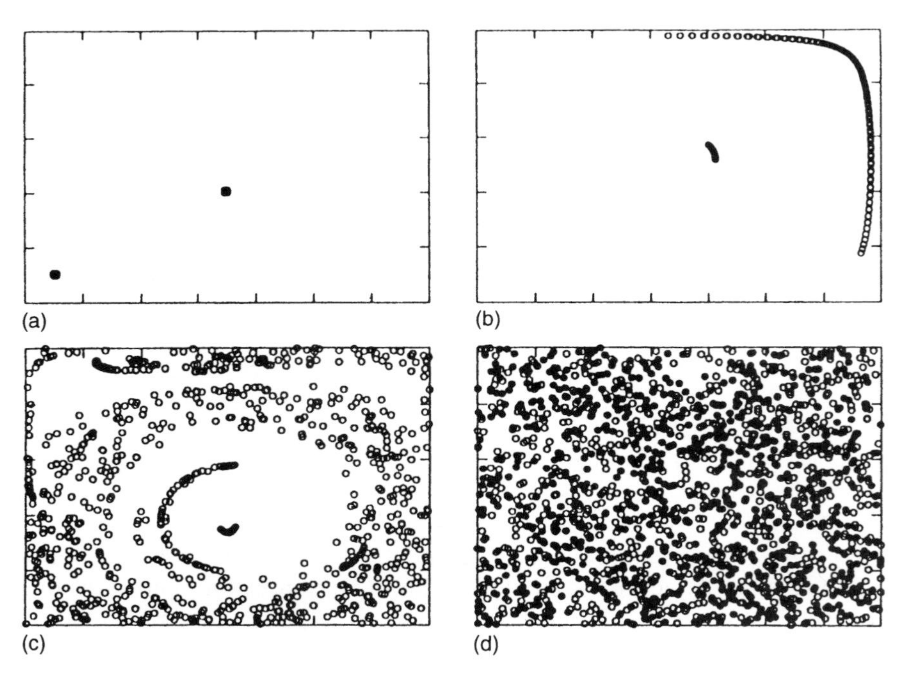

Figure 19.6. Numerical simulation of mixing in 2-D Bénard–Boussinesq convection: (a) initial state, with two groups of 900 markers each; (b) stable convection roll: the markers in the boundary layer were carried along at different distances according to their distance from the edge, and the central region markers remained clustered; (c) soft turbulence: a convection roll remains, but it is not stable and a new hot plume appeared in the middle; markers initially in the boundary layer are dispersed, the ones in the centre remain clustered in one of the motionless spots; (d) 'hard' turbulence: the two groups of markers are completely scattered. From Kellogg, 1993.

Even with such chaotic convection, some motionless spots, not mixing with the surrounding medium, remain for long periods. This is not incompatible with the idea of different geochemical 'reservoirs'.

Finally, Yuen *et al.* (1993) studied convection with very high values of Ra, when a phase change opposes it at a certain level (exothermal for a downwelling thermal and endothermal for an upwelling plume). This is indeed the case at 660 km in the Earth (see Section 17.3). Convection then tends to stratify into two layers. But now and then, temporarily, it becomes single-layered. Solheim and Peltier (1994) arrived to the same result with a model of spherical, axisymmetrical layers.

It was stated in Section 17.3 that a single-layer convection and a two-layer convection seemed to currently coexist on Earth. Could we be the witnesses of one of these transitions from one to the other?

19.10 D″ LAYER AND HOT SPOT ORIGIN

In 1949, the seismologist Bullen had found close to the core a thin mantle layer with slower seismic velocities, which he called the D″ layer. It was interpreted at the time as a layer with a different chemical composition. (For example, a chemical reaction of the core iron with mantle silicates would have formed FeO and FeSi).

After Morgan (1971) suggested that the plumes at the origin of hot spots were starting from this level (he saw in plumes the driving mechanism for the entire mantle convection), Stacey argued that the D″ layer was a thermal boundary layer linked to single-layer convection in the entire mantle. In 1983, with Loper, he presented a model now become classic, the *Stacey–Loper model* (Stacey and Loper, 1983).

According to this theory, there would be genuine 'chimneys' of diameter R, where the temperature would be higher by ΔT than elsewhere, at the same level, in the mantle. As in any chimney, the Archimedes thrust $\pi R^2 \rho \alpha \Delta T g$ per height unit must balance the viscous friction at the periphery: $\tau(R) = 2\pi R \eta (du/dr)$. The resulting volume rate is given by the Poiseuille formula:

$$\frac{\dot{M}}{\rho} = \pi R^4 \rho g \alpha \frac{\Delta T}{4\eta} \tag{19.33}$$

The speed must be high enough that heat does not dissipate by thermal conduction during the ascent. A jacket hot enough to limit this conduction, but not enough to be dragged along, forms around the chimney. ΔT must be high for a good draught; the D″ layer feeding the chimney must therefore be very pronounced and its viscosity $\eta_{D''}$ much lower than the lower mantle viscosity (η_i). According to Stacey and Loper (1983), we need $\eta_i/\eta_{D''} > 10^4$ and $\Delta T \approx 840°$. Then $2R = 24$ km at the start, 15 km at the top of the upper mantle. (With diameters so small, the chimneys and their hot jackets remain undetectable through seismology.) The average upwelling speed is 0.5 m/year, the volume of a hundred chimneys (with low-pressure densities) is 17 km^3/year. Hot-spot basalts would exude from these hot rocks (the most intense hot spot, Hawaii, emits 0.5 to 1 km^3/year of lavas, on average).

The heat transferred from the core to the surface would be 1.6 TW. This value is low, but Stacey and Loper (1993) chose a very low conductivity in the D″ layer: $K = 1.2\,W\,m^{-1}\,K^{-1}$. Anyway, the current consensus is that the contribution of the core to geothermal flux is very small, around 10%.

The study of the core–mantle boundary and the D″ layer has greatly progressed during the last years, since the installation of the new worldwide seismic networks. A topography of a few kilometres was found at the core–mantle boundary, and a general ellipticity different from the one of the geoid. There follows a purely mechanical effect, which is the coupling of the core and mantle rotations. This question was briefly mentioned in Section 17.4 and will not be developed further.

According to Kendall and Shearer (1994), the thickness of the D″ layer is irregular: between 150 and 350 km, with an average of 260 km. According to previous authors, it did not even exist in some regions.

Wysession *et al.* (1994) detected lateral seismic velocity variations in the D″ layer. They found the velocities are smaller by 3% below the Moluccas, larger by 3 to 5% around, below the NW Pacific, Southeast Asia and Australia. They interpret the low velocity as resulting from a plume in formation, and the high velocities by the presence of cold plates reaching down to there. But they do not exclude the contribution of a different chemical composition. The D″ layer may be locally depleted in iron, silica SiO_2 having exuded from the liquid core (it would be as a very high-pressure form of stishovite).

Note that Lenardic and Kaula (1994) infer the opposite. They assume that the D″ layer is of chemical origin, and that a cold plate can only lie on its top. The very high ensuing thermal gradient would then lead to the formation of a plume.

Larsen and Yuen (1997) performed accurate numerical computations, considering not only the steady state but also simulating the evolution with time. They find that the ascending velocity of a hot plume is 10 times higher by adopting a power law ($n = 3$) viscosity for the upper mantle, rather than a Newtonian viscosity (this can go up to 50 times if one considers the maximum velocity). But their model is in 2 dimensions, and only deals with a rectangle of upper mantle. They follow the hypothesis, supported by Anderson (1995) of plumes starting in the hot limit layer around 660 km. The relative fixity of hot spots remains then unexplained. And, as with the Stacey–Loper model, this does not explain why the appearance of a hot spot is often accompanied by a considerable emission of basalts, with no common measure with later volcanism (see Section 5.9).

I personally think that the fixity of hot spots and the initial strong emission of basalts may be explained if we assume the existence of a very viscous layer between 670 and 750 km deep (hypothesis of Pari and Peltier, 1995). I shall call it a 'hypo-lithosphere' (see the next section). As lower mantle convection is very slow, it would form a single shell in the entire Earth and therefore would not participate to this convection. The large basaltic magma pockets would have formed below, and Archimedes thrust would have finally enabled them to cross the hypo-lithosphere. The hot-spot chimneys would be remainders of these events, and would be relatively fixed because corresponding to these holes in the hypo-lithosphere. If this hypothesis were true, the subducting plates penetrating into the lower mantle would cross the hypo-lithosphere in fixed locations in the hot-spot system of reference.

19.11 EXPLOITATION OF SEISMIC TOMOGRAPHY DATA

From the publication of the first seismic tomography results for the lower mantle, it was found that the cold high-velocity regions were not correlated with the actual positions of subduction zones, but with the positions they had 110 to 180 Ma ago, at the time of the Pangaea, assuming that the hot spots had remained fixed since (Richards and Engelbretson, 1992).

(Past position depends on the choice of the frame of reference. Subduction zones on the American west coasts remained fixed relative to these continents, but the continents moved in the absolute HS frame.)

Ricard *et al.* (1993) showed that the lower mantle is a memory of past subductions. They represent seismic velocity anomalies with spherical harmonics to degree 15, with a resolving power (smallest wavelength) still of 1,300 km. This blur makes the following model valid, despite its simplicity. Each plate portion corresponding to a 5 Ma period is supposed to have gone down vertically (in the HS system) at the same velocity than the horizontal portion moves. By penetrating into the lower mantle, more viscous, the velocity is reduced along a fixed ratio, to be adjusted, and the same for all plates.

When this ratio ranges between 3 and 4, a very significant correlation appears with the structure provided by lower mantle tomography. This correlation is higher than the correlation between tomography results published by different authors. The density anomalies obtained with this model also explain 72% of gravity anomalies (represented by spherical harmonics up to degree 8). Up to 96% can be explained by the model of Ricard *et al.* (1993) when removing the harmonics of degree $n = 2$ and order $m = 1$, which are highly perturbed by glacio-isostasy.

It remains to estimate the viscosity ratio that leads to a reduction of velocities by a factor 3–4. Richards (in Sabadini *et al.*, 1991) performed 2-D numerical experiments (rectangular domain, Newtonian viscosity, steady state). He affirms that velocity decreases as the logarithm of viscosity. Ricard *et al.* (1993) inferred from this that $\eta_i/\eta_S = 30$ to 40.

Pari and Peltier (1995) interpreted seismic tomography data in a different and more sophisticated manner. They draw from them a density distribution in the mantle (always expressed with spherical harmonics, functions of the radius), and, from there, for different viscosity models, they calculate the velocities without introducing the heat equation. Viscosities are adjusted to find a proper geothermal flux at the bottom of the lithosphere, as well as the geoid undulations. Glacio-isostatic rebound, used to determine the average viscosity of the upper mantle, is not used except as a test of the adjusted model.

Pari and Peltier (1995) cannot find the large geoid wavelengths unless they increase the lower mantle viscosity below 1,200–1,400 km only. This fits the

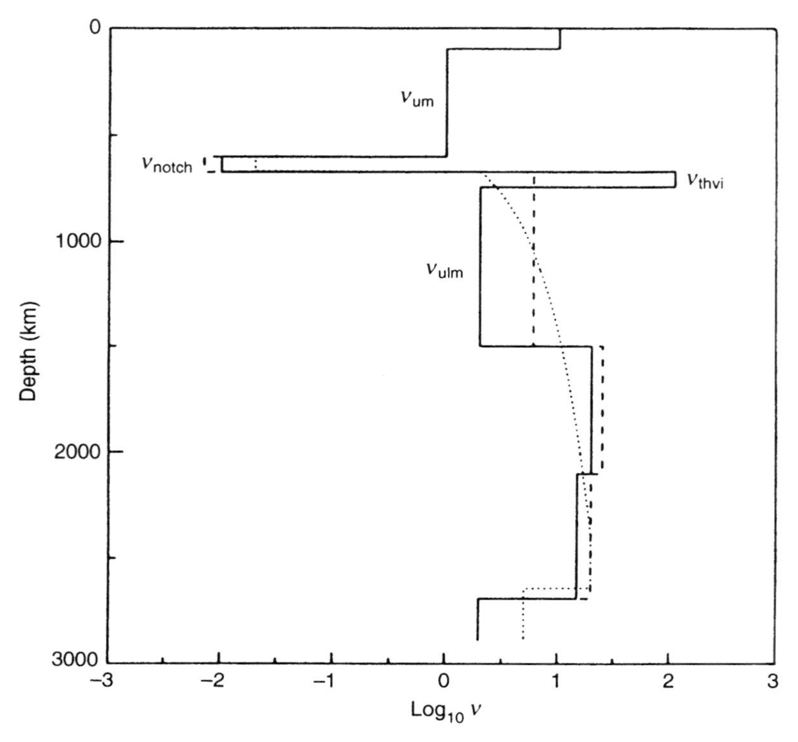

Figure 19.7. Viscosity profiles tested by Pari and Peltier to match the constraints of the problem (see text for details). From Pari and Peltier, 1995.

predictions of Poirier and Libermann (1984) mentioned in Section 18.8. A thin layer with very low viscosity must be introduced in the upper mantle to find the shorter-wavelength geoid undulations. It cannot be the oceanic LVZ (the fit must be for continents as well, and in particular the Canadian shield). It can therefore only be a thermal limit layer at the bottom of the upper mantle. Figure 19.7 shows viscosity profiles matching the imposed constraints. Between 100 and 600 km deep, Pari and Peltier (1995) still assume a uniform viscosity $\eta_s = 10^{21}$ Pa s (in the figure, the 0 of the logarithmic scale).

The upper layer of the model, 100 km thick, has a high fluidity: $\eta = 10^{22}$ Pa s. It can be justified by arguing that it did not include only the lithosphere, but also part of the oceanic LVZ. In the middle of the lower mantle, viscosity increases by a factor 10, from 2×10^{21} Pa s to 20×10^{21} Pa s. The thickness of the D″ layer (200 km) and its viscosity ($\eta_{D''} = 2 \times 10^{21}$ Pa s) were assumed *a priori*.

The largest viscosity oscillation around the 670-km discontinuity corresponds to the two superimposed boundary layers. It would drop to 10^{19} Pa s between 600 and 670 km, and increase to 10^{23} Pa s between 670 and 750 km. With an interface between the two mantles at 2,130 K and $H^* = 465$ kJ/mol, this corresponds to temperature differences with the heart of convection, where the temperature follows an adiabatic profile, +320°C and −450°C respectively. The latter seems too large, and the 10^{23} Pa s viscosity exaggerated. But this is only an arbitrary assumption made by Pari and Peltier (1995). They could satisfy the constraints of the problem, without introducing this second very viscous boundary layer. At 670 km, at equal temperature, the post-spinel phase of the lower mantle should anyway be 100 to 1,000 times more viscous than the spinel phase of the upper mantle. If η_s between 100 and 600 km, and η_i between 750 and 1,500 km only differ by a factor 2, this would be because the latter is much hotter. This 650°C-offset of the two geotherms is also found in the model of Jeanloz and Morris (Figure 19.4).

Thus, despite all the points still needing clarification, a quite consistent image of thermal convection in the mantle starts emerging. Let us sum it up.

Convection in the upper mantle is chaotic and unpredictable. This explains the scattering of the subducted oceanic crust, the sudden changes in the location of mid-ocean ridges, the formation of back-arc basins by short-lived mid-ocean ridges. This convection transfers the heat to evacuate (excluding the heat from continental crust radioactivity) toward the mid-ocean ridges, which are the big heat sinks. The upper mantle ensures most of the return current, facilitating the existence of a lower boundary layer. This return current may cause deep earthquakes in subducted plates (see Section 12.12). The geometry of this convection is linked to the geometry of the plates, reacts to the collision of continents, etc.

The lower mantle convection is separate, but not independent, because of the penetration by many subducted plates. Velocities are important only in the upper third, because the viscosity below becomes 20 to 30 times larger, the effect of pressure prevailing over the effect of temperature. The analogue of the lithosphere around 700 km must be a single shell, fixed in the hot-spot reference frame, and not a mosaic of plates. The lower part being almost motionless, the relative positions of the 'chimneys' at the origin of hot spots only evolve very slowly.

It is not excluded that, at large intervals, there may have been single-layer mantle convection episodes in the Earth's history. But we may be sceptical about the grandiose speculations which were aroused by this possibility. Geological history is very incomplete for the Palaeozoic, and only just beginning for the Precambrian. But it seems to exclude the periodic formation of a Pangaea and its break-up, according to a fixed scenario. Many bifurcations punctuate the path of the Earth's evolution, and although it obeys the laws of physics, Nature likes fantasy.

Appendix

Mathematical complements

A.1 INVARIANTS OF SPACE FUNCTIONS AND VECTOR FIELDS

In geophysics, one studies scalar functions of space coordinates (e.g. temperature) or vectors whose three components are functions of the coordinates (e.g. gravity or the magnetic field). When these functions have partial derivatives, we can define other scalars of vectors which are the same in any system of reference, and are called the *invariants* of the function or vector field. They give its local and intrinsic properties. For a function, the useful invariants are the *gradient* (a vector) and the *Laplacian* (a scalar). For a vector field, the useful invariants are the *curl* (a vector), the *divergence* (a scalar), and the *vector Laplacian* (a vector).

Recall first the definitions of the scalar and vector products of two vectors **a** and **b**, with Cartesian components a_x, a_y, a_z and b_x, b_y, b_z, respectively:

$$\text{Scalar product:} \qquad \mathbf{a} \cdot \mathbf{b} = a_x b_x + a_y b_y + a_z b_z \tag{A.1}$$

$$\text{Vector product:} \qquad \mathbf{a} \wedge \mathbf{b} = \begin{bmatrix} a_y b_z - a_z b_y \\ a_z b_x - a_x b_z \\ a_x b_y - a_y b_x \end{bmatrix} \tag{A.2}$$

In particular, $\mathbf{i}, \mathbf{j}, \mathbf{k}$ denoting the unit vectors on the x-, y- and z-axes respectively:

$$\mathbf{k} = \mathbf{i} \wedge \mathbf{j}; \qquad \mathbf{i} = \mathbf{j} \wedge \mathbf{k}; \qquad \mathbf{j} = \mathbf{k} \wedge \mathbf{i} \tag{A.3}$$

The direction of vectorial products depends on whether the trihedron $(\mathbf{i}, \mathbf{j}, \mathbf{k})$ is right-handed (as usual) or left-handed. The *work* of the vector ***a*** from A to B along a curve (C) is the sum of the infinitesimal scalar products of the small segments $d\mathbf{s}$ along (C):

$$\int_A^B \mathbf{a} \cdot d\mathbf{s} = \int_A^B (a_x\, dx + a_y\, dy + a_z\, dz) \tag{A.4}$$

If (C) is a closed curve (B = A), the position of A in (C) is not important; only the direction of the trajectory is important. Work is then called *circulation* of ***a*** on (C), and written:

$$\oint \mathbf{a} \cdot d\mathbf{s} \tag{A.5}$$

Consider a continuous surface (S), bounded by a contour (C). Let $\boldsymbol{n}$ be the normal unit vector, always on the same side (one-sided surfaces as Moebius strips are excluded). The *flux* of the vector $\boldsymbol{a}$ through (S) is the scalar:

$$\varphi = \iint_S \mathbf{a} \cdot \mathbf{n}\, dS \tag{A.6}$$

Gradient. The gradient of a function $f(x, y, z)$ is a vector with components equal to the x-, y- and z-partial derivatives respectively. Introducing the symbolic vector ∇, it reads:

$$\mathbf{grad}\, f = \nabla f = \begin{bmatrix} \partial f/\partial x \\ \partial f/\partial y \\ \partial f/\partial z \end{bmatrix} \tag{A.7}$$

(Beware: the vector ∇ can only be used in Cartesian coordinates, and even then the application of the rules of multiplication can lead to mistakes.)

The work of $\mathbf{grad}\, f$ from A to B equals the increase of f, whatever path has been followed. Consequently: (1) *the circulation of a gradient is always zero*; (2) *the force lines of* $\mathbf{grad}\, f$ *are normal to any surface* $f = a$ *constant* (i.e. they cross them perpendicularly).

The gravity field from a material point, the electrostatic field due to an electric charge, the excitation vector due to an isolated magnetic pole are all gradients of potential $-C/r$, C being a constant and r being the distance to the acting point. These fields are radial, of intensity C/r^2. It is easily demonstrated that the flux out of a closed surface is $4\pi C$ when the acting point is inside the surface, and zero when it is outside. With many acting points, the fields are adding vectorially, but the potentials and fluxes are adding algebraically. Consequently:

- there are functions (the respective potentials) such that any gravity field, any electrostatic field, etc., are its gradients;
- when a field results from centrifugal radial force Cr^{-2}, the flux out of a closed surface equals $4\pi \times$ (sum of all Cs for the points inside the surface) (*Gauss theorem*).

Curl. The *curl* of vector $\boldsymbol{a}$ is the vector:

$$\mathbf{curl\, a} = \nabla \wedge \mathbf{a} = \begin{bmatrix} \partial a_z/\partial y - \partial a_y/\partial z \\ \partial a_x/\partial z - \partial a_z/\partial x \\ \partial a_y/\partial x - \partial a_x/\partial y \end{bmatrix} \tag{A.8}$$

Consider the surface (S) bounded by a contour (C) and let us link the positive direction on (C) with the direction of $\boldsymbol{n}$ by the right-handed screw convention: one advances in the direction of n by turning in the direction of (C) , as with a corkscrew. It can be demonstrated that the circulation of a vector along (C) equals the flux of its curl through (S) (*Stokes theorem*):

$$\oint_C \mathbf{a} \cdot d\mathbf{s} = \iint_S \mathbf{curl\ a} \cdot \mathbf{n}\, ds \tag{A.9}$$

The curl of any gradient is always zero:

$$\mathbf{curl}(\mathbf{grad}\, f) = \nabla \wedge \nabla \cdot f = 0 \tag{A.10}$$

Divergence. The divergence of a vector $\boldsymbol{a}$ is the scalar:

$$\text{div}\ \mathbf{a} = \nabla \cdot \mathbf{a} = \frac{\partial a_x}{\partial x} + \frac{\partial a_y}{\partial y} + \frac{\partial a_z}{\partial z} \tag{A.11}$$

The divergence of a curl is always zero:

$$\text{div}(\mathbf{curl\ a}) = \nabla \cdot \nabla \wedge \mathbf{a} = 0 \tag{A.12}$$

The flux of a vector out of a closed surface (S) limiting a volume (V) equals the sum of its divergence in the entire volume (V) (*Ostrogradski theorem*):

$$\iint_S \mathbf{a} \cdot \mathbf{n}\, ds = \iiint_V \text{div}\ \mathbf{a}\, dV \tag{A.13}$$

In particular, when $\boldsymbol{a} = \mathbf{grad}$, we must sum in (V) the *Laplacian* of f:

$$\text{div}(\mathbf{grad}\, f) = \nabla^2 f = \frac{\partial^2 f}{\partial x^2} + \frac{\partial^2 f}{\partial y^2} + \frac{\partial^2 f}{\partial z^2} \tag{A.14}$$

When the flux of a vector coming out of a closed surface is zero, this vector is said to be *conservative*. If we consider a tube whose walls are force lines (*force tube*), the flux through a cross-section of the force tube will be the same anywhere along the tube. (To prove it, consider the closed surface formed by two cross-sections and the portion of tube between them.)

Divergence and curl of a product (scalar × vector)

The expressions with Cartesian coordinates are used to demonstrate the following identities:

$$\begin{cases} \text{div}(f\mathbf{a}) = f\ \text{div}\ \mathbf{a} + \mathbf{grad}\, f \cdot \mathbf{a} \\ \nabla \cdot (f\mathbf{a}) = f(\nabla \cdot \mathbf{a}) + (\nabla f) \cdot \mathbf{a} \end{cases} \tag{A.15}$$

$$\begin{cases} \mathbf{curl}(f\mathbf{a}) = f\ \mathbf{curl\ a} + \mathbf{grad}\, f \wedge \mathbf{a} \\ \nabla \wedge (f\mathbf{a}) = f(\nabla \wedge \mathbf{a}) + (\nabla f) \wedge \mathbf{a} \end{cases} \tag{A.16}$$

Vector gradient and vector Laplacian. The vector gradient $\nabla \boldsymbol{a}$ (not to be confused with $\nabla \cdot \boldsymbol{a}$) is neither a vector nor a scalar, but a matrix:

$$\nabla a = \begin{bmatrix} \dfrac{\partial}{\partial x} \\ \dfrac{\partial}{\partial y} \\ \dfrac{\partial}{\partial z} \end{bmatrix} [a_x \quad a_y \quad a_z] = \begin{bmatrix} \dfrac{\partial a_x}{\partial x} & \dfrac{\partial a_x}{\partial y} & \dfrac{\partial a_x}{\partial z} \\ \dfrac{\partial a_y}{\partial x} & \dfrac{\partial a_y}{\partial y} & \dfrac{\partial a_y}{\partial z} \\ \dfrac{\partial a_z}{\partial x} & \dfrac{\partial a_z}{\partial y} & \dfrac{\partial a_z}{\partial z} \end{bmatrix} \tag{A.17}$$

Multiplying the vector ∇ on the left by this matrix, we get a genuine vector, with the Cartesian components $\nabla^2 a_x, \nabla^2 a_y, \nabla^2 a_z$. This is the *vector Laplacian*, which should be written $\nabla \boldsymbol{a} \nabla$ but is noted $\nabla^2 \boldsymbol{a}$. The following identity shows the vector Laplacian is an invariant relative to any change in the coordinate axes, like the gradient, the divergence and the curl.

$$\begin{cases} \nabla^2 \mathbf{a} = \mathbf{grad}(\text{div}\ \mathbf{a}) - \mathbf{curl}(\mathbf{curl\ a}) \\ \nabla^2 \mathbf{a} = \nabla(\nabla \cdot \mathbf{a}) - \nabla \wedge (\nabla \wedge \mathbf{a}) \end{cases} \tag{A.18}$$

A.2 SPHERICAL HARMONICS

Any function defined *on a circle*, and therefore periodic function of the angle φ to the centre, can be decomposed uniquely in a Fourier series:

$$F(\varphi) = \sum_{m=0}^{+\infty} A_m \cos m\varphi + B_m \sin m\varphi \tag{A.19}$$

Similarly, any function defined *on a sphere* can be decomposed uniquely in a double series of *(surface) spherical harmonics*. Noting θ the colatitude and φ the longitude, we get the decomposition:

$$F(\theta, \varphi) = \sum_{n=0}^{+\infty} \sum_{m=0}^{n} N_{nm} P_{nm}(\cos\theta) \cdot [(-1)^m A_{nm} \cos m\varphi + B_{nm} \sin m\varphi] \tag{A.20}$$

A_{nm} and B_{nm} are coefficients with the dimensions of F. The P_{n0} are the Legendre functions, the other P_{nm} are the associated Legendre functions. N_{nm} is a normalization factor. Calling Y_i the surface harmonic $(-1)^m N_{nm} P_{nm} \cos\varphi$ or $N_{nm} P_{nm} \sin m\varphi$, and $dS = d(\cos\theta)/d\varphi$ the infinitesimal area on the unit sphere (S), we have:

$$\begin{cases} \displaystyle\iint_S Y_i Y_j \, dS = 0 \qquad (i \neq j) \\ \displaystyle\iint_S Y_i^2 \, dS = 1 \end{cases} \tag{A.21}$$

The Y_i are *orthogonal* and *normalized*. There follows the following value of the A_{nm} and B_{nm} (designated by C_i), proving that the decomposition in spherical harmonics is always mathematically possible and is unique, although, in practice, it cannot be used to determine the coefficients C_i:

$$C_i = \iint_S FY_i \, dS \tag{A.22}$$

n is the *degree* of the harmonic and m is its *order*. If we perform again the decomposition in spherical harmonics with a different pole for the spherical coordinates, the new A_{nm} and B_{nm} are linear expressions of the old coefficients, *with the same degree*. The P_{nm} of lower degrees and their normalization factors are:

$$\begin{cases} P_{00} = 1; \qquad P_{10} = \cos\theta; \qquad P_{11} = \sin\theta \\[2ex] N_{00} = \sqrt{\dfrac{1}{4\pi}}; \qquad N_{10} = N_{11} = \sqrt{\dfrac{3}{4\pi}} \\[2ex] P_{20} = \frac{3}{2}\cos^2\theta - \frac{1}{2}; \qquad P_{21} = 3\sin\theta\cos\theta; \qquad P_{22} = 3\sin^2\theta \\[2ex] N_{20} = \sqrt{\dfrac{5}{4\pi}}; \qquad N_{21} = \sqrt{\dfrac{5}{12\pi}}; \qquad N_{22} = \sqrt{\dfrac{5}{48\pi}} \end{cases} \tag{A.23}$$

A.3 ERROR FUNCTION

It can be demonstrated by integrating $\exp[-(x^2 + y^2)]$ on the entire plane (x, y), in Cartesian coordinates and in polar coordinates, that:

$$\int_0^{+\infty} e^{-\zeta^2} \, d\zeta = \frac{\sqrt{\pi}}{2} \tag{A.24}$$

This result explains the definition of the *error function* $\operatorname{erf} x$ and the *complementary error function* $\operatorname{erfc} x$, used mainly in statistics:

$$\begin{cases} \operatorname{erf} x = \dfrac{2}{\sqrt{\pi}} \displaystyle\int_0^x e^{-\zeta^2} \, d\zeta \\[2ex] \operatorname{erfc} x = 1 - \operatorname{erf} x = \dfrac{2}{\sqrt{\pi}} \displaystyle\int_x^{+\infty} e^{-\zeta^2} \, d\zeta \end{cases} \tag{A.25}$$

The limit value for $x \to +\infty$ is reached very fast, as shown by the few values below:

x	0	0.477 ...	1.163 ...	1.321 ...	$+\infty$
Erf x	0	0.5	0.9	0.99	1
Erfc x	1	0.5	0.1	0.01	0

An integration by parts shows that:

$$\int_0^{+\infty} \operatorname{erfc} x \, dx = \frac{1}{\sqrt{\pi}} \tag{A.26}$$

A.4 FOURIER TRANSFORM

Imaginary numbers and the identity $e^{i\omega x} = \cos \omega x + i \sin \omega x$ are assumed to be known to the reader. We write z^* the imaginary conjugate of z.

The Fourier transform (FT) of a function $f(x)$ can be defined in two ways, respectively written $F(\nu)$ and $\bar{f}(\omega)$. The new variable $\nu = 1/\lambda$ is the *wave number* (frequency, if x is the time), and the new variable $\omega = 2\pi\nu$ is the *pulsatance*. The first definition is used in signal processing and makes the properties of the FT easier to remember. The second definition is used more often in this book, and makes the writing easier when derivatives of $f(x)$ intervene.

Any function $f(x)$ indefinitely differentiable and decreasing towards ∞ faster than any negative power of $|x|$, can be expressed as:

$$f(x) = \int_{-\infty}^{+\infty} F(\nu) \, e^{i2\pi\nu x} \, d\nu \tag{A.27}$$

It is demonstrated that:

$$F(\nu) = \int_{-\infty}^{+\infty} f(x) \, e^{-i2\pi\nu x} \, dx \tag{A.28}$$

Note that if the FT of $f(x)$ is $F(\nu)$, the FT of $F(x)$ is $f(-\nu)$. Writing $2\pi\nu = \omega$, the equations (A.27) and (A.28) become:

$$\begin{cases} f(x) = \displaystyle\int_{-\infty}^{+\infty} \left[\frac{1}{2\pi} F\left(\frac{\omega}{2\pi}\right)\right] e^{i\omega x} \, d\omega = \int_{-\infty}^{+\infty} \bar{f}(\omega) \, e^{i\omega x} \, d\omega \\ \bar{f}(\omega) = \displaystyle\frac{1}{2\pi} F\left(\frac{\omega}{2\pi}\right) = \frac{1}{2\pi} \int_{-\infty}^{+\infty} f(x) \, e^{-i\omega x} \, dx \end{cases} \tag{A.29}$$

Looking for the FT of a signal is performing its *spectrum analysis*. It can be done when the signal is periodic and therefore the integral of (A.28) is not converging, but the FT in this case is not a real function any more. It is a *distribution*, a mathematical being that will not be introduced here.

If $f(x)$ is real and symmetrical in x, $F(\nu)$ and $\bar{f}(\omega)$ are real, and $F(-\nu) = [F(\nu)]^*$, $\bar{f}(-\omega) = [\bar{f}(\omega)]^*$. The equations (A.27) and (A.28) then become:

$$\begin{cases} f(x) = 2\displaystyle\int_0^{+\infty} F(\nu) \cos 2\pi\nu x \, d\nu = 2\int_0^{+\infty} \bar{f}(\omega) \cos \omega x \, d\omega \\ F(\nu) = 2\displaystyle\int_0^{+\infty} f(x) \cos 2\pi\nu x \, dx \\ \bar{f}(\omega) = \displaystyle\frac{1}{\pi}\int_0^{+\infty} f(x) \cos \omega x \, dx \end{cases} \tag{A.30}$$

If $f(x)$ is real and antisymmetrical in x, then $F(\nu)$ and $\bar{f}(\omega)$ are purely imaginary and:

$$\begin{cases} f(x) = 2i \int_0^{+\infty} F(\nu) \sin 2\pi\nu x \, d\nu = 2i \int_0^{+\infty} \bar{f}(\omega) \sin \omega x \, d\omega \\ F(\nu) = -2i \int_0^{+\infty} f(x) \sin 2\pi\nu x \, dx \\ \bar{f}(\omega) = -\frac{i}{\pi} \int_0^{+\infty} f(x) \sin \omega x \, dx \end{cases} \quad \text{(A.31)}$$

Parseval theorem and FT of a convolution

Most of the FT properties derive from *Parseval theorem*. Representing FTs with upper-case letters, this theorem states that:

$$\int_{-\infty}^{+\infty} f(x)[g(x)]^* \, dx = \int_{-\infty}^{+\infty} F(\nu)[G(\nu)]^* \, d\nu \quad \text{(A.23)}$$

Let us write: $[G(\nu)]^* = H(\nu)\, e^{i2\pi\nu t}$. According to equation (A.27), and taking the imaginary conjugates:

$$[g(x)]^* = \int_{-\infty}^{+\infty} [G(\nu)]^* \, e^{-i2\pi\nu x} \, d\nu = \int_{-\infty}^{+\infty} H(\nu)\, e^{i2\pi\nu(t-x)} \, d\nu = h(t-x) \quad \text{(A.33)}$$

The Parseval theorem reads then:

$$f \otimes h = \int_{-\infty}^{+\infty} f(x)h(t-x) \, dx = \int_{-\infty}^{+\infty} F(\nu)H(\nu)\, e^{i2\pi\nu t} \, dt \quad \text{(A.34)}$$

The term on the left-hand side is the *convolution* of functions f and h (here, a function of t). Its FT is FH, the product of the TFs of f and h.

2-D Fourier transforms

These results extend to a function of two variables $f(x, y)$, indefinitely differentiable and decreasing towards ∞ faster than any negative power of $r = (x^2 + y^2)^{1/2}$.

$$\begin{cases} f(x, y) = \int_{-\infty}^{+\infty} d\nu_x \int_{-\infty}^{+\infty} d\nu_y F(\nu_x, \nu_y) \exp[i2\pi(\nu_x x + \nu_y y)] \\ F(\nu_x, \nu_y) = \int_{-\infty}^{+\infty} dx \int_{-\infty}^{+\infty} dy\, f(x, y) \exp[i2\pi(\nu_x x + \nu_y y)] \end{cases} \quad \text{(A.35)}$$

The 2-D convolution of the two functions $f(x, y)$ and $h(x, y)$ is the following function of x' and y':

$$f \otimes h = \int_{-\infty}^{+\infty} dx \int_{-\infty}^{+\infty} dy\, f(x, y)h(x' - x, y' - y) \quad \text{(A.36)}$$

It can be written:

$$f \otimes h = \int_{-\infty}^{+\infty} d\nu_x \int_{-\infty}^{+\infty} d\nu_y F(\nu_x, \nu_y) H(\nu_x, \nu_y) \exp[i2\pi(\nu x' + \nu_y y')] \tag{A.37}$$

It follows that the FT of the convolution $f \otimes h$ is the product of the Fourier transforms of f and h.

We can write $\omega_x = 2\pi\nu_x$ and $\omega_y = 2\pi\nu_y$ to define the Fourier transform:

$$\bar{f}(\omega_x, \omega_y) = \frac{1}{4\pi^2} F\left(\frac{\omega_x}{2\pi}, \frac{\omega_y}{2\pi}\right) \tag{A.38}$$

A.5 IMPULSE RESPONSE AND LAPLACE TRANSFORM

Impulse response

The *unit step* function (Heaviside function) $\Upsilon(t)$ is 0 when $t < 0$ and 1 when $t > 0$. Any function $f(t)$ null for $t = -\infty$ may be considered as the sum of increments $df\,\Upsilon(t')$ every dt', from $t' = -\infty$ to $t' = t$:

$$f(t) = \int_{-\infty}^{t} \left.\frac{df}{dt}\right|_{t=t'} \Upsilon(t')\, dt' \tag{A.39}$$

If the response of a *linear* system to an excitation $\Upsilon(t)$ is $R(t)$, its response to $f(t)$ is given by the *convolution*:

$$\frac{df}{dt} \otimes R = \int_{-\infty}^{t} \left.\frac{df}{dt}\right|_{t=t'} R(t - t')\, dt' \tag{A.40}$$

We assume that $R(t)$ tends toward 0 for $t \to \infty$ (the system is stable for this type of excitations). Integrating by parts, we find:

$$\frac{df}{dt} \otimes R = \int_{-\infty}^{t} f(t') \left.\frac{dR}{dt}\right|_{t=t'} dt' = f \otimes \frac{dR}{dt} \tag{A.41}$$

The derivative dR/dt is known with more accuracy than the derivative df/dt, because R results from a purely mathematical calculation, whereas f results from field measurements with many errors. dR/dt is the *impulse response* of the system. It is the response to a *Dirac distribution* $\delta(t)$, which must be considered in these mathematical operations as the derivative of $\Upsilon(t)$.

Laplace transforms

Laplace transforms are interesting to calculate convolutions. The Laplace transform of a function $f(t)$ is the function of s defined by the following integral:

$$\mathfrak{L}f(t) = \tilde{f}(s) = \int_{0}^{+\infty} e^{-ts} f(t)\, dt \tag{A.42}$$

The Laplace transform exists only if this integral is converging for any value of s. This is the case if $f(t)$ equals 0 for $t < t_1$ and remains finite for $t > t_1$.

Once $\tilde{f}(s)$ is known, $f(t)$ can be obtained through integration of $e^{ts}f(s)$ along specific curves in the complex plane (they must be at the right of any singularity of $f(s)$).

The similarity of the Laplace and Fourier transforms (with $\omega = is$) explains why the main properties are the same (with a slight difference for the derivatives):

$$\begin{cases} \mathfrak{L}(c_1 f_1 + c_2 f_2 + \ldots) = c_1 \tilde{f}_1 + c_2 \tilde{f}_2 + \ldots \\ \mathfrak{L}(f \otimes g) = \tilde{f}\tilde{g} \\ \mathfrak{L}\left(\dfrac{df}{dt}\right) = s\tilde{f} - f(+0) \end{cases} \tag{A.43}$$

Thus, the equation $df/dt = -\alpha f$ is expressed with Laplace transforms as:

$$\tilde{f} = \frac{f(+0)}{s + \alpha} \tag{A.44}$$

Tables of Laplace transforms tell us that:

$$\mathfrak{L}(\Upsilon(t)\, e^{-\alpha t}) = \frac{1}{s + \alpha} \tag{A.45}$$

And therefore $f = f(+0)\Upsilon(t)\exp(-\alpha t)$. When the resulting Laplace transform is not in the tables, or is obtained numerically, inversion is always possible, to go from $f(s)$ to $f(t)$, by a numerical quadrature in the complex plane.

Synoptic table of the Earth's history.

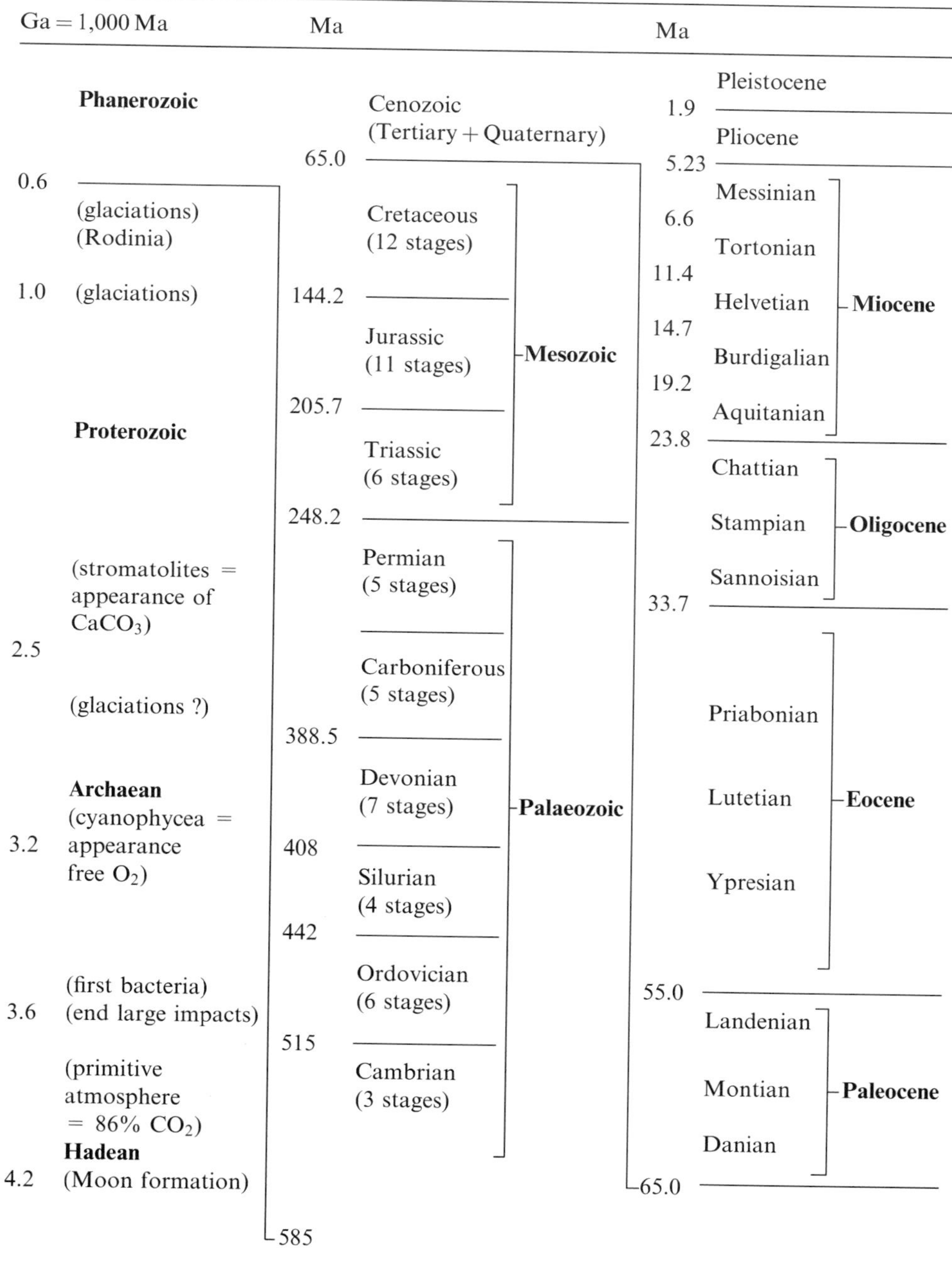

4.55 Earth formation (= isotopic homogenization)

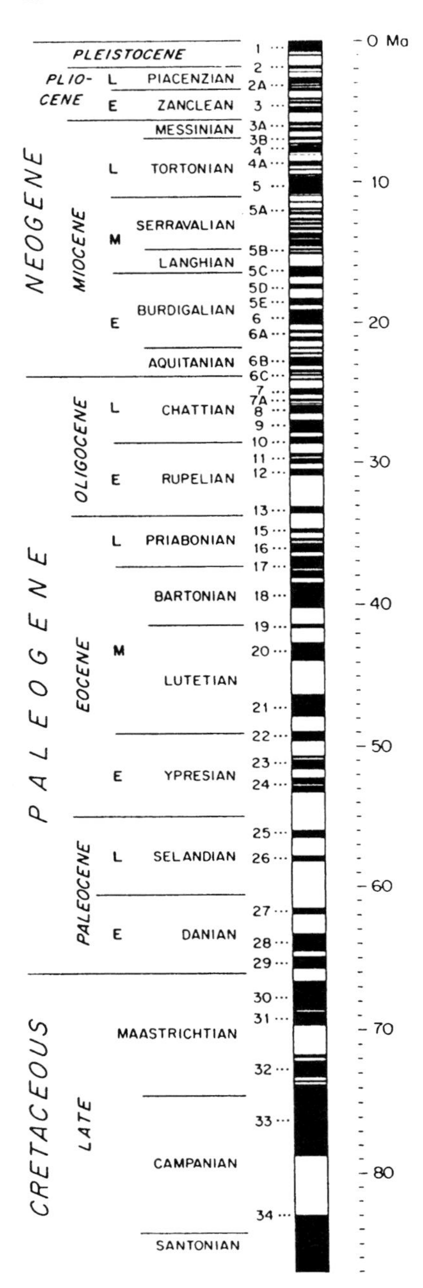

L = Late (Upper), E = Early (Lower). The Lower Oligocene divides into Stampian and Sannoisian; the Lower Palaeocene (Selandian) into Landenian and Montian.

Figure A.1. Palaeomagnetic time scale (chrons), compared to the radio-chronologic time scale and the stratigraphic scale (stages, based on marine fossils and microfossils). From Cande and Kent, 1992.

Bibliography

Abramowitz and Stegun, *Handbook of Mathematical Functions*, Dover, 1965
Albarède and Condomines, *La Géochimie*, PUF 'Que sais-je' no. 759, 1976
Allègre, C., *La dérive des continents*, Scientific American – Pour la Science, 1980
Allègre and Michard, *Introduction à la géochimie*, PUF, 1973
Anderson, D.L. in Sabadini, Lambeck and Boschi (eds.), *Glacial Isostasy, Sea Level and Mantle Rheology*, NATO ASI Series C-334, Kluwer Acad. Publ., pp. 379–401, 1991
Anderson, D.L., *Review of Geophysics*, **33** (1), 125–149 AGU, 1995
Anderson and Bass, *Nature*, **320**, 321–328, 1986
Anderson and Spitzler, *Physics of the Earth and Planetary Interiors*, **4**, 62–64, 1970
Argenio *et al.*, in *Géologie des chaînes alpines issues de la Téthys*, BRGM, 1980
Argus and Gordon, *Journal of Geophysical Research*, **95** (B11), 17,315–17,324, 1990
Argus and Gordon, *Geophysical Research Letters*, **18**, 2,039–2,042, 1991
Atkinson, *Journal of Geophysical Research*, **89** (B6), 4,077–4,114, 1984
Atwater, *Bulletin of the Geological Society of America*, **81**, 3,513–3,536, 1970
Auboin, *Geosynclines*, Elsevier, 1965
Audrey-Charles *et al.*, *Géologie des chaînes alpines issues de la Tethys*, BRGM: Orleans, 1980
Autran and Peterlongo, Massif Central, in *Géologie des Pays Européens: France, Belgique, Luxembourg*, Dunod, 1980
Avouac *et al.*, *Journal of Geophysical Research*, **98** (B4), 6,755–6,804, 1993
Ballèvre *et al.*, in Roure, Heitzmann and Polino (eds), *Deep Structure of the Alps*, Mémoire de la Société Géologique de France, no. 156, 1990, pp. 227–236
Banerdt and Sammis, *Physics of the Earth and Planetary Interiors*, **41**, 108–124, 1985
Baranzagi and Isacks, *Journal of Geophysical Research*, **76**, 8,493–8,516, 1971
Bard *et al.*, *Nature*, **345**, 405–410, 1990
Barrell, *Journal of Geology (University of Chicago)*, **22**, 655–683, 1914
Bergé, Pomeau and Vidal, *L'ordre dans le Chaos*, Hermann: Paris, 1984
Berger, *Review of Geophysics*, **26** (4), 624–657, 1988
Bina and Helffrich, *Journal of Geophysical Research*, **99** (B8), 15,863–15,890, 1994

Bina and Silver, *Geophysical Research Letters*, **17**, 1,153–1,156, 1990
Biot, *Geological Society of America Bulletin*, **72**, 1,595–1,620, 1961
Blanpied *et al.*, *Geophysical Research Letters*, **18**, 609–612, 1992
Bottinga and Allègre, *Philosophical Transactions of the Royal Society of London*, A, **288**, 501-525, 1978
Bowen, *The Evolution of Igneous Rocks*, Princeton University Press, 1928
Bureau des Longitudes; *La Terre, les eaux, l'atmosphère*, Gauthiers-Villars, 1977
Busse, in: Peltier ed., *Mantle Convection*, Gordon & Breach Sci. Publ., pp. 23–95, 1989
Caffee and Hudson, *Lunar and Planet. Science*, **18**, 145–146, 1987
Cagniard, *Annales de Géophysique*, **9**, 95–125, 1953
Calais, E. *et al.*, Evolution paléogéographique et structurale du domaine Caraïbe du Lias à l'Actuel: 14 étapes pour 3 grandes périodes, *C.R. Acad. Sci. Paris*, **309** (II), 1,437–1,444, 1989
Cande and Kent, *Journal of Geophysical Research*, **97** (B10), 13,917–13,951, 1992
Cande and Kent, *Journal of Geophysical Research*, **100** (B4), 6,093–6,095, 1995
Cara, Thèse d'Etat, Univ. Paris-VI, France, 1978
Carbotte and McDonald, *Journal of Geophysical Research*, **99** (B7), 13,609–13,632, 1994
Carlson, *Review of Geophysics*, **32** (4), 337–361, 1944
Carlson and Johnson, *Journal of Geophysical Research*, **99**, B2, 3,201–3,214, 1994
Carter, *Review of Geophysics and Space Physics*, **14**, 301–360, 1976
Carter, N.L., M.C. Tsenn, *Tectonophysics*, **136**, 27–63, 1987
Castelnau *et al.*, *Journal of Geophysical Research*, **101** (B6), 13,851–13,868, 1996
Cathles, *The Viscosity of the Earth's Mantle*, Princeton Univ. Press, 1975
Chapple and Forsyth, *Journal of Geophysical Research*, **84**, 6,729–6,749, 1979
Charlesworth, *The Quaternary Era*, E. Arnold: London, 1957
Chase, *Geophysical Journal of the Royal Astronomical Society*, **29**, 117–122, 1972
Chase, *Earth and Planetary Science Letters*, **37**, 355–368, 1978
Chemenda *et al.*, *Earth and Planetary Science Letters*, **132**, 225–232, 1995
Chen *et al.*, *Journal of Geophysical Research*, **98** (B12), 21,927–21,941, 1993
Chen *et al.*, *Journal of Geophysical Research*, **99** (B2), 2,909–2,924, 1994
Chen, Courtillot *et al.*, *Journal of Geophysical Research*, **98** (B12), 21,927–21,941, 1993
Chester *et al.*, *Journal of Geophysical Research*, **98**, 771–786, 1983
Chopin, *Contrib. Mineral. Petrol.*, **86**, 107–118, 1984
Choy and Boatwright, *Journal of Geophysical Research*, **100** (B9), 18,205–18,228, 1995
Christensen, *Geophysical Journal of the Royal Astronomical Society*, **77**, 343–384, 1984
Christensen and Hofmann, *Journal of Geophysical Research*, **99** (B10), 19,867–19,884, 1994
Christensen and Mooney, *Journal of Geophysical Research*, **100** (B6), 9,761–9,788, 1995
Corchete *et al.*, *Journal of Geophysical Research*, **100** (B12), 24,133–24,146, 1995

Cormier and McDonald, *Journal of Geophysical Research*, **99** (B1), 543-564, 1994
Coulomb and Jobert (eds), *Traité de Geophysique interne, 1. Sismologie et pesanteur*, 1973; *2. Magnétisme et géodynamique*, 1976, Masson.
Courtillot, *Tectonics*, **1**, 239–250, 1982
Courtillot and Besse, *Science*, **237**, 1,140–1,147, 1987
Cunningham *et al.*, *Journal of Geophysical Research*, **100** (B5), 8,257–8,266, 1995
Daly, *Igneous Rocks and the Depths of the Earth*, McGrawHill, 1933
Dalziel, *Annual Review of Earth and Planetary Science*, **20**, 501–526, 1992
Dana, *American Journal of Science*, **5**, 423–443, 1873
Dansgaard, *Meddelelser om Grönland*, **165**, 1961, 120 pp.
Davies, *Journal of Geophysical Research*, **85** (B5), 2,517–30, 1980
Davies, *Journal of Geophysical Research*, **93** (B9), 10,451–10,488, 1988
Debelmas, *Géologie de la France*, **2**, Doin, 1974
DeMets *et al.*, *Geophysical Journal International*, **101**, 425–478, 1990
Dercourt *et al.*, Geological evolution of the Tethys belt from the Atlantic to the Pamirs since the Lias, *Tectonophysics*, **123**, 241–315, 1986
Detrick *et al.*, *Nature*, **292**, 142–143, 1981
Dewey and Burke, *Journal of Geology* (University of Chicago), **81**, 683–692, 1973
Dickinson and Snyder, *Journal of Geophysical Research*, **84**, 561–572, 1979
Dieterich, *Journal of Geophysical Research*, **84**, 2,161–2,168, 1979
Dixon *et al.*, *Phil. Mag.*, **45**, 519–529, 1982
Doucouré *et al.*, *Journal of Geophysical Research*, **101** (B5), 11,291-11,303, 1996
Drury and Urai, *Tectonophysics*, **172**, 235–253, 1990
Duncan and Richards, *Review of Geophysics*, **29** (1), 31–50, 1991
Dunlop, *Journal of Geophysical Research*, **100** (B2), 2,161–2,174, 1995
Dupré and Allègre, C., *Nature*, **286**, 17–22, 1980
Duval, J. *Glaciology*, **27**, 129–140, 1981
Dziewonski *et al.*, *Physics of the Earth and Planetary Interiors*, **10**, 12–48, 1975
Dziewonski and Anderson, *Physics of the Earth and Planetary Interiors*, **25**, 297–356, 1981
Ehlers (ed), *Glacial deposits in NW Europe*, Balkema, 1983
Emiliani, *Journal of Geology*, **66**, 264–275 and 572–577, 1958
Engdahl *et al.*, *Tectonophysics*, **37**, 95–116, 1977
Engdahl and Scholz, *Geophysical Research Letters*, **4**, 473–476, 1977
Enkin *et al.*, *Journal of Geophysical Research*, **97** (B10), 13,953–13,985, 1992
EOS Trans. AGU, **61** (4), 1980
Fallon and Dillinger, *Journal of Geophysical Research*, **97** (B5), 7,129–7,136, 1992
Farrell and Clark, *Geophysical Journal of the Royal Astronomical Society*, **46**, 647–667, 1976
Fleitout and Yuen, *Journal of Geophysical Research*, **89**, B11, 9,227–9,244, 1984
Forsyth, *Journal of Geophysical Research*, **90** (B14), 12,623–12,632, 1985
Forsyth and Uyeda, *Geophysical Journal of the Royal Astronomical Society*, **43**, 163–200, 1975
Gansser, *Geologie des chaines alpines issues de la Thetys*, BRGM: Orléans, 1980
Gast, *Geochemica and Cosmochemica Acta*, **32**, 1,057–1,087, 1968

Gignoux, *Bulletin de la Société Géologique de France*, 18, 739–761, 1948
Girdler and Styles, in: Pilger and Rösler (eds), *Afar between Continental and Oceanic Rifting*, Vol. 2, Schweizerbartsche Verlagsbuchhandlung, Stuttgart, 1976
Glatzmaier and Roberts, *Nature*, **377**, 203–209, 1995
Glatzmaier and Schubert, *Journal of Geophysical Research*, **98** (B12), 21,969–21,976, 1993
Goguel, *Traité de tectonique*, Masson & Cie, 1952
Gordon and Stein, Global tectonics and space geodesy, *Science*, **256**, 333–342, 1992
Gradstein *et al.*, *Journal of Geophysical Research*, **99** (B12), 24,051–24,074, 1994
Granlund, E., *Sveriges Geologi*, 2nd edn, 1949
Gratier, Ménard and Arpin, in Courand, Dietrich and Park (eds), *Alpine tectonics*, 1989
Green and Ringwood, *Contrib. Mineral. and Petrol.*, **15**, 103–190, 1967
Gripp and Gordon, *Geophysical Research Letters*, **17**, 1,109–1,112, 1990
Grosswald, *Quaternary Research*, **13**, 1–32, 1980
Grunow, *Journal of Geophysical Research*, **98**, (B8), 13,815–13,833, 1993
Gubbins and Master, *Adv. In Geophys.*, **21**, 1–50, 1979
Guinot, in Coulomb and Jobert (eds), *Traité de géophysique interne*, 1, Masson, p. 529–564, 1973
Gutenberg and Richter, *Seismicity of the Earth and associated phenomena*, 2nd edn, Princeton University Press, 1954
Gzovsky, *Geophysical Journal of the Royal Astronomical Society*, **14**, 331–339, 1967
Haarmann, *Die Oscillationstheorie*, Ferdinand Enke, Stuttgart, 1930
Hager, *Journal of Geophysical Research*, **89** (B7), 6,003–6,015, 1984
Hales, *Earth and Planetary Science Letters*, **6**, 31–34, 1969
Hall, *Natural History of New York*, Appleton, 1859
Hallam, *Nature*, **225**, 139–144, 1970
Hallet *et al.*, *Permafrost and Periglacial Processes*, **2**, 283-300, 1991
Hamilton, in Talwani and Pitman (eds), *Island arcs, deep sea trenches and back-arc basins*, Maurice Ewing Series no. 1, AGU, 1977
Hanks, *Geophysical Journal of the Royal Astronomical Society*, **23**, 173–189, 1971
Harper, *Rev. Aquatic Sci.*, **1**, 319–336, 1989
Haug, *Bulletin de la Société Géologique de France*, **28**, 617–711, 1900
Heiskanen and Vening Meinesz, *The Earth and its gravity field*, McGraw-Hill, 1958
Herrin *et al.*, *Bulletin of the Seismological Society of America*, **58**, 1,196–2,141, 1968
Hey *et al.*, *Journal of Geophysical Research*, **85** (B7), 3,647–3,658, 1980
Hickman *et al.*, *Journal of Geophysical Research*, **100** (B7), 12,831–12,840, 1995
Higashi *et al.*, *J. of Glaciology*, **21**, 589–620, 1978
Hill, *The mathematical theory of plasticity*, Clarendon Press: Oxford, 1980
Hirn *et al.*, *Phil. Trans. Roy. Soc.* London, A **326**, 17–32, 1988
Hobbs, *Tectonophysics*, **92**, 35–69, 1983
Hofmann, *Ann. Rev. Earth Planet. Sci.*, **16**, 543–603, 1988
Houston and De Bremaecker, *Journal of Geophysical Research*, **80**, 762–750, 1975
Imbie *et al.*, in Berger *et al.* (eds), *Milankovitch and climate*, Reidel, 1984, pp. 269–305

Irving, *Nature*, **270**, 304–309, 1977
Isacks and Molnar, *Review of Geophysics and Space Physics*, **9**, 103–174, 1971
Isacks *et al.*, *Journal of Geophysical Research*, vol. 73, p. 5,855–5,899, 1968
Ito *et al.*, *Journal of Geophysical Research*, **100** (B4), 5,901–5,910, 1995
Jackson and Pollack, *Journal of Geophysical Research*, **89**, 10,103–10,108, 1984
James and Ivins, *Geophysical Research Letters*, **22** (8), 973–976, 1995
Jamison and Cook, *Journal of Geophysical Research*, **85** (B4), 1,833–1,838, 1980
Jaoul, Thèse d'Etat, Univ. Paris Orsay, 1980
Jarvis and Peltier, in Peltier (ed.), *Mantle convection*, Gordon & Breach Sci. Publ., 1989
Jaupart and Parsons, *Physics of the Earth and Planetary Interiors*, **39**, 14–32, 1985
Javoy, *Geophysical Research Letters*, **22**, 2,219–2,222, 1995
Jeanloz, in: Peltier (ed), *Mantle convection*, Gordon & Breach Sci. Publ., 203–259, 1989
Jeanloz and Morris, *Ann. Rev. Earth Planet. Sci.*, **14**, 377, 1986
Jeffreys, *The Earth*, 3rd edn., Cambridge University Press, 1952
Jeffreys and Bullen, Seismological tables, *British Assoc. Adv. Sci.*, Gray-Milne Trust, 1940
Jolivet *et al.*, *Journal of Geophysical Research*, **99** (B11), 22,237–22,259, 1994
Jordan, *Journal of Geophysics*, **43**, 473–496, 1977
Jordan *et al.* in Peltier (ed.), *Mantle convection*, Gordon & Breach, 1989, pp. 97–201
Jurdy, *Journal of Geophysical Research*, **84** (B12), 6,796–6,802, 1979
Jurdy *et al.*, *Journal of Geophysical Research*, **100** (B9), 17,965–17,975, 1995
Justice *et al.*, *Geophysical Research Letters*, **9**, 1,005–1,008, 1982
Kao and Chen, *Journal of Geophysical Research*, **100** (B7), 9,881–9,903, 1995
Kaufman and Royden, *Journal of Geophysical Research*, **99** (B8), 15,723–15,739, 1994
Kellogg, *Advances in Geophysics*, **34**, 1–33, 1993
Kendall and Shearer, *Journal of Geophysical Research*, **99** (B6), 11,575–11,590, 1994
Kerkhove and Desmons; *Géologie des pays européens: France, Belgique, Luxembourg*, Dunod, 1980
Kovacheva, *Geophysical Journal of the Royal Astronomical Society*, **61**, 57–64, 1980
Krishnamurti, *Journal of Fluid Mechanics*, **42**, 295–320, 1970
Kukla, *Trans. Nebraska Acad. Sci.*, **6**, 57–93, 1978
Lago and Rabinowicz, *Geophysical Journal of the Royal Astronomical Society*, **77**, 461–482, 1984
Lambeck *et al.*, *Geophysical Journal International*, **103**, 451–468, 1990
Larsen and Yuen, *Geophysical Research Letters*, **24** (16), 1,995–1,998, 1997
Lemoine, *Revue de Géographie physique et Géologie dynamique*, **2**, 205–230, 1959
LeMouël, in Coulomb and Jobert (eds), *Traité de géophysique interne, 2, Magnétisme et géodynamique,* Masson, p. 161–200, 1976
Lenardic and Kaula, *Journal of Geophysical Research*, **99** (B8), 15,697–15,708, 1994
LePichon, *Journal of Geophysical Research*, **73**, 3,661–3,697, 1968
LePichon and Angelier, *Tectonophysics*, **60**, 1–42, 1979
LePichon *et al.*, *Journal of Geophysical Research*, **100** (B7), 12,675–12,690, 1995

Lerch *et al.*, *Journal of Geophysical Research*, **99** (B2), 2,791–2,839, 1994

Lewis and Sandwell, *Journal of Geophysical Research*, **100** (B1), 379–400, 1995

Lincoln and Schlanger, *Journal of Geophysical Research*, **96** (B4), 6,727–6,752, 1991

Lisowski *et al.*, *Review of Geophysics*, US Nat. Report 1987–90, p. 865, 1987

Liu and Ringwood, *Earth and Planetary Science Letters*, **28**, 209–211, 1975

Lliboutry, *Traité de Glaciologie, 2: Glaciers, variations du climat, sols gelés*, Masson & Cie: Paris, 1965

Lliboutry, *Journal of Geophysical Research*, **74** (B7), 6,525–6,540, 1969

Lliboutry, *Nature*, **250** (5464), 298–300, 1974

Lliboutry, *Very slow flows of solids*, M. Nijhoff/Kluwer Academic Publ., Dordrecht, 1987

Lliboutry, *Sciences géométriques et télédétection*, Masson, 1992

Lliboutry, *International Journal of Plasticity*, **9**, 619–632, 1993

Logatchev in Pilger and Rösler (eds), *Afar between continential and oceanic rifting*, Schweizerbartsche Verlagsbuchhandlung, Stuttgart, 1976.

Lundgren and Giardini, *Journal of Geophysical Research*, **99** (B8), 15,833–15,842, 1994

McCalpin and Nishenko, *Journal of Geophysical Research*, **101** (B3), 6,233–6,253, 1996

McClain and Atallah, *Geology*, **14**, 574–576, 1986

McConnell, *Journal of Geophysical Research*, vol. 70, p. 5,171–5,188, 1965

McGarr, *Journal of Geophysical Research*, **85**, 6,231–6,238, 1980

McKenzie, *Geophysical Journal of the Royal Astronomical Society*, **18**, 1–34, 1969

McKenzie and Parker, *Nature*, **216**, 1,276, 1967

McKenzie and Weiss, *Geophysical Journal of the Royal Astronomical Society*, **42**, 131–174, 1975

Madariaga and Perrier, *Les tremblements de Terre*, Presses du CNRS, 1991

Malahoff and Handschumacher, *Journal of Geophysical Research*, **76**, 6,265–71, 1971

Mattauer, *Les déformations des matériaux de l'écorce terrestre*, Hermann, 1973

Means and Jessel, *Tectonophysics*, **127**, 67–86, 1986

Menard, *Marine geology of the Pacific*, McGraw-Hill, 1964

Mercer, *Quaternary Research*, **6**, 125–166, 1976

Mercer and Sutter, *Palaeogeography, Palaeoclimatology, Palaeoecology*, **38**, 185–206, 1982

Minster *et al.*, *Geophysical Journal of the Royal Astronomical Society*, **36**, 541–576, 1974

Minster and Jordan, *Journal of Geophysical Research*, **83**, 5,331–5,354, 1978

Mitrovica, *Journal of Geophysical Research*, **101** (B1), 555–569, 1996

Molina, Garza and Zijderveld, *Journal of Geophysical Research*, vol. 101 (B7), p. 15,799–15,818, 1996

Molinari *et al.*, *Acta Metallurgica*, **35** (12), 2,983–2,994, 1987

Molnar *et al.*, *Geophysical Journal of the Royal Astronomical Society*, **40**, 383–420, 1975

Molnar and Gipson, *Journal of Geophysical Research*, **101** (B1), 545–553, 1996

Molnar and Tapponnier, *Science*, 189, p. 419–426, 1975

Monnet, Ph.D. thesis, Univ. Claude Bernard, Lyon, 1983
Mooney and Meissner in Fountain *et al.* (eds), *Continental Lower Crust*, Elsevier, 1992
Morgan, *Journal of Geophysical Research*, **73**, 1,952–1,982, 1968
Morgan, *Nature*, **230**, 42, 1971
Morgan, *Bull. Am. Assoc. Petrol. Geol.*, **56**, 203–213, 1972
NAT studying group, *Journal of Geophysical Research*, **90**, 10,321–10,341, 1985
Néel, *Cahiers de Physique*, **17**, 47–50, 1943
Néel, *Cahiers de Physique*, **25**, 3–20, 1944
Nerem *et al.*, *Journal of Geophysical Research*, **99** (B2), 2,791–2,839, 1994
Nerem *et al.*, *Journal of Geophysical Research*, **99** (B8), 15,053–15,074, 1994
Nicolas and Poirier, *Crystalline plasticity and solid state flow in metamorphic rocks*, Wiley, 1976
Nicolas *et al.* in Roure, Heitzmann and Polino (eds), *Deep structure of the Alps*, Soc. Géologique de France, Mémoire No. 156, 1990, pp. 15–27
Nisbet and Walker, *Earth and Planetary Science Letters*, **60**, 105–113, 1982
Nye, *Proc. Roy. Soc. London* A, **207**, 554–572, 1951
O'Connor and Duncan, *Journal of Geophysical Research*, **95**, 17,475–17,502, 1990
O'Neill and Jeanloz, *Journal of Geophysical Research*, **99** (B10), 19,901–19,915, 1994
Pari and Peltier, *Journal of Geophysical Research*, **100** (B7), 12,731–12,751, 1995
Peltier, *Review of Geophysics and Space Physics*, **12**, 649–669, 1974
Peltier, *Advances in Geophysics*, **24**, 1–146, 1982
Peltier, *Science*, **265**, 195–201, 1994
Peltier and Jiang, *Journal of Geophysical Research*, **101** (B2), 3,269–3,290, 1996
Percival *et al.* in Fountain *et al.* (eds), *Continental Lower Crust*, Elsevier, 1992
Perram *et al.*, *Journal of Geophysical Research*, **98** (B8), 13,835–13,850, 1993
Plafker, *Journal of Geophysical Research*, **77**, 901–925, 1972
Poirier *et al.*, *Physics of the Earth and Planetary Interiors*, **32**, 273–287, 1983
Poirier and Libermann, *Physics of the Earth and Planetary Interiors*, **35**, 283–293, 1984
Poli and Schmitt, *Journal of Geophysical Research*, **100** (B11), 22,299–22,314, 1995
Pollack *et al.*, *Review of Geophysics*, **31** (3), 267–280, 1993
Poupinet, *Earth and Planetary Science Letters*, **43**, 149–161, 1979
Rapp and Pavlis, *Journal of Geophysical Research*, **95** (B13), 21,885–21,911, 1990
Ratschbacher *et al.*, *Journal of Geophysical Research*, **99** (B10), 19,917–19,945, 1994
Raymo, in Kukla and Went (eds), *Start of a glacial*, NATO ASI Series I-3, Springer Verlag, p. 207–223, 1992
Ricard *et al.*, *Journal of Geophysical Research*, **98** (B12), 21,895–21,909, 1993
Richard and Allègre, *Earth and Planetary Science Letters*, **47**, 65–74, 1980
Richards and Engelbretson, *Nature*, **355**, 437–440, 1992
Ringwood in Hurely (ed.), *Advances in Earth science*, MIT Press, Cambridge, Mass., pp. 287–356, 1966
Ritzwoller and Lavely, *Review of Geophysics*, **33**(1), 1–66, 1995
Robin, *J. Glaciology*, **2**, 523–533, 1954
Roeloffs and Lagbein, *Rev. of Geophysics*, **32** (3), 315–336, 1994

Rogers, *J. of Geology*, **104**, 91–107, 1996
Rothé, *New Scientist*, 11 July 1968
Rudwick and Fountain, *Review of Geophysics*, **33** (3), 267–309, 1995
Ruina, *Journal of Geophysical Research*, **88**, 10,359–10,370, 1983
Rutland, *Nature*, **233**, 252–255, 1971
Sabadini *et al.* (eds), *Glacial isostasy, sea level and mantle rheology*, NATO ASI Series C-334, Kluwer Acad. Publ., 1991
Sacks *et al.*, *Tectonophysics*, 56, p. 101–110, 1979
Sauramo, *Geol. Fören. Förhandl.*, 60, p. 397–404, 1944
Savostin *et al.*, *Tectonophysics*, **123**, 1–35, 1986
Schlich, R., *Le Courrier du CNRS*, no. 17, 1975
Schopf (ed), *Earth's earliest biosphere*, Princeton Univ. Press, 1983
Schreyer, W., *Journal of Geophysical Research*, **100** (B5), 8,353–8,366, 1995
Schubert *et al.*, *Journal of Geophysical Research*, **85**(B5), 2,531–2,538, 1980
Sclater *et al.*, *Review of Geophysics and Space Physics*, **18**, 269–311, 1980
Shaw and Suppe, *Journal of Geophysical Research*, **101** (B4), 8,623–8,642, 1996
Sleep, *Journal of Geophysical Research*, **95**, 6,715–6,736, 1990
Smith *et al.*, *Journal of Geophysical Research*, **95** (B13), 22,013–22,041, 1990
Solheim and Peltier, *Journal of Geophysical Research*, **99** (B4), 6,997–7,018, 1994
Solomon and Gollub, *Physical Review* A, **43**, 6,683–6,693, 1991
Sonett *et al.*, *Science*, **273**, 100–104, 1996
Song, *Review of Geophysics*, **35** (3), 297–313, 1997
Song and Richards, *EOS, Transactions AGU*, **77** (31), 1996
Soudarin and Cazenave, *Geophysical Research Letters*, **22**, 469–472, 1995
Stacey and Loper, *Physics of the Earth and Planetary Interiors*, **33**, 45–55 and 304–317, 1983b
Stauder, *Advances in Geophysics*, **9**, 1–76, 1962
Stein and Stein, *Journal of Geophysical Research*, **99** (B2), 3,081–3,095, 1994
Stevenson in Peltier (ed.), *Mantle convection*, Gordon & Breach Sci. Publ., 1989, pp. 817–873
Stixrude *et al.*, *Science*, **257**, 1,099–1,101, 1992
Stone and Runcorn, eds, *Flow and creep in the solar system*, NATO ASI Series C-391, Kluwer Acad. Publ., 1993
Stoyko, *Annales Guébhard*, **46**, 293, 1970
Suess, *Das Antlitz der Erde*, 1871-1901
Sugimura and Uyeda, *Island arcs, Japan and its environs*, Elsevier, 1972
Suyenaga, *Journal of Geophysical Research*, **84**, 5,599–5,604, 1979
Sykes, *Journal of Geophysical Research*, **72**, 2,131–2,137, 1967
Tait *et al.*, *Journal of Geophysical Research*, **99** (B2), 2,897–2,907, 1994
Talwani *et al.*, *Journal of Geophysical Research*, **70** (2), 341–352, 1965
Talwani, Windisch and Langseth, *Journal of Geophysical Research*, **76**, 473–517, 1971
Talwani and Pitman (eds), *Island arcs, deep-sea trenches and back-arc basins*, Maurice Ewing Series no. 1, AGU, 1977

Talwani *et al.* (eds), *Deep drilling in the Atlantic Ocean: continental margins and palaeo-environment*, Maurice Ewing series no. 3, AGU, 1980
Tamaki and Joshima, *Journal of Geophysical Research*, **84** (B9), 4,501–4,510, 1979
Tapponnier and Molnar, *Nature*, **264** (5,584), 319–324, 1976
Tapponnier *et al.*, Mesozoic ophiolites, sutures and large-scale tectonic features in Afghanistan, *Earth and Planetary Science Letters*, **52**, 355–371, 1981
Tardy *et al.* in Roure, Heitzmann and Polino (eds), *Deep structure of the Alps*, Soc. Géol. de France, Mémoire no. 156; Soc. Géol. Suisse, Mémoire no. 1; Soc. Geologica Italiana, Vol. speciale no. 1, 1990, pp. 217–226.
Taylor and McLennan, *Review of Geophysics*, **33** (2), 241–265, 1995
Thellier, Magnétisme interne, in Goguel (ed), *Géophysique*, Encyclopédie de la Pléiade, Gallimard, 1971, pp. 234–376.
Udías and Channell (eds), *Evolution and tectonics of the western Mediterranean and surrounding areas*, IGN: Madrid, 1980
Uyeda, in: Talwani and Pitman (eds), *Island arcs, deep-sea trenches and back-arc basins*, Maurice Ewing Series no. 1, AGU, 1977
Uyeda and Kanamori, *Journal of Geophysical Research*, **84**, 1,049–1,061, 1979
Valet and Laj, *Earth and Planetary Science Letters*, 54, p. 53–63, 1981
Van der Voo and Boessenkool, *Journal of Geophysical Research*, **78** (23), 5,118–5,127, 1973
Van der Voo, *Review of Geophysics*, **28** (2), 167–206, 1990
Vassiliou *et al.*, *J. Geodynamics*, **1**, 11–28, 1984
Vaughan and Coe, *Journal of Geophysical Research*, **86**, 389–404, 1981
Vernick *et al.*, *Journal of Geophysical Research*, **99** (B12), 24,209–24,219, 1994
Verosub and Roberts, *Journal of Geophysical Research*, **100** (B2), 2,175–2,192, 1995
Vine, *Science*, **154**, 1,405–1,415, 1966
Vogt, *Journal of Geophysical Research*, **86**, 950–960, 1981
Waddington and Walder (eds), *The physical basis of ice-sheet modelling*, IAHS Publ. No. 170, 1987
Walcott, *Review of Geophysics and Space Physics*, **10**, 849–884, 1972.
Wallace, *Review of Geophysics and Space Physics*, **10**, 849–884, 1972
Wallace, *Geophysical Research Letters*, **22** (16), 2,331, 1995
Wang *et al.*, *Journal of Geophysical Research*, **96** (B3), 4,413–4,421, 1991
Wei, *Geophysical Research Letters*, **22**, 957–960, 1995
Weeks, *Journal of Geophysical Research*, **98** (B10), 17,637–17,648, 1993
Weertman, *Phil. Trans. Roy. Soc. London*, A.**288**, 9–26, 1978
Wesson *et al.*, *Proceedings of the conference on tectonic problems of the San Andreas fault system*, Stanford University Press, 1973
Westaway, *Journal of Geophysical Research*, **98** (B12), 21,741–21,772, 1993
Westaway, *Journal of Geophysical Research*, **100**(B8), 15,173–15,192, 1995
Wezel, *Mem. Soc. Geol. Ital.*, **24**, 531–568, 1982
Wilcock *et al.*, *Journal of Geophysical Research*, **100** (B12), 24,147–24,165, 1995
Wood, *Annual Review of Earth and Planetary Science*, **16**, 53–72, 1988
Wu, *Geophysical Journal International*, **114**, 417–432, 1993
Wysession *et al.*, *Journal of Geophysical Research*, **99** (B7), 13,667–13,684, 1994

Yeats and Prentice, *Journal of Geophysical Research*, **101** (B3), 5,847–5,453, 1996
Yiou *et al.*, *Geophysical Research Letters*, **22** (16), 2,179–2,182, 1995
Yuen *et al.* in Stone and Runcorn (eds), *Flow and creep in the solar system*, NATO ASI Series C, no. 391, Kluwer Acad. Publ., 1995, pp. 147–173
Ziegler and Crowell in McElhinny and Valencio (eds), *Paleoreconstruction of the continents*, AGU Geodynamic series **2**, 31–37 and 45–49, 1981
Zindler and Hart, *Annual Review of Earth and Planetary Science*, **14**, 493–571, 1986
26e Congrès Géologique International, *Géologie des Pays Européens: Espagne, Grèce, Italie, Portugal, Yougoslavie*, Bordas/Dunod, 1980

Index